Practical Aspects of Flow Assurance in the Petroleum Industry

Emerging Trends and Technologies in Petroleum Engineering

Series Editor: *Abhijit Y. Dandekar*

Wax Deposition: Experimental Characterizations, Theoretical Modeling, and Field Practices
Zhenyu Huang, Sheng Zheng, and H. Scott Fogler

Hydraulic Fracturing
Michael Berry Smith and Carl Montgomery

Unconventional Oil and Gas Resources: Exploitation and Development
Usman Ahmed and D. Nathan Meehan

Petroleum Fluid Phase Behavior: Characterization, Processes, and Applications
Raj Deo Tewari, Abhijit Y. Dandekar, and Jaime Moreno Ortiz

Practical Aspects of Flow Assurance in the Petroleum Industry
Jitendra Sangwai and Abhijit Dandekar

For more information about this series, please visit:
https://www.routledge.com/Emerging-Trends-and-Technologies-in-Petroleum-Engineering/book-series/CRCEMETRETEC

Practical Aspects of Flow Assurance in the Petroleum Industry

Jitendra Sangwai

Abhijit Dandekar

CRC Press
Taylor & Francis Group
Boca Raton London New York

CRC Press is an imprint of the
Taylor & Francis Group, an **informa** business

First edition published 2023
by CRC Press
6000 Broken Sound Parkway NW, Suite 300, Boca Raton, FL 33487-2742

and by CRC Press
2 Park Square, Milton Park, Abingdon, Oxon, OX14 4RN

ISBN: 978-0-367-49074-4 (hbk)
ISBN: 978-0-367-54952-7 (pbk)
ISBN: 978-1-003-09130-1 (ebk)

DOI: 10.1201/9781003091301

Typeset in Times
by SPi Technologies India Pvt Ltd (Straive)

Contents

Series Preface

This petroleum engineering book series includes works on all aspects of petroleum science and engineering, but with special focus on emerging trends and technologies that pertain to the paradigm shift in the petroleum engineering field. It deals with the increased exploitation of technically challenged and atypical hydrocarbon resources that are receiving a lot of attention from today's petroleum industry, as well as the potential use of advanced non-traditional or non-conventional technologies such as nanotechnology in diverse petroleum engineering applications. These areas have assumed a position of prominence in today's petroleum engineering field. However, although scientific literature exists on these emerging areas in the form of various publications, much of it is scattered and highly specific. The purpose of this book series is to provide a centralized and comprehensive collection of reference books and textbooks covering fundamentals but paying close attention to these emerging trends and technologies from the standpoint of the main disciplines of drilling engineering, production engineering, and reservoir engineering.

Given the dwindling supply of easy to produce conventional oil, rapidly climbing energy demands, the sustained ~$100/bbl oil price, and technological advances, the petroleum industry is increasingly in pursuit of E & P of atypical or unconventional and technically challenged oil and gas resources, which may eventually become the future of the petroleum industry. Unconventional resources typically include (1) coal bed methane or CBM gas; (2) tight sands gas in ultra low permeability formations; (3) shale gas and shale oil in very low permeability shales; (4) oil shales; (5) heavy and viscous oils; (6) tar sands; and (7) methane hydrates. Compared to the world's proven conventional natural gas reserves of 6600+ trillion cubic feet (TCF), the combined CBM, shale gas, tight sands gas, and methane hydrate resource estimates are in excess of 730,000 TCF[1-3]. Similarly, out of the world's total 9 to 13 trillion barrels of oil resources, the conventional (light and medium oil) is only 30%, whereas heavy oil, extra heavy oil, tar sands, and bitumen combined make up the remaining 70%[4]. In addition shale-based oil resources worldwide are estimated to be between 6 and 8 trillion barrels[5]. As a case in point, shale-based oil production in North Dakota has increased from a mere 3000 barrels/d in 2005 to a whopping 400,000+ barrels/d in 2011[6]. Even the most conservative technical and economic recovery estimates of the unconventional resources represent a very substantial future energy portfolio that dwarfs the conventional gas and oil reserves. However, to a large extent these particular resources, unlike the conventional ones, do not fit the typical profile and are to some extent in the stages of infancy, thus needing a different and unique approach from the drilling, production, and reservoir engineering perspective.

The petroleum engineering academic and industry community also is aggressively pursuing nanotechnology with the hope of identifying innovative solutions for problems faced in the overall process of oil and gas recovery. In particular, a big spurt in this area in the last decade or so is evident from the significant activities in terms of research publications, meetings, formation of different consortia, workshops, and dedicated sessions in petroleum engineering conferences. A simple literature search for a keyword *nanotechnology* on www.onepetro.org, managed by the Society of Petroleum Engineers (SPE), returns over 250 publications dating from 2001 onwards with the bulk of them in the last 5 or 6 years. Since 2008, SPE also organized three different applied technology workshops specifically focused on nanotechnology in the E & P industry. An Advanced Energy Consortium (AEC) with sponsorships from some major operators and service companies also was formed in 2007 with the mission of facilitating research in "micro and nanotechnology materials and sensors having the potential to create a positive and disruptive change in the recovery of petroleum and gas from new and existing reservoirs". Companies such as Saudi Aramco have taken the lead in taking the first strides in evaluating the potential of employing nanotechnology in the E & P industry. Their trademarked Resbots™ are designed for deployment with the injection fluids for in-situ reservoir sensing (temperature, pressure, and fluid type) and intervention, eventually leading to more accurate reservoir characterization once fully developed. Following successful laboratory core flood tests, they conducted the industry's first field trial of reservoir nanoagents[7].

The foregoing is clearly a statement of the new wave in the petroleum engineering field that is being created by emerging trends in unconventional resources and new technologies. The publisher and its series editor are fully aware of the rapidly evolving nature of these key areas and their long-lasting influence on the current state and future of the petroleum industry. The series is envisioned to have a very broad scope that includes but is not limited to analytical, experimental, and numerical studies and methods and field cases, thus delivering readers in both academia and industry an authoritative information source of trends and technologies that have shaped and will continue to impact the petroleum industry.

Abhijit Dandekar
University of Alaska Fairbanks

REFERENCES

1. http://www.eia.gov/analysis/studies/worldshalegas/
2. Kawata, Y. and Fujita, K., Some Predictions of Possible Unconventional Hydrocarbons Availability Until 2100, Society of Petroleum Engineers (SPE) paper number 68755.
3. http://www.netl.doe.gov/kmd/cds/disk10/collett.pdf
4. https://www.slb.com/~/media/Files/resources/oilfield_review/ors06/sum06/heavy_oil.ashx
5. Biglarbigi, K., Crawford, P., Carolus, M. and Dean, C., Rethinking World Oil-Shale Resource Estimates, Society of Petroleum Engineers (SPE) paper number SPE 135453.
6. Mason, J., http://www.sbpipeline.com/images/pdf/Mason_Oil%20Production%20Potential%20of%20the%20North%20Dakota%20Bakken_OGJ%20Article_10%20February%202012.pdf
7. Kanj, M.Y., Rashid, M.H. and Giannelis, E.P., Industry First Field Trial of Reservoir Nanoagents, Society of Petroleum Engineers (SPE) paper number SPE 142592.

Preface

Chapter 1 serves as the introduction to the topic of flow assurance by first defining it in terms of "what is it" and its relevance given the (challenging) oil and gas field operating and transport environments. Next, the "big five" solid phases, that is, hydrates, waxes, asphaltenes, scales, and naphthenates that potentially impact the production of oil and gas, and the transport pipelines are briefly introduced. The summary of various flow assurance activities and the organization of the book segues into the subsequent chapters.

Chapter 2 provides details on the chemistry of reservoir fluids and the chemical and physical properties of different reservoir fluids. The important reservoir engineering properties are also discussed in brief. This is followed by the fundamentals of fluid phase behavior, with a focus on pure system and multicomponent hydrocarbon fluid phase behavior. A detailed discussion on the phase behavior of five different reservoir fluids has also been provided. The fluid sampling methods relevant for flow assurance studies have been discussed. This includes the details on well conditioning, types of fluid sampling methods, and special considerations to be taken during fluid sampling of reservoir fluids for flow assurance studies.

Chapter 3 focuses on laboratory studies and characterization of petroleum reservoir fluids, which basically serves as a forerunner to the flow assurance data collection process. Compositional measurements and typical fluid studies such as constant composition expansion, differential liberation, constant volume depletion, and separator tests are covered with sufficient details. Commonly used heavy-end characterization calculation methods also are covered. Some of the numerical examples included demonstrate the processing of laboratory data to determine fluid properties, as well as heavy-end splitting and grouping. The chapter concludes with produced water characterization in terms of salinity and physical properties and the role of salinity from the standpoint of hydrates, scale, naphthenates, and corrosion.

Chapter 4 provides an introduction to various organic and inorganic solid deposits such as hydrates, paraffin wax, asphaltenes, and naphthenate and their characterization techniques. A brief introduction to corrosion and scale has also been provided. Various types of corrosion are briefly discussed. Organic solid phase envelopes for hydrates, asphaltene, and waxes are discussed briefly and finally, a composite solid phase envelope of the Gulf of Mexico has been presented to highlight the use of phase envelope in the flow assurance.

Chapter 5 focuses on the assessment of the solid organic and inorganic phases in the flow assurance problems. Hydrates- and flow-related rules of thumb essential for mitigation of flow assurance issues are discussed. Empirical methods for hydrates conditions and asphaltene assessment are covered. Details on various experimental techniques for fluid studies on hydrates, wax, and asphaltenes are provided. This is followed by equations of state, and other relevant modeling methods for organic solids, such as asphaltenes, hydrates, and wax, are elaborated. Finally, other flow assurance issues, such as emulsion formation and fouling, are briefly discussed.

Chapter 6 details the deepwater oil and gas environment and related challenges. The deepwater environment, such as temperature, density, and salinity, is briefly discussed. An overview of the various offshore structures or platforms and deepwater station positioning systems is provided. Various elements of subsea production, processing, and separation systems are detailed. Also, a deployment window for subsea processing is discussed.

Chapter 7 is devoted to the management and control of solids phases in flow assurance. Mechanisms of inorganic and organic solid deposition are discussed in detail. Various intrusive and nonintrusive methods for locating flowline blockage have been discussed. Remediation methods for organic and inorganic solids, including hydrate, wax, asphaltenes, and scales, are covered. Dosing rates for various chemicals used for wax, asphaltene, hydrates, and scales are discussed briefly. Finally, production chemicals inventory is discussed, including the strategy and real-time chemical inventory and usage platform, along with advantages of real-time chemical monitoring system.

Chapter 8 basically highlights the potential influence of the flow assurance solids on flow in the porous media and in bulk, that is, pipe transport, depending on the prevailing temperature and pressure conditions. The concept of formation damage, wellbore skin, Darcy's law in terms of skin, and productivity index are covered. Given the (somewhat) omnipresence of asphaltenes from the reservoir to surface, notable case studies of asphaltene deposition and their analysis/interpretation are also included. Toward the end topics of relevance to flow assurance in long-distance pipelines, such as gel strength and drag-reducing agents, are covered.

Chapter 9 is dedicated to fluid mechanics, which can be construed as a backbone of flow assurance. The chapter begins with the fundamental mechanical energy balance equation that relates the pressure gradient to friction and hydrostatic and kinetic terms, respectively. The solution of this equation for single-phase and mixed systems basically allows the determination of pressure at a given location in a flow system, which is of interest to the engineer, since, for example, this can be mapped with potential solid(s) forming conditions. A variety of pertinent fluid mechanics topics covered in this chapter include static, as well as dynamic, cases vis-à-vis shut-in and flowing pressure gradients for single and mixed systems. Additionally, commonly employed empirical and mechanistic multiphase models for pressure gradient prediction also are described to the practical extent possible. Several step-by-step solved numerical examples are included throughout much of the chapter. Finally, other peculiarities of flow assurance, such as slugging and water hammer, are also summarized at the end.

Chapter 10 is mostly devoted to case studies that have been reported in the literature on a given type of flow assurance solid encountered in operations, the diagnosis of the problem, the mitigation strategy (including the screening of inhibitor in some cases), and the treatment effectiveness. The documented case studies are summarized in the form of graphs and tables for quick "lessons learned". Additional topics of interest closely linked to flow assurance include health, safety, and environment; economics; the status of technology development; and operating company perspectives. The chapter and the book end with a framework or workflow that a flow assurance engineer can use as a "toolbox".

Acknowledgments

I would like to dedicate this book to my late grandmother, Godavari B. Sangwai, for her care and love. I sincerely acknowledge my research scholars, mainly Siddhant Prasad, Satyajit Chowdhury, Umasankar Behera, Chandan Sahu, Ganesh Kumar, Gomathi Rajalakshmi, and Mayank Rakesh, for their constant support during this work. I sincerely thank my PhD supervisor, Prof. Santosh K Gupta of IIT Kanpur, for his continued guidance and love throughout my academic endeavors. I am thankful to my parents, Mr. Shital Sangwai and Ms. Vasudha Sangwai, and my wife, Priti, and my children, Saket and Akshit, for making my life enjoyable. I sincerely acknowledge the contribution made by several scientists and authors in the field that has made this work possible.

Jitendra S. Sangwai
Chennai, India

I would like to dedicate this work to my late grandfather, Purushottam V. Dandekar. I thank my parents, Yashwant and Sunanda Dandekar, for their blessings and my wife, Mrudula; son, Shamal; and daughter, Rama, for their support during the course of writing Chapters 1, 3, and 8–10 of this book.

Finally, once again it is my honor to gratefully acknowledge the works of numerous scientists and engineers and their publications in the literature pertaining to the subject covered in this book.

Abhijit Y. Dandekar
Fairbanks, Alaska

Author Biographies

Dr. Jitendra S. Sangwai is currently working as a *full* professor at the Department of Chemical Engineering and is associated with the Petroleum Engineering Program of the Department of Ocean Engineering at the Indian Institute of Technology Madras. He holds an M.Tech and a PhD in chemical engineering from IIT Kharagpur and IIT Kanpur, respectively. He worked with Schlumberger for a brief period before joining academia. Dr. Sangwai's research interest lies mainly in gas hydrates, enhanced oil recovery, rheology of complex fluids, etc. He has published approximately 140 international journal papers, 90 conference publications and filed 20 patents. He has graduated 20 PhD and several master's-degree students. Dr. Sangwai is the recipient of the Society of Petroleum Engineers' Distinguished Achievement Award for Petroleum Engineering Faculty of the South Asia and Pacific region in 2017, the National Award for Technology Innovation from the Gov. of India (2016 and 2018), the Young Faculty Recognition Award for excellence in teaching and research, Institute Research and Development Awards (both at Early- and Mid-Career level) and Shri. J. C. Bose Patent Award from IIT Madras, and SPE Regional Service Award (2015). Dr. Sangwai has been highlighted as 'One among 25 Emerging Investigators' and Top 1% Highly Cited Author recognition by the American Chemical Society Journals.

Abhijit Dandekar is a professor and the chair of petroleum engineering at the University of Alaska Fairbanks. He earned a B.Tech (chemical engineering) and a PhD (petroleum engineering) from Heriot-Watt University, UK. He has been a visiting faculty in China, Denmark, India, Nigeria and South Africa. He is a Society of Petroleum Engineers (SPE) distinguished member, recipient of SPE Western North America Region Distinguished Achievement Award for Faculty, and a Fulbright specialist. He has authored more than 100 refereed and conference papers in petroleum engineering; a widely adopted textbook, *Reservoir Rock and Fluid Properties*; and a book, *Petroleum Fluid Phase Behavior*. He is also the editor for Emerging Trends and Technologies in Petroleum Engineering book series (Taylor & Francis).

1 Introduction to Flow Assurance

1.1 WHAT IS IT?

Simply speaking, *flow assurance* is a self-explanatory term that came into being in the early 1990s. Although the exact reference for its origin is not known, Brazilian oil company Petrobras is credited with coining the Portuguese term *Garantia de Fluxo*, which is literally translated as "guarantee the flow" or "flow assurance". Since oil and gas flow in the entire infrastructure takes place through pipes, valves, and fittings of different dimensions, flow assurance includes all issues important to maintaining the flow from sandface or bottomhole to sales. Although flow through porous media is the precursor and shares some common issues, it is not part of flow assurance. However, some topics pertaining to subsurface flow called formation damage are covered in one of the chapters.

1.2 OPERATING ENVIRONMENT AND CONSEQUENCES

Let us first examine the operating environment and the variables and the resulting consequences that play a major role in flow assurance. Figure 1.1 shows a basic schematic of oil and gas production in onshore and offshore environments. In the case of an onshore environment, the distance between the wellhead and a production separator can be as short as less than a mile, whereas in an offshore environment, the produced fluids may have to flow over a distance of several miles in challenging terrain. For example, the Stones field[1] in the Gulf of Mexico, offshore Louisiana, is one of the record holders, with a reservoir depth of 26,500 ft below sea level (17,000 ft below the mudline or seabed), which means a water depth of 9,500 ft. Additional challenges include a very rough terrain, furrowed sea bottom and close to escarpment. Production from this deep high-pressure, high-temperature (HPHT) field is shipped via a floating production storage and offloading (FPSO) system and a gas export line that faces approximately 3,000 ft of elevation change.

Next, the operating environment needs to be placed in perspective by considering the variation in temperature as a function of depth, such as the one shown in Figure 1.2 for the Gulf of Mexico, the Arabian Sea and the Bay of Bengal. Even though surface temperatures are as high as 70–80°F, temperatures at the depth of 7,000–10,000 ft are in the vicinity of 40°F because of the fact that sunlight is unreachable to such depths. Therefore, as shown in Figure 1.3, it is not difficult to understand the drop in both the pressure and temperature as the HPHT-produced fluids travel through the flowline and riser system (also refer to Figure 1.1) into a separator. Despite the drop in temperature (indirectly pressure also) that is usually minimized by insulating the flowline and riser systems, the prolonged exposure to low surrounding temperatures is conducive to phase change in the flow infrastructure. The resulting phase change is

DOI: 10.1201/9781003091301-1

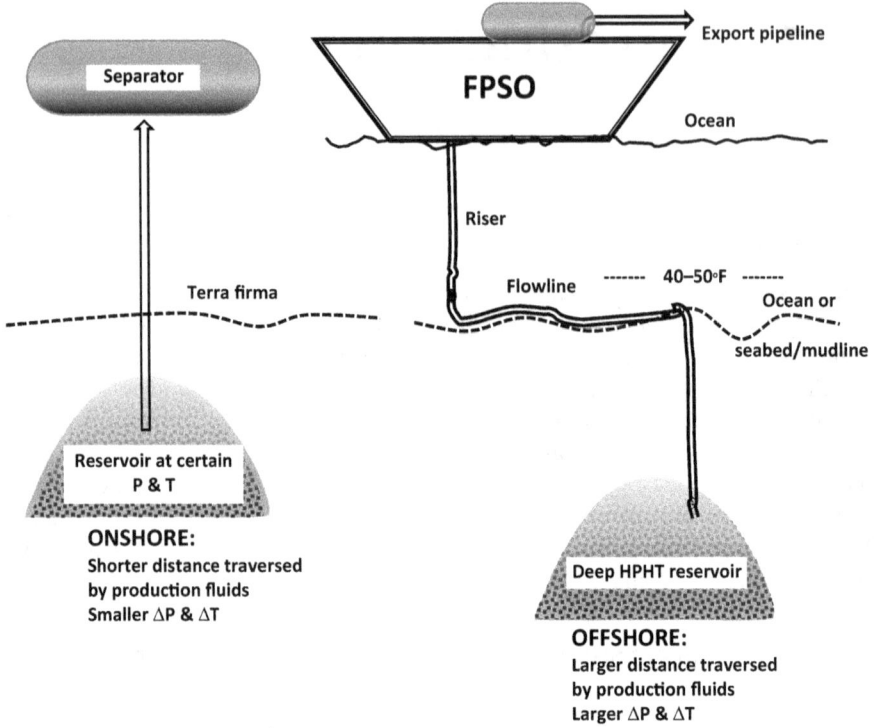

FIGURE 1.1   Schematic of onshore and offshore operating environments (see discussion in Section 1.2).

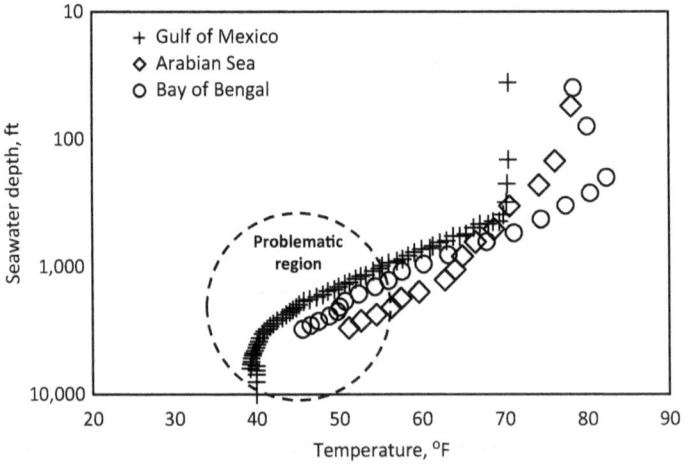

FIGURE 1.2   Seawater temperatures vs. depth for Gulf of Mexico, Arabian Sea and the Bay of Bengal. Plot constructed based on data reported on respective Government websites.

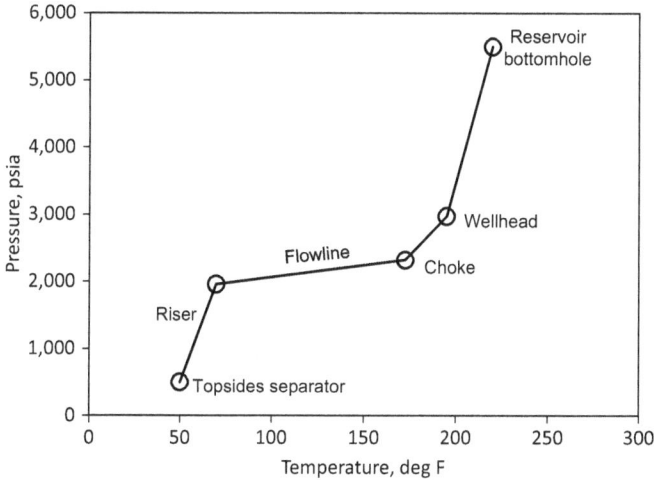

FIGURE 1.3   An example of a typical drop in pressure and temperature from HPHT reservoir conditions to topsides separator.

due to the change in pressure, temperature, and the composition of the produced hydrocarbons as well as the water phase. Although the formation of a gas phase from liquid and vice versa is not that problematic, the precipitation of a solid phase is the most worrisome from flow assurance standpoint. What are then the consequences? The precipitated solid phase may eventually deposit on walls of the flowline or any pipe for that matter, altering the geometry and thus blocking the flow and interrupting the production operations. The important question to which the flow assurance engineer seeks answers for is, at what location in the flow infrastructure will the pressure and temperature drop to a certain value at which solids are likely to precipitate and perhaps deposit from a given produced stream (composition-specific)? This scenario is depicted in Figure 1.4.

Operating conditions such as those encountered in offshore installations and to a relatively lesser extent in onshore also manifests other ancillary flow assurance challenges. These include corrosion, oil–water emulsions, continuously altering fluid

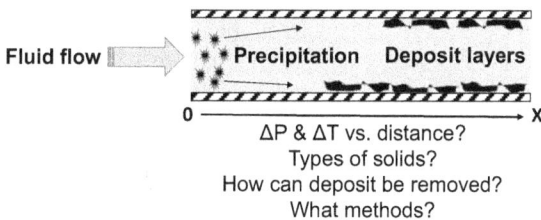

FIGURE 1.4   Operating environment change and consequences – precipitation or flocculation likely leading to deposition and flow impairment.

properties, start-ups and shutdowns, transient conditions, and multiphase flow. This brings up key questions such as friction factors, liquid holdups, slugging, superficial velocities, flow regimes, Reynold's numbers, heat capacities, and overall heat-transfer coefficients.

1.3 THE "BIG FIVE"

The commonly recognized solid phases are (1) ice-like gas hydrates, (2) candle-like waxes, (3) tar-like asphaltenes, (4) saltlike scale, and (5) soap-like naphthenates. These are nicknamed the "big five" by Gudmundsson.[2] The organic solid phases, such as gas hydrates, waxes and asphaltenes, originate from the five reservoir fluids,[3] namely, dry gases, wet gases, gas condensates or retrograde gases, volatile oils, and black oils. Additionally, heavy oils and extra-heavy oils are generally responsible for the formation of waxes and asphaltenes. Inorganic solids, such as scale, are attributed to formation water, whereas naphthenates are linked to both the water and the hydrocarbon liquid phase given their amphiphilic nature.[2] Flow interruptions or blockage caused by gas hydrates can be sudden or instantaneous, much like a heart attack or failure in a human body, whereas those caused by other solid phases tend to be somewhat gradual over time, akin to slower death by diseases such as cancer.

1.4 OIL AND GAS TRANSPORT PIPELINES

These generally include midstream operations, such as long-distance pipelines that carry crude oil and natural gas. Other smaller magnitude examples are short-distance wet-fuel gas lines that need to deliver gas at a specific pressure, which can be achieved by a throttling process across an expansion valve that may lead to rapid cooling due to Joule-Thomson effect.[4] A classic example of a long-distance pipeline is the 800-mile-long Trans-Alaska Pipeline System (TAPS)[5] that carries all the crude oil produced on the North Slope of Alaska to the ice-free port of Valdez in southern Alaska. TAPS is not heated, but the only protection against the ambient temperatures of −40 to −50°F in winter is the 4-in.-thick fiberglass insulation. A typical drop in the temperature from inlet to outlet can be as much as 50–60°F over a distance of 800 miles. Such type of drastic change in the flow conditions leads to the formation of wax, necessitating mechanical cleaning via frequent pig (pipeline inspection gauge) deployment. Sloan[4] has discussed a case of a gas pipeline, in Louisiana's Dog Lake field, getting exposed to low winter ambient temperatures posing a risk of gas hydrate formation and blockage. As an example, Figure 1.5 shows a typical scenario that may lead to the formation of gas hydrates due to the throttling process in a fuel line.[4] Therefore, flow assurance challenges in such types of midstream or process operations are generally confined to one particular solid phase; however, the fundamentals of their formation (phase change, etc.) are the same as those in the case of upstream production operations.

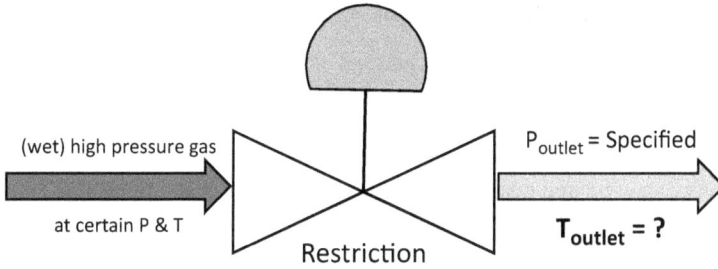

FIGURE 1.5 Throttling of a high pressure (wet) gas across a restriction to a specified outlet pressure that may reduce the outlet temperature, due to Joule-Thomson cooling effect, to gas hydrate formation conditions. Schematic drawn on the basis of scenario depicted by Sloan.[4]

1.5 FLOW ASSURANCE ACTIVITIES

By nature, flow assurance is a truly multidisciplinary activity that involves not just one particular engineering discipline but also relies on concepts from chemical, mechanical, and electrical engineering supported by expertise in chemistry. The variety of engineering components can be grouped under four domains, namely, (1) fluid sampling, (2) thermal and hydraulic modeling, (3) system design, and (4) operating strategies. The different activities under these domains are summarized in Figure 1.6, which also is the backbone of this book.

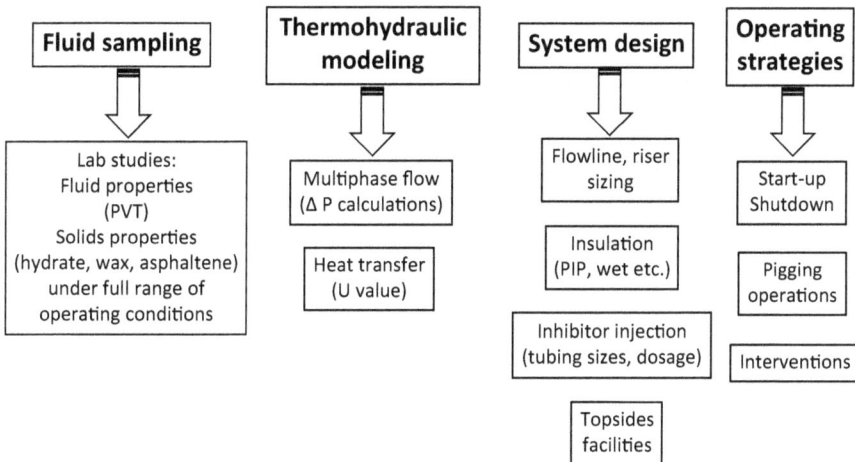

FIGURE 1.6 Typical flow assurance activities.

1.6   ORGANIZATION OF THE BOOK

Based on the foregoing, the significance of flow assurance, particularly in a deep offshore operating environment where the industry is focusing, should be amply clear. The specialized nature of flow assurance also has led to the creation of a new discipline called flow assurance engineering. Although many excellent texts that are very subject-specific and deal with particular solid phases and fluid mechanics exist, currently there is a lack of a book that covers the practical aspects of flow assurance that an engineer can refer to as a guide.

Starting with the introduction to flow assurance in this chapter the book follows a logical sequence that first deals with the overview of petroleum reservoir fluids, their phase behavior, compositional characteristics, and the routine pressure–volume–temperature tests. The organic and inorganic solid phases are defined next, which is followed by empirical methods and specific laboratory tests conducted for their assessment. The deepwater oil and gas environment is described next, which segues into the motivation and control of solids in flow assurance. Fluid mechanics in flow assurance and the role of solid phases are covered in later chapters. The final chapter of the book is dedicated to documented flow assurance case studies and a road map or toolbox that an engineer can use.

REFERENCES

1. Hoffman, J., Clausing, K., Robinson, S., Subramanian, P., & Zummo, A. (2017). The Stones Project: Subsea, Umbilical, Riser and Flowline Systems. In *Offshore Technology Conference* (p. 20). Offshore Technology Conference. https://doi.org/10.4043/27569-MS
2. Gudmundsson, Jon S. 2018. *Flow Assurance Solids in Oil and Gas Production.* Boca Raton, FL: CRC Press, Taylor & Francis Group.
3. McCain, William D. 1990. *Properties of Petroleum Fluids* (2nd Edition). Tulsa, OK: Pennwell Books.
4. Sloan, E. 2000. Dendy and Society of Petroleum Engineers (U.S.). *Hydrate Engineering.* Monograph / SPE; Henry L. Doherty series, vol. 21; Monograph (Society of Petroleum Engineers (U.S.)), v. 21; Monograph (Society of Petroleum Engineers (U.S.)), Henry L. Doherty series.2973142. Richardson, Tex.: Society of Petroleum Engineers.
5. https://www.alyeska-pipe.com/assets/uploads/pagestructure/TAPS_PipelineFacts/ editor_uploads/Final%20Fact%20Book%202019.pdf, accessed March 9, 2020.

2 Overview of Petroleum Reservoir Fluids

2.1 INTRODUCTION

Petroleum reservoir fluids include complex hydrocarbon fluids, such as crude oil and gas, present in the subsurface reservoirs. As the exploration and production activities are increasingly being carried out under harsh conditions (like offshore reservoirs), therefore, it is necessary to understand the reservoir fluid type and properties to design suitable production equipments (surface, subsurface, refinery, and equipments related downstream components, etc.) and develop strategies to mitigate flow assurance issues, especially during offshore oil and gas production.[1] Also, information on the reservoir fluid properties is critical for reserves forecasting and auditing, enhanced oil recovery methods, and reservoir management. The fluid type is generally determined during the early stage of reservoir planning (using sampling techniques). The composition and physicochemical properties of reservoir fluids play an important role in designing and optimizing oil and gas production methods, helping in efficient reservoir management. There needs to be utmost care during the measurement and prediction of various reservoir fluid properties, which otherwise will lead to high uncertainties in the reservoir volume and recovery prediction, thus affecting the efficient strategies for flow assurance management.[2-4] Before production, static data on the fluid property measurements are available, but once production commences, the dynamic data on variation in fluid composition on production timescale become apparent and needful. The main objective of this chapter is to discuss the silent features of five reservoir fluids, their chemical and physical properties, and phase behavior fundamentals. Subsequently, a discussion on the sampling of these hydrocarbon fluids concerning the flow assurance perspective is provided.

2.2 CHEMICAL AND PHYSICAL CHARACTERISTICS OF RESERVOIR FLUIDS

Petroleum reservoir fluid is a complex and multicomponent mixture, of various hydrocarbons ranging from the simplest (C_1 or methane) to large and complex, as heavy as C_{200}.[5] It also contains some heterocyclic compounds (compounds of oxygen, nitrogen, and sulfur), along with some trace metals, such as nickel, vanadium, and zinc, which are present in the heavier fraction, such as asphaltene. The physicochemical properties of reservoir fluids (oil, gas, water) are discussed in the following sections.

DOI: 10.1201/9781003091301-2

2.2.1 Chemistry of Reservoir Fluids

2.2.1.1 Chemical Properties of Gases and Crude Oil

Gas produced from the reservoir primarily include hydrocarbon gases, but it can also contain a mixture of hydrocarbon and non-hydrocarbon gases with varying concentrations. Hydrocarbon gases include alkanes with carbon numbers C_1 to C_5, a smaller amount of hexane and higher hydrocarbons. Non-hydrocarbon gases consist of CO_2, nitrogen, and hydrogen sulfide (H_2S).

Crude oil contains complex hydrocarbons (paraffins, naphthenes, aromatics, resins, asphaltenes), along with heteroatoms, namely, nitrogen, oxygen, and sulfur. It also contains trace metals, including nickel, vanadium, and zinc. These are discussed in the following sections.

2.2.1.1.1 Paraffins

Paraffins, or saturated hydrocarbons, have a structural formula of C_nH_{2n+2}, where n is the carbon number. They contain sp^3 hybridized carbon atoms linked with hydrogen atoms either in a straight chain (n-paraffins) or having one or more side chains (iso-paraffins). At standard pressure and temperature conditions, paraffins less than C_5 are gaseous while C_5 to C_{15} remain liquid. In contrast, paraffin with C_{15} is very viscous and probably solid waxes at surface conditions. Figure 2.1 shows the structural representation of the normal and iso-paraffins.

2.2.1.1.2 Naphthenes

Naphthenes have a structural formula of C_nH_{2n}. The properties of naphthene compounds are similar to that of paraffin with the same carbon number. Naphthenes or cycloalkanes form closed ring structures containing sp^3-hybridized carbon atoms connected with hydrogen atoms. Figure 2.2 shows the structural representation of cyclopentane and cyclohexane.

2.2.1.1.3 Aromatics

Aromatic hydrocarbons contain one or more rings with sp^2-hybridized carbon atoms connected to one hydrogen atom each. In petroleum reservoir fluids, the polyaromatic compounds contain at least two rings. Figure 2.3 shows the chemical structure of the aromatics compound, namely, benzene and naphthalene.

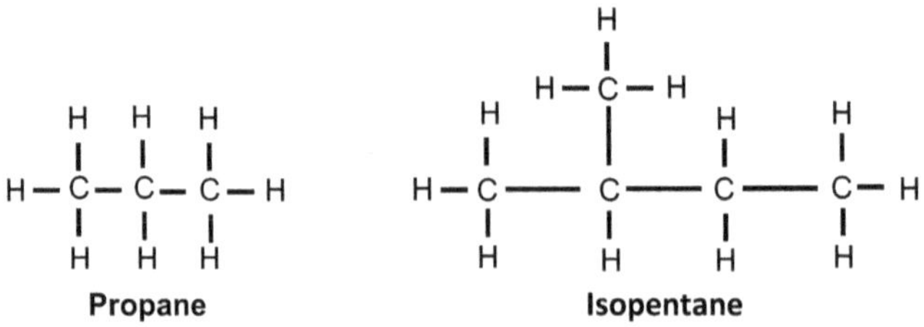

FIGURE 2.1 Chemical structure of normal and iso-paraffins.

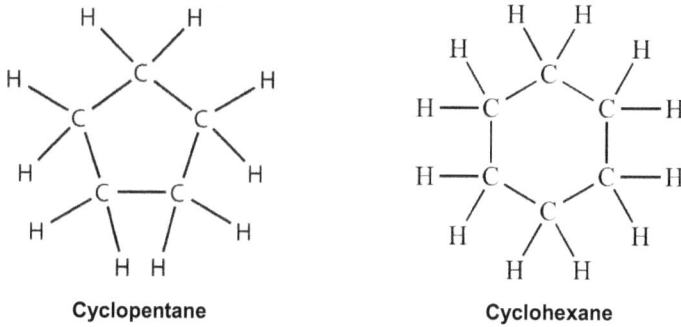

FIGURE 2.2   Chemical structure of cycloalkanes or naphthenes.

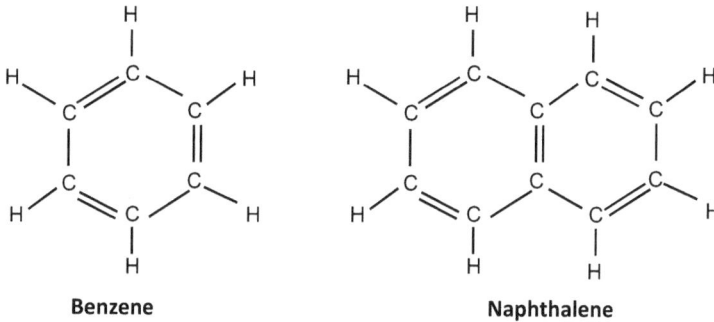

FIGURE 2.3   Chemical structure of aromatics.

2.2.1.1.4 Asphaltenes

Asphaltenes are the heaviest and most polar fractions of crude oil. They contain poly-nuclear aromatic core, aliphatic side chains, heteroatoms, and trace amounts of heavy metals. They are insoluble in alkanes (viz., n-pentane and n-heptane) but soluble in aromatics (e.g., toluene and benzene). The composition and properties of asphaltenes are inconsistent and vary with the source. The presence of non-flocculated asphaltenes is responsible for the black hue of some crude oils and residuals. Figure 2.4 shows the two-dimensional molecular structure of asphaltene from the Ecuadorian oil proposed by Billups et al.[6]

2.2.1.1.5 Resins

Resins, like asphaltenes, are polar components of crude oil; however, they have a lower molecular weight and are less polar than the latter. Resins have a higher hydrogen/carbon (H/C) ratio, have fewer heteroatoms, and help keep asphaltenes stable in the crude oil at reservoir conditions. They dissolve in normal alkanes and aromatic solvents but not in liquid propane. Resins are shiny, sticky, and dark brown.

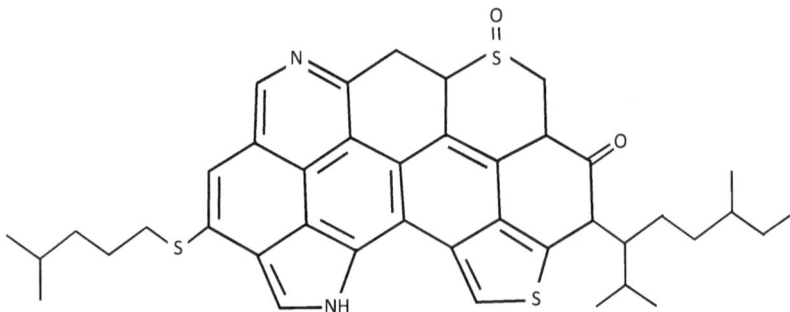

FIGURE 2.4 Molecular structure of asphaltene extracted from the Ecuadorian oil.[6] Reproduced from Billups et al.[6] with permission from American Chemical Society (Copyright American Chemical Society, 2019).

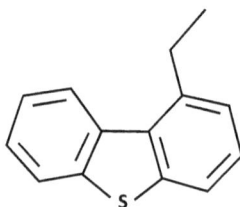

FIGURE 2.5 Molecular structure of the resin molecule from the Athabasca oil sand.[7] Reproduced from Murgich et al.[7] with permission from American Chemical Society (Copyright American Chemical Society, 1999).

Murgich et al.[7] proposed a two-dimensional molecular structure of the resin molecule from the Athabasca sand oil (see Figure 2.5).

2.2.1.1.6   Heterocompounds

The reservoir gas contains varying amounts of nonhydrocarbon components (impurities), such as carbon dioxide (CO_2), oxygen (O_2), hydrogen sulfide (H_2S), sulfur, nitrogen (N_2), helium, argon, and hydrogen. The crude oil contains acids, esters, ketones, phenols, and alcohol. Commonly found acids in young and immature crude oils include fatty acids, isoprenoids, naphthenic, and carboxylic acids. The presence of steranes in various crudes is a key indicator of their organic origin.

2.2.1.1.7   Trace Metals

Crude oil also contains trace metals such as iron, aluminum, calcium, magnesium, copper, lead, tin, antimony, zinc, silver, nickel, chromium, molybdenum, and vanadium. The most significant trace elements, however, are vanadium, nickel, and iron. These trace metals have such a low concentration that their value is stated in parts per million. Trace metal concentrations in crude oil are inversely related to the American Petroleum Institute (API) gravity of crude oil.

The percentage of paraffinic, naphthenic, and aromatic constituents in a reservoir fluid is commonly called the PNA distribution. Water is also an essential constituent of the reservoir fluid. A vast amount of water available in the reservoir is generally located in the water region found below the oil and gas zone due to the limited miscibility of water with hydrocarbon. Typically, this water is highly saline water with a salinity of even more than 30,000 parts per million (ppm) at certain places.

2.2.1.2 Chemical Properties of Formation Water

2.2.1.2.1 Formation Water Salinity

Formation water salinity measures the dissolved salts in the water and is determined by measuring the electrical conductivity. It is generally expressed in the unit of ppm or gram per liter (g/L) or percentage (%). Salts dissolved in the formation water are dissociated into their ions. Sodium (Na^+) and chloride (Cl^-) ions are predominant in the formation water, while magnesium (Mg^{2+}), sulfate (SO_4^{2-}), calcium (Ca^{2+}), and potassium (K^+) are other significant ions. Fresh water has a salinity of approximately 200 ppm, while saturated water has a salinity of around 400,000 ppm. Seawater has a salinity of approximately 35,000 ppm.

2.2.2 Physical and Reservoir Engineering Properties of Fluids

The physical properties of reservoir fluids, including gases, crude oil, and formation water, are described next.

2.2.2.1 Physical Properties of Gases

Understanding the relationship between volume (V), pressure (P), temperature (T), and different chemical and physical behavior of gases is an essential precursor to solving gas reservoir engineering problems. The various properties of gases comprise the following.

2.2.2.1.1 Apparent Molecular Weight (M_a)

The apparent molecular weight of a gas mixture is the contribution of molecular weights of individual components, along with their mole fractions in the mixture, and is given by

$$M_a = \sum y_i M_i \qquad (2.1)$$

where M_a is the apparent molecular weight, y_i is the mole fraction, M_i is the molecular weight, and i stands for the ith component.

2.2.2.1.2 Standard Volume (V_{sc})

The volume occupied by 1 pound mole of gas at standard pressure (14.7 psia) and temperature (60°F) is known as the standard volume.

For an ideal gas, the following relationship holds,

$$PV = ZnRT \tag{2.2}$$

Hence, from the ideal gas equation, the standard volume of gas can be calculated. P is pressure, V is volume and T is the temperature. The standard volume of 1 pound mole of gas at 14.7 psi pressure and 60°F temperature is given by

$$V_{sc} = 379.4 \frac{scf}{lb - mole} \tag{2.3}$$

2.2.2.1.3 Density (ρ_g)

The density of a gas mixture is computed by substituting the apparent molecular weight in the equation of state as

$$\rho_g = \frac{m}{V} = \frac{PM_a}{ZRT} \tag{2.4}$$

where ρ_g is the density of the gas mixture (lb/ft³) and M_a is the apparent molecular weight.

2.2.2.1.4 Specific Volume (V)

The volume occupied by the unit mass of an ideal gas is known as its specific volume. It is also defined as the inverse of the density.

$$V = \frac{v}{m} = \frac{RT}{PM_a} = \frac{1}{\rho_g} \tag{2.5}$$

where V equals the specific volume (ft³/lb) and ρ_g is gas density (lb/ft³).

2.2.2.1.5 Specific Gravity (γ_g)

The ratio of the density of gas to the density of air at the same temperature and pressure conditions is known as specific gravity.

$$\gamma_g = \frac{\rho_g}{\rho_{air}} \tag{2.6}$$

$$\gamma_g = \frac{\rho_g}{\rho_{air}} = \frac{M_a}{28.96} \tag{2.7}$$

2.2.2.1.6 Gas Compressibility Factor (Z)

The gas compressibility factor, also known as the "Z-factor" or "gas deviation factor," is a measure of deviation from the ideal gas behavior at a specific pressure and temperature conditions. It gives the relationship among volume, pressure, and temperature.

$$Z_{(P,T)} = \frac{PV_{\text{elevated condition}}}{PV_{\text{standard condition}}}$$ (2.8)

$$PV = ZnRT$$ (2.9)

It is the proportion of the actual volume of n-moles of gas at a given pressure and temperature condition to the ideal volume of the same number of gas moles at identical pressure and temperature condition. Therefore, Z is given as

$$Z = \frac{V_{\text{actual}}}{V_{\text{ideal gas}}} = \frac{V}{(nRT)/P}$$ (2.10)

2.2.2.1.7 Compressibility of Natural Gases (C_g)

Fluid compressibility is extremely important in reservoir engineering calculations. It is dependent on the pressure and temperature conditions. Gases are highly compressible, while compressibility of liquid is less or sometimes considered constant. The change in volume per unit volume with a unit pressure drop at a constant temperature is known as isothermal gas compressibility. It can be expressed as

$$C_g = -\frac{1}{V}\left(\frac{\partial V}{\partial P}\right)_T = \frac{1}{P} - \frac{1}{Z}\left(\frac{\partial Z}{\partial P}\right)_T$$ (2.11)

where C_g is the coefficient of isothermal gas compressibility (psi^{-1}). C_g can be obtained using the equation of state. For ideal gases, the gas compressibility factor (Z) is 1. Therefore, C_g can be expressed as

$$C_g = \frac{1}{P}$$ (2.12)

The relationship between isothermal gas compressibility (C_g), isothermal pseudo-reduced compressibility (C_{pr}), and pseudo-reduced pressure (P_{pc}) is given as

$$C_{pr} = C_g P_{PC}$$ (2.13)

2.2.2.1.8 Gas Formation Volume Factor (B_g)

The gas formation volume factor relates the gas volume at reservoir conditions to that under standard temperature and pressure (STP) conditions, that is, 14.7 psi and 60°F.

It is also known as the ratio of the actual volume of gas under reservoir conditions to the volume occupied by the same quantity of gas at STP conditions. It can be mathematically represented by

$$B_g = \frac{V_{reservoir}}{V_{surface}} = \frac{(V)_{P,T}}{(V)_{sc}} = \left(\frac{P_{std.\ condition}}{T_{std.\ condition}}\right)\frac{ZT}{P} \tag{2.14}$$

where B_g is the gas formation volume factor (ft³/scf), $V_{P,T}$ (ft³) is the volume of gas at P and T at reservoir condition and V_{sc} (scf) is the volume of gas at STP conditions.
    Therefore,

$$B_g = (0.02827)\frac{ZT}{P} \tag{2.15}$$

where Z is the gas compressibility factor.

2.2.2.1.9 Expansion Factor (E_g)

The expansion factor is the reciprocal of the gas formation volume factor. It can be represented by

$$E_g = \frac{(V)_{sc}}{(V)_{P,T}} = \frac{1}{B_g} \tag{2.16}$$

2.2.2.1.10 Gas Viscosity (μ_g)

The gas viscosity represents the internal fluid friction or resistance to the gas flow. The velocity gradient is high when the gas viscosity between the layers is low. When the gas viscosity increases, the fluid flow decreases, and the velocity gradient also decreases. Gas viscosity is defined as the ratio of the shear force per unit area to local velocity. The unit of gas viscosity (μ_g) is denoted by centipoises, poise, and micro poise.
    1 poise = 100 centipoises = 6.72×10^{-2} lb mass/ft-sec = 2.09×10^{-3} lb-sec/ft².
Also, gas viscosity (μ_g) is the function of pressure, temperature and mole fraction:

$$\mu_g = (P,T,y_i) \tag{2.17}$$

2.2.2.2 Physical Properties of Crude Oil

2.2.2.2.1 Oil Gravity (γ_o)

The ratio of oil density to that of water at a given pressure and temperature is known as oil-specific gravity. It is written as

$$\gamma_o = \frac{\rho_o}{\rho_w} \tag{2.18}$$

where γ_o is the specific gravity of the crude oil, ρ_o is the density of crude oil (lb/ft³) and ρ_w is water density (lb/ft³). The API distinguishes crude oil in terms of API gravity. It depicts the density of crude oil in comparison to water. It can be calculated using oil-specific gravity at STP conditions. It can be given mathematically as

$$\text{API} = \frac{141.5}{\gamma_o} - 131.5 \qquad (2.19)$$

Based on API gravity, crude oil can be divided into four categories:

a) Light crude oil (API gravity > 31.1° API)
b) Medium crude oil (22.3° API < API gravity < 31.1° API)
c) Heavy crude oil (10° API < API gravity < 22.3° API)
d) Extra-heavy crude oil (API gravity < 10° API)

2.2.2.2.2 Specific Gravity of the Solution Gas (γ_g)

The weighted average of specific gravities of the separated gases from the individual setup is used to determine the specific gravity of the solution gas. It can be expressed by

$$\gamma_g = \frac{\sum_{i=1}^{n} \left(R_{sep}\right)_i \left(\gamma_{sep}\right)_i + R_{st}\gamma_{st}}{\sum_{i=1}^{n} \left(R_{sep}\right)_i + R_{st}} \qquad (2.20)$$

where n is the number of separators, R_{sep} stands for separator gas–oil ratio (GOR; scf/STB), R_{st} is the GOR from the stock tank (scf/STB), γ_{sep} is the separator gas gravity, and γ_{st} is stock tank gas gravity.

2.2.2.3 Gas Solubility (R_s)

The volume of gas that is developed from the crude oil in standard cubic feet to the volume of produced oil following gas evolution in the stock tank barrel is known as gas solubility and is shown in Figure 2.6. Pressure, temperature, API gravity

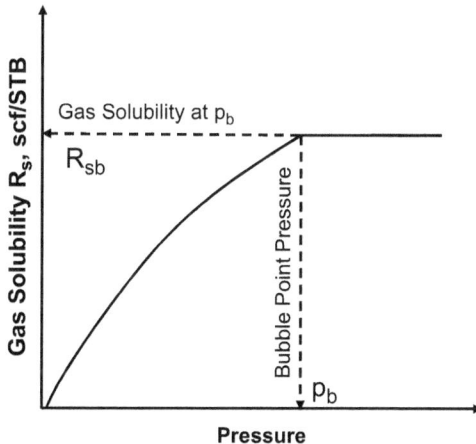

FIGURE 2.6   Schematic of the gas solubility versus pressure diagram.

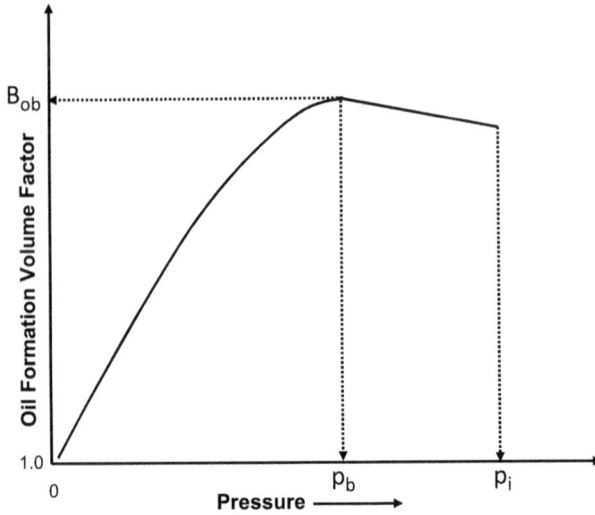

FIGURE 2.7 Schematic of the oil formation volume factor versus pressure.

and gas gravity affect the gas solubility. R_{sb} represents the gas solubility at the bubble point (P_b).

$$R_s = \frac{\text{volume of gas evolved from oil in scf}}{\text{volume of produced oil following gas evolution in STB}} \tag{2.21}$$

2.2.2.2.4 Oil Formation Volume Factor (B_o)

The oil formation volume factor is the ratio of the volume of crude oil along with the gas at reservoir conditions and the volume of crude oil at surface conditions. The graph of the oil formation volume factor versus pressure is shown in Figure 2.7. It is represented mathematically as

$$B_o = \frac{V_{o\ \text{at reservoir}}}{V_{o\ \text{at surface}}} = \frac{(V_o)_{P,T}}{(V_o)_{sc}} \tag{2.22}$$

where B_o is the oil formation volume factor (bbl/STB), $(V_o)_{P,T}$ and $(V_o)_{sc}$ is the volume of oil under reservoir conditions (bbl) and at STP conditions (STB), respectively.

2.2.2.2.5 Crude Oil Density (ρ_o)

Crude oil density is the ratio of the mass of crude oil to its volume at a particular pressure and temperature. Crude oil can be characterized concerning its density as its property changes with density, which can be calculated using the material balance equation as

$$\rho_o = \frac{62.4\gamma_o + 0.0136 R_s \gamma_g}{B_o} \tag{2.23}$$

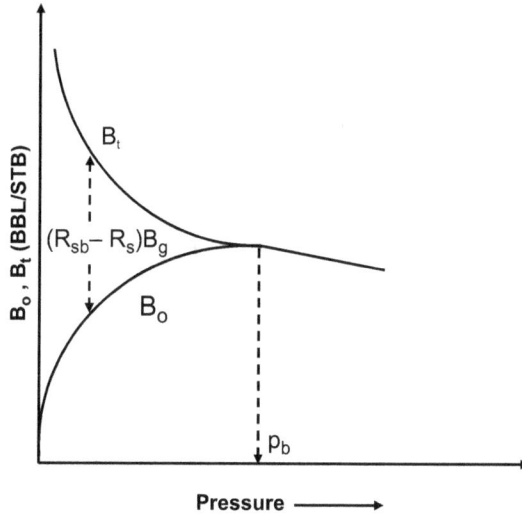

FIGURE 2.8   Schematic of the B_t and B_o with reservoir pressure.

where ρ_o is the oil density (lb/ft³), γ_o is the specific gravity of the stock tank oil and R_s is gas solubility (scf/STB).

2.2.2.2.6   Total Formation Volume Factor (B_t)

The total formation volume factor is the ratio of the total volume of hydrocarbon mixture at specified reservoir conditions to the volume of oil at STP conditions. It illustrates the connection between volume and pressure for a hydrocarbon mixture beneath the bubble point pressure. Figure 2.8 depicts the change in the total formation volume factor (B_t) and the oil formation volume factor (B_o) as a function of pressure. P_b is the bubble point pressure. It is mathematically expressed as

$$B_t = \frac{\left(V_o\right)_{P,T} + \left(V_g\right)_{P,T}}{\left(V_o\right)_{sc}} \tag{2.24}$$

where B_t is the total formation volume factor (bbl/STB), $(V_o)_{P,T}$ is the volume of the oil at P and T (bbl), $(V_g)_{P,T}$ (ft³) is the volume of the liberated gas at P and T, $(V_o)_{sc}$ is the volume of the oil at STP conditions (STB). In terms of gas solubility (R), the B_t is given by

$$B_t = B_o + \left(R_{sb} - R_s\right)B_g \tag{2.25}$$

2.2.2.2.7   Crude Oil Viscosity (μ_o)

The crude oil viscosity varies with pressure, temperature, and dissolved gas in the crude oil. Figure 2.9 indicates that the viscosity of oil increases with a fall in pressure below the bubble point pressure due to the release of dissolved gases. The oil viscosity increases as the pressure rises above the bubble point pressure due to liquid

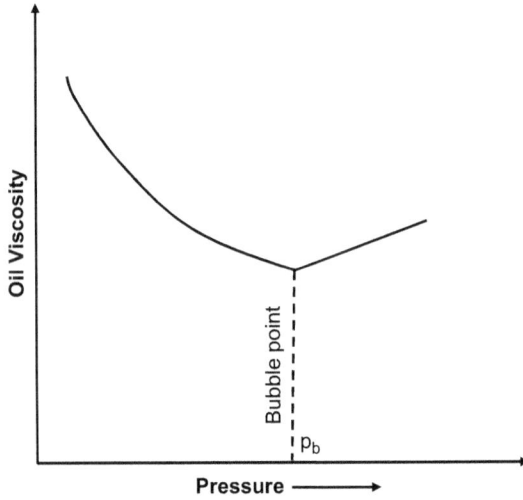

FIGURE 2.9   Schematic of the oil viscosity versus pressure.

compression. Crude oil viscosity will be minimal at saturation pressure (bubble point pressure). Also, the crude oil viscosity reduces with an increase in temperature. The property and type of crude oil change with increased crude oil viscosity and increased friction between fluid layers.

2.2.2.3   Physical Properties of Formation Water

Formation water often gets produced with oil and gas from the reservoir. Several properties of formation water are of interest for reservoir engineering calculation, as discussed.

2.2.2.3.1   *Isothermal Compressibility of Water (C_w)*

Formation water isothermal compressibility is the change in water volume per unit water volume with a unit pressure drop at a constant temperature. It can be expressed mathematically as

$$C_w = -\frac{1}{V_w}\left(\frac{\partial V_w}{\partial P}\right)_T \tag{2.26}$$

where C_w is the water isothermal compressibility (psi^{-1}), V_w is the volume of water, and P is the pressure. Formation water, or brine, is considered a slightly compressible liquid with a very low or constant C_w.

2.2.2.3.2   *Water Formation Volume Factor (B_w)*

The ratio of the volume of water (with dissolved gas) at reservoir condition to the volume of water at surface condition is known as the water (brine) formation volume factor.

$$B_w = \frac{V_{w\ at\ reservoir}}{V_{w\ at\ surface}} \quad \text{bbl / STB} \tag{2.27}$$

2.2.2.3.3 Water Viscosity (μ_w)

Water viscosity depends on the pressure, temperature, and dissolved solids. It is directly proportional to salinity, pressure and inversely proportional to temperature. Therefore, water viscosity increases with the pressure and salinity and decreases with temperature.

2.2.2.3.4 Gas Solubility in Water (R_{sw})

The solubility of gases in water does not affect the viscosity of water. Gas in the water phase behaves differently than in hydrocarbons. The pressure is higher when the gas is soluble in water, maintaining high reservoir pressure. The production rate also increases when the gas is dissolved into the water.

2.2.2.3.5 Water Density (ρ_w)

The density of water is always higher than the density of gas and oil. Therefore, water tends to dip downward, and thus, the buoyancy force occurs in the reservoir, helping reservoir fluid flow to the surface.

2.3 PETROLEUM RESERVOIR CLASSIFICATION

Crude oil and natural gas are composed of various hydrocarbons. The difference in their molecular weight means they exist in different states at room temperature. For example, light fractions (C_1–C_4) are present in gaseous form, whereas higher molecular weight alkanes (C_5–C_{15}) exist in the liquid state. A standard approach is required to categorize them into different types based on their chemical and physical characteristics. Oil reservoirs have a reservoir temperature less than the critical temperature of the reservoir fluid (i.e., oil). In contrast, gas reservoirs have a temperature greater than the critical temperature of reservoir fluids. Hydrocarbon reservoirs are classified into the following five main types: dry gas, wet gas, gas condensate, volatile oil, and black oil.[8, 9] Various reservoir fluids are represented by a schematic as shown in Figure 2.10, and their typical composition is given in Table 2.1. The phase behavior of these reservoir fluids is discussed in Section 2.4.

2.3.1 COMPOSITION OF OIL AND GAS RESERVOIR FLUIDS

Gas reservoirs, by name, are in the gaseous state at reservoir pressure and temperature conditions. The critical temperature of these fluids is less than the prevailing reservoir conditions (discussed in relevance to the phase diagram in the subsequent section). They primarily have a high content of lighter and intermediate hydrocarbons (C_1–C_6). Dry gas has mainly C_1 and C_2 and other hydrocarbons in a very small amount. On the other hand, gas condensate contains more C_4 and

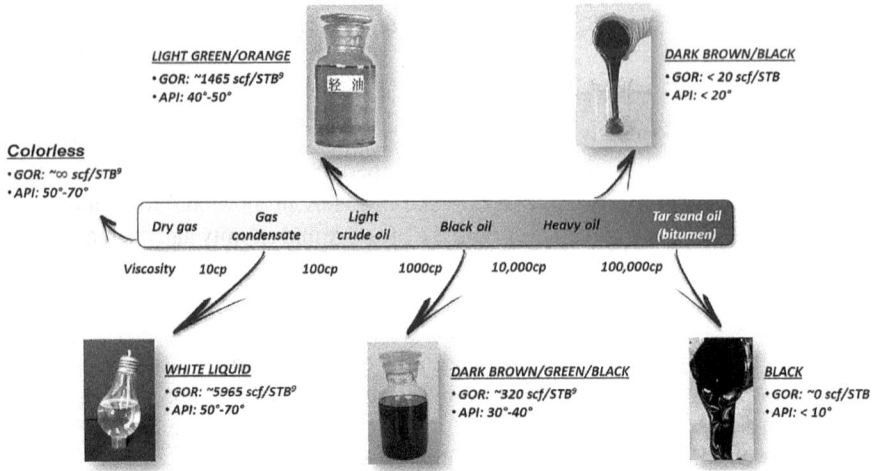

FIGURE 2.10   Representation of various reservoir fluids and their properties.[9]

TABLE 2.1

Cronquist Reservoir Classification and Typical Chemical Composition of Reservoir Fluid[8, 9]

		Black Oil	Volatile Oil	Gas Condensate	Wet Gas	Dry Gas
	CO_2	0.02	1.82	2.37	1.41	0.10
	N_2	0.34	0.24	0.37	0.25	2.07
Composition (mol %)	C_1	**34.62**	**57.6**	**73.19**	**92.46**	**86.12**
	C_2	4.11	7.35	7.8	3.18	5.91
	C_3	1.01	4.21	3.55	1.01	3.58
	iC_4	0.76	0.74	0.71	0.28	1.72
	nC_4	0.49	2.07	1.45	0.24	-
	iC_5	0.43	0.53	0.64	0.13	0.50
	nC_5	0.21	0.95	0.68	0.08	-
	C_6	1.16	1.92	1.09	0.14	-
	C_{7+}	**56.4**	**22.57**	**8.21**	**0.82**	-

C_5 components than wet and dry gas. The C_{7+} content in gas condensate is also significantly higher than wet gas (~8.21 mol% for gas condensate and ~0.82 mol% for wet gas, Table 2.1).[8, 9] Oil reservoirs consist of black oil and volatile oil, both of which have a relatively lower content of C_1. In addition, markedly higher C_{7+} content in black oil (~56.4 mol%) and volatile oil (~22.57 mol%) is visible compared to gas reservoirs.[8, 9]

2.3.2 TAR SAND AND BITUMEN

Another unconventional crude oil reserve includes tar sand and extra heavy oil. Tar sand, also called oil sand or bituminous sand, is a bitumen-infused deposit. It is dense, viscous, and generally immobile under reservoir conditions. The API gravity of tar sand bitumen lies in the range of 5° API to 10° API.[10] The reservoir viscosity of tar sand is greater than 10,000 cP. Tar sand also contains significant concentrations of sulfur, nitrogen, oxygen, and heavy metals. Tar sand is more viscous and less volatile than conventional and heavy oil. The amount of bitumen in the Canadian Athabasca tar sand deposits is estimated to be at least 2 trillion barrels (2×10^{12} bbl). Likewise, the amount of bitumen in the Venezuelan Orinoco tar sand deposits is about 2 trillion barrels (2×10^{12} bbl).[11] Alberta, Canada, is home to 81% of the world's known recoverable tar sand bitumen.

Extra-heavy oil is bitumen in a nearly solid state, unable to flow freely under ambient conditions. Bitumen derived from tar sands is often classified as extra-heavy oil.[10] Heavy oil primarily contains more asphaltene, resins, and trace amounts of heavy metals. The API gravity of extra-heavy oil remains less than 10° API. The reservoir viscosity of extra-heavy oil lies between 100 cP to 10,000 cP. Natural bitumen and extra-heavy oil are the by-products of massive amounts of conventional oils that have been produced and degraded, primarily by bacteria.[12] Bitumen and extra-heavy oil are chemically and texturally similar to the residuum produced by refinery distillation of light oil.

2.3.3 EXAMPLES OF WORLDWIDE OIL AND GAS RESERVOIRS

Following are a few examples of reservoirs worldwide based on the reservoir fluid types (as in Table 2.1).

a) Dry gas (non-associated) reservoirs:[13, 14]
   - Hugoton Gas Field: US
   - Hassi R'Mel gas field: Algeria
   - Groningen gas field: The Netherlands
b) Wet gas reservoirs:
   - Tiguentourin, Hassi Ouan Abecheu, Hassi Ouan Taredert and Hassi Farida fields: ("In Amenas Gas Project"), Algeria[15]
   - Marcellus Shale: Marcellus, New York[16]
c) Gas condensate reservoir:
   - South Pars/North Dome: Offshore Persian Gulf
   - North Belut field: Offshore (West Natuna Sea), Indonesia[17]
   - Arun field: Onshore North Sumatra, Indonesia[17]
   - Santa Barbara field: Venezuela[17]
   - Urengoyskoye gas and condensate field: northern West Siberia Basin, Russia[18]

d) Black oil reservoir:
  • Digboi, Naharkatiya, Moran-Hugrijan oil field: Assam, India
  • Bombay High: Offshore Arabian Sea
e) Heavy oil reservoir:
  • Carabobo field, Orinoco belt: Venezuela
f) Bitumen reserves:
  • Athabasca, Cold Lake, Peace River: Canada

2.4 PHASE BEHAVIOR FUNDAMENTALS

The term *phase* refers to any homogeneous and physically well-defined quantity of matter, separated by certain boundaries from the rest of the mixture. For example, water, ice, and water vapor are three different phases. At certain conditions of pressure and temperature, either one, two or all three phases can coexist. Based on the pressure and temperature conditions in the reservoir and production systems, petroleum fluids may exist in gas, liquid, and solid phases. In the oil, the solids, such as waxes, may also exhibit various kinds of crystalline structures. The phases of the systems and their properties do change with pressure and temperature conditions. The terminology of phase behavior is refereed to infer this phase-change behavior in the fluid system.[2, 19]

The hydrocarbon fluids exist in the reservoir in all three phases at various pressure and pressure conditions. Most of the time, the solid phase behavior of hydrocarbon fluids (such as wax, asphaltene, and hydrates) is not given enough attention. The solids in the petroleum fluids and their phase behavior often result in flow assurance challenges. Thus, operating pressure and temperature conditions are of prime importance to predict the corresponding phase of the reservoir fluid to extract the hydrocarbon fluids effectively and easily. The distinct phase diagram of various hydrocarbon fluids can be inferred from experimental or mathematical techniques concerning various conditions of pressure, temperature, chemistry, and composition. Understanding the respective phase diagram and corresponding reservoir fluid properties is important for engineers to estimate the recovery of hydrocarbons and mitigate flow assurance issues. Phase diagrams also help in modeling the fluid flow behavior of the reservoir fluids in porous media, wellbore, and surface facilities. The phase behavior of hydrocarbon fluids is vital for predicting and estimating various parameters. These include multiphase flow nature, gas void fraction, liquid hold up, pressure drop in the wellbore and tubular, wax asphaltene deposition, hydrate formation, and so on.[1] The details on various phase diagrams are discussed subsequently. To understand the phase behavior of complex reservoir fluids, one must understand the single-component and binary-system phase behavior.

2.4.1 Phase Diagram of a Pure System

Pure component phase diagrams are the simplest to construct as only one component is present in the system. According to the Gibbs phase rule, since the system has at least one phase, the number of degrees of freedom is two, and thus, two-dimensional

phase diagrams are suitable for understanding the phase behavior. The pure component phase diagram information is important to understand the more complex multicomponent hydrocarbon systems as often observed in the petroleum industry. This
section briefly discusses the phase behavior of pure water to understand the phase
behavior concepts and then learn about the bubble point, dew point and critical points
associated with phase diagrams.

2.4.1.1 Phase Diagram of Pure Water

Water can be present in three different states, such as liquid water (L), water vapor
(V), and ice (S). A schematic of the phase behavior of water is shown in Figure 2.11.
If the system under consideration is at a one, two or three-phase equilibrium condition, then the system will have zero, one or two degrees of freedom. Hence, the phase
diagram can thus be represented by point, line or region in the P-T diagram. The
three curves, SV, LV and SL, represent solid–vapor, liquid–vapor, and solid–liquid
phase equilibrium, respectively. The space within the curves represents one phase
region (and has two degrees of freedom). The P and T conditions may vary without
changing the phase type and nature (solid, liquid, or vapor). All three-phase meet at
the triple point and thus have a zero degree of freedom, meaning the P and T conditions are fixed at the triple point. A critical point also has a zero degree of freedom. At
this condition, the properties of the vapor phase and liquid phase become identical.
Above the critical point, the fluid cannot be vaporized due to increased temperature
or liquified by increasing pressure. The region beyond the critical point is referred to
as supercritical fluids.

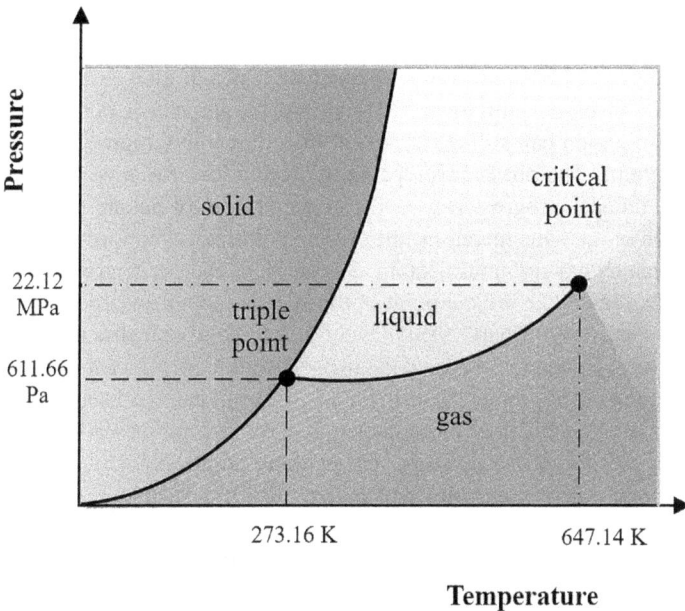

FIGURE 2.11   Schematic of the phase behavior of pure water.

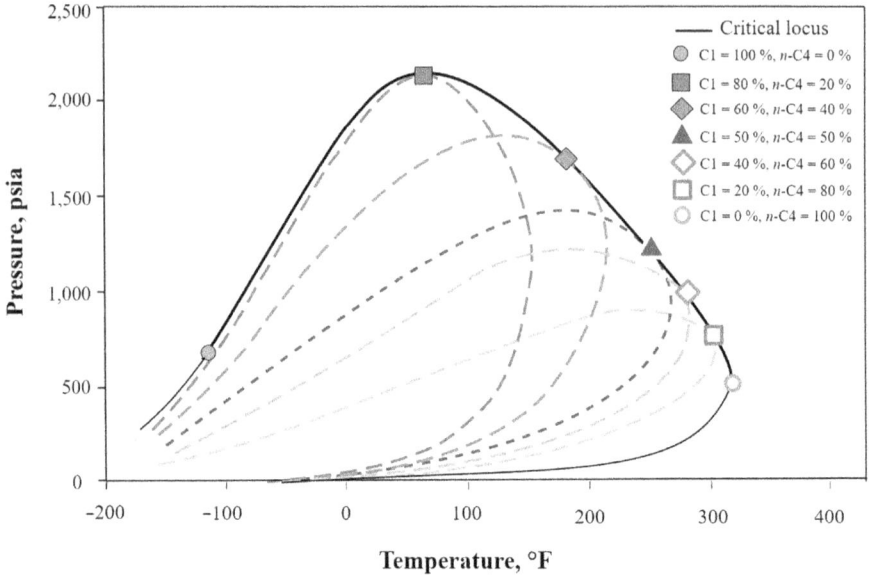

FIGURE 2.12 Phase envelope of a binary system containing methane and n-butane at various compositions.

2.4.2 PHASE DIAGRAM OF BINARY AND TERNARY SYSTEMS

Initially, it is vital to understand the concepts of the phase behavior of binary component systems. Figure 2.12 depicts the phase envelopes of methane and normal butane mixture with varying compositions.[1] As methane concentration decreases in the system, the phase envelopes shift toward the right. As the percentage of methane reduces, the two-phase region can still exist even at the comparably higher temperatures on the P-T diagrams. The phase envelope shifts from a lower to a higher temperature, and the size reduces progressively as the concentration of butane increases. As the concentration of methane increases, the critical point moves toward the critical point of pure methane on a certain trajectory. Similarly, as the concentration of butane in the system increases, the critical point of the system moves toward the critical point of butane. As shown in Figure 2.13, the addition of heavier hydrocarbon (n-decane) also affects the size and the shape of the phase envelop.[1] A noticeable shifting of the critical point toward the right side of the phase diagram can be observed. Thus, it can be inferred that as the number of components in the system increases, the change in shape and size of the phase envelope will be more noticeable. Crude oil has a wide variety of hydrocarbon molecules and organic and nonorganic gases, which affect the phase behavior of various hydrocarbon systems. The phase behavior of complex hydrocarbon fluids is estimated using complex experimental and analytical methods, primarily based on the types, quantity, composition, and chemistry of components present in the hydrocarbon fluid. For detailed information on the phase diagram of

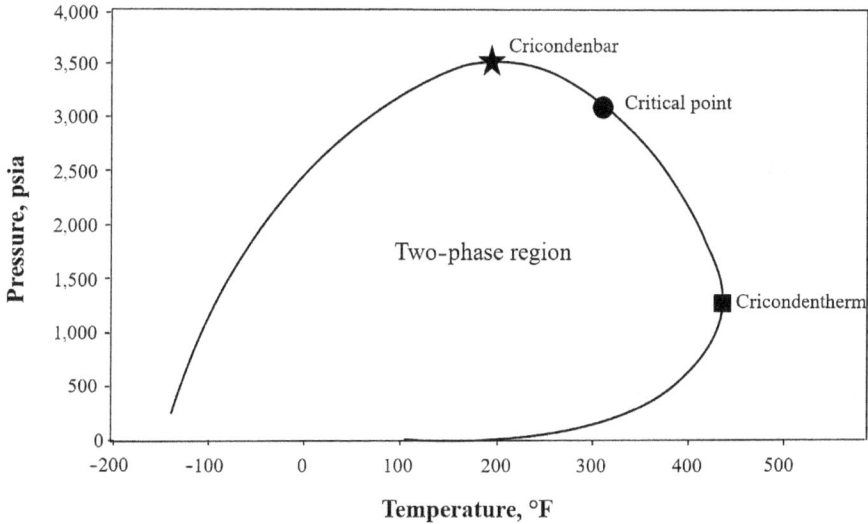

FIGURE 2.13 Phase envelope of a ternary system consisting of 70 mol% methane, 20 mol% n-butane, and 10 mol% n-decane.

multicomponent systems, readers can refer to other texts.[1, 2, 19, 20] We discuss further details on the phase behavior of hydrocarbon fluids in the subsequent section.

2.4.3 PHASE BEHAVIOR OF RESERVOIR FLUIDS

The reservoir fluid phase behavior has implications in exploration, production, transportation strategy and surface facilities design. These hydrocarbon fluids are attributed to their respective categories based on the fluid analysis done in the lab, which involves the reservoir fluid's pressure–volume–temperature (PVT) analysis. Each of these reservoir fluids is produced using specific production techniques. The most important factors in classifying the five types of reservoir fluids are their differences in composition (as discussed in Section 2.3). The size and shape of each fluid's phase diagram vary based on the reservoir fluid type.

Typically, the reservoir hydrocarbon fluids are classified based on their nature at the prevailing reservoir temperature and pressure conditions in the phase behavior curve. The reservoir fluids are expected to flow from reservoir to wellbore and separator, affecting their state from the reservoir to the surface. Thus, the information on phase behavior change during the production operations is vital when defining the classification of reservoir fluids. Production profile data, such as GOR, API gravity, and fluid appearance in standard conditions, are primarily used for easy classification. Laboratory experiments also provide accurate information on the reservoir fluid type and their various properties based on phase behavior. However, it should not be forgotten that the best practice to classify the reservoir fluid is to distinguish them with the help of phase diagrams. The phase behavior of five types of reservoir fluids is discussed in the following section.

2.4.4 CONSTRUCTION OF PHASE DIAGRAM

The phase envelope of the reservoir fluid consists of the bubble point and dew point curve determination at various temperatures (saturation pressures, bubble point, or dew point), which can be done by using extensive laboratory experiments or the robust equation-of-state (EOS) models. Once the bubble point and dew point pressure conditions at various constant temperatures are determined, a phase envelope can be constructed on the cartesian coordinates with pressures (P) on the y-axis and temperatures (T) on the x-axis. These phase diagrams are commonly referred to as P-T diagrams. The typical procedure adopted while plotting the P-T diagram using laboratory experiments and the state equation is briefly discussed here.

A schematic of the PVT cell is shown in Figure 2.14. PVT cells are used in the laboratory for dew point and bubble point pressure estimation of various hydrocarbon fluids. The pressure inside the cell is maintained at reservoir conditions by a hydraulically operated piston. The temperature is regulated by water or oil bath or a suitable method. The cell is also equipped with a sapphire window for visualizing the hydrocarbon fluid phase behavior change with change in the P-T conditions. The general practice is to use live reservoir fluids for phase envelope determination. However, a recombined sample of gas and oil from the wellhead or separator is used to get representative reservoir fluid samples. For this, a separate recombinant cell is used, or they can be recombined based on existing GOR in the PVT cell.

Figure 2.15 shows the schematic of measurement of bubble point pressure for a crude oil sample. Once the live reservoir fluid sample or recombined fluid sample is transferred into the PVT cell at reservoir pressure (typically at higher pressure), the temperature of the PVT cell is maintained at the desired condition at which condition the bubble point is to be measured. Sufficient time is allowed to attain the thermal equilibrium. Once the sample attains a homogeneous single phase in the cell by applying predefined pressure and temperature conditions, cell pressure is decreased gradually in stages (isothermal conditions).

Fluid-phase change is observed and recorded at each stage using a video camera and cathetometer attached to the PVT cell. The cathetometer accurately estimates the gas–oil interface level, which is used to calculate the exact gas and liquid amounts present at each stage during the gradual pressure decrease. Information is then used to estimate the single- and two-phase regions of the phase envelop involving tie-lines. The fluid is allowed to be in thermodynamic equilibrium at each step during the gradual pressure decrease to avoid any experimental errors. The pressure is recorded as the bubble point once the first gas bubble is observed (see Figure 2.15). Similarly, the dew point pressure is recorded when the first dew of liquid is observed. After one such experiment, the same sample is loaded to a single-phase condition at another temperature. The preceding procedure is repeated to generate bubble and dew point loci, and the phase envelope is constructed. PVT cells are also commonly used in flow assurance studies, including asphaltene phase behavior prediction. PVT cell experiments are time-consuming and tedious. Also, they are not available easily at many locations, restricting their use.

In PVT studies, a true representative of reservoir fluids may not be obtained sometimes due to some hurdles, including unfeasible sampling due to time constraints,

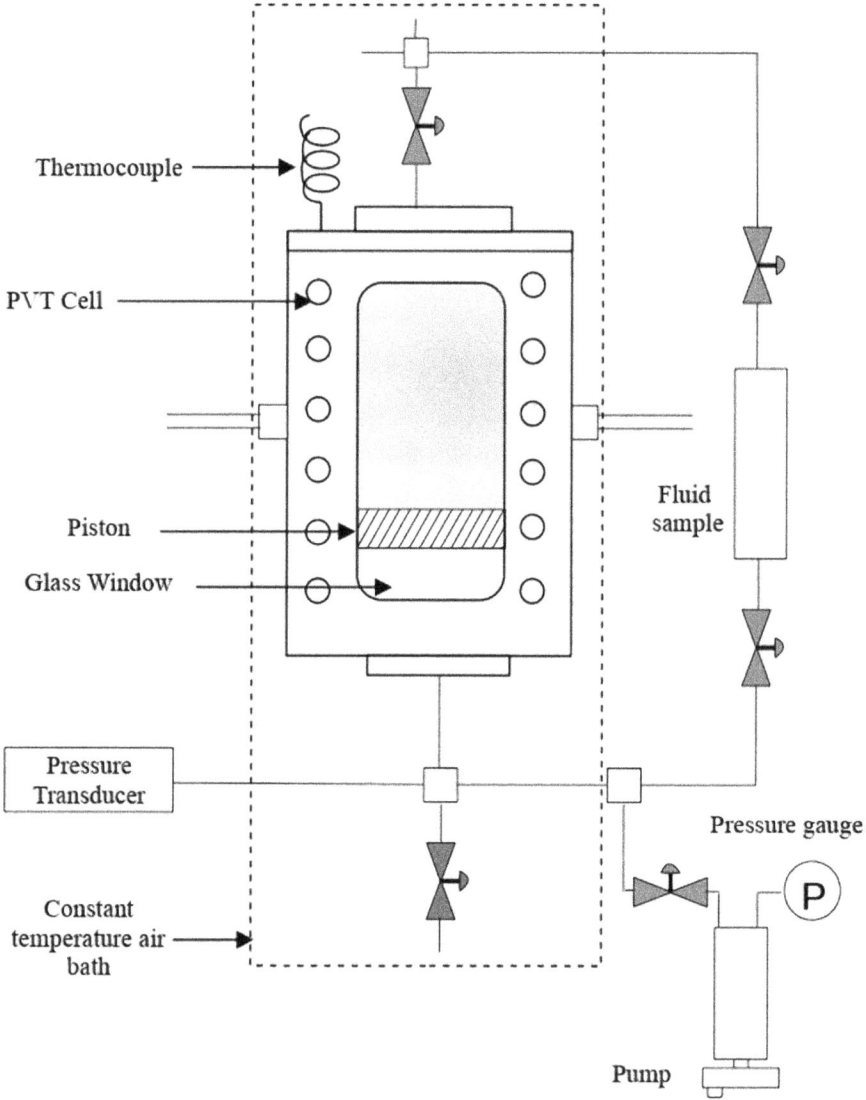

FIGURE 2.14 Typical pressure–volume–temperature (PVT) cell employed to carry out various PVT studies on hydrocarbon fluids and develop phase envelop experimentally.

doubtful reservoir or recombinant samples, erroneous laboratory results, and so on. The bubble point and dew point pressures are estimated from the various EoS models in such situations. The reservoir fluid composition must be known if EoS models are to be used. EoS models are not only useful in the construction of phase envelopes of hydrocarbon fluids but also in the estimation of crude oil and gas compositions using flash calculations, API gravity, the viscosity of the hydrocarbon fluids, and many other reservoir fluid properties of the equilibrium oil and gas phases, relevant for flow assurance studies.

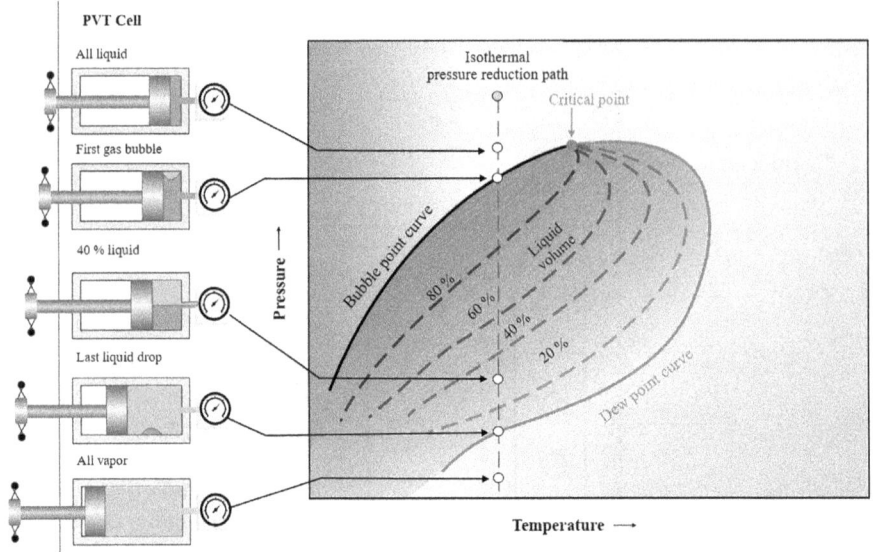

FIGURE 2.15 Schematic representation of measurement of bubble point pressure in a PVT cell.

2.4.5 SALIENT FEATURES OF THE PHASE BEHAVIOR OF FIVE RESERVOIR FLUIDS

2.4.5.1 Black Oil

Black oils are one of the most widely found reservoir fluids. As the name suggests, they are dark-colored, sometimes resembling black. Ordinary black oils generally contain more than 20% C_{7+} fraction. As a result, their phase envelopes are broader than other petroleum reservoir fluids. A typical black oil phase diagram is shown in Figure 2.16.[21] The critical point of the black oil system lies on the phase envelope much farther than the reservoir temperature, toward the right side.

Points 1, 2, and 3 indicate the reservoir conditions, the isothermal pressure reduction path intersection with bubble point curve and separator conditions, respectively. The pressure and temperature change during the production, leading to an actual pressure reduction path, is shown in Figure 2.16. The crude oil exists in the liquid phase only between points 1–2 and can dissolve more gas at these conditions. It is an undersaturated condition, and the corresponding reservoirs are called undersaturated reservoirs. As the pressure declines and point 2 is reached on the bubble point curve, the crude oil gets saturated and tends to release gas instead of further dissolution. Hence, the reservoir conditions from this point onward correspond to the saturated reservoirs (within the phase envelope) and have lower bubble points due to higher critical temperatures. Lower bubble point results in a low GOR (low shrinkage oils). The amount of gas and oil in equilibrium beyond point 2 (on lines 2–3) at each stage is represented by the corresponding quality lines within the phase envelope. The free gas fraction is subtracted from the volumetric estimation of the remaining crude oil at 1. The oil–gas quality lines lie closer to the bubble point at

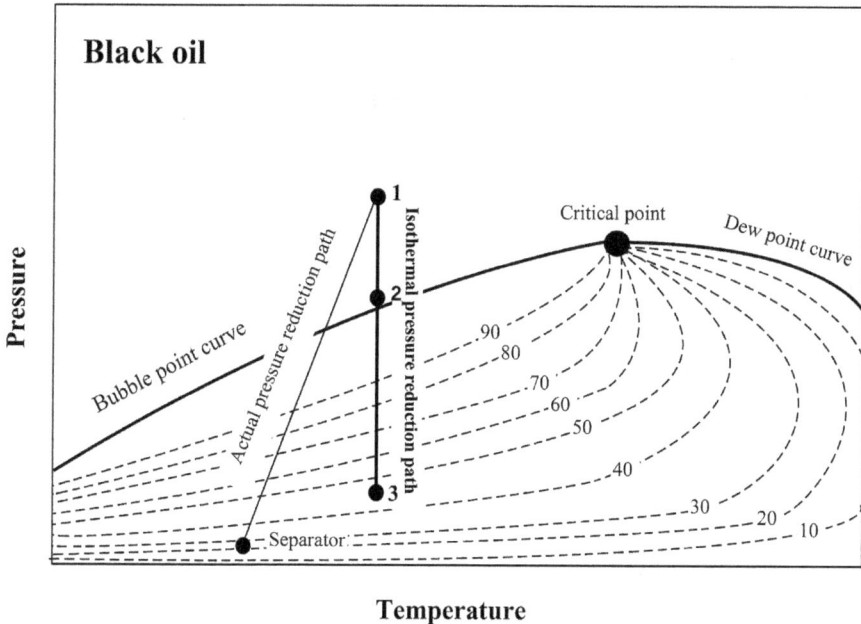

FIGURE 2.16   Schematic of the phase behavior of ordinary black oil.

separator conditions (or point 3), which is operated preferably to yield more crude oil at separator conditions due to the low-shrinkage behavior. The following black oil properties are widely accepted in the absence of a perfect definition. Generally, the API gravity of these crude oils ranges from 40° to 50°, and the GOR is approximately 320 scf/STB.[9, 22] The B_o or formation oil volume factor is less than 2 rb/STB (rb: reservoir barrel).[22] As ordinary black oils have a large fraction of higher molecular weight hydrocarbons. Therefore, they appear dark, although green and brownish colors may also be encountered occasionally.[2]

2.4.5.2   Volatile Oil

Figure 2.17 shows the schematic representation of the typical phase behavior of volatile oil. The reservoir conditions are closer to the cricondentherm of the volatile oil phase envelope. Also, they are located at a lower pressure and temperature as compared to the black oils. Due to the wide variety of hydrocarbon, the phase envelope covers a comparatively higher pressure range, leading to higher bubble point pressure than the ordinary black oils. The critical point of volatile oil lies much closer to the reservoir temperatures than ordinary black oils; hence, they are also known as near-critical crude oils. Volatile oils contain a relatively higher number of lighter hydrocarbons than ordinary black oil. They are also referred to as high-shrinkage oil due to more gas dissolved in the oil.

In Figure 2.17, the straight-line 1–2–3 represent the isothermal pressure depletion path. Due to the high-shrinkage behavior of the oil, the quality lines are relatively closely packed and closer to the bubble point curve. It means there will be many

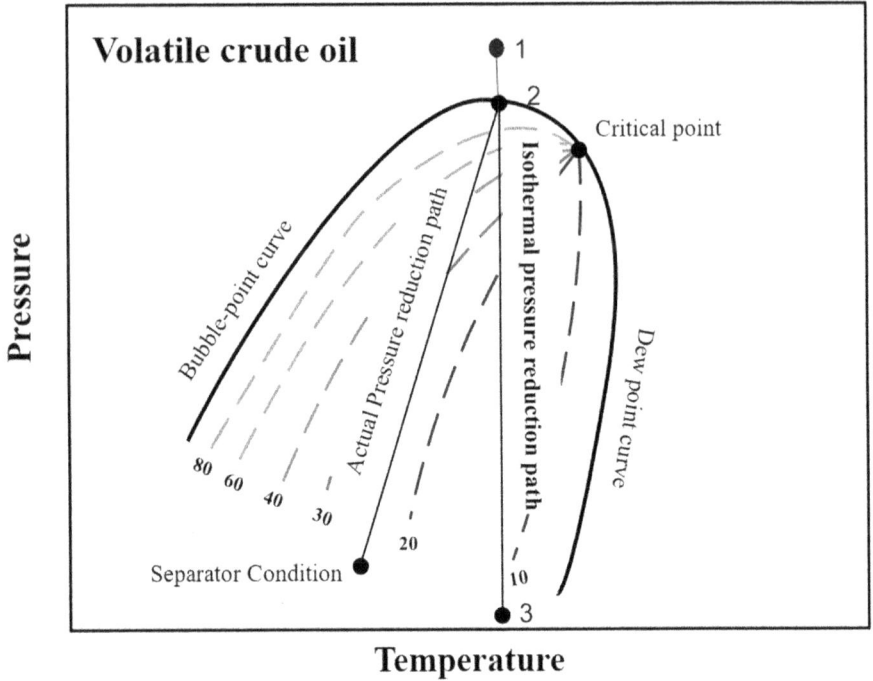

FIGURE 2.17   Schematic of the phase behavior of volatile oil.

quality lines intersecting line 2–3 once the pressure gets reduced below the bubble point curve.[22, 23] As the quality lines shift closer to the bubble point curve, the surface separation condition happens to be at lower iso-vols (quality lines), resulting in less surface liquid recovery. However, the separator gas can further be condensed to improve the oil recovery. The volatile crude oil has API gravities above 40°.[9, 22] The GORs typically lies between 2000–3000 scf/STB,[22] although a value of approximately 1465 scf/STB has also been reported.[9] The oil formation volume factor values are generally 2 rb/STB or higher.[2] One of the characteristic of volatile oils is that produced gases contain a significant quantity of liquid, contributing to oil production at the surface during depletion at the later stages.[9]

2.4.5.3   Retrograde Gas Condensate

Figure 2.18 shows the schematic of the phase behavior of a typical retrograde gas condensate reservoir. Retrograde gas condensate systems generally contain a higher amount of methane (about 75–85% of the total composition), the rest being C_2 to C_7 hydrocarbon and sometimes, not more than 12.5% heavier C_{7+} fractions. As the gas condensates contain lesser hydrocarbon components than black and volatile oils, their phase envelops are also smaller than the latter two. The critical point shifts farther down and towards decreasing temperature on the phase envelop. The reservoir conditions fall well within the gas phase, typically between the critical point and cricondentherm.[2, 23]

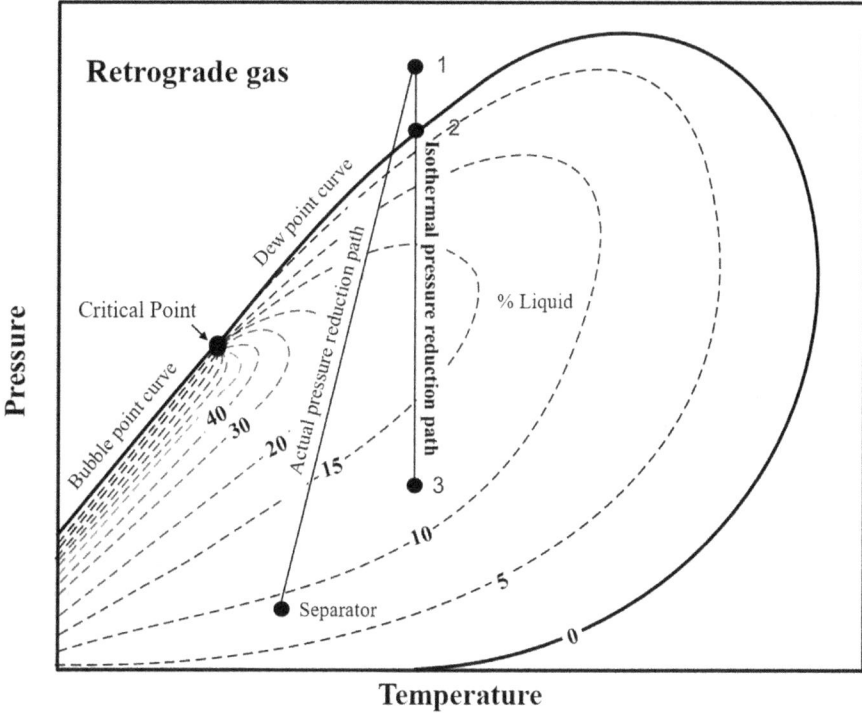

FIGURE 2.18   Schematic of the phase behavior of retrograde gas condensate.

The vertical line 1–3 represents the isothermal pressure depletion path lying between the critical point and cricondentherm. Point 1 corresponds to the initial vapor phase of the hydrocarbon fluid. As the isothermal pressure depletion of the reservoir commences, the vapor phase keeps on expanding until point 2, also known as retrograde dew point. A further reduction in pressure from point 2 causes heavier molecules in the vapor gas to lose their kinetic energy, and thus, they agglomerate and precipitate from the gas. Subsequent depletion of the pressure farther down on the envelope causes the condensed liquid to start evaporating slowly. The pressure depletion path crosses each tie-line twice in this process, known as the retrograde condensation phenomenon. The two-phase will exist simultaneously in the reservoir pores in the retrograde condensate region, one as gas and the other as liquid condensates. The free liquid saturation in pores increases with liquid dropout. This liquid saturation remains an immobile phase until a critical saturation. The separator condition falls farther from the bubble point curve, and it produces less liquid and more gas, which gets condensed at the separator. Hence, the GCR (gas-to-condensate ratio) is generally used for gas condensate systems rather than the GOR. The phase envelope size reduces beyond the critical point, and cricondentherms move closer to the reservoir temperature with an increase in GCR, leading to a very little precipitation of condensate liquid in the reservoir. Typical hydrocarbons fluids with this kind of behavior are termed *lean gas condensates*. However, the reservoir fluids having lower gas condensates ratios will result in more liquid dropout just after passing dew point pressure. They are typically known as rich gas condensates. Due to liquid dropout in gas condensate reservoir, the recombined sample of precipitated condensates at separator and produced gas at the surface does not resemble the exact hydrocarbon fluid composition present in the reservoir,[24] thus making the downhole fluid sampling the preferred choice for gas condensate reservoirs, especially during the later production stages of the reservoir. The GCR of typical gas condensate reservoirs ranges from 3,000–150,000 scf/STB, and the API gravity of condensates ranges from 50–70°.[9, 22] The high API gravity condensates look watery white or slightly colored.[19, 22]

2.4.5.4 Wet Gas

As discussed earlier, this system predominantly contains abundant methane molecules, a few intermediate hydrocarbon compounds, and C_{7+} fraction is less than 1 wt%. Figure 2.19 shows the typical phase envelop of a wet gas system. Wet gas exhibits a much smaller phase behavior envelope than other fluid types, which is shifted toward lower temperatures relative to retrograde gas condensates. The critical point is placed on the left side of the envelope at a lower temperature. The reservoir (pressure and temperature) lies farther away from the phase envelope than the cricondenbar and cricondentherm (in the gas phase). The isothermal pressure reduction line (1–2) never intersects the phase envelope during the pressure depletion.[23] The separator conditions are located inside the phase envelope, close

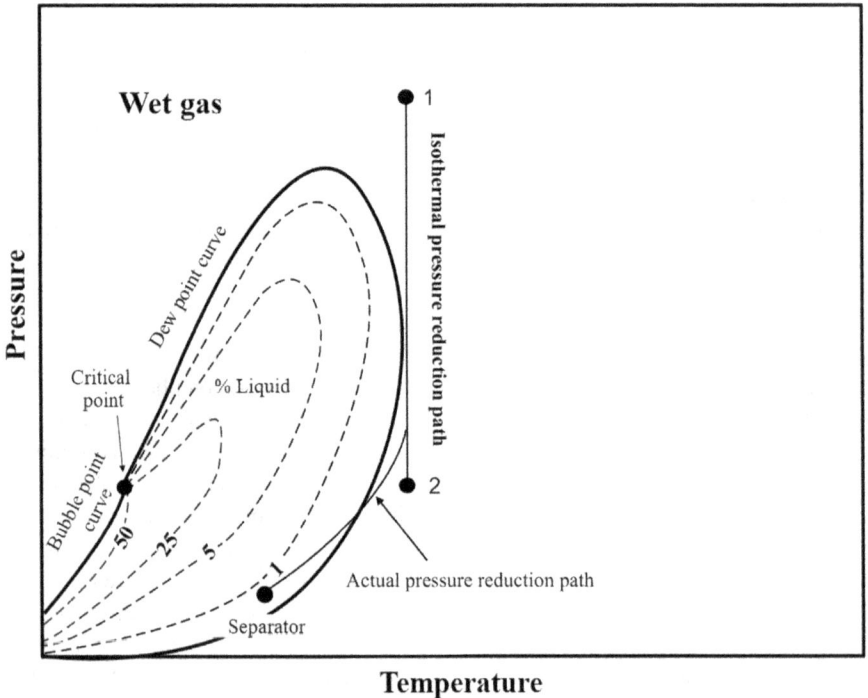

FIGURE 2.19   Schematic of the phase behavior of wet gas.

to the dew point curve. As the wet gas flows from the reservoir to the separator, some liquid is produced as it crosses the corresponding quality line in the phase envelope. The separator liquid and gas can be recombined to form the representative reservoir wet gas sample as there is no liquid dropout within the reservoir. In practice, when the producing GCR is approximately 60,000 to 100,000 scf/STB, the hydrocarbon fluids are defined as wet gas. Typical, API gravity of wet gas ranges from about 50–70° API, and generally, they have a watery white appearance.[2, 19]

2.4.5.5 Dry Gas

Figure 2.20 shows the schematic of the phase envelope of the dry gas system. Dry gas predominantly contains methane. Hence, the phase envelope is thin, simplest compared to all other types of reservoir fluids, and lies along with critical at much lower temperatures than the wet gas system. The reservoir and separator conditions fall well within the gas phase, producing only gases at surface conditions. The isothermal pressure depletion line (1–2) never intersects the phase envelope of the dry gas reservoir. Hence, practically no condensates will be produced. The dry reservoirs do not contain heavier hydrocarbons, and the GOR of dry gas is greater than 100,000 scf/STB,[9, 19] and API gravity ranges from 86–104°.[25]

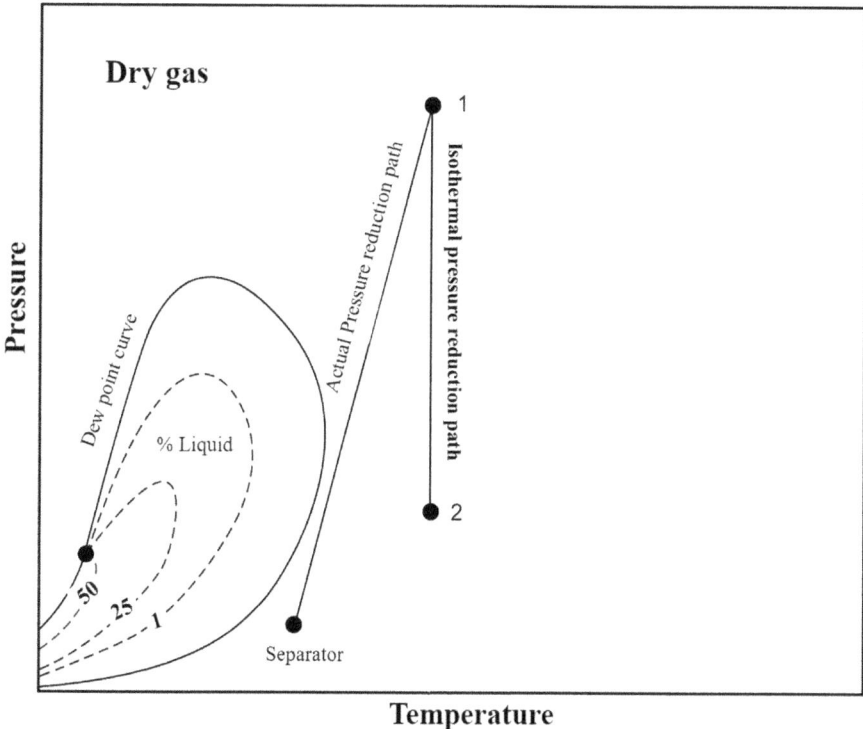

FIGURE 2.20   Schematic of the phase behavior of dry gas.

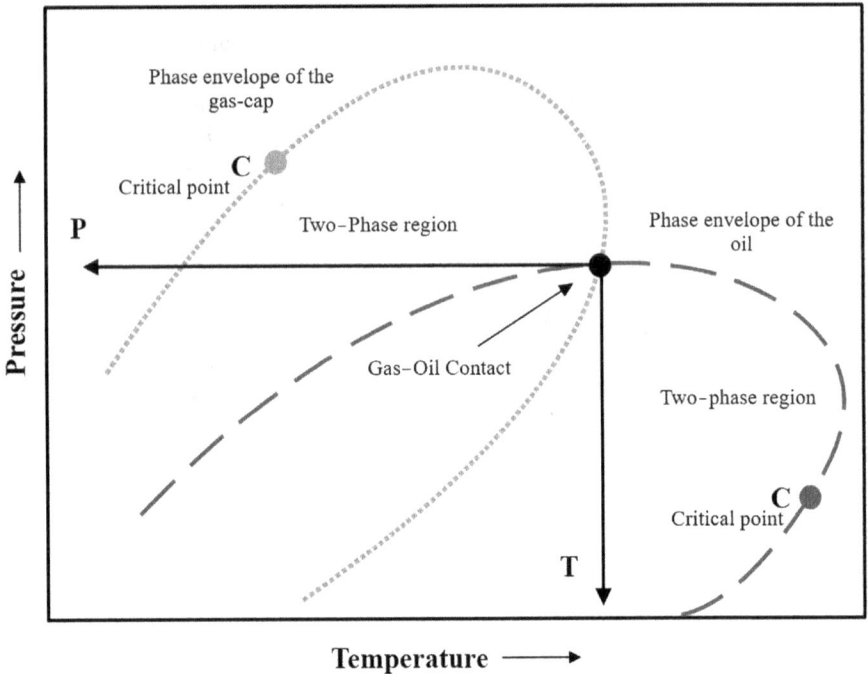

FIGURE 2.21 Intersection of equilibrium of the oil and gas phase envelopes in saturated petroleum reservoirs.

2.4.5.6 Saturated Petroleum Reservoirs

In such a system, the reservoir pressure is less than the bubble point pressure of the entire hydrocarbon fluid, and as a result, both gas and liquid phases are present. The gas evolves out of the crude oil and migrates upwards (upper part of the reservoir), forming a gas cap over the oil column. Therefore, two-phase envelopes attributing to both the phases must be considered.[20] Figure 2.21 shows the schematic representation of a saturated hydrocarbon reservoir. The saturated reservoir contains both the phase envelopes intersecting at one common point, corresponding to the gas–oil contact in the reservoir. Assuming that both phases are in equilibrium, the bubble point pressure of oil will be equal to the dew point pressure of gas in the overridden gas cap. In such a reservoir, the production strategy is to exploit the compressibility of the gas columns to improve the oil recovery potential of the reservoir. The positions of both the critical points are dependent on the hydrocarbon fluid composition.

2.5 FLUID SAMPLING FOR FLOW ASSURANCE

Fluid sampling plays a vital role in understanding the type and quality of the reservoir fluid. As different types of reservoir fluids pose different flow assurance challenges, petroleum engineers must have detailed information about the reservoir fluid analysis to mitigate the flow assurance issues. During fluid sampling, the testing engineer should ensure that the collected fluid sample is a true representative of the reservoir fluid. Fluid sampling is the most crucial aspect of PVT and reservoir fluid property studies. It entails securing a representative reservoir fluid sample,[1,3,26] either from the subsurface or a surface location. The fluid sampling influences the quality of the fluid property data obtained from laboratory investigation substantially. Fluid sampling is vital to the well testing operations. All measurements and studies become invalid if the collected reservoir fluids samples do not represent the reservoir conditions. The information collected from fluid sampling studies is used for a variety of reservoir engineering studies (PVT, core analysis, reservoir analysis, enhanced oil recovery, etc.) and to assess the flow assurance studies (flow potential, water chemistry and scales, precipitation of wax, asphaltenes, hydrates, etc.). Fluid sampling also helps in the early detection of corrosive and dangerous components in the production streams, such as CO_2, H_2S, Hg, and others.[26]

As stated previously, the fluid samples can be obtained either from surface locations such as separators, wellhead, pipelines, stock tanks, wellbores, or subsurface. It is crucial to ensure that the fluid sample remains representative of the reservoir fluid during handling and storage (sampling bottles) until all the necessary estimations have been finished. Problems are aggravated for remote fluid sampling locations, such as offshore and harsh atmospheric conditions. The fluid samples containing corrosive gases (e.g., CO_2, H_2S) and liquids should be carefully handled, and characterization should be done as soon as possible. Although the trained fluid sampling specialist adopts careful sampling procedures, there will never be a guarantee that the sample under investigation is genuinely representative of the reservoir condition.[27] It is, therefore, a good practice to collect as many fluid samples as possible during fluid sampling and testing for reliability before such samples are used for further studies. The best possible estimates of reservoir fluid properties during the entire life of a reservoir are made by multiple fluid sampling and testing.

2.5.1 WELL CONDITIONING

The well commences production immediately after the drilling and completion are finished. This initial phase, also known as the cleanup flow period, involves pushing the drilling fluids away from the wellbore and near-wellbore region. This could result in the drawdown below the saturation pressure, affecting various reservoir types differently (discussed later),[28] and governing the fluid sampling type and procedure. Well conditioning plays a crucial part in collecting surface and subsurface samples and is discussed later.[29]

2.5.1.1 Undersaturated Oil Reservoirs

The fluid sampling from this type of reservoir is simplest in theory. In the highly undersaturated reservoir, only the virgin reservoir fluid flows through the wellbore. However, the reservoir pressures could be in drawdown below the saturation pressure (see Figure 2.22a) due to cleanup or other reasons. The resulting two-phase must be vacated from the wellbore by flowing it at a lower rate, thereby minimizing the drawdown (see Figure 2.22b). The bottomhole samples must be acquired in single-phase

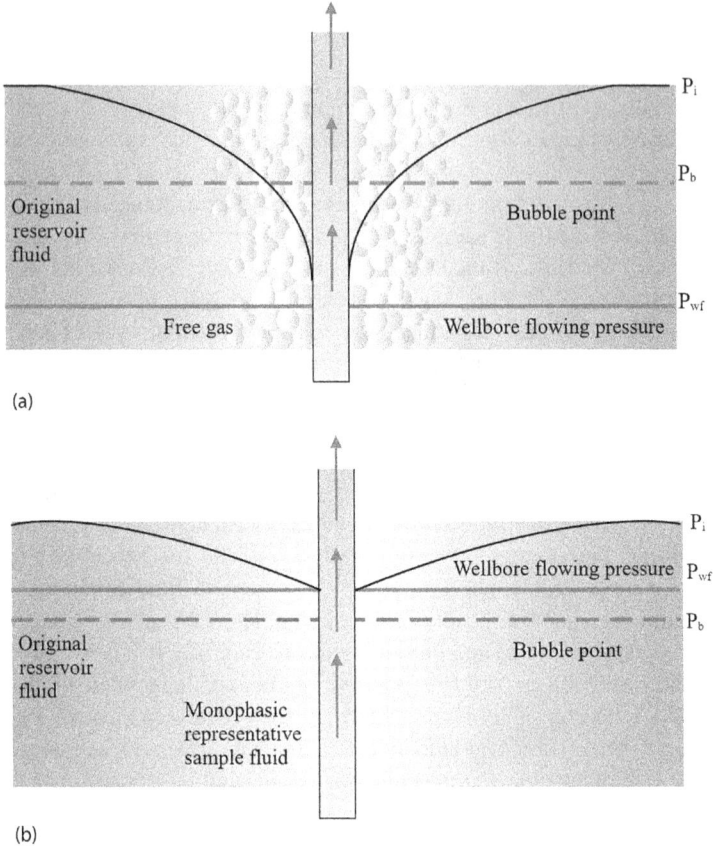

FIGURE 2.22 Well condition fluid sampling procedure: (a) Pressure drawdown results in a non-representative sample, (b) well conditioning by reducing pressure drawdown giving a single-phase representative sample.

fluid while the surface samples should suffice as long as the two-phase fluid has a similar ratio as the reservoir fluid. The sampling from such reservoirs is started after ensuring a stable separator GOR.

2.5.1.2 Saturated Oil Reservoirs

As flowing a saturated reservoir will result in the drawdown below the saturation pressure, sampling is complex for these reservoirs. The well should be closed to allow the pressure to build up, allowing more gas dissolution. A stabilized GOR can be achieved by flowing the well at lower flow rates. The bottomhole sampling will lead to an error, and thus, sampling from the separator is preferred. However, if only bottomhole sampling is available, it should be taken at a trickle flow to minimize the drawdown.

2.5.1.3   Gas Condensate Reservoirs

As long as the condensate reservoirs remain undersaturated, the condensate will remain in solution. The GOR stabilization method may not be suitable due to possible liquid dropout, which may change its phase near the wellbore, resulting in a rich-gas phase zone compared to the original reservoir fluid (Figure 2.22a). The fluid sampling from these reservoirs will not represent the reservoir fluid accurately unless the liquid dropout is significantly less and limited close to the wellbore zone (Figure 2.22b), allowing fast dispersal of rich gas to the surrounding zone in a short time.[20, 30] There are several methods for achieving the minimum required flow rate, including an industry nomogram.[31] The best approach is to select a small tubing diameter before the test to achieve a minimum lift velocity with a low flow rate, reducing the drawdown.[32]

2.5.2   Methods of Fluid Sampling

The sampling methods are categorized into the following broad groups:

* Subsurface (bottomhole/downhole) sampling
* Surface sampling
* Subsea sampling

2.5.2.1   Subsurface Sampling

Subsurface sampling, commonly called downhole or bottomhole sampling, collects the representative live reservoir fluid sample at the in situ pressure and temperature conditions. A suitable candidate must be selected based on the productivity index, skin factor, and stabilized in situ pressure conditions with minimal drawdown. A new well is always preferred to void free gas saturation in the flow stream. Downhole sampling is also done on wells during their lifetime to analyze the formation fluids (water, oil, gas) and rock properties. Special care is taken to acquire specific fluid from the production zone when the formation fluids contain multiphase streams (oil, water, gas). Subsurface sampling is helpful to locate oil–water–gas contact zone and provide information for casing and perforation operation during the well completion. To estimate the depth and height of gas–oil contact (GOC) and oil–water contact (OWC) in the well tubing, samples are collected at different depths and analyzed for multiple phases.[1, 33] Series of productivity tests are conducted to infer the bottomhole flowing pressure. Based on the productivity test data, candidate wells are selected having the highest flowing downhole pressure at the stabilized flow rate of reservoir fluid. A new well must be flowed for enough time to remove any contamination due to mud invasion and ensure stabilized flow until a constant GOR is observed at the surface condition.[34]

Figure 2.23 shows a schematic of a subsurface sampling operation and the tool.[35] In this method, a sampler is run in the wellbore at a particular depth to collect the sample. The sampler acquires single-phase reservoir fluid samples from the properly conditioned wells and can handle much higher pressures and temperatures. Undersaturated reservoirs are ideal for bottomhole sampling. In contrast, this

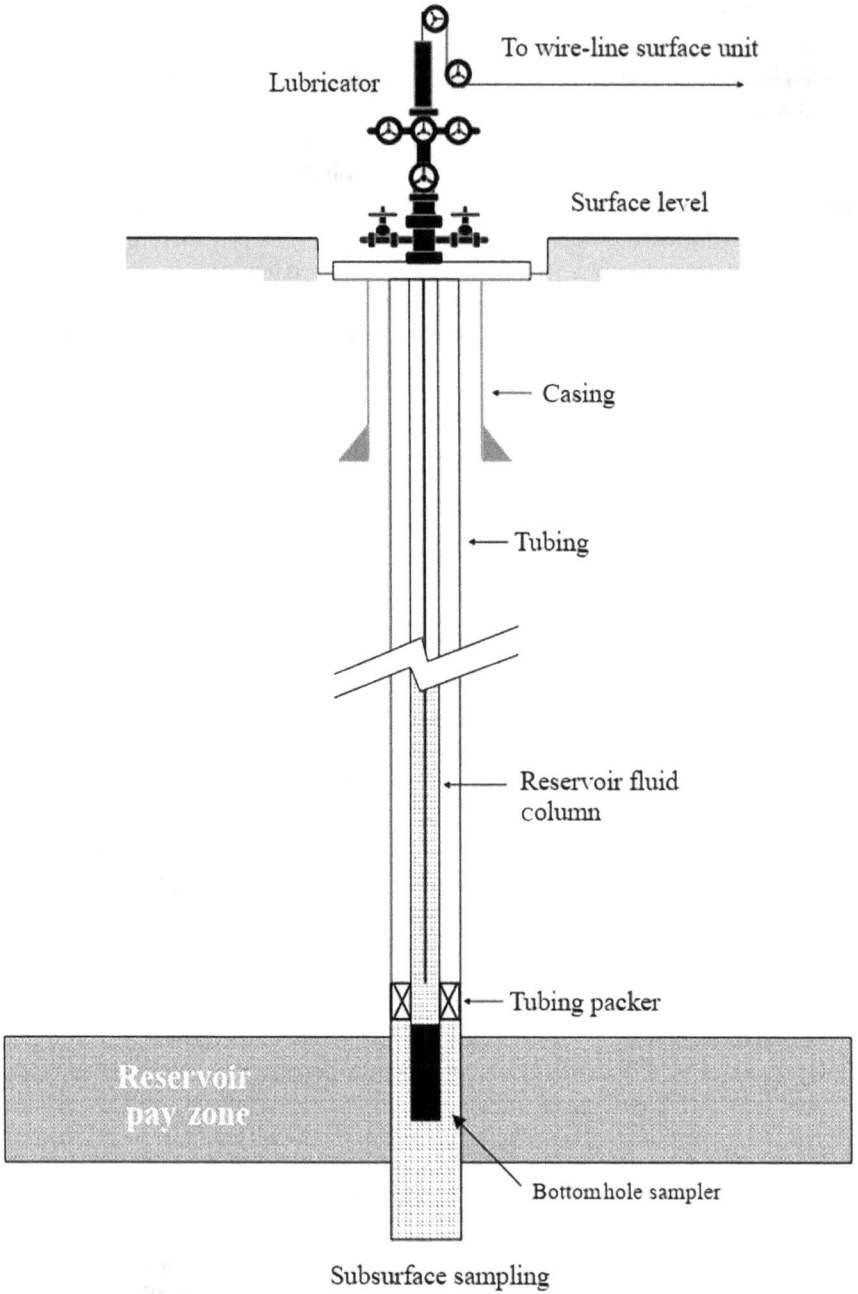

To wire-line surface unit

Lubricator

Surface level

Casing

Tubing

Reservoir fluid column

Tubing packer

Reservoir pay zone

Bottomhole sampler

Subsurface sampling

FIGURE 2.23 Schematic of the downhole or subsurface sampling tool and operation.

sampling method may not be recommended for saturated and depleted gas condensate reservoirs.[36] After the reservoir fluid sample is retrieved from the candidate well,
simple tests such as opening and closing pressure, bubble point pressure, and fluid
compressibility are measured for quality checks. Only a small volume of the reservoir sample fluid can be acquired using this method (typically 600 cm³). General
practice is acquiring many samples and comparing their saturation pressures at ambient temperature on the well site to check the sample. Downhole sampling is time-
consuming, costly, and requires producing wells to shut down during the sampling
operation.[35] For studies related to asphaltene deposition, wax deposition, in situ
emulsion formation, scales, and corrosion, downhole samples are preferred to avoid
losing these components during fluid flow from the reservoir to the surface.

2.5.2.1.1 Tools for Subsurface Sampling

2.5.2.1.1.1 *Downhole Sampler* Figure 2.24 shows a schematic of the bottomhole sampler.[37] The downhole tool consists of oil and air chambers, a piston-type
system separating the two chambers, and a triggering system with a closing

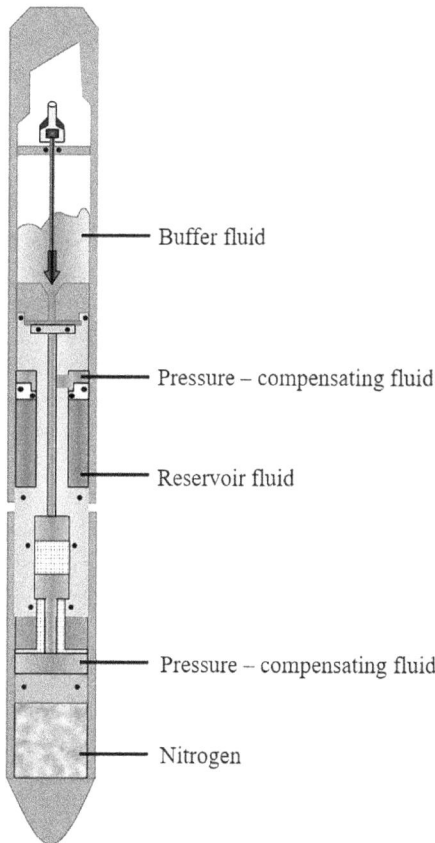

FIGURE 2.24   Schematic of the bottomhole sampler.

mechanism. Once the tool reaches the desired location in the well using a slick wire-line, reservoir fluid enters the oil chamber slowly (to avoid flashing) due to a hydraulic mechanism. The sample chamber closes automatically upon filling with oil. The tool is then pulled up, and the sample is then transferred to sample bottles using a mobile transfer bench.[37]

2.5.2.1.1.2 Transfer Bench   The sampled fluid from the downhole tool is transferred to the sample bottle using a transfer bench and sent to the laboratory for further investigation. The transfer bench setup consists of a hydraulic transfer pump, control valves, and transportation box.[37]

2.5.2.1.1.3 High-Pressure Single-Phase Sample Cylinder   Figure 2.25 shows a schematic of the high-pressure, single-phase sample cylinder.[37] The sample cylinder or bottle has a chamber and a piston-type mechanism (separating the sample and hydraulic fluid) to pressurize the sample at desired pressure conditions. The bottle comes in different materials of construction (stainless steel or titanium) based on the type of fluid to be stored and transferred.[37]

2.5.2.2   Surface Sampling

Surface sampling is the easiest and most economical way of acquiring fluid samples at the separator conditions once the producing GOR at the surface is stabilized (after well conditioning). Figure 2.26 shows a schematic of a typical three-phase separator with liquid and gas sampling technique.[35] The oil and gas samples are collected simultaneously, and recombination of separator oil and gas samples is carried out to obtain representative reservoir fluid samples, which can be used for further PVT studies. Several samples are acquired as per the need. As the gas is more compressible than oil, more gas samples are collected at the separator. The information, such as flowing bottomhole pressure, shut-in, separator and stock tank conditions, the specific gravity of fluids, the volume of sample and separator oil, GOR, and so on, are recorded for further analysis in the laboratory. Specific care must be taken for fluid containing wax and asphaltene to avoid deposition in the sampling chamber. In such cases, the temperature of sample bottles is maintained at a higher temperature than the operating conditions to avoid any possible deposition.

The sample cylinder used for surface sampling contains a floating-type piston arrangement (similar to subsurface sample cylinders). These cylinders are rated for high-pressure, and high-temperature (HPHT) applications (15,000 psi and 300–400°F). The piston isolates the sample from the hydraulic fluid and is leakproof.

FIGURE 2.25   Schematic of a high-pressure single-phase sample cylinder.

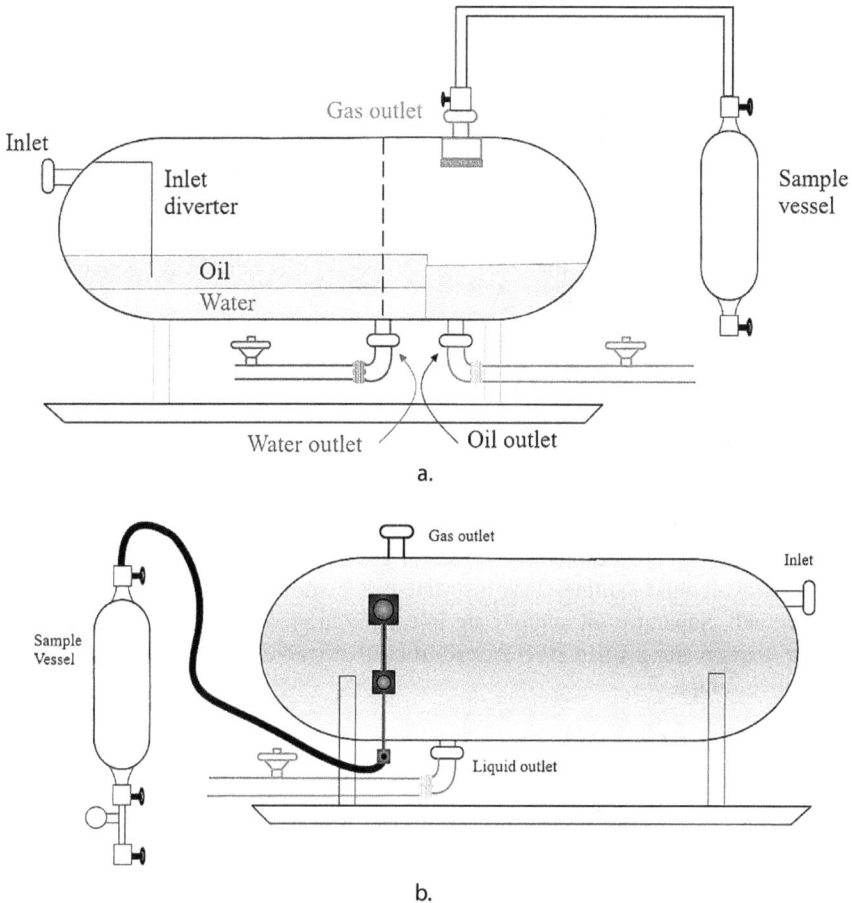

FIGURE 2.26   Schematic of the separator: (a) gas sampling; (b) liquid sampling.

During sampling, the piston arrangement is employed to suck the sample in the cylinder displacing part of the hydraulic fluid in the chamber. The oil sample bottles have few liters of capacity, while the gas sample bottle can acquire up to 20 L of sample.[1, 33, 34]

Surface sampling is preferred for gas condensates and saturated oil reservoirs. One major problem during surface sampling is the carryover of oil to the gas stream for (lean) gas condensate fluids. The carryover liquid may or may not be collected during separator gas sampling, resulting in the potential error in measuring fluid composition. In such conditions, isokinetic sampling or mini-laboratory (Thorton) sampling methods are suggested.[30]

Wellhead sampling is preferred for single-phase fluids at wellhead conditions (early production life), such as saturated oil, wet gas, or dry gases. During the early production life, wellhead sampling can be used as most fluid remains in a single phase at the wellhead. The wellhead or surface samples might give incorrect information on the wax or asphaltene content for flow conditions within the deposition envelope (inside the wellbore). Wellhead samples can be used for further studies if the wellhead temperature is above the wax appearance temperature (WAT).[1, 30]

The third type of surface sampling technique, known as split-phase sampling, is used for a two-phase wellhead sample or a volume of liquid entrained in the gas stream.[38] A stable GOR is first achieved, and then both phases are mixed inside a mixing head. Following this, a probe collects the sample at a specific GOR. This type of sampling is particularly used for condensate reservoirs having a low CGR. Although it has been there for more than 60 years, it is not always economical or better than separator sampling.

2.5.2.2.1 Tools for Surface Sampling

PVT surface separator sampling kit and recombination cell apparatus are used for surface sampling. The sampling kits mainly contain oil and gas sample bottles, valves, leak detectors, pressure, and temperature sensors to collect separator gas and oil samples safely and easily. Care is taken to avoid leakage during sampling and that the valves are tightly closed after the procedure is over.[37] Recombination cell is used in the laboratory to recombine the separator gas and oil sample to get the representative reservoir fluid sample. This instrument is essentially an HPHT rocking-type pressure cell. Separator oil and gas are injected at a predefined ratio and rocked for several hours under a fixed HPHT condition, homogenizing the reservoir fluid representative sample.[37]

2.5.2.3 Subsea Sampling

Subsea sampling refers to retrieving reservoir fluid samples from permanent subsea installations such as production trees or manifolds for testing purposes.[39, 40] It enables gathering of valuable data (including tracer detection, scale, and fluid composition) by using the subsea multiphase flow meters and improves the allocation process, reservoir management, well diagnostic capability, recovery, and timely action for mitigating flow assurance issues.

2.5.2.3.1 Role of Subsea Sampling in Reducing PVT Uncertainties

Subsea meters are typically configured with the PVT information gained from the sample during drilling, but these data are rarely updated due to the lack of representative samples. Also, fluid properties, including reservoir fluid densities, water conductivity, oil permittivity, and mass attenuation may change throughout the well's life cycle, reducing the meter's accuracy.[41] This may increase the uncertainty as high as ±30%.[42] Some multiphase flow meters can measure water fractions with uncertainties below 0.01% in real time, which is vital to mitigate the flowline plugging. Thus, accurate knowledge of fluid properties and correct method selection is of utmost importance. Measuring the multiphases without separation is crucial for marginal field development. They are the only option for fluid sampling from depleted reservoirs to accurately update, and characterize the reservoir.[42] Subsea flow multiphase flow meters are based on venturi or a pressure differential device. The subsea sampling uses a multi-energy gamma-ray system (simple integration) or a combination of electromagnetic sensors and a single gamma-ray system.[42]

2.5.2.3.2 *Various Subsea Sampling Techniques*

2.5.2.3.2.1 Remotely Operated Vehicle–Assisted Subsea Fluid Sampling A major advantage of remotely operated vehicles' (ROVs') assistance is the ability to obtain samples independent of a large platform in an extreme deep-water environment, resulting in huge financial savings. A more complex sampling system can be designed by mounting sampling equipment onto ROV. Accordingly, two different sampling systems have been designed by Oceaneering Deepwater Technical Solution (DTS), namely, flow-through and single-draw systems. A flow-through system depends on the pressure differential to drive the produced fluids through the system, obtained by selecting sampling points via venturi and blind trees.[39, 43] Venturis offers a consistent pressure difference, while the advantage of blind tree lies in its ability to offer well mixed-multiphase or phase-biased samples. Single-draw systems pull a fixed volume of fluid into a sample chamber and maintain a constant pressure as long as power and communications remain uninterrupted. Intelligent sampling systems employ various electrical sensors to monitor pressure, temperature, and flow rate, which prevent flow assurance issues.[39]

2.5.2.3.2.2 Focused Sampling The focused sampling uses wireline tools to take the sample from two production zones between formation and sampling tool, namely, sample zone and guard zone.[44] A schematic of the conventional and focused sampling probe is shown in Figures 2.27. It contains a dual flowline for the sample and guard flow instead of a single flowline for conventional sampling (Figure 2.27b and c). It also has a pressure gauge along with a separate pump for each flowline. This technique can separate the filtrate contamination from the virgin reservoir fluid.

2.5.2.3.2.3 Digital Fluid Sampling Digital fluid sampling is an advanced fluid sampling technique developed especially for deep water. It is facilitated by reservoir fluid geodynamics (RFG), which combines PVT analysis, geochemical fingerprinting, and reservoir geology with downhole fluid data (DFA), enabling an enhanced reservoir fluids description and optimized data acquisition. In this regard, some vital issues defining the reservoir concerns include the impact of biogenic on reservoir fluid behavior, baffling and fault compartmentalization, real-time prediction of reservoir fluid properties, and flow assurance risk assessment.[45] Optical density and asphaltene content can be correlated to aid digital sampling for different geochemical and PVT scenarios. A typical reservoir fluid cyclic workflow for geodynamics is given in a schematic in Figure 2.28.[45]

Digital fluid sampling techniques enable predicting critical reservoir fluid properties in real time to develop and validate the reservoir models. They are an alternative to the physical fluid samples, which are needed for the model calibration and correlations. It has been successfully employed in a deepwater reservoir in the Gulf of Mexico. The data set consisted of 154 DFA sampling stations in 26 different wells, of which 120 DFA high-quality stations were admissible for RFG implementation and integration, with 130 PVT reports and 58 geochemical fingerprint analyses.

The quality of the fluid sample collected at various locations determines the success of further studies on the sample. Thorough quality checks are performed on the

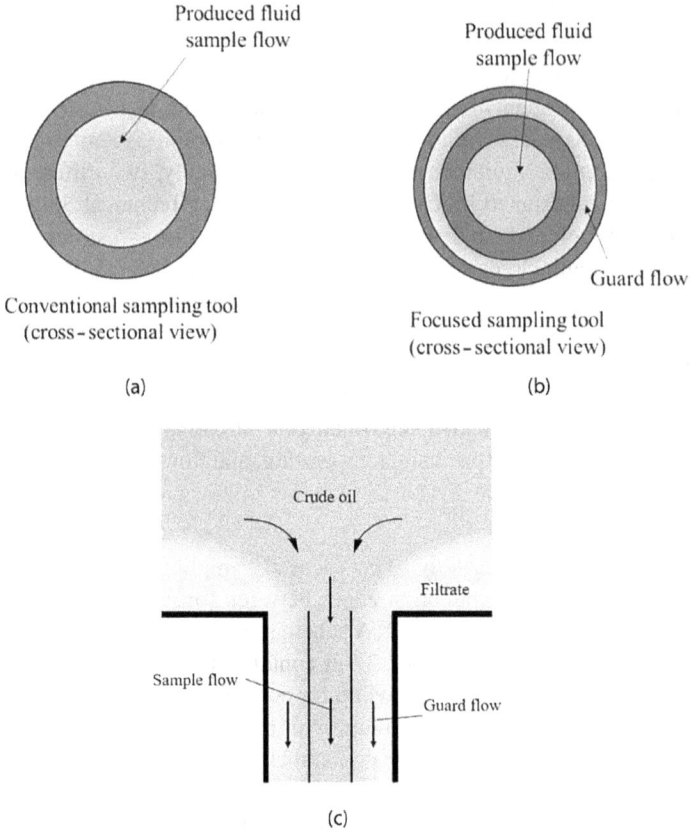

Produced fluid
sample flow

Produced fluid
sample flow

Guard flow

Conventional sampling tool
(cross-sectional view)

Focused sampling tool
(cross-sectional view)

(a)

(b)

Crude oil

Filtrate

Sample flow

Guard flow

(c)

FIGURE 2.27   (a) Conventional sampling tool; (b) focused sampling tool; (c) filtrate invasion in focused sampling tool.

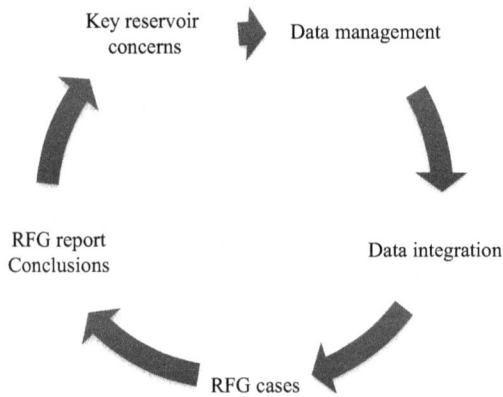

Key reservoir
concerns

Data management

RFG report
Conclusions

Data integration

RFG cases

FIGURE 2.28   Cyclic workflow of reservoir fluid geodynamics.

sample collected at bottomhole or surface conditions before carrying further studies.[46] Some quality-check methods include measuring the sampler opening and closing pressure, bubble point pressure, and so on.[1, 47] Separator samples should only be used when the recombination of separator gas and liquid resembles the representative reservoir fluid sample. Generally, bottomhole samples are preferred over surface samples if the gas and oil properties, and flow rates at the surface conditions are in question. However, if the surface conditions are reliable, the surface sampling procedure can provide statistically valid GORs measured over a long sampling period. Whenever possible, separator liquid and gas samples should be taken simultaneously and in the same locations to have the identical sampling conditions for both fluids.

2.5.3 SPECIAL CONSIDERATION FOR FLOW ASSURANCE

2.5.3.1 Waxy and Asphaltenic Crude Oil

The solid organics such as asphaltene and waxes are typically soluble in the crude oil at the reservoir conditions and contribute to the overall composition of the reservoir fluid. During the production of reservoir fluid, the pressure and temperature of the fluid decrease up to the surface, affecting the solubility of organic solids. It also causes precipitation of asphaltene and waxes in the wellbore, tubing, and separator, leaving the bulk of these heavier hydrocarbons out of the reservoir fluid. Any sampling done at the separator will now result in a sample fluid with less solid organic components. Hence, the recombination of separator gas and liquid will not result in a representative reservoir fluid sample. It may be rendered invalid for further analyses, including wax and asphaltene deposition studies, phase behavior studies, and reservoir fluids, and may affect any flow assurance mitigation strategies. Thus, great care must be considered during sampling such fluid. To sample such fluids at surface conditions, large diameter, short distance, and heated sampling lines should be used. For oils with the potential to deposit wax in the wellbore, downhole samples are collected using single-phase sampling tools (as shown in Figure 2.24).

Problems associated with asphaltene deposition are difficult to predict. A low asphaltene concentration fluid may be more problematic than a high asphaltene concentration. Once asphaltenes are separated from the bulk phase, it is very difficult to rehomogenize them. Therefore, using a single-phase sampler above the onset pressure is important to avoid separation from the bulk phase. Sample bottles with movable mixing balls for agitation (as shown in Figure 2.25) can be used during the collection and transportation of the sample. The samples in these bottles are typically agitated at reservoir conditions for more than 24 hours to get them back to the reservoir conditions giving a possibility to recover the original reservoir fluid sample from the bottle/cylinders.[27]

2.5.3.2 Sampling of Crude Oil Emulsions

Crude oil emulsions are another flow assurance element. The emulsion can be strong and difficult to break, making their sampling necessary to avoid various flow assurance issues. Obtaining emulsions samples is necessary to check various emulsion braking systems such as de-emulsifiers. Emulsions samples are preferably collected at in situ sampling locations, where the representative emulsion phase exists instead of the bottom or top of the sampling location where the free water or oil phase may exist, affecting the measurements. It is suggested to take the sample from the pipe at mid-height using a "quill" sampling probe. It is also better to choose the location with high turbulence and velocity to avoid issues with gravity segregation.

To avoid emulsification at a pressurized location, a floating piston–type sample bottle can be used. Even though not present within the pipeline or separator, the emulsion may form during sampling due to erroneous procedure. One must also ensure that emulsion formation is avoided during the sampling process due to the high flow rate. The emulsion can be strong or weak depending on the composition, and operating conditions. Samples should be carefully preserved to avoid disturbance to their initial state of stability.[27]

2.5.3.3 Corrosive Reservoir Fluid

The sampling quality may be affected due to the presence of heavy organic solids (viz., asphaltenes and waxes), nonorganic corrosive components (H_2S and CO_2), or mishandling of the sample bottle during transfer. During transportation to the lab, the actual H_2S content may reduce due to its corrosion potential, making it necessary to carry on-site measurements. The online measurement process uses sniffer or Drager tubes for H_2S, CO_2, and other hydrocarbon gases. Alternatively, inert gas bags may be used to collect the gas samples. It is recommended to do on-site measurement using two different techniques to estimate the H_2S reliably. Also, on-site or lab analyses should be finished within 24 hours to reduce the corrosive effects of H_2S on the sampling container. Special sampling bottles made of Inconel are generally employed for sampling highly corrosive reservoir fluids.[48] Portable gas chromatographs are gaining interest for quick on-site estimation of hydrocarbon and non-hydrocarbon gas composition. Measuring sulfate-reducing bacteria (SRB) is recommended as they can contaminate the reservoir fluid, affecting the measurement and posing risk during their handling. Novel instrumentation and portable analytical equipment have made performing on-site or in situ measurements of reservoir fluid properties relatively easy.[27]

2.5.4 POTENTIAL ERRORS DURING FLUID SAMPLING

Figure 2.29 provides possible locations and potential errors during fluid sampling.[26] Possible errors could be due to the inappropriate sampling post drilling and completion operation resulting in contamination from drilling fluid, mud filtrate, and completion fluids. Excessive drawdown may result in phase separation, and thus the collected two-phase sample may not represent the in situ reservoir fluid. As discussed earlier, well conditioning is a prerequisite for such a situation. In reservoirs with multiple pay zones, the mixing of reservoir fluid from a different pay zone may lead to inaccurate samples. Intermittent flows and two-phase flows are not suitable for proper sampling procedures; hence, the flow rate should be stabilized for the correct fluid sample. Some reservoir fluids may contain highly reactive and corrosive components such as non-hydrocarbon gases (CO_2 and H_2S), which can result in poor sampling if the correct materials are not used. During transfer and transportation, care

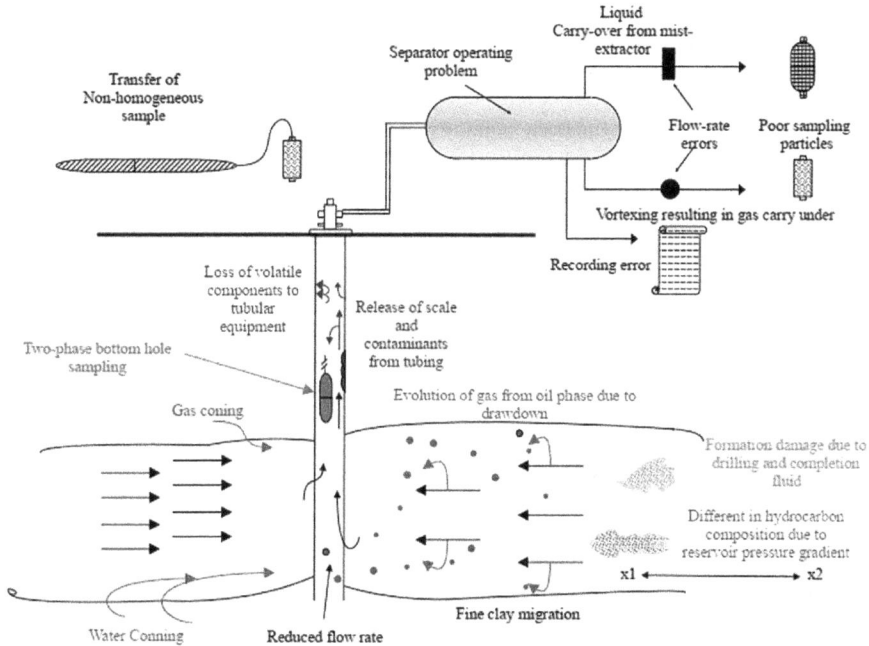

FIGURE 2.29 A schematic view of wellsite sampling and measurement errors.

should be taken that there is no leakage. The fluid with asphaltene and wax deposition potential should be preserved at reservoir conditions or above WAT and asphaltene onset pressure. For the surface sample, the flow rate should be stabilized and under equilibrium. Incorrect separator size, high flow rates, wax and asphaltene deposition inside the separator, and so on could lead to errors. In addition, wax and asphaltene tend to stabilize oil–water emulsions. The separator sample may not provide accurate information if the emulsion is strong and stable. As disused earlier, the liquid or gas carry over to other streams should be avoided in the separator. The potential errors could result from inaccurate data recording, such as fluid flow rate, GOR, pressure, and temperature during recombination and analyses of fluid samples.[26]

REFERENCES

1. Dandekar, A. Y. (2013). *Petroleum Reservoir Rock and Fluid Properties* (2nd Edition). CRC Press, Taylor & Francis Group, Boca Raton, FL, USA.
2. McCain, W. D. (1990). *The Properties of Petroleum Fluids* (2nd Edition). PennWell Publishing Co., Tulsa, OK.
3. Standing, M. B. (1981). *Volumetric and Phase Behavior of Oil Field Hydrocarbon Systems* (2nd Edition). Society of Petroleum Engineers of AIME, Dallas, TX.
4. Speight, J. G. (2016). *Introduction to Enhanced Recovery Methods for Heavy Oil and Tar Sands* (2nd Edition). Gulf Professional Publishing, Elsevier, Oxford, UK.
5. Pedersen, K. S., Christensen, P. L., & Shaikh, J. A. (2014). *Phase Behavior of Petroleum Reservoir Fluids* (2nd Edition). CRC Press, Taylor & Francis Group, Boca Raton, FL.

6. Billups, W. E., Verma, M., Brinson, B. E., Vishnyakova, E., Alemany, L. B., & Shammai, M. (2019). Birch reduction of asphaltenes, synthesis of hydroasphaltenes. *Energy & Fuels*, 33 (9), 8040–8044. https://doi.org/10.1021/acs.energyfuels.9b01298

7. Murgich, J., Abanero, J. A., & Strausz, O. P. (1999). Molecular recognition in aggregates formed by asphaltene and resin molecules from the Athabasca oil sand. *Energy & Fuels*, 13 (2), 278–286. https://doi.org/10.1021/ef980228w

8. Cronquist, C. (1979). Evaluating and producing volatile oil-reservoirs. *World Oil*, 5, 159.

9. Morton-Thompson, D., & Woods, A. M. (1992). *Development Geology Reference Manual*. American Association of Petroleum Geologists, Tulsa, OK.

10. Speight, J. G. (2005). Natural Bitumen (Tar Sands) and Heavy Oil. In *Coal, Oil Shale, Natural Bitumen, Heavy Oil and Peat*, Vol. 2. EOLSS Publishers, Oxford, UK.

11. Speight, J. G. (2020). *The Refinery of the Future* (2nd Edition). Gulf Professional Publishing, Elsevier, Oxford, UK.

12. Meyer, R., & Attanasi, E. D. (2004). Natural Bitumen and Extra-Heavy Oil. In *2004 Survey of Energy Resources*, Clarke, A., & Trinnaman, J. A. (Eds.). Elsevier Science & Technology Books.

13. Tiratsoo, E. N. (1979). *Natural Gas* (3rd Edition). Gulf Publishing Company. Huston, TX.

14. Gibson, R. Gibson Oil Statistics; http://www.gravmag.com/oilold.html#gasfields (accessed December 3, 2021)

15. Tsourakis, F. In Amenas Gas Project: the largest wet gas development project in Algeria; https://www.algeria-summit.com/post/in-amenas-gas-project-the-largest-wet-gas-development-project-in-algeria (accessed July 26, 2021)

16. Kargbo, D. M., Wilhelm, R. G., & Campbell, D. J. (2010). Natural gas plays in the Marcellus shale: challenges and potential opportunities. *Environmental Science & Technology*, 44 (15), 5679–5684. https://doi.org/10.1021/es903811p

17. Rahimzadeh, A., Bazargan, M., Darvishi, R., & Mohammadi, A. H. (2016). Condensate blockage study in gas condensate reservoir. *Journal of Natural Gas Science and Engineering*, 33, 634–643. https://doi.org/10.1016/j.jngse.2016.05.048

18. Besta, S. What are the largest natural gas fields in Russia?; https://www.nsenergybusiness.com/features/largest-natural-gas-fields-in-russia (accessed July 26, 2021).

19. Ahmed, T. (2018). *Reservoir Engineering Handbook* (5th Edition). Gulf Publishing Co. Huston, TX.

20. Danesh, A. (1998). *PVT and Phase Behaviour of Petroleum Reservoir Fluids* (1st Edition, Vol 47). Elsevier, Amsterdam, The Netherlands.

21. Luna, P., & Hidalgo, A. (2014). Mathematical modeling and numerical simulation of two-phase flow problems at pore scale. In *Proceedings of the 2014 Madrid Conference on Applied Mathematics*. Madrid, Spain.

22. Moses, P. L. (1986). Engineering applications of phase behavior of crude oil and condensate systems (includes associated papers 16046, 16177, 16390, 16440, 19214 and 19893). *Journal of Petroleum Science and Technology*, 38, 715–723, https://doi.org/10.2118/15835-PA

23. Hadi, P. (2010). Field Operational Problems Due to Condensate Formation in Retrograde Gas Reservoirs. M.S. Thesis, Petroleum University of Technology, Iran.

24. Danesh, A., Henderson, G. D., & Peden, J. M. (1991). Experimental Investigation of Critical Condensate Saturation and Its Dependence on Interstitial Water Saturation in Water-Wet Rocks. *SPE Reservoir Engineering*, 6(3), 336–342. https://doi.org/10.2118/19695-PA

25. Viswanathan, B. (2017). Natural Gas. In *Energy Sources: Fundamentals of Chemical Conversion Processes and Applications*. Elsevier, Amsterdam, The Netherlands. pp. 59–79. https://doi.org/10.1016/B978-0-444-56353-8.00003-4

26. Ezekwe, N. (2010). *Petroleum Reservoir Engineering Practice* (1st Edition). Pearson, Westford, Massachusetts.
27. PEH:Fluid Sampling – PetroWiki; https://petrowiki.spe.org/PEH:Fluid_Sampling#Reservoir-Fluid_Type (accessed August 1, 2021)
28. Bon, J., Sarma, H. K., Rodrigues, J. T., & Bon, J. G. (2007). Reservoir-fluid sampling revisited - A practical perspective. *SPE Reservoir Evaluation & Engineering*, 10 (06), 589–596. https://doi.org/10.2118/101037-PA
29. Lawrence, J. J., Chorneyko, D. M., Smith, C. K., & Nagarajan, N. (2008). Representative reservoir fluid sampling: Challenges, issues, and solutions. In *International Petroleum Technology Conference*. Kuala Lumpur, Malaysia, December 3-5. (IPTC-12401-MS). https://doi.org/10.2523/IPTC-12401-MS
30. Whitson, C. H. (1998). *Fluid Sampling & Laboratory Data*. Norsk Hydro.
31. Turner, R. G., Hubbard, M. G., & Dukler, A. E. (1969). Analysis and prediction of minimum flow rate for the continuous removal of liquids from gas wells. *Journal of Petroleum Technology*, 21(11), 1475–1482. https://doi.org/10.2118/2198-PA
32. Preparing for fluid sampling; https://petrowiki.spe.org/Preparing_for_fluid_sampling#cite_note-r2-2 (accessed December 3, 2021).
33. Amyx, J. W., Bass, D. M., & Whiting, R. L. (1960). *Petroleum Reservoir Engineering: Physical Properties* (1st Edition). McGraw-Hill, Texas, USA.
34. Dake, L. P. (1998). *Fundamentals of Reservoir Engineering* (1st Edition, Vol. 8). Elsevier, Amsterdam, The Netherlands.
35. Todd, A. C. *Reservoir Engineering*. Heriot-Watt University.
36. Towler, B. F. (1989). Reservoir engineering aspects of sampling of sampling of saturated oils for PVT analysis. *Society of Petroleum Engineers* (SPE 19438-MS).
37. Vinci Technologies. (2012). *PVT Tool*. Catalogue of Products.
38. Williams, J. M. (1994). Getting the best out of fluid samples. *Journal of Petroleum Technology*, 46 (09), 752–752. https://doi.org/10.2118/29227-PA
39. Mancini, C. S., & Turnbull, J. (2011). Introduction to Remotely Operated Vehicle Assisted Subsea Sampling. In *Offshore Technology Conference*. Houston, Texas, USA, May 2–5. (OTC-21324-MS). https://doi.org/10.4043/21324-MS
40. Pinguet, B., Theron, B., & Leong, Y. S. (2014). Representative subsea sampling: A key enabler for flow assurance and better metering uncertainty. In *International Petroleum Technology Conference*. Kuala Lumpur, Malaysia, December 10–12. (IPTC-17963-MS). https://doi.org/10.2523/IPTC-17963-MS.
41. Kelner, E. (2012). An ROV-deployed deepwater subsea sampling system. In *Offshore Technology Conference*. Houston, Texas, USA, April 30–May 3. (OTC-23412-MS). https://doi.org/10.4043/23412-MS.
42. Pinguet, B. G., Vethe, E., Smith, M. T., Smith, G., Sbordone, A., & Nighswander, J. A. (2012). Subsea sampling: Reducing uncertainty from PVT by utilizing representative sampling subsea. In *Offshore Technology Conference*. Houston, Texas, USA, April 30–May 3. (23340-MS). https://doi.org/10.4043/23340-MS
43. Hall, G. (2011). Subsea liquid sampling using flow-through technique. In *Offshore Technology Conference*. Houston, Texas, USA, May 2–5. (OTC-21787-MS). https://doi.org/10.4043/21787-MS.
44. Dong, C., Del Campo, C., Vasques, R., Hegeman, P., & Matsumoto, N. (2005). Formation testing innovations for fluid sampling. In *17th Annual Deep Offshore Technology International Conference and Exhibition*. Brazil, November 8–10.
45. Gelvez, C., Cedillo, G., Soza, E., Gonzalez, D., Slotnick, B. S., Moreno, S., Pineda, W., Saidian, M., Mullins, O. C., Paul, S., Cañas, J., & Kulkarni, A. (2021). Digital fluid sampling in deep water reservoirs using reservoir fluid geodynamics: The beginning of the

digital fluid sampling revolution. In *SPWLA 62nd Annual Logging Symposium*. Society of Petrophysicists and Well Log Analysts (SPWLA). Online. May 17–20. https://doi.org/10.30632/SPWLA-2021-0010.

46. Paredes, J. E., Perez, R., Perera, L. M., & Larez, C. J. (2014). Screening criteria and methodology for quality check and PVT selection for reservoir studies. In *SPE Annual Technical Conference and Exhibition*. Amsterdam, The Netherlands, October 27–29. (SPE-170695-MS). https://doi.org/10.2118/170695-MS.

47. Afanasyev, V., Theuveny, B. C., Guieze, P., & Bastos, V. (2008). Sampling with multiphase flowmeter in Northern Siberia - condensate field experience and sensitivities. In *SPE Russian Oil and Gas Technical Conference and Exhibition*. Moscow, Russia, October 28–30. (SPE-115622-MS). https://doi.org/10.2118/115622-MS

48. Schlumberger. Single-Phase Sample Bottle - Schlumberger https://www.slb.com/-/media/files/testing-services/product-sheet/single-phase-sample-bottle.ashx (accessed December 3, 2021).

3 Laboratory Studies and Characterization of Petroleum Reservoir Fluids

3.1 INTRODUCTION

A simple application of Gibb's[1] phase rule to petroleum reservoir fluids tells us the three degrees of freedom, namely, pressure, temperature, and composition. These three independent intensive variables and the chemistry (paraffinic, naphthenic, aromatic, etc.) basically define the fluid-phase behavior and properties. For example, for a five-component system to exist in a single phase, six degrees of freedom are required ($F = C - P + 2 = 5 - 1 + 2 = 6$) – two being pressure and temperature and the remaining four being the molar composition of the four components (the fifth is automatically specified since mole fractions add up to 1). It is important to recognize the fact that among the three degrees of freedom, the composition is a given and is much like the signature or DNA of a reservoir fluid. Therefore, we first consider general compositional measurements followed by specialized analysis, such as high-temperature gas chromatography (HTGC) and saturates–aromatics–resins–asphaltenes (SARA) given their significance in flow assurance. We then examine the effect of pressure and temperature, in conjunction with composition, on the phase behavior and properties of petroleum reservoir fluids. These laboratory data obtained through conventional pressure–volume–temperature (PVT) tests are frequently used to calibrate the equations-of-state (EOS) models that are extended or applied to flow assurance studies. Finally, also covered in this chapter is numerical fluid characterization and produced water analysis.

3.2 BASIC COMPOSITIONAL MEASUREMENTS

Although one can measure compositions under a variety of conditions, the discussion here is confined to the overall original or feed composition of reservoir fluids, which is one of the primary inputs to EOS models. Reservoir fluid composition can be broken down into three parts: (1) well-defined, discretely identified components that include the non-hydrocarbons and everything up to n-pentane; (2) pseudo components or single carbon number (SCN) fractions, such as C_6 to C_{19} or C_{29}; and (3) the plus fraction, such as C_{20+} or C_{30+}. Every reservoir fluid description is customized or user-defined according to the applications; for example, the plus fraction can be a C_{7+} or a C_{15+} or C_{25+} and so forth. Note that the well-defined components are named

DOI: 10.1201/9781003091301-3

as such because their properties are universal regardless of the fluid; however, the pseudo fractions and particularly the plus fractions are unique not only for a fluid type (black oil, volatile oil, etc.) but also within the fluid type (gas condensate A vs. B). Therefore, the fluid composition is much like DNA. Table 14.3 in Dandekar[2] shows the breakdown of a black oil composition into these three parts.

For any compositional measurements, a physical sample of the reservoir fluid is necessary. This could be in the form of a bottomhole single-phase sample or two separate companion samples from a separator, that is, the separator gas and liquid, respectively. The separately analyzed samples are then numerically recombined per the appropriate gas-to-oil ratio (GOR) to obtain the molar composition of the "live fluid". A somewhat similar approach is applied to the bottomhole sample, known as the blowdown method. The live sample is flashed to simulated standard conditions, and the separated gas and liquid are analyzed and then recombined per the separation GOR to achieve the overall composition. This type of methodology is necessary given the limitations of handling live-sample high-pressure conditions in a gas chromatograph (GC) and laboratory-scale distillation units. However, from a flow assurance standpoint, the most ideal scenario would be the capture of a live fluid under in situ conditions and then a direct compositional analysis in that state to avoid phase splits (especially solid organic phases) that would likely introduce uncertainties in modeling. Although high-pressure GC techniques have been proposed,[3, 4] their widespread applicability for flow assurance is unknown at this point. Therefore, the somewhat traditional approach of compositional analysis is used.

A GC can be used to determine the compositions of both gas and liquid phases; however, it is commonly employed for gas or vapor phases, whereas liquid (degassed or flashed liquid) is distilled in a true boiling point (TBP) apparatus. In both cases, the mixture is discretized into a certain number of components according to their boiling points. A typical GC setup consists of two detectors in a series, namely, the thermal conductivity detector (TCD) and flame ionization detector (FID) for non-hydrocarbon and hydrocarbon components, respectively (although note that the TCD detects both). Also included are a temperature-programmed oven, a packed column, and a data integrator. A TBP distillation unit typically requires a liquid sample of about 100 cc that is heated in a glass flask attached to a glass distillation column. The distilled physical fractions (named TBP or SCN) are collected in a fraction collector. The undistilled material remaining in the distillation flask is termed residue or plus fraction C_{N+} (note that N is one number above the carbon number of the last collected TBP fraction). Both GC and distillation techniques are described in detail elsewhere.[2, 5] As an example, the separately analyzed gas and oil (liquid) sample recombination is shown in Table 3.1

Using the basis of 1 stock tank barrel (STB) oil, 1 STB = 5.615 ft^3 × 52.8 lb/ft^3 = 296.47 lb = 296.47/203.56 = 1.456 lb-mole (203.56 lb/lb-mole is the molecular weight of stock tank oil, assumed as such in column 3, calculated using the summation of column 3 × 4). From the ideal gas equation, n/V = 14.7/(10.732 × (60 + 460)) = 0.002634 lb-mole/scf, which means GOR = 1895.47 × 0.0026/1.456 = 3.428 lb-mole **gas/lb-mole stock tank oil**. Also 1 STB can be written as 1.456/1.456 = 1 lb-mole stock tank oil/**lb-mole stock tank oil**. Since both gas and oil now have a common denominator, columns 2 and 3 can be multiplied by 3.428 and 1, respectively, and then added and normalized to give the recombined fluid composition (see column 5).

TABLE 3.1

Recombination Example Using Separately Analyzed Gas, Oil Composition, GOR, and Stock Tank Oil Density

Component	Separated Gas Mole Fraction	Separated Oil Mole Fraction	MW, lb/ lb-Mole	Recombined Fluid Composition, Mole Fraction
N_2	0.007	0.000	28	0.005
CO_2	0.041	0.000	44	0.032
C_1	0.680	0.000	16	0.526
C_2	0.108	0.000	30	0.084
C_3	0.076	0.000	44	0.059
iC_4	0.013	0.000	58	0.010
nC_4	0.031	0.000	58	0.024
iC_5	0.011	0.000	72	0.009
nC_5	0.015	0.000	72	0.012
C_6	0.018	0.000	86	0.014
C_7	0.000	0.115	96	0.026
C_8	0.000	0.125	107	0.028
C_9	0.000	0.085	121	0.019
C_{10+}	0.000	0.676	250	0.153
	1.000	1.000		1.000

GOR = 1,895.47 scf/STB; stock tank oil density = 52.8 lb/ft^3.

3.3   HTGC AND SARA ANALYSIS

HTGC and SARA analyses are experimental techniques targeted specifically toward the study of paraffin waxes and asphaltenes, respectively. Both measurements are used to obtain the relative quantification or compositional characteristics of these two solid organic phases.

3.3.1   HTGC

In principle, HTGC is similar to normal GC analysis; however, with one major distinguisher and that is the use of temperatures as high as up to 500°C because the main purpose is to obtain compositional characteristics of high carbon number (thus high boiling point) paraffin waxes. Additionally, since the HTGC is used for hydrocarbon detection only, there is no need for a TCD. Although HTGC can be used to analyze any flashed or dead oil sample, it is more commonly applied to compositionally characterize the solid waxy deposit on what is known as a "cold finger". A typical cold finger setup consists of a stainless steel tube of 12 mm in diameter and 72 mm in length dipped in a test sample of oil that is cooled by circulating a coolant inside the finger. The surface temperature of the finger is maintained below the wax appearance temperature of the oil, thereby depositing wax and thus mimicking the wall of an oil pipeline. The cold finger deposit gives an idea of the extent of the flow assurance problem for a given oil. These tests also are frequently employed in

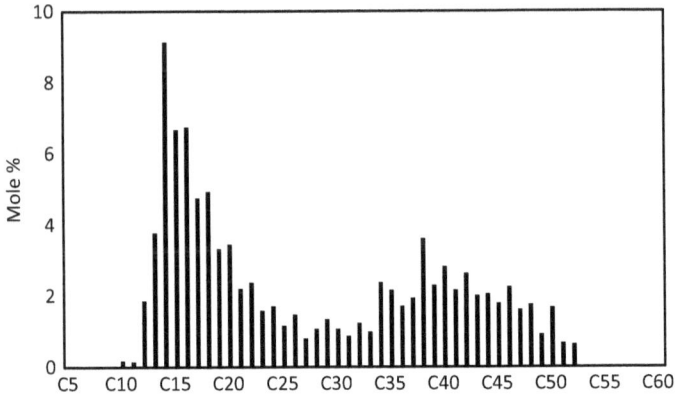

FIGURE 3.1 HTGC of an Alaska North Slope oil sample cold finger deposit. Plot constructed based on data reported by Zhu et al.[6]. Oil API is 28.15 and wax content is 18.2%.

the determination of the efficacy of a given paraffin inhibitor. The deposit is rinsed with methyl ethyl ketone to remove the free oil phase sticking to the deposit, which is dried and weighed prior to HTGC. Cold finger tests and subsequent HTGC are described elsewhere.[6–8] As an example, the molar composition of a cold finger deposit by HTGC for an Alaska North Slope oil sample[6] is shown in Figure 3.1. In Figure 3.2, the HTGC wt% of a cold finger deposit from the South Eagle Ford[8] oil sample and a field deposit[8] is compared. In all three cases, the lower carbon numbers generally represent the oil "trapped" in the deposit.[8] However, in the case of a field deposit, the relatively longer duration over which it forms and the aging process

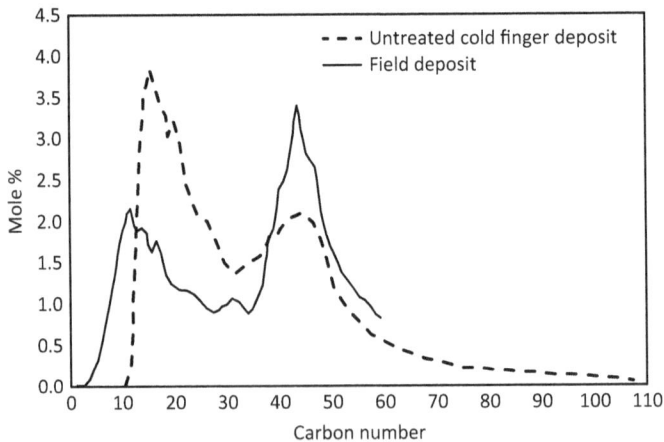

FIGURE 3.2 Comparison of n-paraffin distribution in an untreated cold finger deposit and field deposit. Plot constructed based on data (digitized) reported by Gawas et al.[8]

tend to drive out some of the trapped oil, resulting in higher and harder wax content, which is much more problematic in the flow infrastructure.[8] Therefore, representative cold finger testing conditions and the subsequent HTGC are important.[8]

3.3.2 SARA ANALYSIS

SARA analysis is basically a division of crude oil or condensate in terms of four solubility classes, namely, saturates (S), aromatics (A), resins (R), and asphaltenes (A). Saturates include straight chain, branched, and cycloalkanes or naphthenes; aromatics are those having hexagonal benzene ring structures; and resins are large polar molecules with N, S, and O often attached to their structures.[9] These three solubility classes also are referred to as maltenes.[9] Asphaltenes, on the other hand, contain the largest amounts of N, S, and O along with heavy metals, such as Ni, Cd, and Fe, attached to their structure.[9] The chemical structure of asphaltenes is not universal and is quite complicated from a chemistry standpoint.

A typical reservoir fluid composition in its entirety also contains lighter well-defined components, such as methane, ethane propane, and others, and non-hydrocarbon constituents; however, the four solubility classes, as defined earlier, pertain to the liquid phase. This means that SARA analysis is carried out on flashed/degassed reservoir fluids or stock tank liquids. A brief procedure for obtaining SARA analysis is as follows: The stock tank liquid is mixed with an excess volume of precipitant (n-pentane, hexane or heptane) to separate the asphaltenes since they are not soluble in these n-alkanes. After filtering the asphaltenes, the deasphalted liquid, called maltenes, is separated into resins first (by adding acetone) and then into aromatics and saturates using dimethylformamide.[10] This method is called all-solvent fractionation;[10] other methods are discussed by Speight.[11] The separated SARA is generally reported in wt%.

In flow assurance, SARA analysis is used as one of the criteria or indicators to determine the propensity of a given reservoir fluid in terms of asphaltene precipitation and deposition. Derivative correlations such as colloidal instability index and refractive index that use SARA wt% are employed to judge a given reservoir fluid as far as asphaltene stability or instability is concerned. These are discussed in Chapter 5. The stock tank oil density (specific gravity) or American Petroleum Institute (API) gravity generally correlates reasonably well with SARA wt%. A higher percentage of aromatics, resins and asphaltenes results in higher densities or lower API gravities as is common with heavier fluids, whereas a smaller percentage balanced by higher saturate content is reflected by lower densities or higher API gravities in lighter oils. It should be noted here that the absolute wt% of a given solubility class does not necessarily indicate a problematic or non-problematic fluid, in other words higher wt% of asphaltenes versus lower wt% of asphaltenes; rather, it is the collective influence of each solubility class that determines the problem potential. As an example, Figure 3.3 shows the SARA relationship with API gravity for various stock tank liquids. As seen in Figure 3.3, the data cover a very wide range of API gravities, geographic origins and expected SARA wt% trends that reasonably correlate with the API gravities.

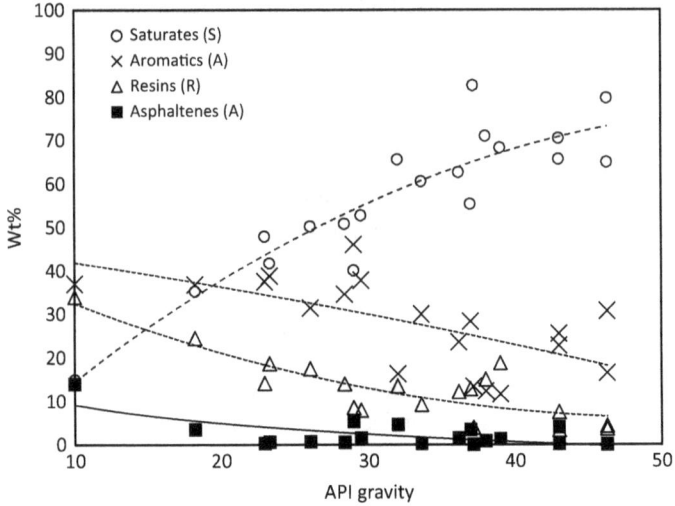

FIGURE 3.3   SARA analysis versus API gravity of stock tank liquids of different geographic origins (North Sea, North America, Europe, North Africa, South America, Middle East). Plot constructed based on data reported in various sources.[12–16]

3.4   CONVENTIONAL PVT TESTS

All conventional PVT tests are carried out in fairly standardized high-pressure high-temperature (HPHT) pressure (P)–volume (V)–temperature (T) equipment in which the pressure can be varied by altering the volume or vice versa and the temperature maintained constant for the most common tests with the exception of a separator test. The basic principle used is pressure depletion conforming to production methods that make use of the reservoir energy. All depletion tests are carried out on bulk reservoir fluids, that is, either bottomhole or recombined separator samples that are properly conditioned but, most important, in the absence of porous media. The underlying assumption is that fluid-phase behavior in bulk approximates the one in porous media. For conventional reservoirs, this works reasonably well; however, this assumption is not valid in nanopores due to the dominant capillary forces.[17] Another simplification, from a practical standpoint, is the absence of water given its immiscibility with hydrocarbons, even though both coexist in the shared pore space.

3.4.1   CONSTANT COMPOSITION EXPANSION

As the name constant composition expansion (CCE) suggests, the "overall" composition of the fluid remains constant throughout the multiple depletion steps. This also means the total mass of the fluid is constant because nothing is removed from the PVT cell. This test also is sometimes referred to as flash liberation or vaporization. The primary purpose of the test is to measure the saturation pressure (bubble or dew point), gas and oil volumes at the test conditions. However, it is a good practice to measure two additional saturation pressures at temperatures lower than the reservoir

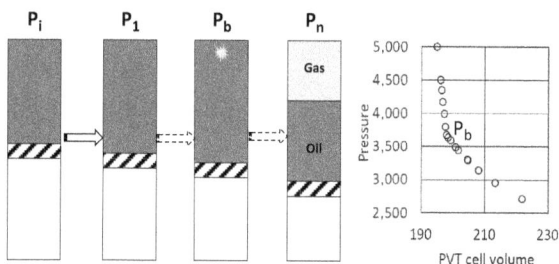

FIGURE 3.4 Schematic of a CCE on an oil sample and the resulting data. Typical CCE starting pressure is approximately 2,500 psi + reservoir pressure.

temperature, such that they represent conditions in the production tubing. These data are also deemed useful in the rigorous tuning of EOS models. Figure 3.4 shows the schematic of a CCE for an oil sample.

Mathematically, the concept of CCE can be readily understood from the molar balance equation,

$$Z_i = \frac{\left(\dfrac{Y_i V_g \rho_g}{MW_g}\right) + \left(\dfrac{X_i V_l \rho_l}{MW_l}\right)}{\left(\dfrac{V_{SP} \rho_{SP}}{MW_{SP}}\right)} \tag{3.1}$$

where i is the component designator ($i = 1$ to n); Z_i is the overall composition (mole fraction); Y_i and X_i are the mole fractions of the gas and liquid phases, respectively; V_g and V_l are the equilibrium phase volumes of gas and liquid, respectively, at a given pressure, ρ_g and ρ_l are the corresponding phase densities; and MW_g (= $\Sigma Y_i MW_i$) and MW_l (= $\Sigma X_i MW_i$) are the molecular weights of gas and liquid phases, respectively, whereas SP denotes single phase. Basically, Equation (3.1) tells us that if Y_i, X_i, V_g, V_l, ρ_g, ρ_l, V_{SP} and ρ_{SP} are measured at various pressures below the dew or bubble point and in single phase, respectively, then the computed Z_i values should be the same. An example of synthetic gas condensate is shown in Figure 3.5.

3.4.2 DIFFERENTIAL LIBERATION AND SEPARATOR TESTS

The differential liberation (DL) test is considered a classical depletion test that mimics the liberation of gas and the corresponding shrinkage of oil in the reservoir below the bubble point pressure.[5] Accordingly, pressure in a PVT cell is reduced in multiple steps below the bubble point (it is assumed that a CCE has been performed before) down to atmospheric pressure. However, at each pressure step, all the liberated or evolved gas (also referred to as free gas) is displaced from the PVT cell, only the equilibrium oil is allowed to be flashed to the next lower pressure and the expulsion of all gas is repeated. Similar to the CCE test, gas and oil volumes are recorded and the liberated gas properties (composition) measured. All other properties/values are

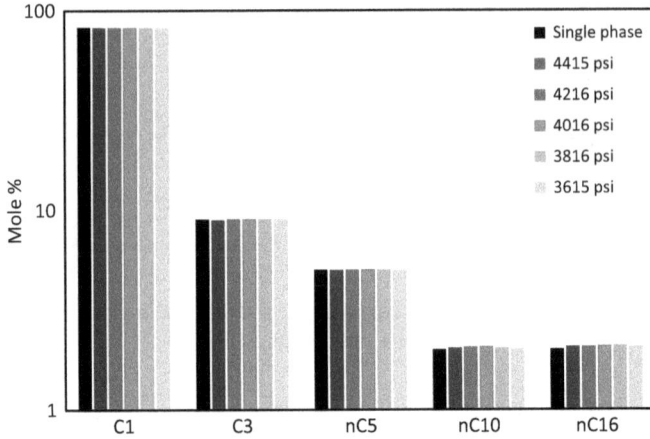

FIGURE 3.5 Consistency of direct compositional measurements and illustration of CCE concept based on the molar balance of Equation 3.1. Note that the five pressures are below the dew point. Plot constructed based on data presented by Danesh et al.[18] for a synthetic five-component gas condensate system.

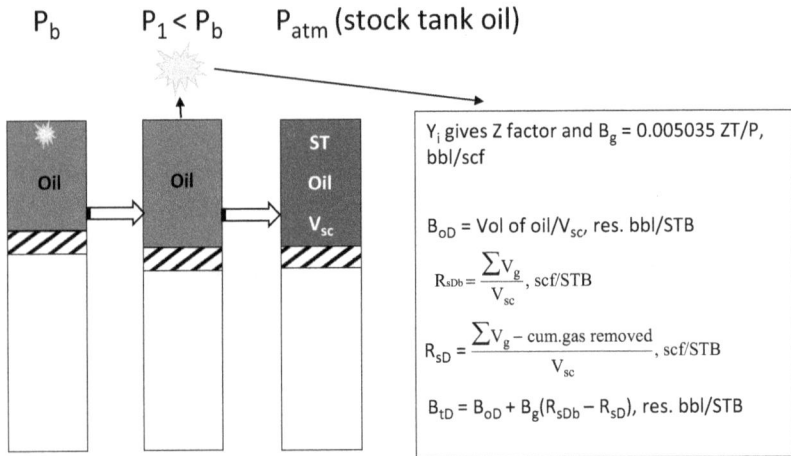

FIGURE 3.6 Schematic of a DL sequence on an oil sample and the resulting data. Test starts with oil at bubble point pressure at reservoir temperature. Variables have been defined in Section 3.4.2.

calculated from volumetric and other data. Densities and viscosities of the oil also are typically measured with an online densitometer and a viscometer, whereas gas density and viscosity are generally calculated. Figure 3.6 depicts a conceptual schematic of the DL test and the resulting data.

Separator tests are typically carried out on black oils to determine the optimum separator pressure. For a black oil, usually two stages of surface separation, that is, a primary or first stage and a secondary or stock tank, are adequate. The primary goal of a separator test is to find out the pressure at which it should be operated in the field

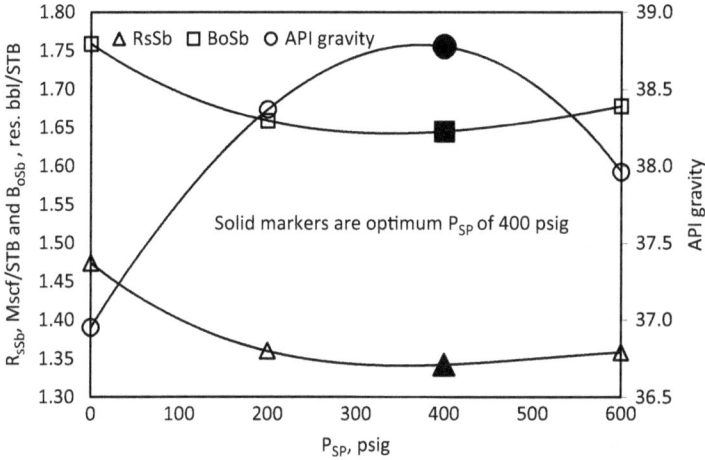

FIGURE 3.7   Identification of optimum primary separator operating pressure at average field temperature of 84°F. Plot constructed based on an exercise problem in Danesh.[5] Note that data at 0 psig means directly one stage separation (stock tank only) is used.

such that the stock tank oil has a maximum in the API gravity and the combined (primary + stock tank) oil formation volume factor and solution GOR are minimal, respectively. Although the tests do not require multiple pressure depletion steps like CCE or the DL does, what is needed is basically a two-step flash, that is, reservoir $\Rightarrow$ $P_{SP} \Rightarrow P_{ST}$, where P_{SP} is the primary separator pressure (unknown, to be determined) and P_{ST} is the stock pressure (near atmospheric). Accordingly, different pressures are tested and an optimum operational P_{SP} determined. Another difference compared to other laboratory tests is the temperature, which, in the case of separator tests, is the average field temperature. Other test details and specifics can be found in McCain.[19] As an example, Figure 3.7 shows the identified optimum conditions for the operation of the primary separator based on the aforementioned criteria.

In material balance and other calculations, such as flow assurance, oil properties are adjusted or corrected as follows.[5, 19]

At pressures above the bubble point,

$$R_s = R_{sSb} \tag{3.2}$$

$$B_o = \left(\frac{V_t}{V_b}\right)_F B_{oSb} \tag{3.3}$$

At pressures below the bubble point,

$$R_s = R_{sD}\left(\frac{R_{sSb}}{R_{sDb}}\right) \tag{3.4}$$

$$B_o = B_{oD} \left(\frac{B_{oSb}}{B_{oDb}} \right)$$

(3.5)

$$B_t = B_{oD} \left(\frac{B_{oSb}}{B_{oDb}} \right) + B_g \left(R_{sDb} - R_{sD} \right) \left(\frac{R_{sSb}}{R_{sDb}} \right)$$

(3.6)

where, B_o, R_s, and B_t are the oil formation volume factor in reservoir barrels per stock tank barrel (res. bbl/STB), the solution GOR in scf/STB, and the total or two-phase formation volume factor in res. bbl/STB, $(V_t/V_b)_F$ is the relative total volume from the CCE or flash liberation (thus the subscript F). The designators S and D denote values from the separator tests and differential liberation, whereas b means the bubble point. Note that B_g (gas formation volume factor in res. bbl/scf) in Equation 3.6 is based on the DL test and represents the value for the free or liberated gas below the bubble point (see Figure 3.6). It should also be noted that above the bubble point, customarily, $B_t = B_o$.

In a typical DL test, the collected data can be readily used to calculate values of B_g, B_{oD}, R_{sD} at various pressure steps and a single value of R_{sDb}. This is illustrated with various equations in conjunction with Figure 3.6. As an example, an exercise problem from Danesh[5] is solved here, and the solution is shown in its entirety. Danesh[5] presented a complete data set of laboratory tests, including reservoir fluid composition and expelled gas composition, which is used for this purpose. Figure 3.8 plots the relative volume $(V_t/V_b)_F$ versus pressure for this oil that shows the bubble point of 4580.7 psia at the reservoir temperature of 195°F. Also plotted in Figure 3.8 is the Y – function, which is an indicator of the relative changes in pressure and volume in

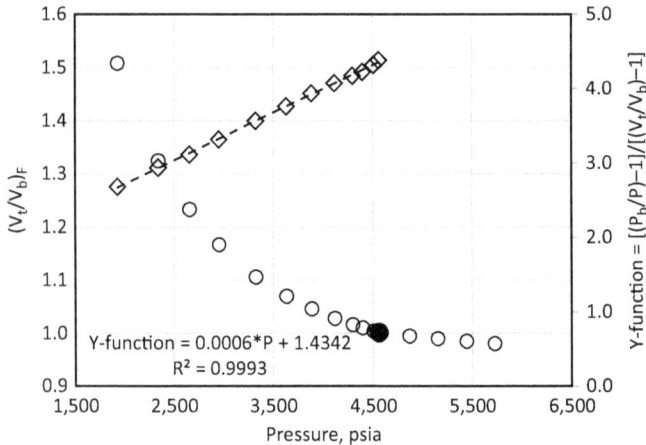

FIGURE 3.8   Relative total volume and Y – function from a CCE test for a reservoir oil showing a bubble point of 4580.7 psia (solid circle) at $(V_t/V_b)_F = 1$ and a Y – function straight line with slight curvature. Plot constructed based on data in Danesh.[5]

the two-phase region. The Y – function versus pressure should yield a straight line or a slight curve, as is the case with these data.

The B_{oD} and R_{sD} values for DL are directly shown in a table; however, B_g values are not reported, but they can be easily calculated. Based on the reported evolved gas compositions, pseudo reduced pressure and temperature allow us to determine the gas deviation factor (Z) by using either the Z-factor chart or methods that have converted the chart into correlations. It should be noted that depending on the gas composition, appropriate correction to the pseudo-critical pressure and temperature should be applied for the presence of non-hydrocarbons and/or C_{7+} fraction (high molecular weight or rich gases). These are discussed, with associated references, elsewhere.[2] Alternatively, the gas composition can be used to determine the molecular weight and, subsequently, the gas gravity, from which the pseudo-critical pressure and temperature are estimated by empirical correlations while keeping the remaining methodology to determine the Z factor the same. Utilizing both approaches, the Z factors determined are nearly the same as the one reported by Danesh.[5] Sample calculation of B_g and B_{tD} at 3,983 psig is as follows:

$$B_g = 0.005035\,ZT\,/\,P = 0.005035\frac{0.889\times(195+460)}{(3983+14.7)} = 0.00073\,\text{res.bbl}\,/\,\text{scf};$$

$$B_{tD} = 1.695 + 0.00073(1541-1261) = 1.900\,\text{res.bbl}\,/\,\text{STB}.$$

Note that the pressure and temperature should always be in absolute units (for field units, psia and °R), and B_g needs to be in consistent units such that B_{tD} results in reservoir barrels per stock tank barrel. The DL data presented by Danesh[5] are extended using B_g and B_{tD} calculations at all pressures, as shown in Table 3.2.

Next, the properties of this oil are adjusted using Equations 3.2 through 3.6. Note that the separator values used in these equations are the ones at optimum primary-stage separator pressure. As seen in Figure 3.7, this value is 400 psig for this oil. Example calculations for 5,151.7 psia ($>P_b$) and 2,721.7 psia ($<P_b$) follow, and Table 3.3 provides the complete table of adjusted values.

At 5,151.7 psia,

$R_s = R_{sSb} = 1,342$ scf/STB (note that the values in Figure 3.7 are in Mscf/STB);

$$B_o = \left(\frac{V_t}{V_b}\right)_F B_{oSb} = 0.9897\times1.645 = 1.628\,\text{res.bbl}\,/\,\text{STB}$$

At 2,721.7 psia,

$$R_s = R_{sD}\left(\frac{R_{sSb}}{R_{sDb}}\right) = 806\times(1342\,/\,1541) = 702\,\text{scf}\,/\,\text{STB}$$

$$B_o = B_{oD}\left(\frac{B_{oSb}}{B_{oDb}}\right) = 1.48\times(1.645\,/\,1.834) = 1.327\,\text{res. bbl}\,/\,\text{STB}$$

TABLE 3.2

Complete Differential Liberation Data Showing the Calculated B_g and B_{tD} Values. Note that the Table of Danesh[5] is Extended Based on Additional Calculation Results

Pressure, psia	B_{oD}, res. bbl/STB	R_{sD}, scf/ STB	Z Factor, Dimensionless	B_g, res. bbl/scf	B_{tD}, res. Bbl/STB
5,726.7	1.798				
5,441.7	1.807				
5,151.7	1.815				
4,864.7	1.825				
4,580.7	1.834	1,541	P_b		
3,997.7	1.695	1,261	0.889	**0.00073**	**1.900**
3,570.7	1.614	1,092	0.865	0.00080	1.973
3,150.7	1.542	939	0.845	0.00088	2.074
2,721.7	1.480	806	0.839	0.00102	2.227
2,291.7	1.422	680	0.843	0.00121	2.467
1,863.7	1.370	564	0.852	0.00151	2.843
1,429.7	1.320	451	0.873	0.00201	3.515
1,000.7	1.271	344	0.897	0.00296	4.810
580.7	1.222	238	0.931	0.00529	8.111
228.7	1.166	135	0.96	0.01384	20.630
14.7	1.059	0	1	0.22435	346.780

TABLE 3.3

Adjusted Values of B_o, R_s and B_t for the Reservoir Oil by Danesh.[5] Note that This Table Represents an Extended Solution of the Exercise Problem

Pressure, psia	B_o, res. bbl/STB	R_s, scf/STB	B_t, res. Bbl/STB
5,726.7	1.613	1,342	
5,441.7	1.620	1,342	
5,151.7	**1.628**	**1,342**	
4,864.7	1.636	1,342	
4,580.7	1.645	1,342	
3,997.7	1.520	1,098	1.699
3,570.7	1.448	951	1.760
3,150.7	1.383	818	1.847
2,721.7	**1.327**	**702**	**1.978**
2,291.7	1.275	592	2.185
1,863.7	1.229	491	2.512
1,429.7	1.184	393	3.096
1,000.7	1.140	300	4.222
580.7	1.096	207	7.096
228.7	1.046	118	17.996
14.7	0.950	0	302.026

Calculation table showing adjusted oil properties, following the previous table.

$$B_t = B_{oD} \left(\frac{B_{oSb}}{B_{oDb}} \right) + B_g \left(R_{sDb} - R_{sD} \right) \times \left(\frac{R_{sSb}}{R_{sDb}} \right)$$
$$= 1.327 + 0.00102 \times (1541 - 806) \times (1342 / 1541)$$
$$= 1.978 \text{ res. bbl / STB}$$

Finally, in DL and other tests, it is quite common to also include oil viscosity and density measurements in particular. Although gas properties do not drastically change with pressure, an order of magnitude or two is quite possible in the case of oils when one compares viscosity at the bubble point versus at atmospheric pressure. This dramatic change in the viscosity from the lowest to highest value is due to the dominant compositional effect or lowering of the solution's GOR from maximum to zero. In flow assurance calculations, it is important to accurately capture this variation, which can be readily realized by examining the dimensionless Reynold's number ($Re = du\rho/\mu$) used in fluid mechanics. Figure 3.9 plots the density and viscosity of a heavy oil, showing more than twice an increase in viscosity at low 100 psia compared to the bubble point of 685 psia. Note the characteristic "V"-shaped trend of both density and viscosity.

3.4.3   CONSTANT VOLUME DEPLETION

The constant volume depletion (CVD) test shares one common feature with the DL test in that the gas is expelled from the PVT cell at pressures below the saturation pressure; however, only a certain volume of gas is removed such that a constant cell

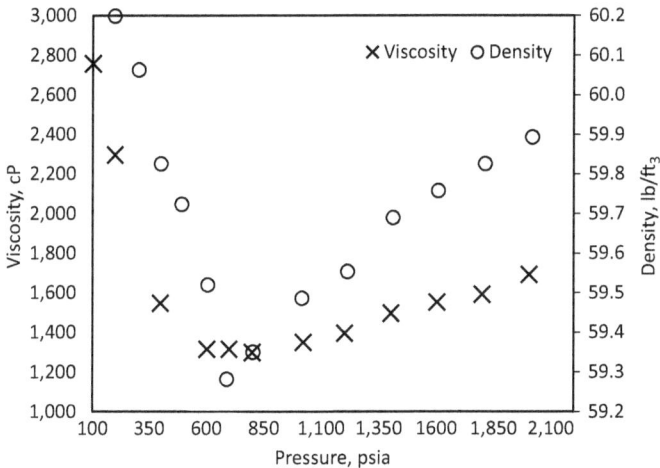

FIGURE 3.9   Heavy oil viscosity and density versus pressure in the single- and two-phase regions at a reservoir temperature of 120°F. Note the low bubble point pressure of about 685 psia, which is typical of heavy oils. Combined plot constructed based on data (digitized) in Quinones-Cisneros et al.[20]

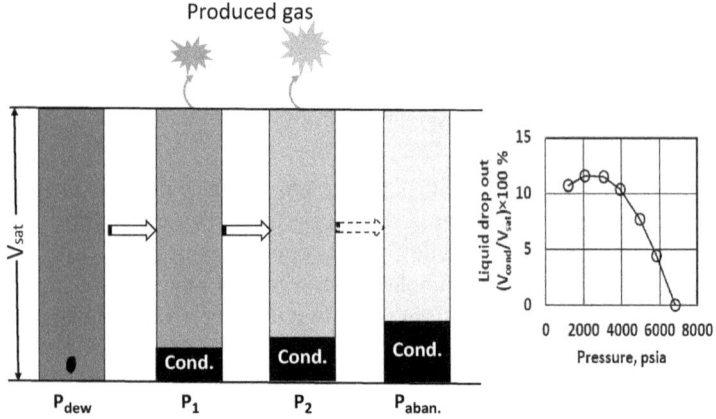

FIGURE 3.10 Schematic of a CVD test on a gas condensate sample and the resulting data. Test typically starts at dew point pressure at reservoir temperature.

volume is maintained throughout the test. This constant cell volume, V_{sat} or volume at the saturation pressure, which, for the most part, refers to the dew point since the test is carried out on gas condensates. The CVD test is also considered for volatile oils because of their unique production characteristics below the bubble point, with much of the liquid production originating from the liberated gas.[21] Unlike the DL test, CVD is generally terminated at relatively higher pressures, which are considered abandonment conditions. Composition of the removed gas is normally reported, which is used in conjunction with other measured data, such as volumetrics, to determine the stagnant retrograde condensate composition from material balance.[2, 5] Also reported in the test is the retrograde liquid dropout, which is defined as retrograde condensate volume/V_{sat}. Similar to other lab data, CVD data are also used to calibrate EOS models. Figure 3.10 shows a schematic of the CVD test.

3.5 FLUID CHARACTERIZATION USING LABORATORY MEASURED DATA

Perhaps the biggest uncertainty in dealing with petroleum reservoir fluids in all cases lies in the heavy end or the plus fraction. The heavy end that we refer to here is C_{7+}. If this fraction is left as is or is not properly characterized, then critical properties and acentric factors derived from such a grossly averaged C_{7+} fraction will likely lead to a poor performance by EOS models and, consequently, highly inaccurate phase behavior and fluid property predictions. Furthermore, even tuning of the C_{7+} properties may not be adequate or satisfactory. Therefore, the primary objective of fluid characterization is to better define the C_{7+} fraction by splitting it, as shown in Figure 3.11.

How is splitting or fluid characterization achieved? The best method, of course, is to physically split either by TBP distillation or GC analysis. However, this is possible only if a physical sample and the facilities are available, which may not always be the case. Therefore, in this section, we cover some of the most commonly used

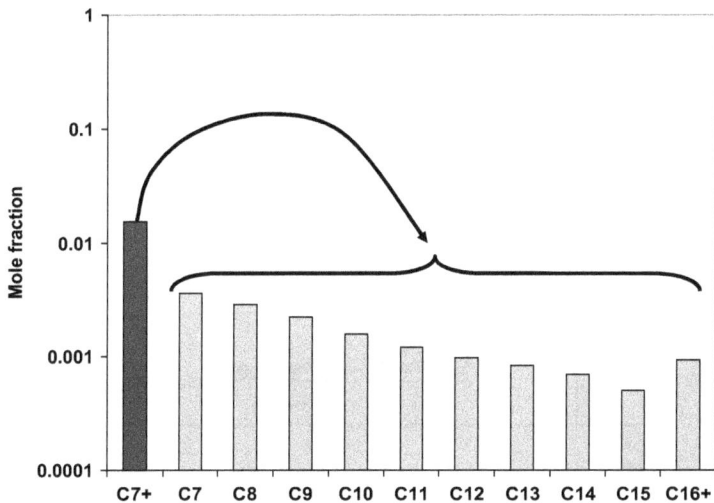

FIGURE 3.11 Concept of splitting or fluid characterization to better define the C_{7+}.

numerical techniques that address the C_{7+} characterization. The principal idea here is to use some measured properties of the C_{7+} fraction such as molecular weight and specific gravity and, in some cases, a measured partial molar distribution if available (e.g., C_7–C_{10}), in which case what is actually characterized is C_{11+}. The molecular weight and specific gravity data are almost always available, which is measured by a freezing point depression apparatus and oscillating U-tube densitometer. Both values are somewhat indicative of how light or heavy a given C_{7+} fraction is and is, of course, reflective of its chemical makeup. Also, the C_{7+} fraction can be construed as a degassed or flashed oil in most cases.

Several numerical methods, which have been discussed elsewhere,[5] are available to split a C_{7+} fraction. The only method covered here is that of Pedersen et. al.[22] which is empirical in concept only and thus does not rely on predetermined or fixed correlation coefficients. Pedersen et. al.[22] proposed an exponential decrease of composition with increasing carbon number, which was based on a large database of fluids from the North Sea and the rest of the world, with measured compositional analysis up to C_{80+}.[5] The empirically conceptual functional form of Pedersen et al. characterization relation is shown by the following equation:

$$Z_{Cn} = \exp\left(A + B_{Cn}\right) \tag{3.7}$$

or, in terms of molecular weight,

$$Z_{Cn} = \exp\left(A + BMW_n\right) \tag{3.8}$$

where Z_{Cn} is the mole fraction of a single carbon number group C_n having a molecular weight of MW_n. A and B are the constants or customized correlation coefficients

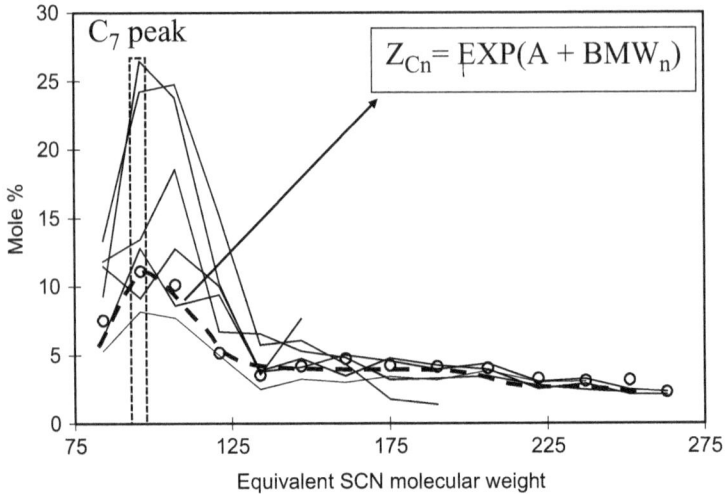

FIGURE 3.12   Molar distribution versus molecular weight of TBP distilled fractions of North Sea oils and condensates. Plot based on data originally presented in Dandekar et al.[23] in the form of mole % versus carbon number. Note the starting of linearity from C_7 peak.

that need to be determined for a given fluid. The exponential concept as described by Equations 3.7 and 3.8 can be readily realized by considering the TBP distillation data of some North Sea oils and condensates, as shown in Figure 3.12.

Although the distribution shown in Figure 3.12 and its mathematical representation are typical of most reservoir fluids, that is not the case as far as other peculiar fluids are concerned.[5] For example, reservoir fluids that have been biodegraded or those that contain an unusual amount of aromatics and naphthenes, which means characterization methods such as the ones described in Equation 3.7 or 3.8 should not be applied since they may lead to an erroneous splitting of the C_{7+} fraction. Krejbjerg and Pedersen[24] presented plots of Z_{Cn} versus C_n for three different oils having API gravities of 28, 18 and 10. The oil with 28° API gravity exhibited a distribution similar to the one shown in Figure 3.12, with the linearity beginning at C_7. However, the other two heavier oils of low API gravities also showed the linearity, albeit beginning at the peak carbon numbers of 11 and 19, respectively. The authors state that the lighter components in the low–API gravity oils may have disappeared over time due to biodegradation. This can be interpreted as follows: If the oil is heavy or of low API gravity, then the concept of exponential decrease may be applied but for characterization of a heavier plus fraction instead of a C_{7+}, which would be the case with typical oils of reasonable API gravities (20+).

Danesh[5] presented an alternate method of determining the constants A and B (Equations 3.7 and 3.8) when limited data, such as only the molecular weight of C_{7+}, are available. In other words, no partial molar distribution is available from which A and B could be calculated. The alternate method is perhaps more practical because quite often having limited data may very well be the case. The basis of the method is two material balance equations of Danesh,[5] which we have generalized as follows:

$$\sum_{Cx}^{C_N} Z_{Cx} = \sum_{Cx}^{C_N} \exp\left(A + BMW_x\right) = Z_{Cx+} \tag{3.9}$$

$$\sum_{Cx}^{C_N} Z_{Cx}MW_x = \sum_{Cx}^{C_N} \exp\left(A + BMW_x\right)MW_x = Z_{Cx+}MW_{x+} \tag{3.10}$$

A value of $x = 7$ in Equations 3.9 and 3.10 means a C_{7+} fraction having a molecular weight of MW_{7+} and a mole fraction of Z_{C7+}. In this case, the summations begin from C_7 to a much higher carbon number, such as C_{45}, but could be extended to even higher values. The generalization shown in Equations 3.9 and 3.10 easily allows the characterization of any plus fraction, for example, C_{19+} in the case of the 10° API oil of Krejbjerg and Pedersen.[24] The working equations are derived by Danesh,[5] and they allow the determination of first B, then A and finally the characterized composition. Again, the generalized forms of the three equations follow:

$$\sum_{Cx}^{C_N} \left(MW_x - MW_{x+}\right)\exp\left(BMW_x\right) = 0 \tag{3.11}$$

$$A = LN\left(Z_{Cx+}\right) - LN\left(\sum_{Cx}^{C_N} \exp\left(BMW_x\right)\right) \tag{3.12}$$

$$Z_{Cx} = \exp\left(A + BMW_x\right) \tag{3.13}$$

Equation 3.11 suggests that the value of B depends only on the molecular weight of the C_{x+} fraction[5] (C_{7+}, C_{10+}, etc.). This means that if generalized molecular weights are used for all fluids for components or SCN fractions x to N, then the value of B will be fixed or unique for every given plus-fraction molecular weight. Note also that the value of B needs to be determined by suitable numerical methods that yield the Equation 3.11 summation approximately equal to 0. Since mole fractions will decrease with an increasing carbon number vis-à-vis molecular weight, the value of B is always negative; however, it is important to keep in mind that sometimes, a false or unrealistic minimum negative B value may be encountered.

Next, we demonstrate the application of Equations 3.11 through 3.13 to the 10°API oil presented in Krejbjerg and Pedersen.[24] As mentioned previously, since the linearity begins at C_{19}, the characterization is basically applied to a C_{19+} fraction. The data reported for this particular oil include descriptions up to C_{36+}. Accordingly, using the mole fractions and molecular weights of all the components from C_{19}–C_{36+},

the C_{19+} molecular weight is calculated using $\sum_{i=19}^{36+} X_i MW_i = MW_{x+} = MW_{19+} = 655.6$.

This value can be obtained after normalizing the C_{19} to C_{36+} mole fractions to 1 or by using them as is but dividing the summation by 0.5696 (C_{19+} mole fraction) – the end result will be the same. Table 3.3 from Krejbjerg and Pedersen[24] shows all the detailed data. Equation 3.11 can now be written for this specific case:

$$\sum_{C19}^{C_{200}} \left(MW_x - 655.6\right)\exp\left(BMW_x\right) = 0 \qquad (3.14)$$

Note that $N = 200$ or C_{200} is chosen in order to be consistent with the value used by the authors. In applying Equation 3.14, molecular weights of all components from C_{19}–C_{200} were estimated from fitting the reported data of the authors. Equation 3.14, written for C_{19} fraction with its molecular weight of 263, would be $(263 - 655.6)$ $\exp(263B)$. Such equations are obviously written for each carbon number all the way up to 200. A particular value of B that would yield the summation of C_{19}–C_{200} for Equation 3.14 to approximately equal to 0 is considered as the converged value. For this particular case, the B value is –0.00247. Table 3.4 shows the results of this calculation and the characterized composition, which is compared with the reported composition of Krejbjerg and Pedersen[24] (see Figure 3.13). Although some differences are evident, overall, the characterization produces a reasonable result.

As described previously, the A and B values can also be obtained by using the partial molar distribution, such as C_{19}–C_{35} in this particular case. Figure 3.14 shows the plot of MW_n versus $LN(Z_{Cn})$ and the straight-line fit resulting in $A = -2.3165$ and

---

TABLE 3.4

Calculation of B Value Using the Pedersen Characterization[22] Limited Data Method[5] for the C_{19+} Fraction of the 10° API Heavy Oil

SCN	MW_n	$(MW_x - 655.6)$ $\exp(-0.00247MW_x)$	$\exp(-0.00247MW_x)$	Z_{Cn}, Mole Fraction
C_{19}	263	−205.2109	0.5226	0.019310
C_{20}	275	−193.1358	0.5074	0.018746
...	...	......	....	......
C_{65}	905	26.7517	0.1071	0.003958
...	...	......	....	......
C_{75}	1045	29.5555	0.0759	0.002803
...	...	......	....	......
C_{95}	1325	25.4659	0.0381	0.001406
...	...	......	....	......
C_{110}	1535	19.9341	0.0227	0.000838
...	...	......	....	......
C_{160}	2234	6.3757	0.0040	0.000149
...	...	......	....	......
C_{200}	2793	2.1722	0.0010	0.000038
		$\Sigma = 3.6E^{-7}$	$\Sigma = 15.4168$	$\Sigma = 0.5696$

$A = LN(0.5696) - LN(15.4168) = \mathbf{-3.2983}$. Values computed in Column 5 are obtained from Equation 3.13.

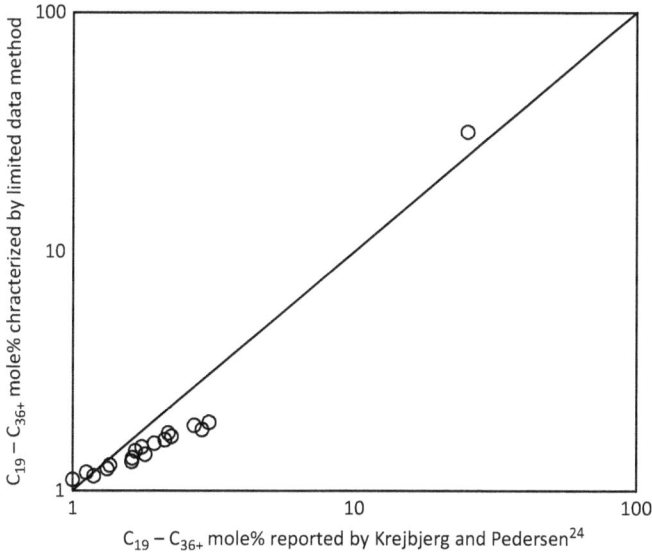

FIGURE 3.13 Comparison of Krejbjerg and Pedersen[24] reported C_{19}–C_{36+} composition and characterized by the limited data method. Note that the C_{36}–C_{200} mole fraction (plotted in %) computed in Table 3.4 are aggregated as C_{36+}.

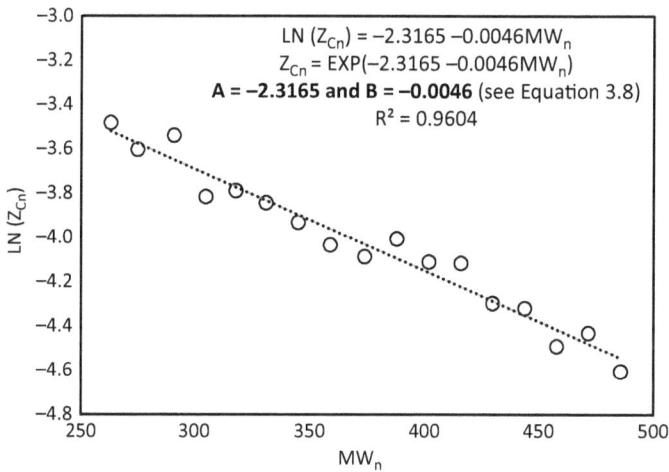

FIGURE 3.14 Determination of A and B values from reported[24] partial molar distribution of C_{19}–C_{35}.

$B = -0.0046$. Clearly, these values are different compared to those from the limited data method, since they reflect the intercept and slope based on the partial molar distribution. However, when the A and B values from Figure 3.14 are used to extend the composition all the way up to C_{200} and the mole fractions C_{19}–C_{200} added, then the sum does not equal 0.5696 (C_{19+} mole fraction). Note that in the summation, C_{19}–C_{35} are the reported values, and the rest are extended by using the calculated A and B. Since material balance equations are employed in the limited data method, the summation of C_{19}–C_{200} automatically equals 0.5696. An alternative approach for the deficit observed using the partial molar distribution is to extend the composition up to some other higher carbon number, for example, C_{199}, and determine the mole fraction

of C_{200+} as follows $C_{200+} = 0.2517 - \sum\limits_{36}^{199} Z_{Cn}$ or $C_{200+} = 0.5696 - \sum\limits_{19}^{199} Z_{Cn}$ (C_{19}–C_{35} mole

fractions being the reported values of partial molar distribution). In this particular example, $C_{200+} = 0.09$, whereas using the limited data method, $C_{200} = 0.000038$ (see Table 3.4). Clearly, this difference is significant, the sensitivity can only be judged after utilizing both sets of compositions in their entirety.

A given plus fraction characterized up to a carbon number as high as 200 means numerous components that need to be handled individually in any application. This is, however, unwieldy and impractical. Therefore, the next step of processing the characterized data is called "grouping" or "lumping" multiple carbon numbers. For example, C_{100} through C_{120} are lumped together and treated as one component. Splitting or characterization first and then grouping may seem like a paradox; however, it is not because both are carried out systematically and not randomly. Lumping offers the advantage of reducing the computational time in large-scale flow assurance and other numerical simulations. The method that is quite commonly used is the one proposed by Whitson.[25] The C_{7+} of a given fluid after characterization is lumped as follows:

$$N_g = \text{Integer}\left[1 + 3.3 \log_{10}(N - 7)\right] \tag{3.15}$$

Using $N = 200$ as the value of the last carbon number in the characterized 10°API oil, we get $N_g = 8$ from Equation 3.15, which means 8 multiple carbon number groups. The grouping is based on the molecular weight cutoff as per the following criteria, applied to the characterized fluid:

$$MW_I = 96\left[\exp\left(\left(\frac{1}{8}\right)LN\left(\frac{2793.3}{96}\right)\right)\right]^I = 96\left[1.524\right]^I \tag{3.16}$$

where I varies from 1 to 8, 96 is the generalized molecular weight of C_7, 8 is the number of groups and 2793.3 is the C_{200} molecular weight (rounded off as 2793). Note that although the characterization was for the C_{19+} fraction (limited data method), the lumping scheme applies to the entire distribution from C_7 to C_{200}. Using Equation 3.16, a table of the 8 groups, as per the molecular weight cutoffs or the upper boundary that groups the components accordingly, can be generated (see Table 3.5).

As seen in Table 3.5, the molecular weight cutoff of 146 means that every component that has molecular weights up to 146 belongs to group 1, that is, C_7–C_{10}, since the molecular weight of C_{10} is 134 and of C_{11} is 147, which automatically falls into

TABLE 3.5
Grouping of the Characterized 10°API Heavy Oil[24]

Group Number (I)	Upper Molecular Weight Boundary/Cutoff	Components in Group
1	146	C_7-C_{10}
2	223	$C_{11}-C_{16}$
3	340	$C_{17}-C_{24}$
4	518	$C_{25}-C_{37}$
5	789	$C_{38}-C_{56}$
6	1203	$C_{57}-C_{86}$
7	1833	$C_{87}-C_{131}$
8	2793	$C_{132}-C_{200}$

group 2. Finally, the consistency of both lumping equations is clear in that the upper boundary of group 8 molecular weight of 2793 is C_{200} (last component). The final step in preparing the complete ready-to-use fluid description involves the determination of lumped group critical properties (T_c, P_c), specific gravity (γ) and the acentric factor (ω). This can be done by first normalizing the group mole fraction to 1 and then applying simple molar mixing rules. Alternatively, group molecular weight and specific gravity can be determined and used in estimating the critical properties and the acentric factor from the many empirical correlations.[5] Example calculations are illustrated in Table 3.6 for group 1.

TABLE 3.6
Example Calculation for Group 1 Properties. Note that the Z_{Cn} Values in Column 2 (and All the Way up to C_{18}) Are from Krejbjerg and Pedersen[24]

Components	Z_{Cn}, mole Fraction	Z_{Cn}, Normalized mole Fraction	MW_n	γ, Dimension-less	T_c, °F	P_c, psia	ω, Dimension-less
C_7	0.0013	0.0722	96	0.722	524.9	445.5	0.280
C_8	0.0032	0.1778	107	0.745	573.5	417.4	0.312
C_9	0.0045	0.2500	121	0.764	625.7	386.6	0.352
C_{10}	0.0090	0.5000	134	0.778	668.9	359.9	0.389
	$\Sigma=0.0180$	$\Sigma=1.0000$					

Group 1 MW = ΣColumn 3 × Column 4 = 123.2 Group 1 γ = (ΣColumn 3 × Column 4)/(ΣColumn 3 × Column 4/Column 5) = 0.766 Group 1 T_c = ΣColumn 3 × Column 6 = 630.8°F Group 1 P_c = ΣColumn 3 × Column 7 = 383 psia Group 1 ω = ΣColumn 3 × Column 8 = 0.358.

TABLE 3.7

Characterized and Subsequently Lumped Fluid Composition of the 10°API Heavy Oil of Krejbjerg and Pedersen[24]

Components or Groups	Mole %	MW	γ, Dimensionless	T_c, °F	P_c, psia	ω, Dimensionless
CO_2	1.44					
C_1	18.72					
C_2	0.14					
C_3	0.03					
iC_4	0.01					
nC_4	0.01					
iC_5	0.01					
nC_5	0.27					
C_6	0.41	84	0.690	458.3	474.5	0.251
C_7–C_{10}	1.80	123	0.766	630.8	383.0	0.358
C_{11}–C_{16}	14.26	189	0.821	808.2	286.6	0.532
C_{17}–C_{24}	16.62	277	0.867	965.9	218.9	0.729
C_{25}–C_{37}	16.77	423	0.905	1,137.9	160.7	1.007
C_{38}–C_{56}	14.25	639	0.941	1,324.7	111.0	1.291
C_{57}–C_{86}	9.92	961	0.983	1,530.5	77.2	1.520
C_{87}–C_{131}	4.31	1442	1.046	1,773.8	47.8	1.629
C_{132}–C_{200}	1.05	2142	1.136	2,132.0	38.5	1.642
	$\Sigma = 100.00$					

Note: Properties of well-defined components can be found elsewhere.[5]

Finally, the complete reservoir fluid composition of this 10°API heavy oil after characterization and lumping is shown in Table 3.7. Note that generally, P_c decreases and T_c increases with the carbon number; however, a reversal in the trend of P_c is seen for the last lumped group, which is likely due to the critical property correlations being out of bounds for high carbon numbers vis-à-vis molecular weights. Therefore, in order to maintain the consistency of the P_c values of the last two lumped groups, these values were based on the Riazi and Daubert[26] correlations, whereas all other values are based on the Twu[27] and Lee–Kesler (for acentric factor)[28] correlations, respectively. As far as the pseudo and plus fractions are concerned, they are normally treated or further split in terms of wax and asphaltene formers and nonformers, respectively, which are discussed in greater detail in Chapter 5.

3.6   PRODUCED WATER CHARACTERIZATION

Given the lack of explicit commercial significance of produced water, which basically originates from the reservoir pore spaces, it is not as rigorously tested in laboratories as hydrocarbons are. Therefore, for the most part, produced water properties are derived from pure water properties by applying a salinity correction factor. Similar to other reservoir fluids, produced water properties also are dependent on composition (salinity), temperature and pressure. Due to the most explicit effect of salinity, it is the most dominant variable affecting water properties and thus is almost always measured in the laboratory as part of produced water characterization. Numerous empirical correlations that apply salinity correction to pure water properties to obtain reservoir engineering parameters are described elsewhere.[5, 19] Therefore, in this section, we only focus on produced water characterization that is of significance in flow assurance, which primarily includes salinity, density, viscosity and water content of natural gas.

3.6.1 PRODUCED WATER SALINITY OR COMPOSITION

Produced water that originates from the reservoir is not pure and always contains dissolved salts, which vary in species and content, and is basically characterized by salinity. The salts typically present are sodium chloride (NaCl), potassium chloride (KCl), calcium chloride ($CaCl_2$), calcium carbonate ($CaCO_3$), magnesium chloride ($MgCl_2$), potassium chloride (KCl), sodium bicarbonate ($NaHCO_3$), sodium sulfate (Na_2SO_4), barium sulfate ($BaSO_4$), calcium sulfate ($CaSO_4$), strontium sulfate ($SrSO_4$) and others. These salts are a combination of cations and anions, that is, positively charged monovalents Na^+, K^+, the divalents Ca^{++}, Mg^{++} and the negatively charged Cl^-, HCO_3^- and SO_4^{2-}, respectively. Note though that the predominant cation and anion in almost every formation water is Na^+ and Cl^-, respectively, or NaCl salt. The concentration of cations and anions is generally reported in mg/liter and is experimentally determined by ion-exchange chromatography. The principles can be found in Fritz and Gjerde.[29] The respective cations and anions can be related to the equivalent salts by stoichiometry. The summary of all the ions represents what is known as total dissolved solids (TDS), expressed in parts per million (ppm). Assuming a water density of approximately 1,000 g/liter; 1 mg/liter = 10^{-3} g/1,000 g = 1 g/10^6 g, which is 1 ppm. Similar to the fact that every petroleum reservoir fluid is compositionally unique, every formation or produced water also has a unique ionic composition that can differ vastly from reservoir to reservoir.[2, 30] For example, the TDS in formation waters can range from 10,000 to 300,000 ppm, whereas seawater is typically 30,000 ppm.[30] Therefore, produced water composition is one of the important measurements carried out on every formation water given its importance in various aspects of flow assurance as outlined next and covered in detail in other pertinent chapters.

3.6.1.1 Gas Hydrate Phase Equilibria and Salinity

Whenever natural gas and water are brought together at the thermodynamically favorable temperature and pressure, gas hydrates form. The water composition or salinity plays an important role in hydrate phase equilibria. For a given gas composition and temperature, if the salinity is varied, then the hydrate equilibrium pressure shifts accordingly. Figure 3.15 shows the gas hydrate equilibrium curves for a synthetic gas mixture of three components (75 mole% methane, 20 mole% ethane, and 5 mole% propane), pure water and salts (NaCl only) of different salinities. As seen in this collection of equilibrium curves, as the salinity increases, the hydrate equilibrium pressure at any given temperature also increases, that is, an upward shift in the typical

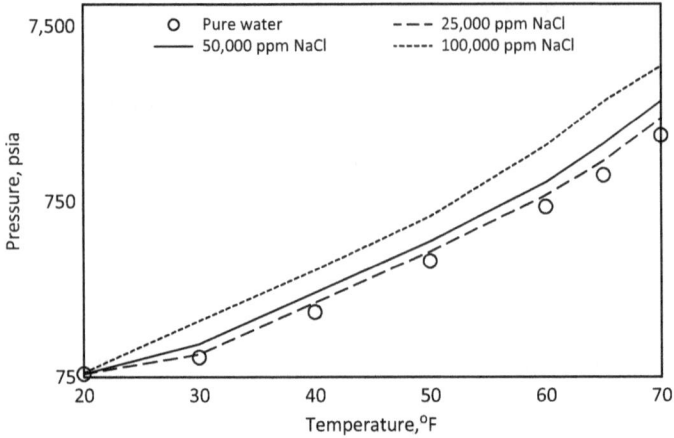

FIGURE 3.15 Collection of hydrate equilibrium curves for a three-component hydrocarbon gas mixture (75 mole% methane, 20 mole% ethane, and 5 mole% propane), pure water, and waters of different NaCl salinities.

operating region of interest. Given the fact that the equilibrium curve represents the equilibria between "liquid water–gas hydrate–gas", an upward shift of the curve simply means an expanded safe region where liquid water and gas are in equilibrium. This is an important practical consequence from a flow assurance perspective. The effect of salt or salinity is basically a thermodynamic hydrate inhibition, which also is the case with methanol and glycol (discussed in Chapter 7).

3.6.1.2   Scaling and Salinity

As rightly stated by Gudmundsson,[9] the composition of the produced water in terms of dissolved salt species is perhaps the prime factor as far as scaling is concerned. Precipitation and scaling occur when the solubility limit of a given species is exceeded and the aqueous phase can no longer retain the salt in a dissolved state. The deposited scale thus may be a combination of different species depending on the requisite alteration in temperature, pressure or both and/or when incompatible produced waters from different formations are mixed.

There are mainly two groups of salts in produced water, that is, salts of high solubility, such as $NaCl$, KCl and $CaCl_2$, and those of relatively low solubility, such as $BaSO_4$, $CaCO_3$, $SrSO_4$ and $CaSO_4$. Clearly, given the lower solubility of the second group, salts that are more likely to precipitate and deposit, commonly referred to as scale, are $BaSO_4$ (barite), $CaCO_3$ (calcite) and $CaSO_4$ (anhydrite). Common salt ($NaCl$) can form crystalline halite, which occurs when the aqueous phase evaporates in production tubulars in HPHT reservoirs.[9] Although, oilfield scaling is not a new problem, its increased importance has emerged due to the exploitation of deeper, offshore HPHT resources and is therefore a significant aspect of flow assurance, which is covered in more detail in Chapter 4.

3.6.1.3   Naphthenates and Salinity

Naphthenates are amphiphilic meaning one end has an affinity to water, whereas the other end has an affinity to a hydrocarbon phase.[9] Therefore, naphthenate formation and subsequent flow assurance problem is a result of coproduction of oil and formation water. Specifically, these are defined or characterized as calcium and sodium naphthenates,[9] respectively, and abbreviated as "CaNs" and "NaNs", respectively. Although coproduction of oil and formation water is perhaps a necessary precondition, it may not always lead to the formation and deposition of naphthenates but may come as a surprise in green- as well as brownfields. Unlike gas hydrate equilibria, currently, no direct correlation between produced water composition and naphthenate formation exists. Field evidence discussed by Gudmundsson[9] suggests that produced waters rich in sodium will form NaNs, whereas those rich in calcium will form CaNs. It is also postulated[9] that the monovalent NaNs are formed relatively easily (note that most produced waters contain high amounts of NaCl compared to other salts) than the divalent CaNs.

3.6.1.4   Corrosion and Salinity

Unlike most flow assurance solids which can be chemically or mechanically removed from the flow infrastructure, corrosion directly impacts the integrity of the flowlines itself by basically altering the inner surfaces (as well as the outer surfaces in bare flowlines offshore), potentially weakening them prone to leaks. Produced water composition and its pH are the determining factors for the extent of corrosion in the flow infrastructure.[9] For the most part, the corrosion issue is a result of CO_2 and water phase behavior in which carbonic acid formation takes place in the flow infrastructure that is conducive to corrosion when CO_2 mixes with water.[30] The solubility of CO_2 in produced water is inversely proportional to salinity. The reduction in CO_2 solubility with increasing salinity is significant as shown by the data of Enick and Klara.[31] For example, at 4,000 psia and 100°F, the ratio of CO_2 solubility in 100,000-ppm and 200,000-ppm salinity brine and fresh water is approximately 0.7 and 0.4, respectively. In a typical producing scenario, the gas coming out of a solution may contain CO_2, which will eventually interact with the coproduced reservoir water, leading to the formation of carbonic acid. In other direct cases, the CO_2 that is part of the injection process (EOR, sequestration) eventually will come into contact with the produced water, again leading to the formation of carbonic acid.

3.6.2   Produced Water Density and Viscosity

Both density and viscosity are important properties needed in hydraulics, which is an integral part of flow assurance. Although the best source for obtaining reliable values is directly measuring a captured sample using a device such as a densitometer or a viscometer, this may not always be feasible. Therefore, the use of empirical correlations is prevalent, since they also offer the convenience of being integrated into various types of flow assurance simulators.

The density of produced water, with respect to the salinity, is obtained by the following equations:[32]

$$\rho_{PW} \text{ at } 60°F \text{ and } 14.7 \text{ psia (standard conditions)}$$
$$= 62.368 + 0.438603S + 0.0016007S^2 \tag{3.17}$$

where ρ_{PW} is the density of produced water in lb/ft³ and S is the salinity in wt%, which can be easily obtained by dividing the TDS (mg/liter or ppm) by 10^4. Equation 3.17 is consistent in that it simply reduces to a density of 62.368 lb/ft³ for $S = 0$, that is, pure water. Produced water density at other conditions (typically higher temperature and pressure) is calculated from

$$\rho_{PWR} = \rho_{PW} / B_W \tag{3.18}$$

where ρ_{PWR} is produced water density in lb/ft³ at other conditions and B_W is the water formation volume factor that has values typically in the vicinity of 1+ res. bbl/STB, and these can be estimated from correlations.[32] Therefore, depending on the actual value of B_W, the produced water density at other conditions will be somewhat lower than at standard conditions.

The empirically produced water viscosity correlations of McCain[32] are valid for salinities up to 26% and pressure up to 15,000 psi; however, the lower temperature limit is 86°F. This may be somewhat marginal in flow assurance applications given the low ambient temperatures in the range of 40–45°F and its impact in cooling the flow stream in an offshore flow line. First, the viscosity of the produced water at atmospheric pressure and given temperature and salinity is calculated by

$$\mu_{PW1} = AT^{-B} \tag{3.19}$$

where, μ_{PW1} and T are viscosity in cP and temperature in °F, respectively, whereas coefficients A and B are related to salinity, S, in wt% by

$$A = 109.574 - 8.40564S + 0.313314S^2 + 0.008722213S^3 \tag{3.20}$$

and

$$B = 1.12166 - 0.0263951S + 6.79461 \times 10^{-4} S^2$$
$$+ 5.47119 \times 10^{-5} S^3 - 1.55586 \times 10^{-6} S^4 \tag{3.21}$$

Note that the substitution of $S = 0$ in Equations 3.20 and 3.21 and $T = 60°F$ in Equation 3.19 will result in pure water viscosity of approximately 1.1 cP (a typical value), which again proves the consistency of the preceding three equations. Next, the produced water viscosity at elevated pressure is obtained from

$$\mu_{PWR} = \mu_{PW1} \left[0.9994 + 4.0295 \times 10^{-5} P + 3.1062 \times 10^{-9} P^2 \right] \tag{3.22}$$

where μ_{PWR} is water viscosity in cP at a given pressure P in psia.

3.6.3 WATER CONTENT OF NATURAL GAS

In 1958, McKetta and Wehe[33] presented a chart that allows the graphical determination of water vapor content of natural gas at a given temperature and pressure. Basically, this chart is used to find out how much water vapor is associated with the gas (expressed in lb mass of water/mmscf of gas; mmscf is million standard cubic feet) at a certain temperature and pressure. In other words, this is akin to the moisture content or humidity of the natural gas. The primary chart is designed for a typical natural gas (mostly methane and thus a gas gravity of ~0.55) and pure water. Since the water content is lowered with the molecular weight of gas (heavier gas gravity) and water salinity, a set of correction factors to the value read from the chart is applied. These two corrections are presented in the form of two insert plots in the water content chart. Sloan[34] expressed the water vapor content chart of McKetta and Wehe[33] in terms of the following correlations:

$$WVC = 10^{C_1 + 0.018T - 0.0000227T^2} \tag{3.23}$$

$$C_1 = 3.0288 - 1.11733 \log_{10}(P) + 0.0515 \left[\log_{10}(P) \right]^2 \tag{3.24}$$

where WVC is water vapor content in lb/mmscf, and T and P are temperature and pressure, respectively, in °F and psia. Equations 3.23 and 3.24 (or the chart), as obviously expected, would result in higher WVC at higher temperatures and vice versa for a given pressure. However, at a given temperature the water vapor content is inversely proportional to pressure, which again is expected.

Whitson and Brulé[30] presented correlations for the insert charts for correcting the WVC for high molecular weight gas and salinity, respectively:

$$C_g = 1 + \frac{\gamma_g - 0.55}{\left(1.55 \times 10^4\right)\gamma_g T^{-1.446} - \left(1.83 \times 10^4\right)T^{-1.288}} \tag{3.25}$$

$$C_s = 1 - 0.002256S^{1.44} \tag{3.26}$$

where C_g and C_s are the respective correction factors, γ_g is gas gravity ($= MW_g/MW_{air}$), MW_g is the gas molecular weight and MW_{air} is the air molecular weight (~28.97), T is the temperature in °F and S is the salinity in wt%. Note that Equation 3.26 is a modified form of the original correlation in terms of salinity in mg/liter or ppm. The consistency of the preceding two correction factor equations is also clear in that both will result in a value of $C_g = 1$ and $C_s = 1$, that is, no correction factor when the gas gravity is 0.55 ($\gamma_{CH4} = 16.04/28.97 \cong 0.55$) and water is pure ($S = 0$). From a practical standpoint, however, some correction will always be necessary given the fact that the gas (natural or otherwise) will seldom be pure methane and produced water will have a nonzero salinity. The corrected water vapor content is thus given by

$$WVC_{corrected} = WVC(\text{Equation } 3.23) \times C_g \times C_s \tag{3.27}$$

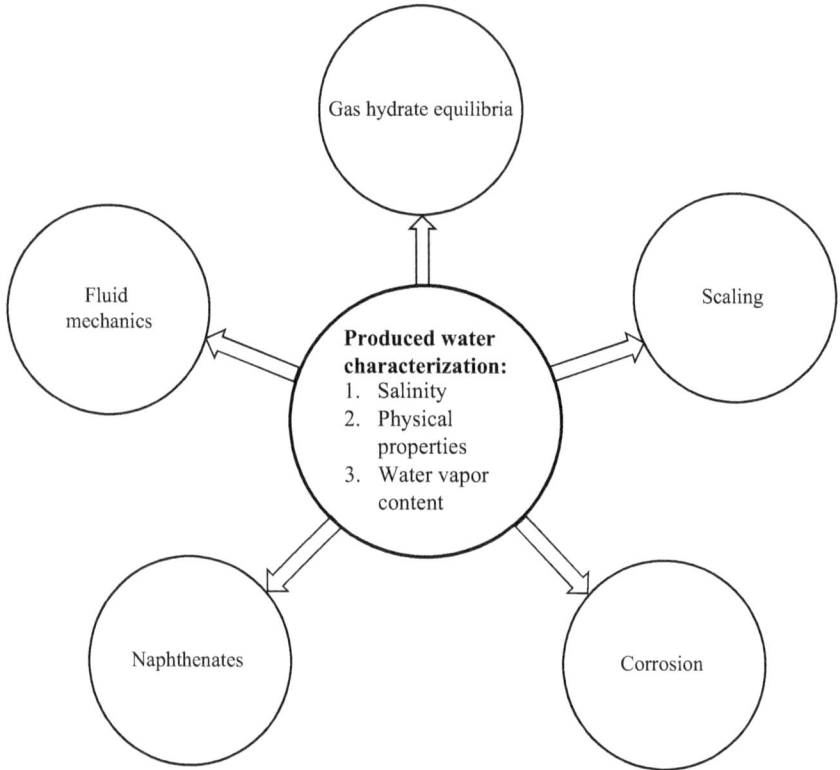

FIGURE 3.16 Significance of produced water characterization in various aspects of flow assurance.

All the previously described WVC equations are strictly valid for gas mixtures that do not contain H_2S and/or CO_2. Therefore, additional correction to Equation 3.27 is necessary for the presence of these two acid gases that classify the gas mixture as sour. Wichert and Wichert[35] presented a nomogram (not shown here) for estimating the sour gas correction factor from a "mole % of H_2S equivalent" concept. For example, even if the gas contains no H_2S but only CO_2, then 70% of whatever CO_2 mole% is present in the gas is considered as H_2S equivalent. This nomogram covers a temperature range of 50 to 350°F and pressure range of 100 to 10,000 psia and H_2S equivalent up to 50 mole%. Since the WVC in a non-hydrocarbon such as H_2S and CO_2 is high, the sour gas correction factor values are more than 1 for the most part and could be as high as 5 if the H_2S equivalent is 50 mole%,[35] in other words, a WVC 5 times the value from Equation 3.23. However, if all three conditions exist for a certain gas, which is high gas gravity, high water salinity and the gas is sour, then the overall correction may be offset because both C_g and C_s are always less than 1, whereas the sourness correction is greater than 1.

The determination of WVC is of practical significance in hydrate calculations carried out to estimate the dosage rate of thermodynamic hydrate inhibitors, such as methanol and glycol. The dosage rate of inhibitors is based on the free water in a flow line, which is a sum of produced water and condensed water. The latter is obtained from the difference between the WVC at higher temperature (start of a flow line) and lower temperature (farther along the flow line or at the end) as the produced stream cools down and water condenses. These two conditions are also termed water in "hot gas" and "cold gas", respectively.[34] Examples of these calculations are shown in Sloan.[34]

3.6.4 SIGNIFICANCE OF PRODUCED WATER CHARACTERIZATION IN FLOW ASSURANCE

Based on the foregoing, in summary, Figure 3.16 illustrates the significance of produced water characterization in flow assurance.

REFERENCES

1. Gibbs, J. W. (2010). Elementary principles in statistical mechanics: Developed with especial reference to the rational foundation of thermodynamics. In *Elementary Principles in Statistical Mechanics: Developed with Especial Reference to the Rational Foundation of Thermodynamics*. https://doi.org/10.1017/CBO9780511686948
2. Dandekar, A. Y. (2013). *Petroleum reservoir rock and fluid properties*. CRC Press.
3. Shah, J., Bostrom, N.-W., Zhdaneev, O., Raghuraman, B., & Paso, K. G. (2009). *High pressure and high temperature chromatography* (Patent No. US2009151426A1). https://worldwide.espacenet.com/patent/search/family/040011234/publication/WO2009079059A1?q=pn%3DWO2009079059A1
4. Danesh, A., & Todd, A. C. (1990). A novel sampling method for compositional analysis of high pressure fluids. *Fluid Phase Equilibria*, 57(1), 161–171. https://doi.org/10.1016/0378-3812(90)80018-7
5. Danesh, A. (1998). *PVT and Phase Behaviour of Petroleum Reservoir Fluids* (A. Danesh (ed.)). Elsevier.
6. Zhu, T., Walker, J. A., & Liang, J. (2008). *Evaluation of Wax Deposition and Its Control During Production of Alaska North Slope Oils*. US DOE Report.
7. Sokhanvarian, K., Diarra, A., Fernandez, J., & Stanciu, C. (2019). Novel Non-Aromatic Non-Ionic Rheology Modifiers for High Paraffinic Crude Oils. In *SPE Annual Technical Conference and Exhibition* (p. 16). Society of Petroleum Engineers. https://doi.org/10.2118/195894-MS
8. Gawas, K., Krishnamurthy, P., Wei, F., Acosta, E., & Jiang, Y. (2015). Study on Inhibition of High-Molecular-Weight Paraffins for South Eagle Ford Condensate. In *SPE Annual Technical Conference and Exhibition* (p. 11). Society of Petroleum Engineers. https://doi.org/10.2118/174817-MS
9. Jon Steinar Gudmundsson. (2018). *Flow Assurance Solids in Oil and Gas Production*. CRC Press/Balkema.
10. M. R. Riazi. (2005). *Characterization and Properties of Petroleum Fractions* (First). American Society for Testing and Materials.
11. Speight, J. G. (2013). *Heavy Oil Production Processes*. Gulf Professional Publishing. https://doi.org/10.1016/C2012-0-00598-4
12. Aske, N. (2002). *Characterization of Crude Oil Components, Asphaltene Aggregation and Emulsion Stability by Means of Near Infrared Spectroscopy and Multivariate Analysis*. Norwegian University of Science and Technology. PhD Thesis.

13. Jamaluddin, A. K. M., Creek, J., Kabir, C. S., McFadden, J. D., Cruz, D., Manakalathil, J., Joshi, N., & Ross, B. (2002). Laboratory Techniques to Measure Thermodynamic Asphaltene Instability. *Journal of Canadian Petroleum Technology*, 41(07), 9. https://doi.org/10.2118/02-07-04

14. Minssieux, L. (1997). Core Damage From Crude Asphaltene Deposition. In *International Symposium on Oilfield Chemistry* (p. 19). Society of Petroleum Engineers. https://doi.org/10.2118/37250-MS

15. del Carmen Garcia, M., & Chiaravallo, N. (2001). *Asphaltenes Deposition Control in Lake Maracaibo Crude Oil Production*. Society of Petroleum Engineers. https://doi.org/10.2118/65009-MS

16. Kabir, C. S., & Jamaluddin, A. K. M. (2002). Asphaltene Characterization and Mitigation in South Kuwait's Marrat Reservoir. *SPE Production & Facilities*, 17(4). https://doi.org/10.2118/80285-PA

17. Tewari, R. D., Dandekar, A. Y., & Moreno, J. (2019). *Petroleum Fluid Phase Behavior: Characterization, Processes, and Applications*. Taylor & Francis Ltd.

18. Danesh, A., Todd, A. C., Somerville, J., & Dandekar, A. (1990). Direct measurement of interfacial tension, density, volume, and compositions of gas-condensate system. *Chemical Engineering Research and Design*, 68(4), 325–330. https://www.scopus.com/inward/record.uri?eid=2-s2.0-0025451940&partnerID=40&md5=334bc1f66422baaaa3095a8604f1cfc2

19. McCain Jr., W. D. (2017). *Properties of Petroleum Fluids* (3rd Edition). PennWell.

20. Quiñones-Cisneros, S. E., Zéberg-Mikkelsen, C. K., Baylaucq, A., & Boned, C. (2004). Viscosity Modeling and Prediction of Reservoir Fluids: From Natural Gas to Heavy Oils. *International Journal of Thermophysics*, 25(5), 1353–1366. https://doi.org/10.1007/s10765-004-5743-z

21. McCain Jr., W. D. (1994). Heavy Components Control Reservoir Fluid Behavior. *Journal of Petroleum Technology*, 46(09), 746–750. https://doi.org/10.2118/28214-PA

22. Pedersen, K. S., Blilie, A. L., & Meisingset, K. K. (1992). PVT calculations on petroleum reservoir fluids using measured and estimated compositional data for the plus fraction. *Industrial & Engineering Chemistry Research*, 31(5), 1378–1384. https://doi.org/10.1021/ie00005a019

23. Dandekar, A. Y., Andersen, S. I., & Stenby, E. H. (2000). Compositional analysis of North Sea oils. *Petroleum Science and Technology*, 18(7).

24. Krejbjerg, K., & Pedersen, K. S. (2006). Controlling VLLE Equilibrium with a Cubic EoS in Heavy Oil Modeling. In *Canadian International Petroleum Conference* (p. 15).

25. Whitson, C. H. (1983). Characterizing Hydrocarbon Plus Fractions. *Society of Petroleum Engineers Journal*, 23(4), 683–694. https://doi.org/10.2118/12233-PA

26. Riazi, M. R., & Daubert, T. E. (1987). Characterization Parameters for Petroleum Fractions. *Industrial and Engineering Chemistry Research*. https://doi.org/10.1021/ie00064a023

27. Twu, C. H. (1984). An internally consistent correlation for predicting the critical properties and molecular weights of petroleum and coal-tar liquids. *Fluid Phase Equilibria*, 16(2), 137–150. https://doi.org/10.1016/0378-3812(84)85027-X

28. Lee, B. I. and Kesler, M. G. (1980). Improve Vapor Pressure Prediction. *Hydrocarbon Processing*, 163–167.

29. Fritz, J. S. and Gjerde, D. T. (2009). *Ion Chromatography, 4th, Completely Revised and Enlarged Edition*. Wiley.

30. Whitson, C. H. and Brulé, M. (2000). *Phase Behavior*. Society of Petroleum Engineers Inc.

31. Enick, R. M., & Klara, S. M. (1992). Effects of CO_2 Solubility in Brine on the Compositional Simulation of CO_2 Floods. *SPE Reservoir Engineering*, 7(02), 253–258. https://doi.org/10.2118/20278-PA

32. McCain Jr., W. D. (1991). Reservoir-Fluid Property Correlations-State of the Art (includes associated papers 23583 and 23594). *SPE Reservoir Engineering*, 6(2), 266–272. https://doi.org/10.2118/18571-PA

33. McKetta, J. J. and Wehe, A. H. (1958). Use this chart for water content of natural gases. *Petroleum Refiner*, 37, 153–154.

34. Sloan, E. D. J. (2000). *Hydrate Engineering*. Society of Petroleum Engineers.

35. Wichert, G. C. & Wichert, E. (2003). New charts provide accurate estimations for water content of sour natural gas. *Oil and Gas Journal*, 64–66.

4 Characterization of Inorganic and Organic Solid Phases

4.1 INTRODUCTION

Crude oil is a complex mixture of hydrocarbons along with various non-hydrocarbons. Hydrocarbon fluids from various reservoirs have different compositions and characteristics. Often, the hydrocarbon fluid composition is of main concern for flow assurance issues related to onshore and offshore reservoirs. As the pressure and temperature change from the reservoir to the surface during fluid flow, heavier hydrocarbon components tend to separate from the bulk hydrocarbon fluid and deposit near the wellbore, in the well tubing, flowlines, risers, pipelines, or in surface separation facilities. Organic solid phases encountered in the oil and gas industry mainly include wax, asphaltene, hydrates, and inorganic solids, such as scales. Even being a lighter hydrocarbon, natural gas can form hydrates with water present in the reservoir fluids in the subsea pipelines.

In the reservoir fluids, the existence of inorganic salts, solids, and water is undesirable, as they can result in serious issues of scale deposition, and fouling and corrosion of the surface and subsea facilities, and can even affect the catalysts in downstream operations.[1] The presence of inorganic salts, waxes and asphaltene can also result in strong and stable oil–water emulsions that become difficult to separate and often lead to flow assurance issues during transportation. Due to the potential impact of the various organic and inorganic solids on the flow assurance issues, it is thus important to understand these solids and their characterization protocols. The conditions under which they form and remain stable are different for different organic solids. The composition of crude oil and gas also plays a role in their stability. The solid organic phase envelope providing insights into their stability under given pressure and temperature conditions is a precursor for developing techniques to avoid their formation and mitigation in onshore and offshore production facilities. In this chapter, we discuss some of the details on organic and inorganic solids, their characterizations, and details on solid organic phase envelope.

4.2 PARAFFINS OR WAXES

Paraffins are saturated hydrocarbons with a general formula as C_nH_{2n+2}, where n refers to carbon number.[2] Paraffins are an essential constituent of any hydrocarbon fluid and can be present with varying molecular weight (and having higher carbon numbers). Typically, paraffins with carbon numbers above 18 or 20 are referred to as

DOI: 10.1201/9781003091301-4

waxes. Petroleum waxes may also contain branched and cyclic paraffins. Wax exhibits a plastic-like solid behavior at surface conditions and melts with an increase in temperature.[3, 4] The presence of varying molecular weight paraffins in waxes impacts various properties, such as crystalline structure, melting point, viscosity, rigidity, and brittleness.[5]

Petroleum waxes are mainly divided into three main categories, namely, paraffin (macrocrystalline), microcrystalline, and semi-microcrystalline waxes. Paraffin or macrocrystalline waxes contain a significant amount of normal (unbranched) paraffins (C_{18}–C_{36}). Normal paraffins are soft, colorless solids obtained from crude oil, coal, and typically contains normal paraffins from carbon numbers C_{20} to C_{40}. These are solids at room temperature and melt above 37°C and boil above 370°C.[6–8] Paraffin waxes have a typical density of 900 kg/m[3, 9] and they are often used to make candles, dyed crayons, electrical insulation, and lubricants.

Microcrystalline wax contains mainly a higher composition of iso-paraffins, which are branched hydrocarbons.[10] The microcrystalline waxes (naphthene and iso-paraffin) crystallize at a higher temperature than normal paraffins, thus depositing early. They have lower carbon numbers and, thus, a finer and rounder crystal structure than macrocrystalline compounds with a higher and extended one.[11] This finer crystal structure is also responsible for their elasticity.[12, 13] They have a higher molecular weight, melting point, density, and viscosity than normal paraffins. The iso-paraffin waxes are produced during the crude oil refining process. The macro-paraffin waxes have better engine combustion characteristics due to higher melting points and molecular weights. Semi-microcrystalline waxes primarily contain naphthene and aromatic waxes with varying crystalline structures based on the carbon number of the compounds. Orthorhombic and triclinic crystalline structures consist of carbon numbers C_{19} to C_{29} and C_{28} to C_{36}, respectively.[14] Normal paraffin, semicrystalline, and microcrystalline waxes are derived from distillates with increasing densities and viscosities (from low viscosity and light distillates to residual oils).[14] Figure 4.1 shows the schematic of different wax crystal structures after crystallization from the bulk fluid. Waxes are soluble in the organic solvent but insoluble in water. In crude oils, wax dissolves under high temperatures. However, they tend to form a solid phase with a decrease in temperature. The deposition of wax in well tubing and pipelines is a major challenge for the oil and gas industry.[13]

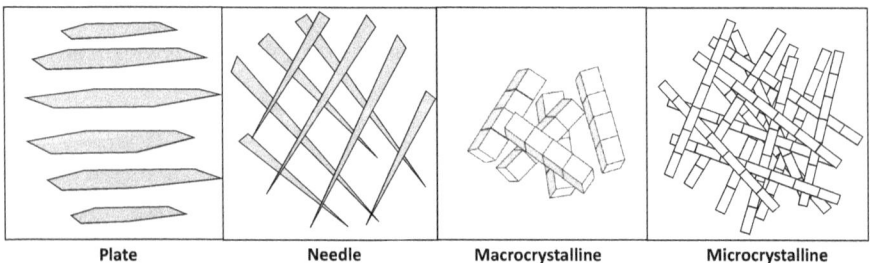

| Plate | Needle | Macrocrystalline | Microcrystalline |

FIGURE 4.1 Schematic of the different wax crystals that may be possible from the crude oil.

4.2.1 Wax Content of Crude Oil

The wax content of crude oil represents the precipitable quantity of wax and compares the waxiness of one crude oil with another. Standard acetone method (UOP method 46-64) and some of its modified versions are used to determine the wax content of the crude oil. Advanced methods such as differential scanning calorimetry (DSC), pulsed nuclear magnetic resonance (NMR), and high-temperature gas chromatography (HTGC) are used at the industrial scale to determine the wax content. Most methods provide wax content at one temperature, while some provide the wax content for differential temperatures. Following are some of the methods to characterize wax content in crude oil.

4.2.1.1 Universal Oil Products 46-64 Method

The wax content differs with crude oil source, and hence it is important to determine the wax content of the crude oil. In Universal Oil Products (UOP) 46-64 method, 2g of given stock tank oil (STO) is dissolved in petroleum naphtha, and then 15 g of fuller's earth is added to the mixture. The mixture is then filtered, which is followed by evaporation. The clarified oil is re-dissolved in the petroleum-naphtha-n-acetone mixture (200 ml with 1:3 petroleum naphtha/n-acetone mixture). The solution is then cooled to 0°F (or desired low temperature), and filtered through a funnel containing a standard filter paper. The wax collected is then cleaned from the mat in a preweighted flask using hot petroleum naphtha. Again, the petroleum naphtha is evaporated, and the wax content is measured.[15]

4.2.1.2 Pulsed NMR

The proton-pulsed NMR method can also quantify wax content (%) precipitated from the crude oil as a function of temperature from −40 to +50°C. In this method, radio frequency excitations are applied to protons that are polarized perpendicular to the magnetic field. After this process, NMR relaxation signals are detected. The signal decay is understood from the supposition of the element obtained from the nuclei from different phases. Protons present in the solid state dominate the decay during the initial stage, while the signal amplitude at the later stage is directly proportional to the protons in the liquid phase. Quantitative information on the percentage of wax present in the crude oil can be determined from the signal information thus obtained.[16]

4.2.1.3 Differential Scanning Calorimetry

Differential scanning calorimetry (DSC) is one of the advanced methods used in flow assurance to investigate the wax content and crystallization potential of crude oil. The procedure involves heating the crude oil sample from low temperature (150°C) to high temperature (80°C) at a rate of 5°C/min. The resulting thermogram exhibits melting endotherms. The heat of dissolution of the wax content is directly proportional to the area of the endotherm. The onset of the exothermic crystallization peak relates to the wax precipitation temperature (WPT). A broad range of waxy crude oils is characterized using DSC. Thermograms of a variety of crude oil may show a broad/modest crystallization peak due to the complexity of the crude oil.[17] For some

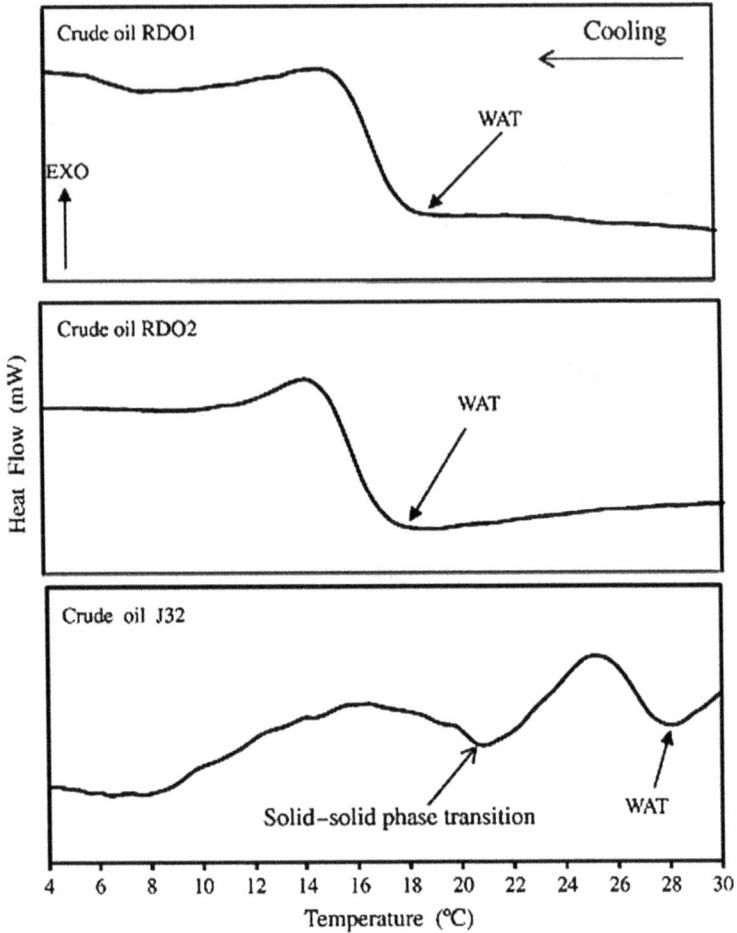

FIGURE 4.2 Exothermic peaks from DSC thermograms of crude oils during cooling of Mexican crude oils (RDO1, RDO2, and J32).[18] Reproduced from Alcazar-Vara and Buenrostro-Gonzalez[18] with permission from Elsevier (Copyright Elsevier, 2011).

cases, the peaks are better defined for model waxy crude oil due to the narrower range of waxy components. Cooling or heating rate may affect the crystallization peak. For a higher cooling rate, say, 10°C/min, supercooling may result in a low value for crystallization peak. A low cooling rate, such as 0.1°C/min, may provide better WPT information.[2] Figure 4.2 shows a sample DSC thermogram for a waxy crude oil.

4.2.1.4 HTGC

HTGC is a method in which a high temperature of up to 430°C can be used to ensure the elution of higher n-alkane with carbon number around C_{100}. HTGC is generally

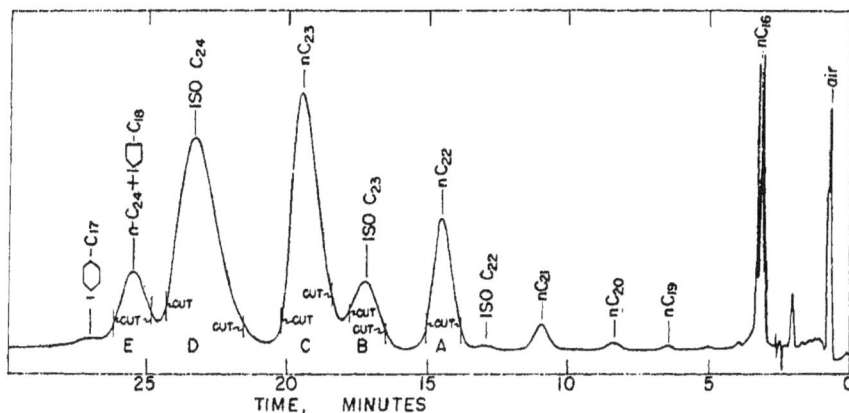

FIGURE 4.3 Chromatogram of a wax concentrate on a microwax column.[19] Reproduced from Czech et al.[19] with permission from American Chemical Society (Copyright American Chemical Society, 1961).

used to determine the nature of alkanes present in crude oil. The ASTM D5442 provides analyses up to carbon number C_{44}, which does not need much higher temperature conditions. The principle involves the separation of alkane based on their boing points, which becomes difficult as the carbon number increases. For acceptable chromatography, nonpolar columns are preferred. Figure 4.3 shows a typical HTGC of a crude oil showing waxy components. Based on the HTGC calibration, a relative area under the curve for each component is converted to its relative abundance in the crude oil sample.

4.3 ASPHALTENE

Asphaltenes are the heavier and most polar compounds present in crude oil with a low economic value. They are dark-colored flaky solids and create carbonaceous products after heating. Asphaltenes are typically insoluble in lighter alkanes and soluble in hydrocarbon solvents, for example, aromatics.[20–22] Asphaltenes precipitation occurs due to solubility change in crude oil resulting from variation in pressure and temperature conditions during transportation. Figure 4.4 shows a schematic of the molecular structure of asphaltenes as reported by Mullins.[23] The asphaltene macromolecule consists of a polynuclear aromatic (PNA) core, along with cycloalkane, branched, and aliphatic side chains. In addition to carbon and hydrogen, the macromolecule also contains heteroatoms such as nitrogen, oxygen, and sulfur, along with trace amounts of heavy metals such as nickel, vanadium, and zinc.[24] The composition and structure of the asphaltene molecules (arrangement of paraffin, olefin, and aromatics) change with the source, resulting in inconsistent properties that can be attributed to the source of the organic matter, pressure, and temperature conditions under which the degradation and transformation of the organic matter occur in the oilfields worldwide.[25] Due to the complex nature of asphaltene, they tend to create

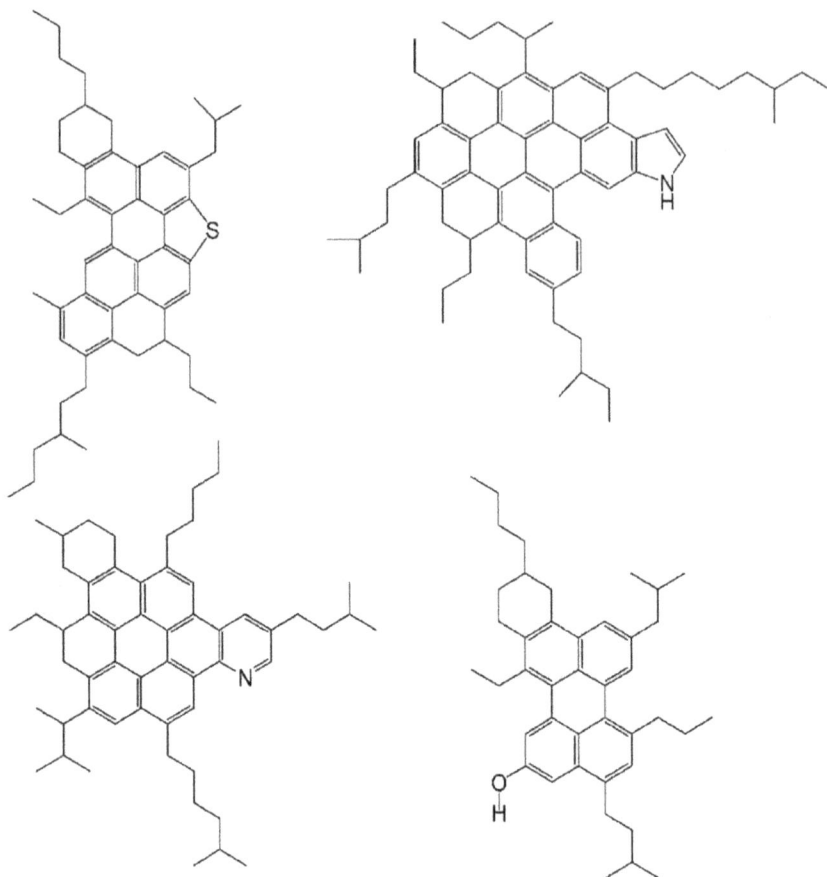

FIGURE 4.4 Proposed structure of the asphaltene molecule. Reproduced with permission from Mullins[23] with permission from American Chemical Society (Copyright American Chemical Society, 2010).

many issues during flow assurance. Thus, asphaltene removal from crude oil will help to reduce the heteroatoms and metal content in the petroleum fraction, thus reducing the complexities in downstream operation.[25]

4.3.1 MODELS FOR ASPHALTENE STRUCTURE

Several researchers have proposed different models to define the structure of asphaltene as they are very complex molecules and their behavior differ significantly with crude oil source. Hence, asphaltenes are generally defined based on their solubility in liquid hydrocarbons (e.g., nC_5 and nC_7 insoluble asphaltene) rather than their specific structure. The archipelago model describes an asphaltene structure containing several aromatic rings connected by aliphatic chains. There is uncertainty in the model due to the unknown number of rings present in the molecule. Another continental

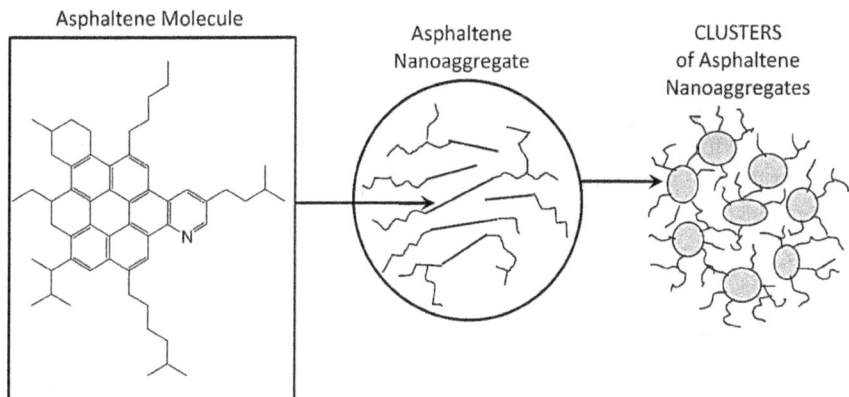

FIGURE 4.5 Schematic of the Yen–Mullins model showing the behavior of asphaltene molecules. Reproduced from Mullins[23] with permission from American Chemical Society (Copyright American Chemical Society, 2010).

model proposes an asphaltene molecule containing several aromatic rings in the middle, with several aliphatic branches. This model is primarily associated with lower molecular weight asphaltene molecules present in light oil.[26]

The Yen and Mullins model is one of the widely accepted models to define the structural behavior of asphaltene molecules. The schematic of the Yen–Mullins model is shown in Figure 4.5. The Yen–Mullins model primarily provides insights into the asphaltene structure based on the type of crude oil. For light crude oils with high American Petroleum Institute (API) gravity, asphaltene consists of a small polycyclic aromatic hydrocarbons (PAH) core with a diameter of 1.5 nm. For such oil, asphaltene concentrations are relatively lower, limiting their size.[26] In the case of black oil, with moderate API gravity, the asphaltene concentration is typically high, resulting in nanoaggregates formation with an average size of 2 nm. These aggregates are slightly bigger than the original molecules. In the case of heavy oil with low API gravity, a significant amount of asphaltene is present, resulting in cluster formation. The cluster can grow over a period with an average diameter of 5 nm due to the agglomeration of various nanoaggregates. This model indicates that the asphaltene will remain either in a single size or nanoaggregates in light oil to heavy oil. On the other hand, it will form aggregates in black to heavy crude oil, thus impacting the API gravity of the crude oil.

4.3.2 Properties of Asphaltenes

4.3.2.1 Aromaticity

Comparing the H/C ratio of an asphaltene to the source indicates the aromaticity trend.[25] For example, the H/C ratio of Alberta heavy oil is 1.5. However, the H/C ratio of an n-heptane asphaltene fraction ranges between 0.98 and 1.26. The lower H/C ratio of the asphaltene fraction indicates a higher aromaticity.[27] Asphaltene aromatic content can vary between 36 and 50% for different sources, based on the different methods such as nuclear magnetic resonance spectroscopy and X-ray photoelectron spectroscopy. The hydrogen deficiency of the asphaltene molecule is responsible for its insolubility and aggregation behavior. Asphaltene can turn into coke due to the thermal cracking process of petroleum, primarily due to the high aromatic content. Heavy crude oil can be upgraded to reduce aromaticity by increasing the H/C ratio (hydrogen addition and carbon rejection), which results in less coke formation.[25]

4.3.2.2   Density and Viscosity

Asphaltene content can affect the density and viscosity of the crude oil, which plays a role during the crude oil production and transportation. The low H/C ratio of the asphaltene fractions results in a high density and viscosity of the fraction than the source sample.[25] The viscosity of the asphaltene fraction can be of the order of 100 times more than the vacuum residue, while the density of the asphaltene fraction can be more than 1000 kg/m³.[27]

4.3.2.3   Molecular Weight

The molecular weight of the asphaltene fraction is important to understand the possibility of crude oil processing for distillation. They may generate enough residue during distillation, affecting the catalyst activity. Most of the vacuum and heavy residues are non-distillates. The boiling point increases with an increase in the molecular weight of the hydrocarbon. The boiling point of asphaltenes fractions of the crude oil can exhibit a wide range.[28] Average molecular weight of asphaltene molecules is determined using various analytical methods. The complex nature and the intermolecular characteristics of asphaltenes can result from varying molecular weights.[25]

4.3.3   CHARACTERIZATION OF ASPHALTENE

The composition and structure of asphaltenes vary with the source as discussed earlier. Thus characterization studies are essential to shed more light on their properties and the effect on flow assurance issues that fall into analytical and spectroscopic methods. The properties commonly studied include the following:

- Molecular weight and structure.
- Functional groups and elemental composition.
- Crystalline parameter of the asphaltene aggregates and their surface morphologies.

Following are some of the methods used to characterize asphaltenes from crude oil.

4.3.3.1   Molecular Weight Distribution

4.3.3.1.1   Vapor Pressure Osmometry

Vapor pressure osmometry (VPO) determines the absolute value for the number-average molecular weight (M_n). It works by taking advantage of the change in the vapor pressure that occurs when solutes are added to the pure solvent. This method is best suitable for lower molecular weight compounds, where aggregation tendency is not present, and results are similar to the mass spectroscopic methods.[29, 30] The M_n for a polymer sample is given by[31]

$$M_n = \frac{K}{\lim\limits_{c \to 0} \dfrac{\Delta V}{c}} \qquad (4.1)$$

K is the calibration constant, ΔV is the bridge imbalance output voltage, and c is the concentration of the polymer–solvent solution.

4.3.3.1.2 Size Exclusion Chromatography

Size exclusion chromatography (SEC) is also called as gel filtration, gel permeation, molecular sieve, and gel exclusion chromatography. SEC is a separation technique used to separate molecules based on size and shape (hydrodynamic radius). The gel allows for the filtering of molecules from a sample based on the molecular size. Thus, larger molecules elute first. It is not a high-resolution technique, and the molecules to be separated must differ by at least twofold in molecular weight.

4.3.3.1.3 Mass Spectrometry Methods

Mass spectrometry (MS) is an analytic method, employing the ionization and mass analysis of organic compounds for determining mass, formula, and structure. A mass analyzer separates ionized masses on the basis of mass to charge ratios (m/z), determined by molecular weight distribution. Various MS techniques are generally used to measure molecular weight distribution of asphaltene, and subfractions as described in the following sections.

4.3.3.1.3.1 Pyrolysis Gas Chromatography–Mass Spectrometry In pyrolysis gas chromatography–mss spectrometry (Py-GC-MS), small fragments are obtained from a sample after heating it and then used for analysis. Chromatography is used to separate the fragments in the same way as that of GC-MS (by flame ionization detection in which organic compound is ionized, burning the compounds in the hydrogen flame).

4.3.3.1.3.2 Matrix-Assisted Laser Desorption Ionization Time-of-Flight Mass Spectroscopy Matrix-assisted laser desorption ionization time-of-flight mass spectroscopy (MALDI-TOF MS) is a soft ionization method in which laser energy is used in a controlled manner to reduce the sample fragmentation associated with traditional MS techniques. The basic principle involves mixing a small amount of the sample with the proton-rich low-molecular-weight matrix allowing easy vaporization. A controlled laser beam is then applied to the analyte, ionizing the sample. Lighter (small m/z) charged ions move quickly to the detector. Thus, the m/z value decides the time of flight. Table 4.1 summarizes the selected studies from the literature regarding molecular weight analyses.

TABLE 4.1

Selected Literature Survey Showing Asphaltene Molecular Weight Ranges for Various Methods

#	Source	Method	Molecular Weight (Da or g/mol)	Literature
1.	Light and heavy crude oils	MALDI-TOF MS, VPO	~1700–4000 (MALDI) ~1400–2500 (VPO)	Acevedo et al.[29]
2.	Middle east, Canadian and Mexican crude oils	SEC, VPO	~5,800–28,000	Durand et al.[32]
3.	Oil sands and vacuum residues	MALDI-TOF	~560–880	Zheng et al.[33]
4.	Vacuum residues	GPC	~1450–1580	AlHumaidan et al.[34]
5.	Atmospheric tower bottom	MALDI-TOF	~886.75	Prasad and Sangwai.[21]

4.3.3.2 Molecular Structure

4.3.3.2.1 NMR Spectroscopy

NMR spectroscopy is used for the determination of structural parameters of the molecule of an unknown organic compound. It can also be used to determine the sample impurity. An external magnetic field is applied, which causes the nuclei of theatoms, in the sample, to spin in a way that is characteristic of the environment of the atom, for example, carbon or hydrogen. This information is then used to delineate the molecular structure. Clutter et al.[35] provided an analytical model to get the structural parameters of crude oil containing a mixture of hydrocarbons based on the ^{1}H and ^{13}C NMR, based on molecular weight values. ^{1}H and ^{13}C NMR spectra for nC_7 asphaltene extracted from Ratawi–Burgan vacuum residue of a Kuwaiti crude oil are given in Figure 4.6a and b. In the generated NMR spectra, the peaks corresponding to various chemical shifts[33, 34] correspond to various structure parameters, which can be found by integrating those peaks.[36] In addition, the distortionless enhancement by polarization transfer (DEPT) spectroscopy is used to distinguish between sp, sp^2, and sp^3 C–H bonds, further refining the molecular structure.[33]

4.3.3.3 Functional Group and Elemental Analysis

4.3.3.3.1 Fourier-Transform Infrared Spectroscopy

In Fourier-transform infrared spectroscopy (FTIR), infrared radiation is absorbed by organic molecules that results in the vibration motions with a frequency specific to a functional group (bond strength and symmetry). The stretching vibrations (resulting from the bond stretching) are greater than the bending vibrations. Stretching frequency follows the order from highest to lowest for triple, double, and single bonds, respectively. The obtained transmittance or absorbance spectrum can be divided into two separate regions. The functional group region lies to the left of about 1500 cm^{-1}. This region represents the vibrations due to bond stretching. The fingerprint region lies to the right of 1500 cm^{-1}, representing bond bending vibrations. The complex nature of the spectra in the fingerprint region makes it difficult to designate the absorption bands. The peaks appearing in the functional group region are then compared with the

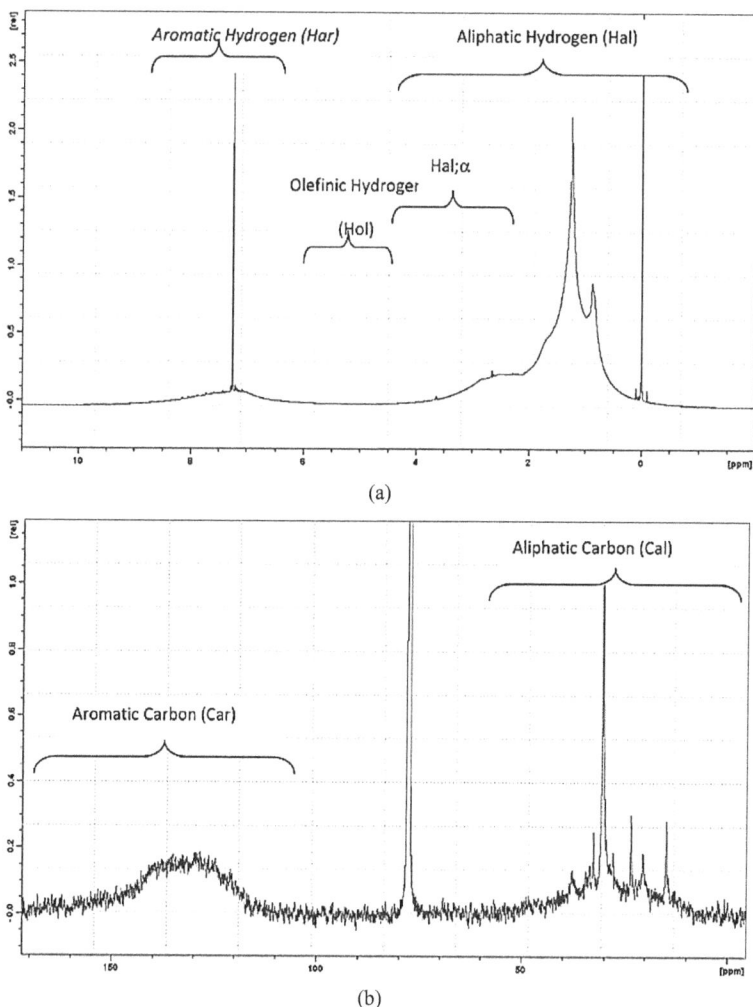

(a)

(b)

FIGURE 4.6 (a) ¹H, and (b) ¹³C NMR spectra for nC₇ asphaltene extracted from Ratawi-Burgan vacuum residue of Kuwaiti crude oil. **Reproduced from AlHumaidan et al.[34] with permission from the American Chemical Society (Copyright American Chemical Society, 2017).**

standard absorption table to determine the presence of the expected functional group in the compound. Typical FTIR absorbance spectra of asphaltene obtained from vacuum distillation feed bitumen and its subfractions[37] are given in Figure 4.7. Various FTIR sample techniques include attenuated total reflection (ATR) and diffuse reflectance (DRIFTS) that can be used depending on the preference. In addition, quantitative analysis can be done by fitting the peaks[38] to gain insights into functional group linkages in the heavy fraction of crude oil (see Figure 4.8). Figure 4.8 shows a FTIR absorbance spectra in a narrow wavenumber range (2750 cm⁻¹ to 3000 cm⁻¹) containing 3 peaks and 1 shoulder peak very close to each other making the spectra convoluted. Thus a deconvolution is necessary for a detailed structural insight.

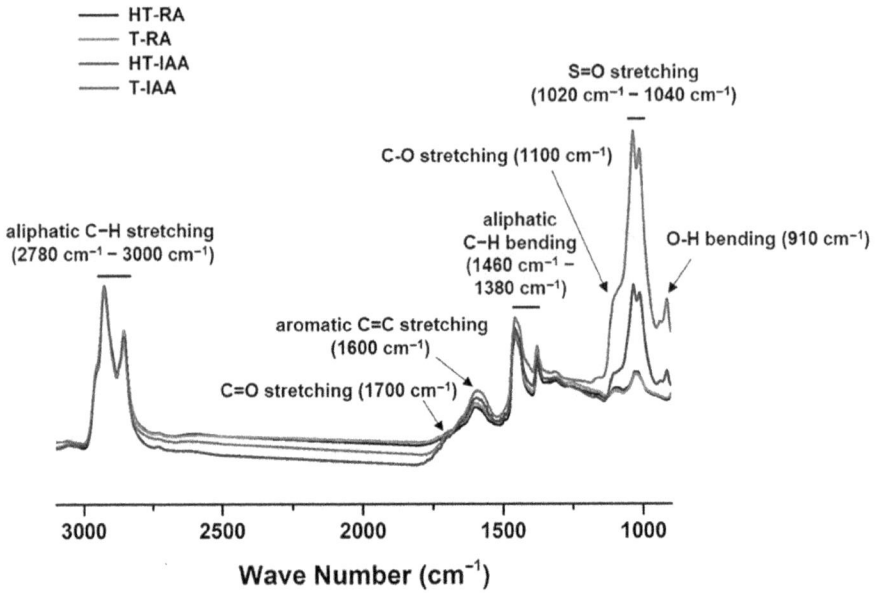

FIGURE 4.7 A typical FTIR absorbance spectra of asphaltene obtained from vacuum distillation feed bitumen and subfractions. Reproduced from Qiao et al.[37] with permission from American Chemical Society (Copyright American Chemical Society, 2017).

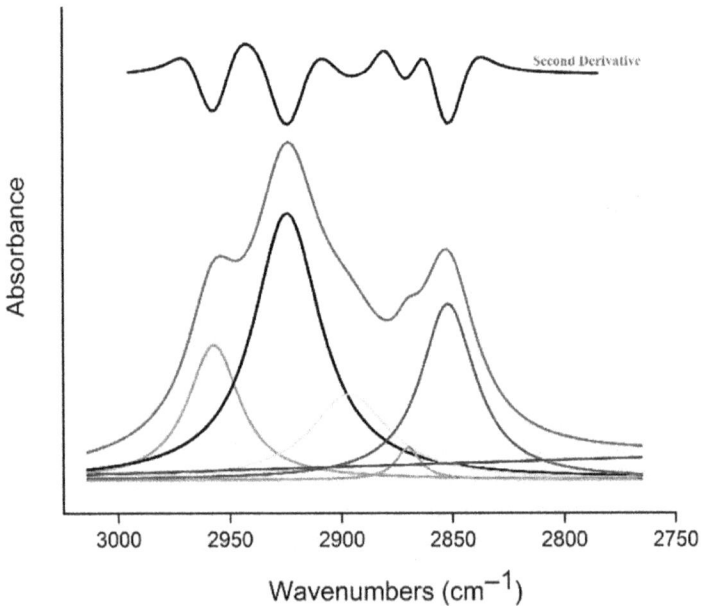

FIGURE 4.8 A typical FTIR curve fitting in terms of the Lorentzian band.[38] Reproduced from Asemani and Rabbani[38] with permission from Elsevier (Copyright Elsevier, 2020).

TABLE 4.2

Typical Elemental Composition of Asphaltenes Extracted from Various Sources[33]

Element (wt%)	Liaohe Vacuum Residue	Vene Vacuum Residue	Buton Oil Sand
C	86.32	82.85	75.06
H	8.3	7.93	7.86
N	2.19	1.82	0.83
S	0.29	5.06	8.35
O (by diff)	2.91	2.35	7.91
Metals (wppm)			
Na	13.6	11.8	8.21
Ni	6.57	4.93	1.16
Zn	0.25	26	1.94

4.3.3.3.2 Organic and Inorganic Elemental Analysis

The organic elemental analysis is concerned with determining the C, H, N, S, and O content of an organic compound based on combustion at high temperature. Upon combustion in the high-temperature furnace, the elements in the specific amount of the sample are converted to CO_2, H_2O, N_2O and SO_2, through which the elemental composition is determined using infrared detectors. Nitrogen is detected by the thermal conductivity (TC) system. For inorganic element analysis, X-ray fluorescence,[39] and atomic absorption spectroscopy[40] are generally used. Typical elemental composition of asphaltenes extracted from oil sand and vacuum residues are given in Table 4.2.

4.3.3.4 Aggregate Crystalline Parameter and Surface Morphology

4.3.3.4.1 Powdered X-ray Diffraction Studies

The structure of asphaltene and the interaction forces play an essential role in understanding their flocculation and deposition phenomena. Powdered X-ray diffraction (PXRD) quantifies the crystalline parameters of asphaltene aggregates based on the peak position and intensity using the Bragg's law.[41-44] The diffraction pattern of asphaltenes is broad, unlike the sharp peak for crystalline compounds. Diffraction peaks of interest are present in the low-angle region (below $\frac{\sin\theta}{\lambda} = 0.2$, λ is the radiation wavelength used in the X-ray diffraction analysis).[41] The γ band, appearing around $2\theta = 18°$ represent the saturate structure and (002) band at around $2\theta = 26°$, represent the aromatic structure or spacing between aromatic layers). The (10) band at $2\theta = 40°$ represents the size of the aromatic sheet. The sample aromaticity can be inferred from the areas under these bands.

4.3.3.4.2 Raman Spectrum Studies

Raman spectroscopy involves the inelastic scattering of photons, also known as Raman scattering. Monochromatic light, usually in near-infrared, near-ultraviolet, or X-rays range interacts with molecular vibrations, phonons, or other excitations,

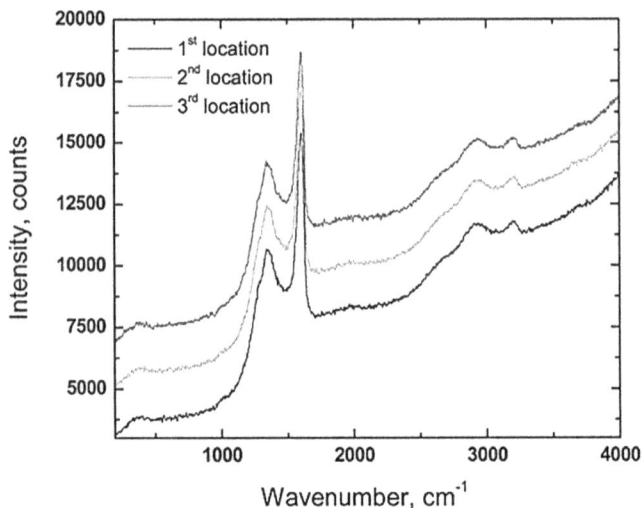

FIGURE 4.9 A Raman shift of asphaltene sample.[45] Reproduced from Abdallah and Yang[45] with permission from American Chemical Society (Copyright American Chemical Society, 2012).

which results in the energy shift of the laser photons. This energy shift represents the vibrational modes in the system. The obtained information is similar to and complemented by information from infrared. Raman shift can be used to deduce structural parameters, such as the size of the aromatic sheet. A typical Raman shift of an asphaltene sample[45] is given in Figure 4.9. Two distinct peak position are visible at wavenumber around ~1250 cm^{-1} and 1500 cm^{-1}. The sharp latter peak suggests a short range ordering of aromatic sheets. Also, it is possible to estimate the aromatic sheet size based on the intensities at these two peak positions.

4.3.3.4.3 Surface Morphology

The surface morphology studies can be done by using a scanning electron microscope (SEM) or a transmission electron microscope (TEM). Apart from qualitative studies, layer diameter can also be quantified by using TEM. An electron gun emits a high-energy electron stream toward a sample, focused using electromagnetic lenses. The interaction of electron stream with the sample leads to the creation of backscattered electrons, secondary electrons, and X-rays. These interactions are then captured to construct a magnified image. A TEM, on the other hand, uses a broad electron beam to create a higher resolution image of a sample's internal structure, providing crystallographic data allowing for the examination of more sample characteristics. SEM and TEM images of the Maya asphaltene sample is given in Figure 4.10a and b.

4.3.3.5 Asphaltene Separation using SARA Analysis

SARA stands for saturates, aromatics, resins, and asphaltenes. The analysis of these four components is called SARA analysis. The objective of the SARA analysis is to differentiate different components of crude oil. The SARA analysis separates

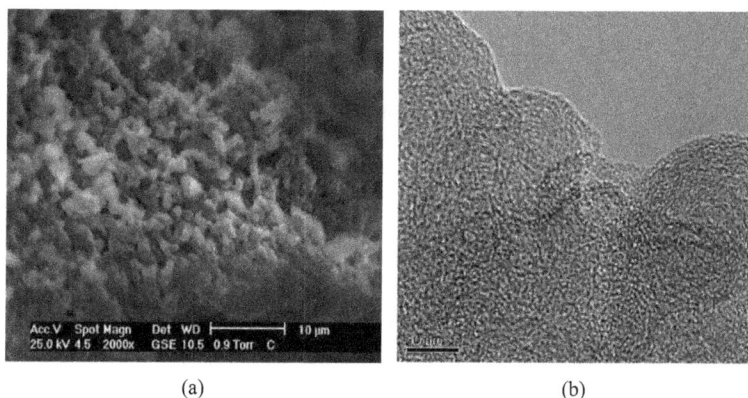

(a)   (b)

FIGURE 4.10   Electron microscopy images of Maya Asphaltene sample[46] (a) SEM; (b) TEM image. Reproduced from Trejo et al.[46] with permission from American Chemical Society (Copyright American Chemical Society, 2009).

the crude oil fractions on basis of their polarity polarizability. The saturates components include linear, branched-chain, and saturated cyclic hydrocarbons. The aromatics contain benzene ring–structured compounds and are polarizable to a certain extent. The resins contain a highly polar end group and a long-alkylated tail. The polar end group of the resin compound contains highly polar compounds (viz., aromatics, naphthenes). The pure resin may be a liquid-like compound or a sticky solid kind of structure. The highly polar end group of the resins also contains heteroatoms such as oxygen, nitrogen, and sulfur. The SARA analysis is performed using three methods, for example, thin-layer chromatography (TLC), high-performance liquid chromatography (HPLC), and gravity-driven chromatographic separation. TLC is the fastest one among them. TLC uses quartz rods coated using silica. In HPLC, the crude oil containing asphaltenes is subjected to deasphalting before using this method. However, HPLC is a time-taking process. The deasphalting is done by adding suitable solvents to the crude oil sample, and the asphaltene is precipitated, and the deasphalted oil is separated.[47]

The procedure for the SARA analysis is outlined in Figure 4.11. In this method, a small amount (~3 g) of crude oil is added to the liquid n-heptane (120 mL) in a round-bottom flask. The mixture is stirred for 2 h at room temperature. The residue deposited at the bottom of the flask is separated after 2–14 h and washed with nC_7. Further purification can be done by Soxhlet extraction as required. The residue is called the asphaltenes, and the deasphalted solution is called the maltenes. The maltenes contain saturates, aromatics, and resins components. The maltenes are then passed through an activated silica gel column. Elution with n-heptane, toluene, and methanol (or dichloromethane) will release saturates, aromatics, and resin (respectively) in the exit solution, separated using a solvent extraction process. Weighing the fraction of the components will provide the mass percentage of each component in the crude oil.[48]

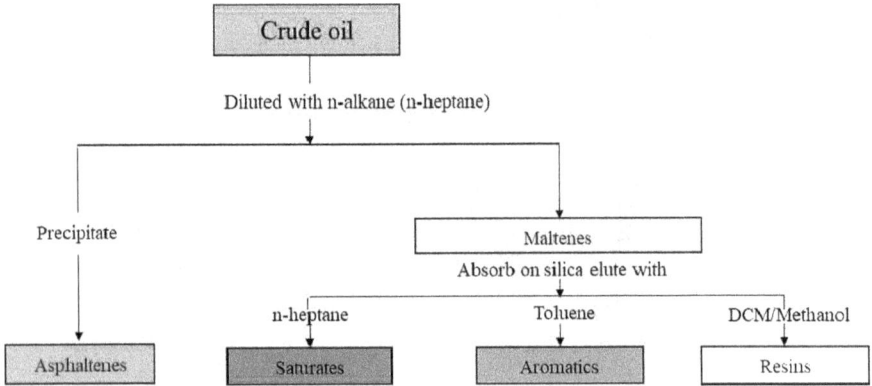

FIGURE 4.11 Schematic of crude oil separation process into asphaltene, saturate, aromatic, resin (SARA) components.

4.4 NAPHTHENATE

Most of the world's crude oil reservoirs are acidic. The acidic nature of crude oil is due to the presence of naphthenic acid. Naphthenic acid reacts with metal ions in the saline water to form metal salts of naphthenic acids known as naphthenates. Naphthenate tends to deposit, thus creating flow assurance issues for the upstream oil and gas industry. Due to the increased production of acidic crude oil, the formation of naphthenate with formation water is very common. Based on the type and quantity of naphthenic acid present in the crude oil and the properties of the formation/reservoir water, the problems with naphthenates can be varied. In recent times, there have been increased cases of issues with metal naphthenate deposition across various locations in West Africa, South America, Asia, and the North Sea.[49]

Naphthenates have the general formula $M(naphthenate)_2$. Here, M stands for metal ions. Metal naphthenates are a complex mixture rather than a specific single component. The metal naphthenates are highly soluble in organic media such as petroleum solvent. Examples of metals naphthenates are copper, sodium, magnesium, calcium, barium, lead, zinc, nickel, and others. Figure 4.12 shows the molecular structure of

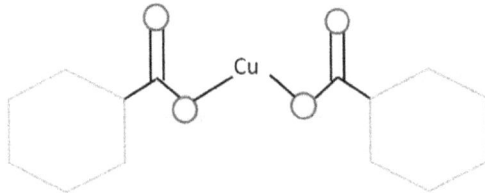

FIGURE 4.12 Molecular structure of copper naphthenate.

TABLE 4.3
Physical and Chemical Properties of Some Metal Naphthenates

Properties	Density (g/cm³)	Flash Point (°C)	Melting Point (°C)	Solubility
Sodium naphthenate	1.059	93.3	125–146	Soluble in water, ethanol, acetone
Calcium naphthenates	0.91–0.97	combustible	>100	Insoluble in water
Magnesium naphthenates	–	combustible	>100	Insoluble in water
Copper naphthenate	1.055	37.8	>100	Insoluble in water

the copper naphthenate. The general formula for copper naphthenate is $Cu(RCOO)_n$ where n is 2.

4.4.1 PROPERTIES OF NAPTHANATES

The physical and chemical properties of some metal naphthenates are given in Table 4.3.

4.4.2 CHARACTERIZATION OF NAPHTHENATES

Naphthenates are characterized by Fourier-transform ion cyclotron resonance mass spectroscopy (FT-ICR-MS). Sodium naphthenate emulsions mainly contain monoprotic saturated carboxylic acids. Similarly, calcium naphthenate consists of teraprotic acid with C_{60}–C_{70} hydrocarbon. Here, the characterization of sodium and calcium is discussed briefly to understand the characterization methods.

4.4.2.1 Sodium Naphthenate

To prepare sodium naphthenate, the oil sample is mixed with brine of 2000 ppm and 0.1 g of $NaHCO_3$. After that, the mixture is agitated for 5 min and allow to settle the solution for 2 days. A dual-phasic mixture of 50% emulsion and 50% oil is obtained, and the sample is collected for analysis. Sodium naphthenate mainly contains low-saturated naphthenic acids. Figure 4.13 shows the mass spectrum of naphthenate extracted by acidified (HCl) toluene. Naphthenates deposits are extracted from two extraction procedure (i.e., top and bottom right of Figure 4.13). However, both the naphthenates shows the presence of the saturated fatty acid suggesting surfactant-like structures [carboxylic head (polar) and aliphatic tail (nonpolar)] enhance the ability to bind monovalent cations in the oil/aqueous brine system. On the contrary, a higher double-bond equivalent (DBE; ~5–7) in naphthenate suggests cycloalkane presence and reduced ability to hold monovalent cation.

4.4.2.2 Calcium Naphthenate

Calcium naphthenates samples are generally collected from the separator due to the deposition of naphthenates on the surface of the separator. The samples are initially

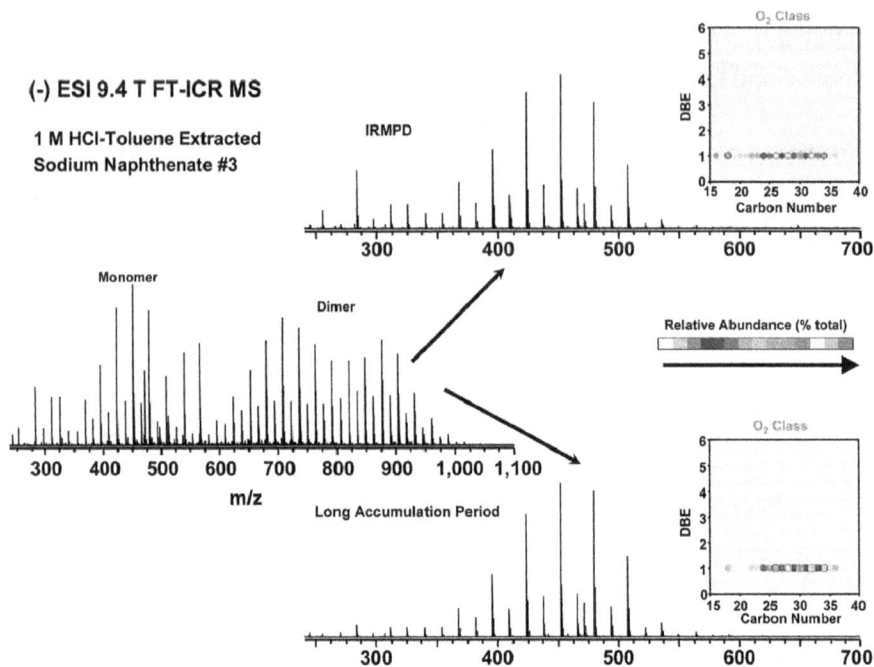

FIGURE 4.13 Mass spectrum of sodium naphthenate (NaNaph) extracted from the acidified toluene (left of the diagram); mass spectra generated (top) by infrared multiple photon dissociation (IRMPD) and during accumulation period (bottom). Insets show isoabundance–color contoured plot of DBE versus carbon number for O_2 class from monomers from NaNaph derived from negative-ion ESI FT-ICR mass spectra (#3 specify sample 3 of sodium naphthenate). Reproduced from Mapolelo et al.[50] with permission from American Chemical Society (Copyright American Chemical Society, 2009).

weighed (say, 100 mg) and washed with toluene to eliminate entrapped crude oil. Then the sample is washed with methylene chloride. Subsequently, the sample is digested with HCl acid followed by toluene extraction and air-drying. The dried samples are used for mass spectrometric analysis. Figure 4.14 shows the isoabundance–color contoured plot of the DBE versus the carbon number for calcium naphthenate deposits. O_8 peaks are shown in abundance for a DBE value of about 10. Tetraprotic acid (ARN acids) presence is confirmed from the DBE value of 10. ARN acids are present in calcium naphthenates but are not present in sodium naphthenates. Hence, calcium naphthenates can be easily detected and differentiated from sodium naphthenates.

4.5   GAS HYDRATES

Natural gas hydrates are crystalline icelike solids and form under high-pressure and low-temperature conditions. Under these conditions, a physicochemical process happens in which guest gas molecules are trapped inside hydrogen-bonded cages formed

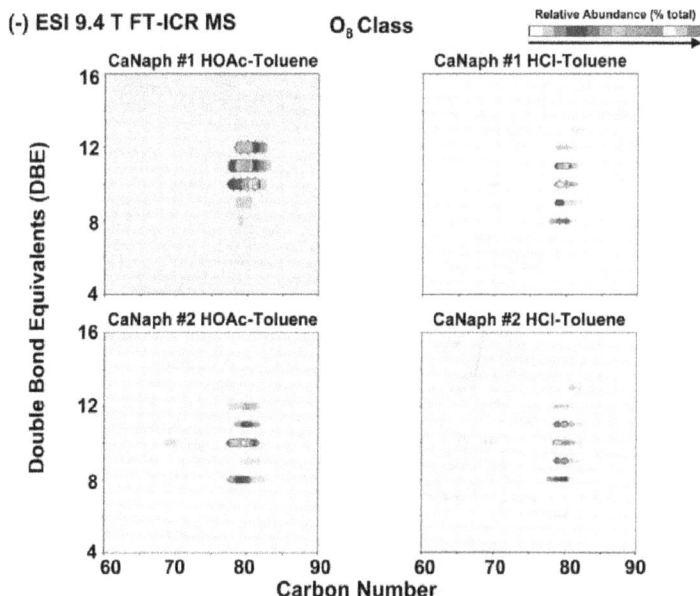

FIGURE 4.14 Isoabundance contoured plots of DBE versus carbon number (O8 class) of calcium naphthenate removed by acidified toluene. Data obtained from ESI FT-ICR mass spectra (negative ion; #1 and #2 indicate the two samples of calcium naphthenate). Reproduced from Mapolelo et al.[50] with permission from American Chemical Society (Copyright American Chemical Society, 2009).

by host water molecules. Once formed, the hydrate structure is stabilized by the van der Waal's interaction between the guest and the cages.[51–53] Gas hydrates are nonstoichiometric compounds, which means there is not necessarily a complete occupancy of gas molecules in the available sites.[54]

4.5.1 Hydrate Structures

Natural gas hydrates have three different crystal structures, cubic structures sI and sII, and hexagonal structure sH as shown in Figure 4.15. The major difference between the hydrate and ice structure is that while ice forms as a pure structure, hydrates will not form without a guest molecule of appropriate size. Structure I (sI) hydrate formation occurs when guest molecules have diameters in the range of 4.2–6 Å, such as CH_4, C_2H_6, CO_2, and H_2S. Smaller molecules, such as N_2 and H_2 ($d < 4.2$ Å), form structure II (sII) hydrate as single guests. Larger molecules, such as C_3H_6 or i-C_4H_8 (6–7 Å), form structure s(II). Molecules in the range of 7–9 Å, such as i-C_5H_{12} or neohexane (2,2-dimethyl butane), can form structure H (sH) when accompanied by smaller molecules such as CH_4, H_2S, or N_2.[54] Details regarding various hydrate structures are provided in Table 4.4.

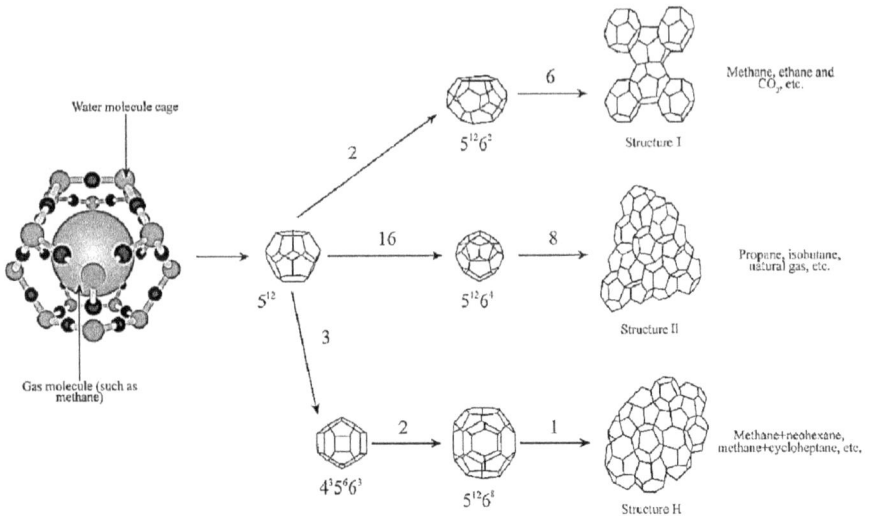

FIGURE 4.15 The five cavity types and the three common unit crystal structures of the clathrate hydrates. Reproduced from Zou[55] with permission from Elsevier (Copyright Elsevier, 2017).

TABLE 4.4

Cage Geometry for Various Hydrate Structures[54]

Hydrate Structure	I		II		H		
Cavity size	Small	Large	Small	Large	Small	Medium	Large
	5^{12}	$5^{12}6^2$	5^{12}	$5^{12}6^4$	5^{12}	$4^35^66^3$	$5^{12}6^8$
Cavities/ unit cell	2	6	16	8	3	2	1
Avg. cavity radius (Å)	3.95	4.33	3.91	4.73	3.94	4.04	5.79
Water molecules/ cavity	20	24	20	28	20	20	36

4.5.2 CHARACTERIZATION OF HYDRATE

There are various characterization techniques for hydrates, some of which are as follow:

a) XRD
b) NMR
c) Raman spectroscopy
d) Neutron diffraction

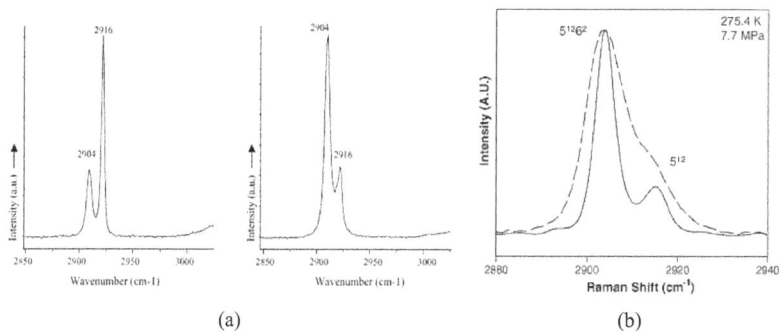

(a)   (b)

FIGURE 4.16 Raman shift of (a) sI (right) and sII (left) structures of methane hydrate; (b)methane hydrate measured near the seafloor showing peaks corresponding to the large ($5^{12}6^{2}$) and small (5^{12}) cages. (a) Reproduced from Schicks et al.[62] with permission from Elsevier (Copyright Elsevier, 2005). (b) Reproduced from Hester et al.[63] with permission from Elsevier (Copyright Elsevier 2007).

4.5.2.1 Raman spectroscopy and X-ray diffraction (XRD)

In situ techniques, namely, XRD, Raman spectroscopy, NMR, and neutron diffraction at low temperature and high-pressure help investigate the clathrate structures, cages occupancies, and their evolution over time, which is essential for studying the kinetic behavior of the hydrate crystals.[56–61] Figures 4.16a and b show the Raman shift for sI and sII structure methane hydrate and deconvoluted peaks corresponding to larger and smaller cages. In Figure 4.16a, structure I and structure II hydrates coexist at the same pressure and temperature, as observed by Schicks et al.[62] The structures sI and sII are distinguished by the relative peak areas corresponding to the C–H stretching band of methane at 2904 and 2916 cm^{-1}. Figure 4.16b shows deconvoluted peaks corresponding to larger $5^{12}6^{2}$ and smaller 5^{12} cages for methane hydrate formed at the seafloor (7.7 MPa and 275.4 K).[63] The Raman shift for hydrate structures can be studied with time in the presence of different additives, as stated earlier, to shed light on their kinetic behavior during various stages of hydrate formation and dissociation. Structural transitions in the hydrate with time can be studied with the help of time-resolved XRD. A time-resolved high-pressure XRD analysis of methane hydrate to study the transition of structure sI to sH and relative intensity of the diffraction lines for sI and sH is shown in Figure 4.17. It is observed from Figure 4.17 that diffraction lines up to about 40 s represent the sI structure. After 40 s, diffraction lines for the sH structure start to appear, and there is a coexistence of sI and sH structures for up to 130 s, after which only the diffraction lines representing sH structure are seen.[64] Apart from the preceding observations, the study also concluded that the transition between sH–FIIhS or sH to a filled-ice Ih structure takes longer than sI–sH, attributing the possible reason for the typical reconstructive transition compared to the cage-recombination mechanism for transition to sH.

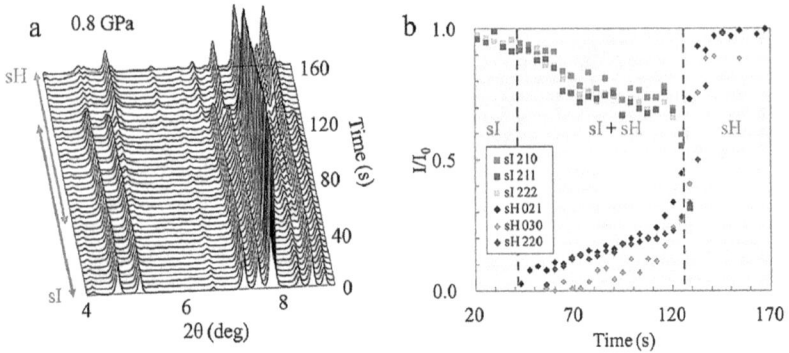

FIGURE 4.17 Time-resolved XRD patterns for sI–sH transition at 0.8 GPa (left, a) and change in the relative intensity of peaks corresponding to sI structure (miller indices 210, 211, and 222) and sH structure (miller indices 021, 030, and 220) (right, b). Reprinted from Hirai et al.[64] with permission from AIP Publishing (Copyright AIP Publishing, 2015).

4.5.2.2 NMR Spectroscopy

NMR spectroscopy can be used for microscale methane hydrate dissociation analysis.[65] In this study, the authors dissociated the methane hydrate (sI) by increasing

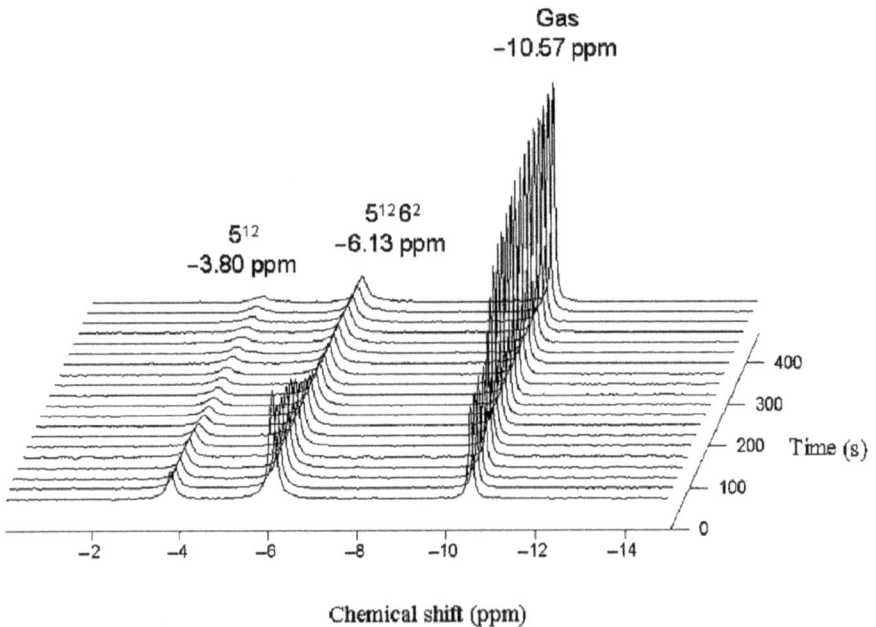

FIGURE 4.18 Time-resolved ^{13}C MAS NMR (100.5 MHz frequency) of sI methane hydrate dissociation from 269 to 271 K (only selected [every fifth] spectra plotted). Reproduced from Gupta et al.[65] with permission from American Chemical Society (Copyright American Chemical Society, 2007).

the temperature from 269 to 271 K. Time-resolved ^{13}C MAS NMR spectra has been recorded with time at a resolution and duration of 5.2048 s and 500 s, respectively (shown in Figure 4.18). It is evident from Figure 4.18 that the intensities of the three peaks at −3.80, −6.13, and −10.57 ppm change with time. The spectrum was then deconvoluted to determine the area of peaks at these shifts, and the ratio of the area of larger to smaller peaks divided by a factor (3), giving cage occupancy.[65]

4.5.2.3 Neutron Diffraction Study

The neutron diffraction study can give insights into structural arrangements of water molecules around methane molecules (hydrate shell size) during various stages of hydrate formation and dissociation.[66] The hydration structure around methane hydrate is measured at 4.5 MPa pressure and temperature range of 4–20°C using the SANDALS diffractometer.[66] The obtained neutron diffraction data is further refined using the empirical potential structure refinement (EPSR) method to shed light on the hydrate formation and decomposition at a molecular level. First-order difference neutron diffraction measurements during methane hydrate formation and dissociation are given in Figures 4.19a and b, respectively. Figure 4.19a shows the presence of two stages of hydrate formation in which the down-peaks are established first, followed by the up-peaks in the 2 Å-1 region. Figure 4.19b shows a smooth transition from 100% hydrate to gas dissolved in water. The first-order difference during decomposition and formation phase can be differentiated by the presence of up-peaks in 2 Å-1, which is unique to the dissociation. A probable cause for the two-step formation could be ordering preference, first for long-range ordering in the initial methane hydrate formation phase followed by the local ordering of the methane molecules.

4.6 CORROSION AND SCALE

Corrosion is a chemical process observed on a metal (refined) surface. The refined metal naturally converts to a more stable form of oxide, hydroxide, carbonate, and sulfide when it reacts with water, hydrogen sulfide, and carbon dioxide in the environment. Oil and natural gas industries often face challenges due to the formation of scales and corrosion. During production, crude oil and natural gas carry many corrosive substances that are responsible for corrosion. Hence, the probability of corrosion formation inside the pipes and fittings is maximal, which results in the degradation of the materials. The resulting degradation, such as reduction in thickness, leads to loss of the mechanical properties (i.e., ductile, impact strength) and subsequently failure or permanent deformation of the materials. The previously mentioned serious issues may force the industries to either replace the components immediately or stop the production for some days.

The deposition of mineral solids containing carbonate/bicarbonate of calcium and magnesium when subjected to heat is called scale. Scale can also be referred to as the deposition of solid corrosion on the surface of the metal. Most of the scaling observed in oil and gas industry in the inner wall surface of the flow pipe is due to the deposition of solid calcium carbonate ($CaCO_3$) and calcium sulfate ($CaSO_4$). Scale deposition in the production pipe tends to decrease its mechanical strength and eventually fails, resulting in loss of profit. A small quantity of scaling on the metal surface can

(a)

(b)

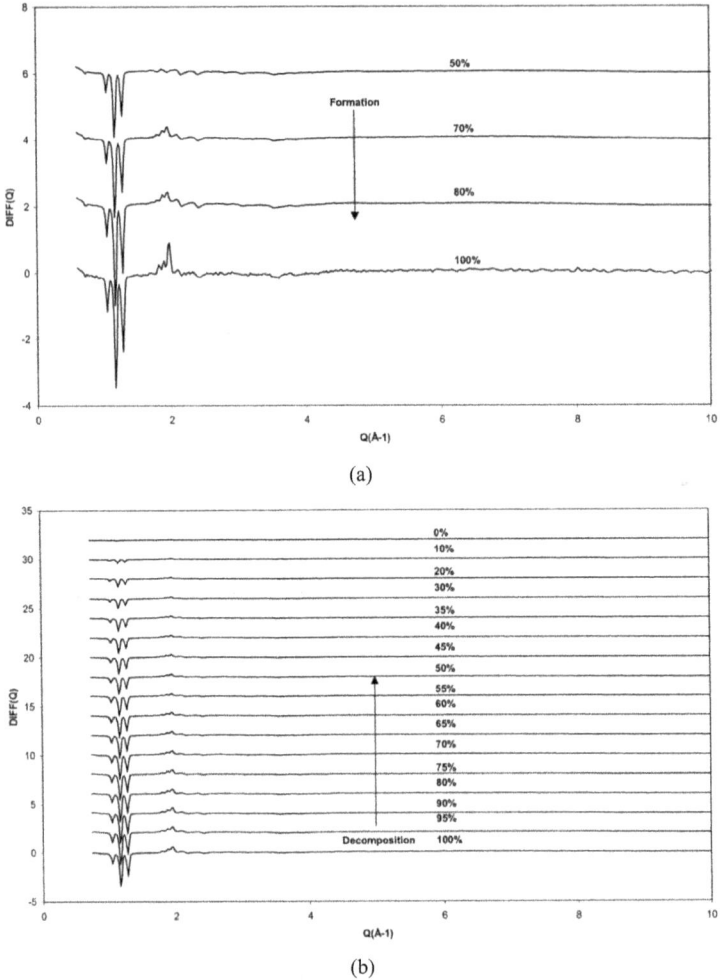

FIGURE 4.19 Neutron diffraction first-order difference measurements at various stages during methane hydrate (a) Formation and (b) Dissociation. Numbers indicate the percentage of hydrate formed. (a and b) Reproduced from Buchanan et al.[66] with permission from American Chemical Society (Copyright American Chemical Society, 2003).

be considered a protective layer if it does not contain any corrosive chemical, hindering corrosion formation. Corrosion and scale in the oil and gas industries are the most challenging problem as both production and surface facilities involve transportation of the oil and gas through pipelines. Therefore, these issues need to be addressed by controlling or preventing them through measure steps (discussed further in Chapter 7).

4.6.1　PRINCIPLES AND PRODUCTS OF CORROSION

Chemical reactions among metals and nonmetals present in the environment result in what is known as a corrosion product. Sometimes, corrosion products help protect

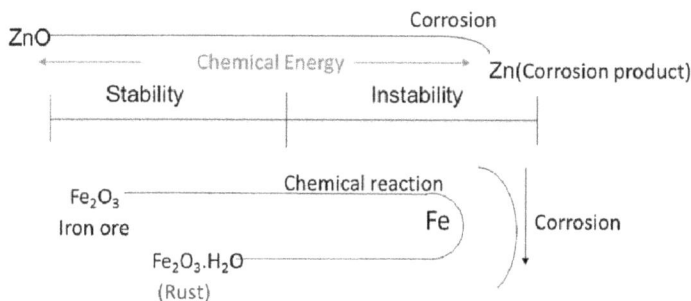

FIGURE 4.20   Variation in energy from metal oxide to metal and corrosion product.

metals from further deterioration. The corrosion product of the zinc protects the zinc metal, and thus the rate of corrosion on zinc is much slower than the iron. This could be one of the reasons why zinc coating is given on the iron metal surface. Figure 4.20 shows the change of the chemical energy during metal extraction from metal ore. The zinc metal is energetically less stable than the iron metal (see Figure 4.20). Hence, when zinc metal is exposed to the atmosphere, it forms a complex compound of zinc hydroxide or zinc carbonate, which partially protects the zinc metal. However, iron rust is nonprotective and nonadherent, allowing oxygen and water access to the metal surface. The reaction of the iron metal with oxygen and water is expressed as

$$2Fe + 3/2O_2 + H_2O \rightarrow Fe_2O_3.H_2O \left(\text{hydrated ferric oxide} \right)/\left(\text{rust} \right).$$

Therefore, the deterioration of iron metal is usually observed even after corrosion products are formed. The stability of the iron metal increases after the formation of the corrosion, but deterioration continues.

4.6.2   ELECTROCHEMICAL THEORY OF CORROSION

Figure 4.21 shows the details of the corrosion process. When metal comes in contact with the electrolyte (corrosive solution), anodic and cathodic zones are created on the metal surface. At the anodic site of the metal, the loss of an electron (oxidation) takes place, and metal converts to metal ions. Similarly, a reduction reaction occurs at the cathodic site, and the hydrogen gas releases, leaving behind hydroxyl ions. These hydroxyl ions react with metal ions to form the metal hydroxide, the first corrosion product. Simultaneously, oxygen and carbon dioxide present in the environment diffuse into the metal, forming a complex compound called rust, for example, $Fe(OH)(HCO_3)$. Finally, the anodic site encounters a loss of metal ions (see Figure 4.21). The details of the anodic and the cathodic reaction are given below with consideration of iron metal.[67]

At the anodic site:

$Fe \rightarrow Fe^{2+} + 2e^-$

At the cathodic site:

$H_2O \rightarrow H^+ + OH^-$

$2H^+ + 2e^- \rightarrow H_2$

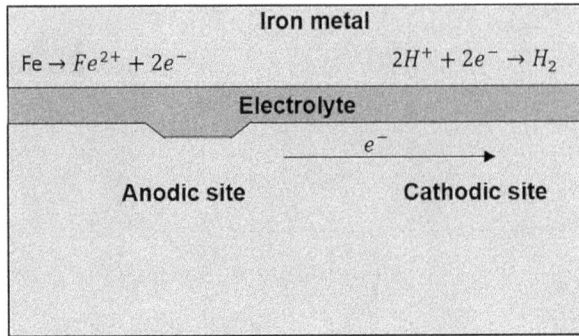

FIGURE 4.21 Schematic representation of iron metal under a corrosion process along with the reactions involved in the formation of rust.

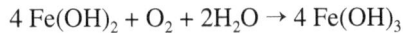

$$Fe^{2+} + 2OH^- \rightarrow Fe(OH)_2$$

$$4\,Fe(OH)_2 + O_2 + 2H_2O \rightarrow 4\,Fe(OH)_3$$

$$Fe(OH)_3 + CO_2 \rightarrow Fe(OH)_2(HCO_3)$$

$$Fe(OH)_2 + CO_2 \rightarrow Fe(OH)(HCO_3)\ (rust)$$

4.6.3 Types of Corrosion

The corrosion in the gas and oil industries leads to damage of the transport pipelines and failure of the parts which may result in the complete shutting down of the plant, as the maintenance cost for corrosion is very high. The different types of corrosion in oil and gas industry are discussed in the following sections.

4.6.3.1 Pitting

The most common type of corrosion that occurs on the surface of metal alloys is called pitting. Pitting corrosion is generally observed in the marine environment of the surface of metal alloys. This is a kind of attack with localized holes on the surface of the metals. The attack spreads rapidly to other parts, leaving some parts remained free from corrosion. Pitting becomes vicious when the metal surface comes in contact with saline water. In addition, other solutions favorable for corrosion contain fluorides and iodides, while sulfides and water enhance the pitting process. As the metal comes in contact with the corrosive solution, the available electrons are given up due to the formation of a complete circuit. The electron moves from one part of the metal surface to another, leading to a tiny hole (Figure 4.22). These tiny holes or pits begin to develop on the metal surface, causing massive damage to the metal surface on a long-term basis. Some oxidizing metal elements (iron, copper, mercury) enable pitting formation even without oxygen. Some metals elements (viz., stainless steel, chromium, passive iron, cobalt, aluminum, copper, and associated alloys) are prone to corrosion. Pitting is considered as a local type of corrosion.

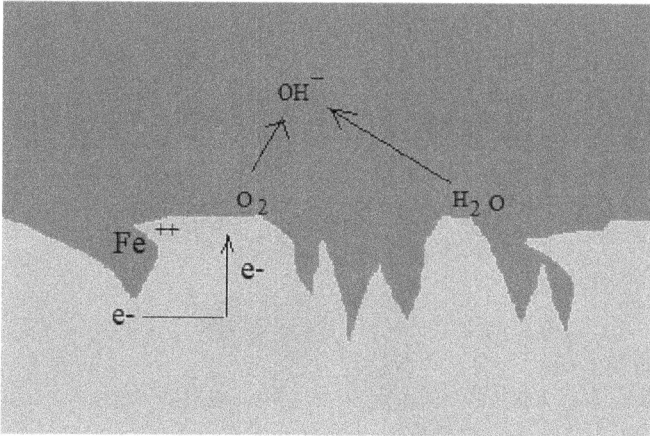

FIGURE 4.22 Pitting corrosion on iron metal surface exposed to seawater.

4.6.3.2 Uniform and Nonuniform Corrosion

In uniform corrosion, corrosion is uniformly distributed on the material. Due to this, a large amount of material can become useless. However, uniform corrosion can be easily detected and prevented quickly. Uniform corrosion is often found in chemical handling equipment. On the other hand, nonuniform corrosion occurs nonuniformly on the surface of the metal due to the inhomogeneous material or the environment. Nonuniform corrosion can be deep, narrow, and cause pinholes on the metal surface. Some of the examples of such corrosion are pitting and crevice corrosion. These types of corrosion are localized, causing limited but severe destruction of the metal. The nonuniform corrosions are classified into various types (viz., spot, point, honeycomb, intergranular, and selective corrosion).

4.6.3.3 Fretting

Fretting corrosion refers to the gradual wear of the metal surface due to contact between the two metal surfaces in motion. This damage on the metal surface occurs due to stress applied on it and the relative motion of the metal surfaces (e.g., by vibration). Severe fretting causes fracture of the metal surface. Fretting damages the condition of the metal surface layer slowly and increases surface roughness, reducing the metals' fatigue strength. For example, wear and tear of the ball bearing system are due to the frettings.

4.6.3.4 Corrosion Due to CO_2/ Sweet Corrosion

Figure 4.23 shows the impact of sweet corrosion on the metal surface. Carbon dioxide is one of the main undesired gasses for corrosion formation in pipelines. CO_2 is soluble in water, and hence it forms carbonic acid. This carbonic acid reacts with iron to form iron carbonate (Fe_2CO_3). Pitting corrosion is an example of corrosion due to CO_2.[68] However, CO_2 corrosion is mostly influenced by other conditions, such as temperature, metal properties, flow condition, and pH and composition of the

FIGURE 4.23 A picture of sweet corrosion on iron surface. Reproduced from Popoola et al.[67] permission from Springer (Copyright Springer, 2013).

aqueous phase. Many mechanisms have been proposed for CO_2 corrosion, but the formation of carbonic acid and bicarbonate ions is widely accepted as the reason behind this sweet corrosion. The chemical reaction mechanism follows:

$$CO_2 + H_2O \rightarrow H_2CO_3$$

$$H_2CO_3 + e \rightarrow H^+ + HCO_3^-$$

$$2H \rightarrow H_2$$

$$Fe \rightarrow Fe^{2+} + 2e$$

$$H_2CO_3 + 2Fe \rightarrow Fe_2CO_3 (\text{iron carbonate}) + H_2$$

4.6.3.5 Corrosion Due to H_2S/Sour Corrosion

Deterioration of the metallic drill string is a major problem in oil and gas industry, which happens due to the presence of hydrogen sulfide (H_2S) gas and water in the system. It is also known as sour corrosion due to the presence of sour gases in the production fluids. Figure 4.24 shows the pipe affected by sour corrosion. The H_2S of the production fluid reacts with water to form a weak acid. Hence, the corrosivity of the production fluid increases with the increase of hydrogen ions in the system. The corrosion product obtained due to sour corrosion is iron sulfide. At a low temperature, iron sulfide forms scale and provides resistance to the slow corrosion process. The chemical reaction can be expressed as follows:

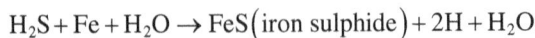

$$H_2S + Fe + H_2O \rightarrow FeS(\text{iron sulphide}) + 2H + H_2O$$

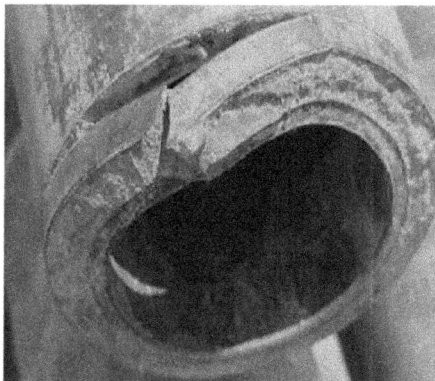

FIGURE 4.24   A picture of sour corrosion on iron surface. Reproduced from Popoola et al.[67] permission from Springer (Copyright Springer, 2013).

4.6.3.6   Oxygen Corrosion

Oxygen corrosion is the most common type of corrosion occuring due to the interaction of oxygen, metal and electrolyte of the environment, and metal oxide is the product of oxygen corrosion. Two sites are created on the metal surface during this corrosion process, known as cathodic and anodic sites. Electron flow from anodic to cathodic area occurs, causing degradation of the metal surface on anodic site. Corrosion that occurs in the underground pipeline of the offshore field region is an example of oxygen corrosion. The pitting is a typical example of oxygen corrosion.

4.6.3.7   Galvanic Corrosion

Galvanic corrosion is an electrochemical process observed on a bimetallic surface. When two metals of different electrochemical potentials are connected in an electrolyte environment, the metal with lesser electrochemical potential behaves as an anode, and the other one serves as a cathode. Galvanic corrosion is generally observed in large offshore structures. Moreover, galvanic corrosion occurs when the anodic area is less than the cathodic area. During the corrosion, the anode loses metal ions to maintain the electron flow through the electrolyte.

4.6.3.8   Crevice Corrosion

Crevice corrosion is confined to a local region. It is generally observed in a gap between two joints where water and oxygen can reach easily. The impact of the corrosion becomes more intense in the presence of chloride in the water. In crevice corrosion, metal is in contact with an aqueous medium. Crevice corrosion occurs due to the concentration difference of the corrodents on the metal surface. Electrochemical potential differences between the metals also result in a selective crevice. At the initial stage, the corrosion is uniform but, over time, converts to the crevice. Oxygen present in the drilling fluid or chloride present in seawater promotes crevice corrosion. However, over time, pitting also affects the shielded part of the metal surface.

4.6.3.9 Erosion Corrosion

Erosion corrosion can be expressed as the degradation of the metallic surface due to the turbulence of the production fluids containing abrasive suspended particles in the flow pipes. In the case of the high turbulence inside the pipelines, the erosion corrosion is directly proportional to the fluid velocity. Erosion corrosion may occur even at mild flow conditions. However, erosion corrosion removes the passive oxide layer from the metal surface. Therefore, erosion increases the rate of corrosion. Holes, groove valleys, and wavy surfaces result from erosion-corrosion in the metal inner surface of the pipeline.

4.6.3.10 Microbial-Induced Corrosion

Microbial-induced corrosion (MIC) is caused due to bacterial activities. Some microbes like *Micrococcus*, *Clostridium*, *Escherichia*, *Bacillus*, *Pseudomonas*, and *Mycobacterium* are found in production fluid. These microbes readily form colonies on the metal surface in favorable environmental conditions. The formation of those colonies are observed in regions on the metal surface that have stagnant water. These microbes generate undesired gases (CO_2, H_2S) and organic acids that corrode the metal surface by increasing toxicity. Some of the bacteria that convert sulfate to sulfide, have a significant contribution to corrosion. The microbial corrosion is identified from the deposition of the black slimy waste materials on the metal surface. MIC may completely block the pipe, causing various flow problems. SEM may help us analyze the MIC corrosion properly.

4.6.3.11 Stress Corrosion

In 1965, stress corrosion was recognized in a gas transmission pipeline in Louisiana. Stress corrosion is a type of localized corrosion that produces cracks on the surface of the metals due to corrosive fluid and stress on the pipe (i.e., bending of the pipe/tensile stress). Stress corrosion is affected by various factors such as the pH of the surrounding fluids, grain boundaries, and patches on the metal surface. It is mainly observed in the bend pipe where tensile force appears. Over time, cracks are created on the metal surface, due to the tensile force at environmental conditions, and consequently, rapture and failure of the metal are observed. It is also reported that cracks are formed due to hydrogen embrittlement. Once the crack develops, oxygen and moisture enter the metal easily and promote corrosion.

4.7 ORGANIC SOLID PHASE ENVELOPES

4.7.1 Wax Precipitation Envelope

Wax precipitation happens when the equilibrium condition of the reservoir fluid is disturbed. If the crystals of the wax form, they may get dispersed in the crude oil and not necessarily deposit. The wax crystals may agglomerate when many such crystals form and come closer in the presence of some nucleating materials. Once they agglomerate, they separate out from the fluid and deposit. Figure 4.25 shows

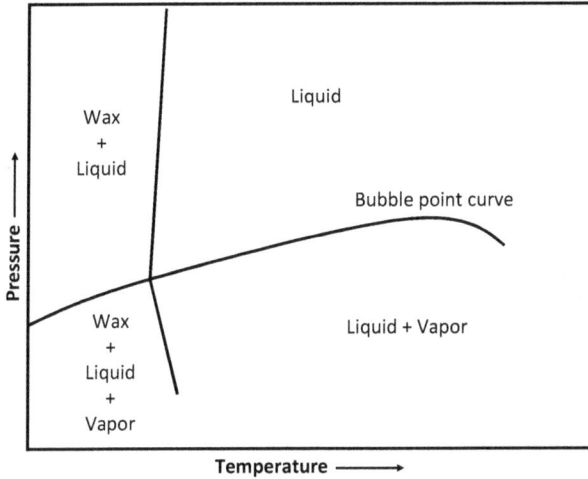

FIGURE 4.25   A schematic of the wax deposition envelope.

a schematic of the wax deposition envelope.[69] The area shows the region for wax precipitation and deposition. As seen in the figure, the line separating wax and liquid from the liquid phase is nearly vertical, thus indicating that the pressure has less impact on the wax deposition. Wax crystallization is more pronounced at lower temperatures, where the wax solubility in the crude oil decreases.[70] This results in increased viscosity of the crude oil. The temperature plays a major role in wax crystallization and thus possibly in its deposition. One of the other reasons for increased wax crystallization and precipitation is a change in the crude oil composition over time due to the production of lighter hydrocarbons from the reservoir.

4.7.2   Asphaltene Precipitation Envelope

Figure 4.26 shows a schematic of the asphaltene precipitation envelope (APE) with pressure and temperature.[71] Precipitation of asphaltene occurs for all pressure and temperature conditions inside the phase envelope. As seen in the figure, we see the upper and lower asphaltene onset pressure (AOP) curve surrounding the saturation (bubble point) curve. Asphaltene is typically present in the dissolved or partly dissolved state along with resins and aromatics in the crude oil at reservoir conditions.[72] At the bubble point, as the crude oil has the highest content of dissolve lighter hydrocarbon gas, which is bad for asphaltene, this makes asphaltene separate out from the crude oil. Asphaltene tends to deposit within the region surrounded by upper and lower asphaltene onset pressure boundary. As the pressure decreases, asphaltene tends to separate out from the bulk fluid. The pressure at which the first molecule of asphaltene separates out from the bulk fluid at a given temperature is called as upper asphaltene onset pressure (upper AOP). The deposition increases till the pressure reaches saturation pressure at a given temperature. With a further decrease in pressure below saturation pressure, some gas will separate out from the crude oil, and thus, the content of light hydrocarbon of the crude oil decreases. Hence, the

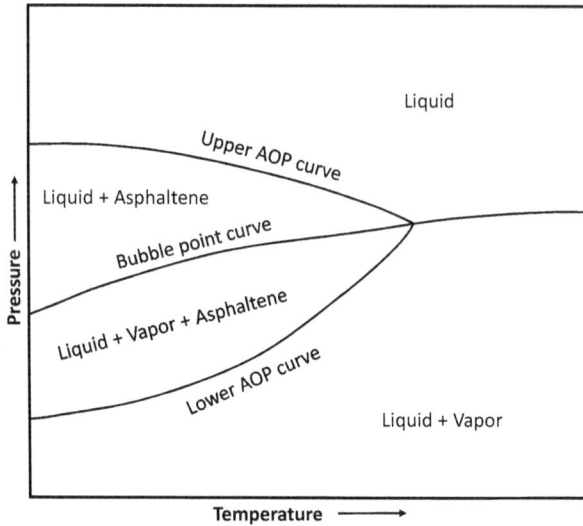

FIGURE 4.26 Schematic of the asphaltene precipitation envelope (APE). AOP: Asphaltene onset pressure.

deposited asphaltene tends to re-dissolve in the bulk fluid and dissolve completely at lower asphaltene onset pressure (lower AOP) boundary conditions. The effect of temperature cannot be clearly visible; however, low-temperature conditions show to increase the asphaltene precipitation. With an increase in temperature, the solubility of resins increases in the n-alkanes, while the asphaltene solubility may decrease, thus reducing the asphaltene in the crude oil.[73] Since a major part of the crude oil contains alkanes, the asphaltene depositions issues are quite frequent, particularly for light crude oil reservoirs. Deposition of asphaltene may also occur during crude oil production, transportation, and storage. Light hydrocarbon gas is often injected near the wellbore and within the reservoir during artificial lift and enhanced oil recovery operations. This affects the stability of asphaltene within the crude oil and thus may aggravate the asphaltene deposition issues.

4.7.3 GAS HYDRATE PHASE ENVELOPE

Gas hydrate phase envelopes represent the pressure and temperature conditions at which hydrates are stable. Hydrate-phase equilibria are well defined and provide boundary conditions within which thermodynamic and kinetic studies are carried out.[54] Hydrate phase diagrams differ from conventional hydrocarbon phase diagrams and are generated through a series of phase stability experiments for varying initial pressures. Figure 4.27a shows a schematic diagram of a typical gas hydrate equilibrium curve. It is divided into a number of regions. The gas hydrate is stable at high pressure and low temperature (top left region). Two quadruple (denoted by points 1 and 2) and one critical point (point 3) exist within the system. At point 1 (lower quadruple point), ice, hydrate, liquid water, and hydrocarbon gas are in

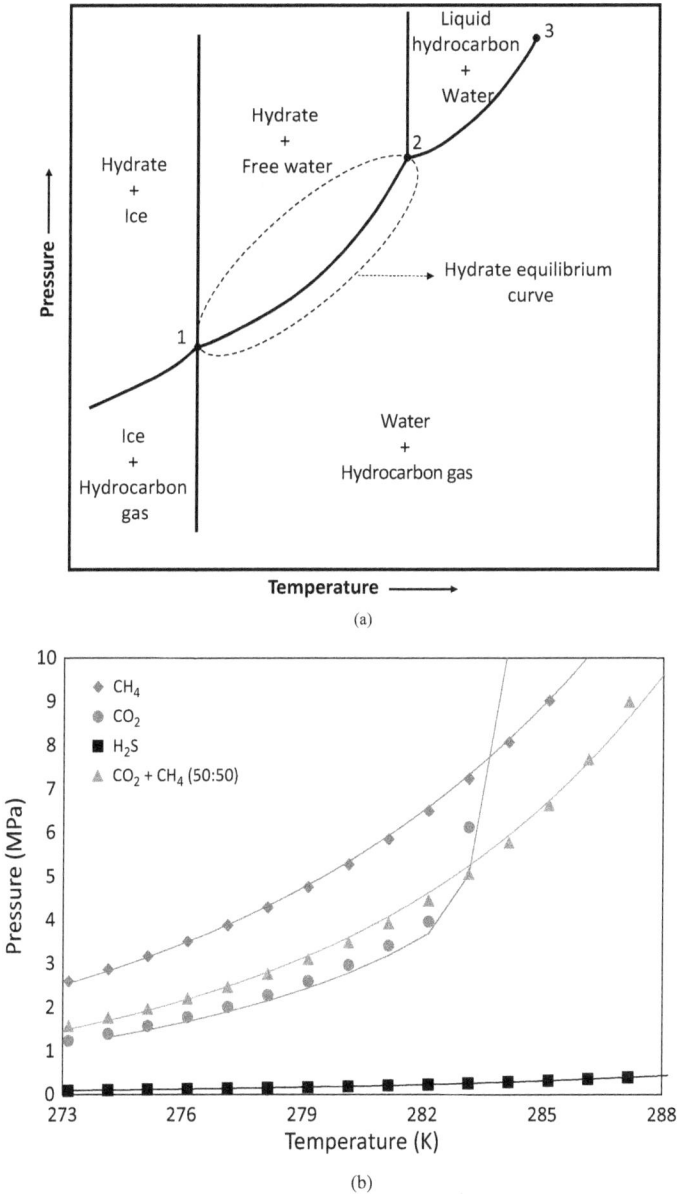

FIGURE 4.27 A schematic of phase equilibrium curve of hydrate of (a) typical hydrocarbon gas and (b) hydrates of CH_4, CO_2, H_2S, and $CH_4 + CO_2$, generated using CSM-GEM.

equilibrium, while at point 2 (upper quadruple point), liquid water, gaseous hydrocarbon, and solid hydrate are in equilibrium. At point 3, liquid hydrocarbon is in equilibrium with liquid water. For practical purposes, segment 1–2, or the loci of hydrate equilibrium points, is considered.

Figure 4.27b shows a phase equilibrium of hydrates of CH_4, CO_2, H_2S, and the mixture of 50:50, CH_4 and CO_2. The phase equilibrium curves for all the systems were calculated using CSM-GEM software. It can be seen that the CH_4 hydrate phase stability envelope is situated at higher pressure and low-temperature conditions. This means methane hydrate is stable at a much higher pressure and lower temperature as compared to other systems. CO_2 hydrate is stable at higher temperatures and lower pressures. H_2S hydrate is the most stable of all the systems. The phase equilibrium curve of hydrate of CO_2 and CH_4 mixture lies between the pure CH_4 and CO_2. This is expected since the mixture contains CO_2, which increases the phase stability envelope or makes it more stable at higher temperatures and lower pressures. This indicates that the phase stability of hydrates of natural gases containing non-hydrocarbon gases will greatly depend on their composition and may form even at lower pressure and higher temperature conditions in subsea pipelines.

4.8   USE OF SOLIDS PHASE ENVELOPES—OPERATING ENVELOPE FOR GULF OF MEXICO

Notz presented a somewhat empirical composite solids phase envelope consisting of equilibrium curves for (1) hydrate, (2) wax, (3) upper asphaltene onset pressure (U AOP), and (4) lower asphaltene onset pressure (L AOP).[74] This is presumably for Gulf of Mexico (GOM)–type fluids and conditions. A modified composite solids phase envelope is depicted in Figure 4.28. As seen in this figure, the hydrate formation curve lies on the extreme left (lower temperatures), followed by the wax equilibrium curve, which is basically composed of two branches that meet or intersect the bubble point curve—the portion starting upward from the bubble point curve represents the solid–liquid equilibria, whereas the one downward represents the solid–liquid–vapor equilibria, terminating on the x-axis, which is the wax appearance temperature (WAT) of the STO. The somewhat narrow region of wax equilibria is depicted by the red elongated box in Figure 4.28. Asphaltenes are represented by two different curves, namely, the U AOP (higher pressure) and the L AOP (lower pressure), respectively. The vapor–liquid equilibrium or the bubble point curve is typically sandwiched between the U AOP and the L AOP (basically this should appear below the brown-colored bubble point curve in Figure 4.28, although the one suggested by Notz[74] is shown as the dashed black curve). Note that, as expected, the U AOP, L AOP and bubble point curves tend to converge at relatively higher temperatures. Some of the modeled cases and their comparison with experimental data, shown by Pedersen et al.[75] indicate this convergence occurring in the vicinity of 3,000 psia and 300°F. Also shown in Figure 4.28 is experimental data on hydrate, wax, U AOP, and bubble point for some GOM fluids and conditions that have been reported in the literature.[76–80] Note that, for the most part, the AOP data reported in the literature appear to be for the U AOP, as reflected in Figure 4.28.

The primary purpose of Figure 4.28 is to basically expand the scope of the composite solids phase envelope such that it can be useful in obtaining a quick "snapshot" of which solids are likely to precipitate (e.g., in steady-state flowing conditions) by mapping or superimposing the key temperature and pressure conditions such as reservoir, wellhead, and the end of flowline. An example case from Notz is depicted

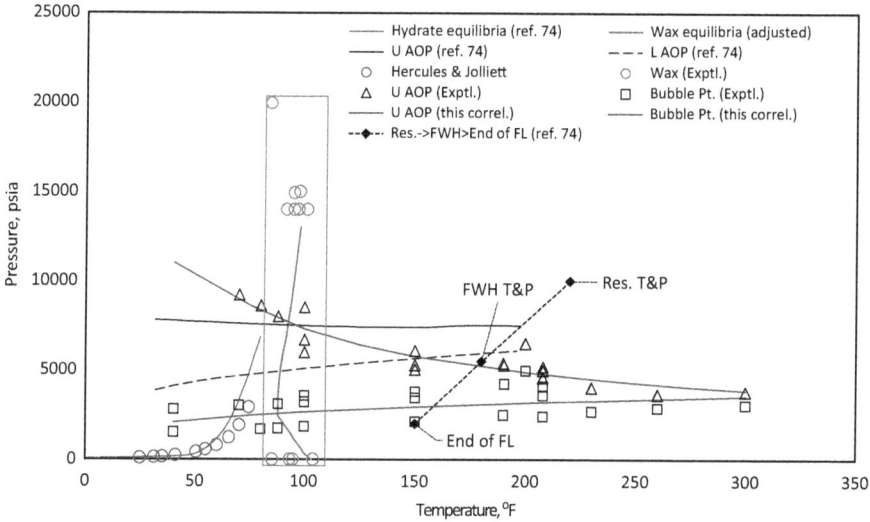

FIGURE 4.28   Composite solids phase envelopes for Gulf of Mexico fluids. Hercules and Jolliett hydrate conditions are from Sloan,[76] wax data is from Gonzalez et al.[77, 78] and U AOP and bubble point data are from Gonzalez et al.[77, 78] and Jamaluddin et al.[79, 80] respectively, while others are self-explanatory as shown in the legend. *Note: No Hydrate; No Wax; No Asphaltene on the right of hydrate and wax equilibrium curves and above the U AOP, whereas hydrate, wax form on the left and asphaltene present/unstable between the U AOP and L AOP (not shown as experimental or correlated), respectively.*

in Figure 4.28, which clearly indicates that neither hydrates nor waxes will form; however, asphaltenes will be unstable through the flowline, beginning about at right past the wellhead. Note that, basically the ending temperatures in the flowline are dependent on the length and the overall heat transfer coefficient.

REFERENCES

1. Enders, M. S. P., Anschau, K. F., Doneda, M., Druzian, G. T., Gomes, A. O., Guimaraes, R. C. L., Flores, E. M. M., & Muller, E. I. (2020). Characterization of inorganic solids present in brazilian crude oil emulsions using scanning electron microscopy (SEM) with energy-dispersive x-ray spectrometry (EDS): Evaluation of the effect of solvents. *Energy & Fuels*, 34(2), 1309–1316. https://doi.org/10.1021/acs.energyfuels.9b03087
2. Roenningsen, H. P., Bjoerndal, B., Hansen, A. B., & Pedersen, B. W. (1991). Wax precipitation from North sea crude oils: 1. Crystallization and dissolution temperatures, and Newtonian and non-Newtonian flow properties. *Energy & Fuels*, 5(6), 895–908. https://doi.org/10.1021/ef00030a019
3. Peters, W., Brandon, H., Jerina, K. L., Wolf, C., & Young, V. L. (2012). *Biomaterials in Plastic Surgery*. Woodhead Publishing Limited, Cambridge, UK.
4. Palou, A., Cruz, J., Blanco, M., Larraz, R., Frontela, J., Bengoechea, C. M., González, J. M., & Alcalà, M. (2014). Characterization of the composition of paraffin waxes on industrial applications. *Energy & Fuels*, 28(2), 956–963. https://doi.org/10.1021/ef4021813

5. Jayalakshmi, V., Selvavathi, V., Sekar, M. S., & Sairam, B. (1999). Characterisation of paraffin waxes by dsc and high temperature GC. *Petroleum Science and Technology*, 17(7–8), 843–856. https://doi.org/10.1080/10916469908949752

6. Freund, M., Csikós, R., Keszthelyi, S., & Mózes, G. Y. (1982). Paraffin Products Properties, Technologies, Applications. In *Development in Petroleum Sciences*, Vol. 14, Mozes G. Y. (Ed.). Elsevier, Budapest, Hungary.

7. Willingham, C. B., Taylor, W. J., Pignocco J. M., & Rossini, F. D. (1945). Vapor pressures and boiling points of some paraffin, alkylcyclopentane, alkylcyclohexane, and alkylbenzene hydrocarbons. *Journal of Research of the National Bureau of Standards (United States)*, 35, 219–244.

8. Chemical Book. Paraffin wax (8002-74-2) https://www.chemicalbook.com/ProductMSDSDetailCB2854418_EN.htm (accessed November 26, 2021).

9. National Physical Laboratory. Kaye and Laby Tables of physical constants https://www.ase.org.uk/resources/kaye-and-laby-tables-physical-constants (accessed November 26, 2021).

10. Joint FAO/WHO Expert Committee on Food Additives. Microcrystalline wax https://www.fao.org/3/W6355E/w6355e0m.htm (accessed November 26, 2021).

11. Bacon, M. M., Romero-Zerón, L. B., & Chong, K. K. (2010). Determining wax type: paraffin or naphthene? *SPE Journal*, 15(4), 963–968. https://doi.org/10.2118/124799-PA

12. HCI wax. Microcrystalline Wax http://www.hciwax.com/index.php/products/microcrystalline-wax.html (accessed November 26, 2021).

13. El-Dalatony, M. M., Jeon, B.-H., Salama, E.-S., Eraky, M., Kim, W. B., Wang, J., & Ahn, T. (2019). Occurrence and characterization of paraffin wax formed in developing wells and pipelines. *Energies*, 12(6), 967. https://doi.org/10.3390/en12060967

14. Mansoori, G. A., Barnes, H. L., & Webster, G. M. (2003). Petroleum Waxes. In *Fuels and Lubricants Handbook: Technology, Properties, Performance, and Testing*, Totten, G. E., Westbrook, S. R., & Shah, R. J., (Eds.). ASTM, West Conshohocken, PA, pp. 525–556.

15. Alghanduri, L. M., Elgarni, M. M., Daridon, J. L., & Coutinho, J. A. P. (2010). Characterization of Libyan waxy crude oils. *Energy & Fuels*, 24(5), 3101–3107. https://doi.org/10.1021/ef1001937

16. Pedersen, W. B., Hansen, A. B., Larsen, E., Nielsen, A. B., & Roenningsen, H. P. (1991). Wax precipitation from north sea crude oils. 2. Solid-phase content as function of temperature determined by pulsed NMR. *Energy & Fuels*, 5(6), 908–913. https://doi.org/10.1021/ef00030a020

17. Paiva, F. L., Calado, V. M. A., & Marchesini, F. H. (2017). On the use of modulated temperature differential scanning calorimetry to assess wax crystallization in crude oils. *Fuel*, 202, 216–226 https://doi.org/10.1016/j.fuel.2017.04.039

18. Alcazar-Vara, L. A. & Buenrostro-Gonzalez, E. (2011). Characterization of the wax precipitation in mexican crude oils. *Fuel Processing Technology*, 92 (12), 2366–2374 https://doi.org/10.1016/j.fuproc.2011.08.012

19. Czech, F. W., Fuchs, R. J., & Antczak, H. F. (1961). Identification of components in paraffin wax by high temperature gas chromatography and mass spectrometry. *Analytical Chemistry*, 33(6), 698–704. https://doi.org/10.1021/ac60174a016

20. Prasad, R. (2008). *Petroleum Refining Technology* (1st Edition). Khanna Publishers, New Delhi, India.

21. Prasad, S. K. & Sangwai, J. S. (2021). Chemical and structural characterisation of nC_7 sphaltenes extracted from atmospheric tower bottom and low waxy crude oil from Indian reservoir. *International Journal of Oil, Gas and Coal Technology*, 26(2), 157–183. https://doi.org/10.1504/IJOGCT.2021.112880

22. Speight, J. G. (2004). Petroleum asphaltenes – part 1: asphaltenes, resins and the structure of petroleum. *Oil & Gas Science and Technology*, 59(5), 467–477. https://doi.org/10.2516/ogst:2004032

23. Mullins, O. C. (2010). The modified Yen model. *Energy & Fuels*, 24(4), 2179–2207. https://doi.org/10.1021/ef900975e

24. Chen, F., Zhu, Q., Li, S., Xu, Z., Sun, X., & Zhao, S. (2018). The function of poly aromatic nuclei structure for adsorption of Vanadyl/Nickel Etioporphyrin on asphaltene/graphene. *Fuel Processing Technology*, 174, 132–141. https://doi.org/10.1016/j.fuproc.2018.02.021

25. Alshareef, A. H. (2020). Asphaltenes: Definition, properties, and reactions of model compounds. *Energy & Fuels*, 34(1), 16–30. https://doi.org/10.1021/acs.energyfuels.9b03291

26. Fakher, S., Ahdaya, M., Elturki, M., & Imqam, A. (2020). Critical review of asphaltene properties and factors impacting its stability in crude oil. *Journal of Petroleum Exploration and Production Technology*, 10, 1183–1200. https://doi.org/10.1007/s13202-019-00811-5

27. Speight, J. G. (1991). *The Chemistry and Technology of Petroleum* (2nd Edition). Marcel Dekker, Inc., New York.

28. Boduszynski, M. M. (1987). Composition of heavy petroleums. 1. Molecular weight, hydrogen deficiency, and heteroatom concentration as a function of atmospheric equivalent boiling point up to 1400.Degree.F (760.Degree.C). *Energy & Fuels*, 1(1), 2–11. https://doi.org/10.1021/ef00001a001

29. Acevedo, S., Gutierrez, L. B., Negrin, G., Pereira, J. C., Mendez, B., Delolme, F., Dessalces, G., & Broseta, D. (2005). Molecular weight of petroleum asphaltenes : A comparison between mass spectrometry and vapor pressure osmometry. *Energy & Fuels*, 19(4), 1548–1560. https://doi.org/10.1021/ef040071+

30. Moschopedis, S. E., Fryer, J. F., & Speight, J. G. (1976). Investigation of asphaltene molecular weights. *Fuel*, 55(3), 227–232. https://doi.org/10.1016/0016-2361(76)90093-4

31. Chanda, M. (2013). *Introduction To Polymer Science And Chemistry a Problem-Solving Approach* (2nd Edition). CRC Press, Taylor & Francis Group, Boca Raton, FL.

32. Durand, E., Clemancey, M., Lancelin, J. M., Verstraete, J., Espinat, D., & Quoineaud, A. A. (2010). Effect of chemical composition on asphaltenes aggregation. *Energy & Fuels*, 24(2), 1051–1062. https://doi.org/10.1021/ef900599v

33. Zheng, C., Zhu, M., Zhou, W., & Zhang, D. (2017). A preliminary investigation into the characterization of asphaltenes extracted from an oil sand and two vacuum residues from petroleum refining using nuclear magnetic resonance, DEPT, and MALDI-TOF. *Journal of Energy Resources Technology*, 139(3), 032905. https://doi.org/10.1115/1.4035746

34. AlHumaidan, F. S., Hauser, A., Rana, M. S., & Lababidi, H. M. S. (2017). NMR characterization of asphaltene derived from residual oils and their thermal decomposition. *Energy & Fuels*, 31(4), 3812–3820. https://doi.org/10.1021/acs.energyfuels.6b03433

35. Clutter, D. R., Petrakis, L., Stenger, R. L., & Jensen, R. K. (1972). Nuclear magnetic resonance spectrometry of petroleum fractions. Carbon-13 and proton nuclear magnetic resonance characterizations in terms of average molecule parameters. *Analytical Chemistry*, 44 (8), 1395–1405. https://doi.org/10.1021/ac60316a002

36. Petrakis, L. & Allen, D. (1987). NMR for Liquid Fossil Fuels. In *Analytical Spectroscopy Library*, (Vol. 1). Elsevier, Amsterdam, The Netherlands.

37. Qiao, P., Harbottle, D., Tchoukov, P., Wang, X., & Xu, Z. (2017). Asphaltene subfractions responsible for stabilizing water-in-crude oil emulsions. Part 3. Effect of solvent aromaticity. *Energy & Fuels*, 31(9), 9179–9187. https://doi.org/10.1021/acs.energyfuels.7b01387

38. Asemani, M. & Rabbani, A. R. (2020). Detailed FTIR spectroscopy characterization of crude oil extracted asphaltenes: Curve resolve of overlapping bands. *Journal of Petroleum Science & Engineering*, 185, 106618. https://doi.org/10.1016/j.petrol.2019.106618

39. Ferreira, S. R., Barreira, F. R., Spinelli, L. S., Leal, K. Z., Seidl, P., & Lucas, E. F. (2015). Comparison between asphaltenes (sub) fractions extracted from two different asphaltic residues: Chemical characterization and phase behavior. *Quimica Nova*, 39(1), 26–31. https://doi.org/10.5935/0100-4042.20150172

40. Leyva, C., Ancheyta, J., Berrueco, C., & Millán, M. (2013). Chemical characterization of asphaltenes from various crude oils. *Fuel Processing Technology*, 106, 734–738. https://doi.org/10.1016/j.fuproc.2012.10.009

41. Yen, T. F., Erdman, J. G., & Pollack, S. S. (1961). Investigation of the structure of petroleum asphaltenes by x-ray diffraction. *Analytical Chemistry*, 33(11), 1587–1594. https://doi.org/10.1021/ac60179a039

42. Andersen, S. I., Jensen, J. O., & Speight, J. G. (2005). X-ray diffraction of subfractions of petroleum asphaltenes. *Energy & Fuels*, 19(6), 2371–2377. https://doi.org/10.1021/ef050039v

43. Alhumaidan, F. S., Hauser, A., Rana, M. S., Lababidi, H. M. S., & Behbehani, M. (2015). Changes in asphaltene structure during thermal cracking of residual oils: XRD study. *Fuel*, 150, 558–564. https://doi.org/10.1016/j.fuel.2015.02.076

44. Kananpanah, S., Kheirkhah, R., Bayat, M., Sadi, M., & Moosavian, M. A. (2017). Comparison of asphaltene structure and morphology under different deasphaltene methods. *Petroleum Science and Technology*, 35(5), 457–464. https://doi.org/10.1080/10916466.2016.1258416

45. Abdallah, W. A. & Yang, Y. (2012). Raman spectrum of asphaltene. *Energy & Fuels*, 26(11), 6888–6896. https://doi.org/10.1021/ef301247n

46. Trejo, F., Ancheyta, J. & Rana, M. S. (2009). Structural characterization of asphaltenes obtained from hydroprocessed crude oils by SEM and TEM. *Energy & Fuels*, 23(1), 429–439. https://doi.org/10.1021/ef8005405

47. Cruz, C. D. L., Márquez, N., Escobar, M. E., & Segovia, S. (1997) An improved chromatographic method for the separation of saturated hydrocarbons, aromatic hydrocarbons, resins and asphaltenes from heavy crude oils. In *213 ACS National meeting*. San Francisco, CA, April 13–17.

48. Acevedo, S., Méndez, B., Rojas, A., Layrisse, I., & Rivas, H. (1985). Asphaltenes and resins from the orinoco basin. *Fuel*, 64(12), 1741–1747. https://doi.org/10.1016/0016-2361(85)90402-8

49. Eke, W. I., Victor-Oji, C., & Akaranta, O. (2020). Oilfield metal naphthenate formation and mitigation measures: A review. *Journal of Petroleum Exploration and Production Technology*, 10, 805–819. https://doi.org/10.1007/s13202-019-00797-0

50. Mapolelo, M. M., Stanford, L. A., Rodgers, R. P., Yen, A. T., Debord, J. D., Asomaning, S., & Marshall, A. G. (2009). Chemical speciation of calcium and sodium naphthenate deposits by electrospray ionization FT-ICR mass spectrometry. *Energy & Fuels*, 23(1), 349–355. https://doi.org/10.1021/ef800642b

51. Buffett, B. A. (2000). Clathrate hydrates. *Annual Review of Earth and Planetary Sciences*, 28, 477–507. https://doi.org/10.1146/annurev.earth.28.1.477

52. Sahu, C., Sircar, A., Sangwai, J. S., & Kumar, R. (2021). Kinetics of methane hydrate formation in the presence of 1-Dodecyl-2-Pyrrolidinone and Tetrahydrofuran in pure water. *Industrial & Engineering Chemistry Research*, 60(20), 7588–7598. https://doi.org/10.1021/acs.iecr.1c00925

53. Sahu, C., Kumar, R., & Sangwai, J. S. (2020). Comprehensive review on exploration and drilling techniques for natural gas hydrate reservoirs. *Energy & Fuels*, 34(10), 11813–11839. https://doi.org/10.1021/acs.energyfuels.0c02202

54. Sloan Jr., E. D., & Koh, C. A. (2007). *Clathrate Hydrates of Natural Gases*. CRC Press, Taylor & Francis Group, Boca Raton, FL.

55. Zou, C., Yang, Z., & Tao, S. (2017). Natural Gas Hydrate. In *Unconventional Petroleum Geology*. Caineng, Z. (Ed.). Petroleum Industry Press, Elsevier, Amsterdam, Netherlands, pp. 391–404.

56. Kirchner, M. T., Boese, R., Billups, W. E., & Norman, L. R. (2004). Gas hydrate single-crystal structure analyses. *Journal of the American Chemical Society*, 126(30), 9407–9412. https://doi.org/10.1021/ja049247c

57. Shin, K., Udachin, K. A., Moudrakovski, I. L., Leek, D. M., Alavi, S., Ratcliffe, C. I., & Ripmeester, J. A. (2013). Methanol incorporation in clathrate hydrates and the implications for oil and gas pipeline flow assurance and icy planetary bodies. *Proceedings of the National Academy of Sciences of the United States of America.*, 110, (21) 8437–8442. https://doi.org/10.1073/pnas.1302812110

58. Haber, A., Akhfash, M., Loh, C. K., Aman, Z. M., Fridjonsson, E. O., May, E. F., & Johns, M. L. (2015). Hydrate shell growth measured using NMR. *Langmuir*, 31(32), 8786–8794. https://doi.org/10.1021/acs.langmuir.5b01688

59. Hassanpouryouzband, A., Joonaki, E., Farahani, M. V., Takeya, S., Ruppel, C., Yang, J., English, N. J., Schicks, J. M., Edlmann, K., Mehrabian, H., Aman, Z. M., & Tohidi B. (2020). Gas hydrates in sustainable chemistry. *Chemical Society Reviews*, 49, 5225–5309. https://doi.org/10.1039/C8CS00989A

60. Cha, M., Shin, K., Lee, H., Moudrakovski, I. L., Ripmeester, J. A., & Seo, Y. (2015). Kinetics of methane hydrate replacement with carbon dioxide and nitrogen gas mixture using in situ NMR spectroscopy. *Environmental Science & Technology*, 49(3), 1964–1971. https://doi.org/10.1021/es504888n

61. Petuya, C., Damay, F., Talaga, D., & Desmedt, A. (2017). Guest partitioning in carbon monoxide hydrate by Raman spectroscopy. *Journal of Physical Chemistry C*, 121(25), 13798–13802. https://doi.org/10.1021/acs.jpcc.7b04947

62. Schicks, J. M., Erzinger, J., & Ziemann, M. A. (2005). Raman spectra of gas hydrates— differences and analogies to ice 1h and (gas saturated) water. *Spectrochimica Acta Part A: Molecular and Biomolecular Spectroscopy*, 61(10), 2399–2403. https://doi.org/10.1016/j.saa.2005.02.019

63. Hester, K. C., Dunk, R. M., White, S. N., Brewer, P. G., Peltzer, E. T., & Sloan, E. D. (2007). Gas hydrate measurements at hydrate ridge using Raman spectroscopy. *Geochimica et Cosmochimica Acta*, 71(12), 2947–2959. https://doi.org/10.1016/j.gca.2007.03.032

64. Hirai, H., Kadobayashi, H., Hirao, N., Ohishi, Y., Ohtake, M., Yamamoto, Y., & Nakano, S. (2015). Time-resolved x-ray diffraction and Raman studies of the phase transition mechanisms of methane hydrate. *Journal of Chemical Physics*, 142(2), 024707. https://doi.org/10.1063/1.4905482

65. Gupta, A., Dec, S. F., Koh, C. A., & Sloan, E. D. (2007) NMR investigation of methane hydrate dissociation. *Journal of Chemical Physics*, 111(5), 2341–2346. https://doi.org/10.1021/jp066536+

66. Buchanan, P., Soper, A. K., Westacott, R. E., Creek, J. L., & Koh, C. A. (2003). In situ neutron diffraction studies of methane hydrate formation and decomposition. *Journal of Chemical & Engineering Data* , 48(4), 778–782. https://doi.org/10.1021/je025622+

67. Popoola, L. T., Grema, A. S., Latinwo, G. K., Gutti, B., & Balogun, A. S. (2013). Corrosion problems during oil and gas production and its mitigation. *International Journal of Industrial Chemistry*, 4, 35. https://doi.org/10.1186/2228-5547-4-35

68. Nalli, K. (2012). Appendix VI: Corrosion and Its Mitigation in the Oil and Gas Industries. In *Process Plant Equipment*. John Wiley & Sons, Inc., Hoboken, NJ, USA, pp. 673–679.

69. Leontaritis, K. J. (1996). The asphaltene and wax deposition envelopes. *Fuel Science & Technology International*, 14 (1–2), 13–39. https://doi.org/10.1080/08843759608947560

70. Fanchi, J. R. (2006). General Engineering. In *Petroleum Engineering Handbook*, (Vol. 1), Lake, L. W. (Ed.). Society of Petroleum Engineers, Richardson, TX.

71. Akbarzadeh, K., Hammami, A., & Kharrat, A. (2007). Asphaltenes – Problematic but rich in potential. *Oilfield Review*, 19(2), 22–43.

72. Mousavi-Dehghani, S. A., Riazi, M. R., Vafaie-Sefti, M., & Mansoori, G. A. (2004). An analysis of methods for determination of onsets of asphaltene phase separations. *Journal of Petroleum Science & Engineering*, 42(2–4), 145–156. https://doi.org/10.1016/j.petrol.2003.12.007

73. Maijoni, A. (2011). *Mechanistic Study of CO_2 Flooding of Asphaltenic Oil Reservoir*. University of Stavanger. Master's Thesis.

74. Notz, P. (2009). *Flow Assurance for Offshore Production*. Petroskills, Houston, TX.

75. Pedersen, K. S., Christensen, P. L., & Shaikh, J. A. (2014). *Phase Behavior of Petroleum Reservoir Fluids*. CRC Press, Taylor & Francis Group, Boca Raton, FL.

76. Sloan, E. D. (2000). *Hydrate Engineering*. Bloys J. B. (Ed.). Society of Petroleum Engineers, US.

77. Gonzalez, D., Gonzalez, F., Pietrobon, M., Haghshenas, M., Shurn, M., Mees, A., Stewart, C., Ogugbue, C., & Duvivier, G. (2016). Strategies to monitor and mitigate asphaltene issues in the production system of a Gulf of Mexico deepwater subsea development. In *Offshore Technology Conference*. Houston, Texas, USA, May 2–5. (OTC-27123-MS). https://doi.org/10.4043/27123-MS

78. Gonzalez, D. L., Mahmoodaghdam, E., Lim, F., & Joshi, N. (2012). Effects of gas additions to deepwater Gulf of Mexico reservoir oil: Experimental investigation of asphaltene precipitation and deposition. In *SPE Annual Technical Conference and Exhibition*. San Antonio, Texas, USA, October 8-10. (SPE-159098-MS). https://doi.org/10.2118/159098-MS

79. Jamaluddin, A. K. M., Joshi, N., Iwere, F., & Gurpinar, O. (2002). An investigation of asphaltene instability under nitrogen injection. In *SPE International Petroleum Conference and Exhibition in Mexico*. Villahermosa, Mexico, February 10–12. (SPE-74393-MS). https://doi.org/10.2118/74393-MS

80. Jamaluddin, A. K. M., Creek, J., Kabir, C. S., McFadden, J. D., D'Cruz, D., Manakalathil, J., Joshi, N., & Ross, B. (2002). Laboratory techniques to measure thermodynamic asphaltene instability. *Journal of Canadian Petroleum Technology*, 41(7), 44–52. https://doi.org/10.2118/02-07-04

5 Assessment of Inorganic and Organic Solid Phases and Other Flow Assurance Issues

5.1 INTRODUCTION

Petroleum engineering as a discipline is quite empirical in nature, primarily because of the fact that we deal with natural materials, such as the reservoir rocks and the (geo-)fluids, that occupy the pore spaces, which are commercially exploited and eventually produced as "finished products" as far as the upstream sector of the oil and gas industry is concerned. This degree of empiricism transpires in flow assurance as well, as manifested by the various rules of thumb and the obviously named empirical methods. The next logical level desired is, of course, the use of a bit more scientifically robust methods such as the equations of state which are used in vapor–liquid, and solid–liquid equilibrium calculations. These models are then fine-tuned based on experimental data for achieving the desired accuracy when predicting the various properties of reservoir fluids, including compressibility factors and density of the fluid. The various models are used for specific reservoir fluid types and composition, such as gases and liquid, as there is no specific equation which is suitable for every fluid. The number of components also increases the complexity of equation-of-states models since the interaction between the components in the various phases affects the thermodynamic properties of the fluid and kinetics. Apart from these phase equilibria models, standardized experimental methods also provide a direct measurement of the parameters, which are useful in tuning the predictions. Hence, modeling gas hydrates is relatively easy as it involves only gas and water as compared to wax and asphaltene, which have a large number of components. Thus, equation of states modeling is vital for phase equilibria calculations and prediction of precipitation of these solid phases to optimize the production strategy and avoid the encountering flow assurance issues. To develop robust models for applications to a variety of flow assurance issues, experimental data are necessary. In the case of waxes and asphaltenes, information on wax appearance temperature, their characteristics, and composition in the crude oil is necessary. In the case of hydrate issues, information on the gas composition and equilibrium pressure and temperature conditions, including formation and dissociation kinetics of gas hydrate, is necessary. Apart from flow wax, hydrates, and asphaltene, there are other issues that are frequently encountered in flow assurance such as emulsion formation, which causes pressure drops across chokes and valves, and fouling caused by chemical reactions or bacterial and algae

DOI: 10.1201/9781003091301-5

colonies. The formation of an emulsion is predominant in the case of a mature reservoir due to increased wax or asphaltene content. These estimations and/or prediction methods for the commonly encountered inorganic and organic solids together with the complementary experimental techniques are covered in this chapter. Also, some flow assurance issues other than wax, hydrates, and asphaltenes are discussed.

5.2   RULES OF THUMB

For the most part, all rules of thumb (RoTs) are based on experience and are basically meant to be guides[1] for further (thorough) action. For example, as stated by Sloan,[1] a flow assurance engineer may determine that additional accuracy is needed in calculating the dosage rates for an inhibitor. It should be noted that RoTs are not intended to replace "absolute truths",[1] but they should be construed as a bit like the first iteration. Given the fact that hydrates perhaps constitute the greatest flow assurance risk, most of the RoTs are for hydrates.[1] In addition, a handful of RoTs for paraffin waxes, flow, and other miscellaneous parameters also exist. All pertinent RoTs are presented in this section.

5.2.1   Hydrates

Perhaps the best compilation of RoTs for hydrates is provided by Sloan,[1] of which the significant ones are summarized in Table 5.1.

5.2.2   Paraffin Waxes

Pedersen et al.,[3] based on an extensive review of wax-related data collected, have proposed the following excellent RoTs for waxy oils: (1) precipitated wax melts at sufficiently high temperatures; (2) higher limit of wax appearance temperature (WAT) is approximately 60°C; (3) WAT reduction ratio is 0.15–0.20°C/mole % dissolved C_1–C_5; (4) if the oil composition is same, WAT increase per atm (pressure) is about 0.02°C; (5) and an upper limit of wax content is 15% for a stable oil.

Other suggested RoTs include that of Gudmundsson[4] for time-based maximum wax deposit thickness in pipelines: one day, less than 1 mm; one week, between 1–10 mm; and for a month, greater than 10 mm. RoTs used in pigging frequency are based on (1) less than 10% of the total cross-sectional area and (2) limiting total wax thickness to 1–4 mm.[5] An RoT-related chemical composition for wax forming groups is typically 80–95% normal paraffins.[6]

5.2.3   Flow-Related and Miscellaneous

According to Notz,[7] the allowable frictional pressure drop from gas condensate and oil pipelines is 10–20 psi/mile and 50–250 psi/mile, respectively, at the design rates. Another RoT pertaining to slug catchers is that they should be 3–5 times the riser volume. The general RoTs that pertain to the oil's American Petroleum Institute (API) gravity and separators[4] are as follows: API > 20° (conventional lighter oils) oil can be separated by gravitation; 20° > API > 15° (medium to heavier oils) more difficult to separate in the conventional tank; and for API < 15° (heavy oils) special measures likely needed due to separation challenges.

TABLE 5.1

Rules of Thumb (RoTs) for Hydrates Compiled and Prepared Based on Listed RoTs in Sloan[1]

RoTs related to hydrate phase equilibria

(1) At $T = 39°F$ and $P > 166$ psig, hydrates will form in a natural gas system if free water is available.

(2) In gas–water systems, hydrates tend to form on the pipe wall, whereas in gas–condensate or oil systems, they frequently form from free water as particles that agglomerate.

(3) It is better to expand a dehydrated gas than a moist or water-saturated gas to prevent hydrate formation.

(4) A larger ΔP at higher inlet temperature is better (the corresponding drop in temperature following the expansion will likely be outside the hydrate equilibrium conditions).

RoTs related to THIs and AAs

(1) At $T = 39°F$ and $P > 1,000$ psia, MeOH lost to vapor phase is 1 lb MeOH/mmscf for every wt% MeOH in the free-water phase.

(2) At $T = 39°F$ and $P > 1,000$ psia, the maximum amount of MEG lost to the gas phase is 0.02 lb/mmscf.

(3) MeOH concentration dissolved in the condensate is 0.5 wt%.

(4) MEG mole fraction in a liquid hydrocarbon at 39°F is 0.03% of the aqueous phase mole fraction of MEG.

(5) MEG injection is used for onshore gas lines when the required MEOH dosage rate exceeds 30 gallons/hour.

(6) Prerequisites for the use of AAs are substantial liquid hydrocarbon phase (oil or condensate) and a maximum water/oil ratio of 40 to 60.

RoTs related to blockage and dissociation

(1) Lack of hydrate blockage $\neq$ lack of hydrates; frequently, hydrates form but flow and are detected in the pigging returns.

(2) Attempts to blow the hydrate plug out of a line by increasing the ΔP lead to more hydrate formation (system farther into the hydrate forming region). In offshore regions, the first step is to inject an inhibitor into the plugged line from the platform to estimate the plug distance. At the same time, any access point can be used for inhibitor injection in onshore settings.

(3) The hydrate dissociation process is usually days or weeks, regardless of the applied techniques.

Miscellaneous RoTs

(1) Agglomeration of individual hydrate particles results in an open hydrate mass with porosity in excess of 50% and is permeable to gas flow (permeability–length ratio of $8.7{:}11 \times 10^{-15}$ m). Owing to these characteristics, although the pressure is transmitted, liquid flow is substantially impeded – this is somewhat akin to what is known as the Klinkenberg slippage effect[2] in porous media.

(2) Costs associated with MeOH loss can be substantial if the total fraction of either the vapor or the liquid hydrocarbon phase is very large relative to the aqueous phase.

AA = antiagglomerant; MeOH = methanol; MEG = monoethylene glycol; THI = thermodynamic hydrate inhibitor.

5.3   EMPIRICAL METHODS

Similar to the RoTs, empirical methods also are based on experience; however, the main difference is the latter are "data-based". In some cases, this may be expressed or cast in the form of fitted equations or correlations that contain coefficients, while others may require calculating a certain variable or parameter and checking its placement on a deterministic map, table, or diagram.

5.3.1 HYDRATE FORMATION CONDITIONS

In the case of hydrates, perhaps the best example is the graphical method[8] to determine the hydrate formation conditions using the gas gravity (recall $\gamma_g = MW_g/28.97$, where MW_g is the gas molecular weight, calculated from composition). Basically, the chart is a collection of six hydrate Pressure–Temperature (PT) condition curves for γ_g ranging from 0.55 (methane) to as high as 1.0. Although the chart offers great simplicity in determining the hydrate forming conditions for a given gas gravity, caution needs to be exercised that it was generated for sweet gases and thus is not recommended if substantial amounts of CO_2, N_2, and H_2S are present. Therefore, as recommended by Sloan,[1] it should be used to obtain the first estimate of hydrate PT conditions. Many empirical correlations also exist that allow the calculation of PT conditions for sweet natural gases (no inhibitor effect). The following is an example of one such empirical equation proposed by Towler and Mokhatab,[9] which expresses the hydrate formation gas gravity charts:

$$T = 13.47\ln(P) + 34.27\ln(\gamma_g) - 1.675\left[\ln(P) \times \ln(\gamma_g)\right] - 20.35 \qquad (5.1)$$

where T is the hydrate formation temperature in °F, P is pressure in psia, and γ_g is gas gravity (dimensionless).

5.3.2 ASPHALTENE STABILITY ASSESSMENT

Primarily, four different graphical empirical methods or screening criteria indicate asphaltene stability in crude oil from the standpoint of flow assurance issues. Two methods are based on the undersaturation degree (density–pressure relationship). The other two are based on compositional analysis of the oil (usually, stock tank) in terms of saturates (S), aromatics (A), resins (R), and asphaltenes (A). A subcategory of one of the SARA-based methods uses the colloidal instability index (CII) as a determinant, a substitute for the graphical version. Figure 5.1 shows what is known as the de Boer[10] plot, which divides the oil density at reservoir conditions versus the reservoir pressure and the bubble point pressure difference (or undersaturation) in three regions, namely, (1) severe problems, (2) mild or no problems, and (3) no problems.

The "de Boer plot" basically indicates severe asphaltene issues for the light or highly undersaturated crude oils. In principle, the Oilphase–Schlumberger asphaltene stability index is somewhat similar,[11] which basically offers a binary choice based on density undersaturation. If the difference between the initial oil density and the one at bubble point; when $(\rho_{oi}-\rho_{ob})$ is >0.025, the system is considered as unstable, whereas when $(\rho_{oi}-\rho_{ob}) < 0.025$, the system is stable. The SARA-based screening criteria are depicted in Figure 5.2a. The first approach uses only the asphaltene versus resin or asphaltene–resin ratio approach.[11] The second approach uses a combination of saturates + asphaltenes versus aromatics + resins.[12] For example, if the oil contains relatively higher asphaltene content (or saturates and asphaltenes combined) but is counterbalanced by the resin (or aromatics and resin combined), then the oil

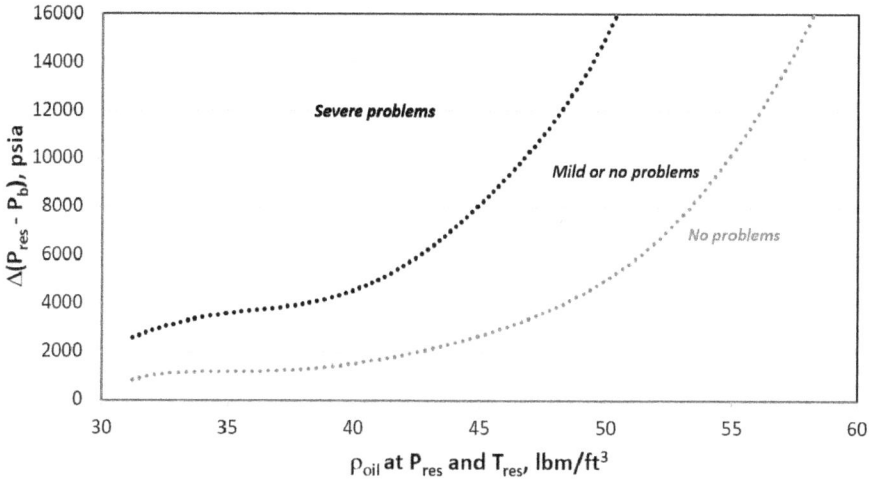

FIGURE 5.1 Reconstructed de Boer plot of oil density versus degree of undersaturation. Plot represents polynomially fitted expressions of values in de Boer et al.[10]

may be considered as stable or manifesting minor or no problems from the standpoint of asphaltenes. Yen et al.[12] proposed the CII, which is basically the numerical interpretation of the second plot in Figure 5.2b, as a screening criterion to determine asphaltene problem severity:

$$CII = \frac{Saturates + Asphaltenes}{Aromatics + Resins} \tag{5.2}$$

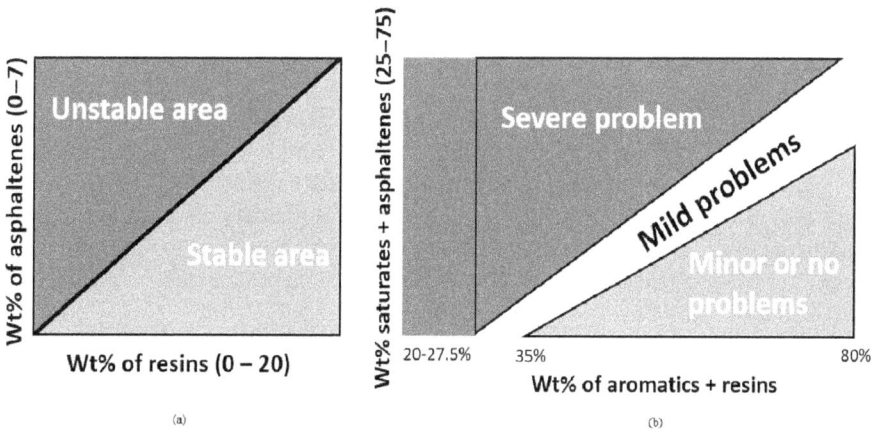

FIGURE 5.2 SARA-based screening for asphaltene assessment. Shown here are conceptual representations of criteria proposed by (a) Jamaluddin et al.[11] and (b) Yen et al.[12]

TABLE 5.2

SARA Analysis of Referenced Crudes and their Asphaltene Assessments Compared with Available Field Evidence

Crude	S, wt%	Ar, wt%	R, wt%	As, wt%	CII	Field Reports or Experience
Oil 6 Lake Maracaibo (Venezuela)[14]	52.8	37.8	7.9	1.5	1.19	*Frequent asphaltene plugging*
Oil 7 Lake Maracaibo (Venezuela)[14]	34.3	30.3	32.1	3.3	0.60	*No asphaltene related issue*
Marrat (Kuwait)[15]	71.0	12.3	15.1	0.8	2.62	*Extensively documented asphaltene issues*
Hassi–Messaoud (Algeria)[13]	70.5	25.5	3.3	0.15	2.45	*Extensively documented asphaltene issues*
Boscan (Venezuela)[13]	15.0	37.0	34.0	14.0	0.41	*Well-known trouble-free oil*
Oil A (Middle East)[16]	68.3	11.6	18.8	1.3	2.29	*Not stated explicitly but construed as problematic given the detailed experimental evaluation*
Oil B (Gulf of Mexico)[16]	65.6	16.3	13.5	4.6	2.36	*Not stated explicitly but construed as problematic given the detailed experimental evaluation*

SARA wt% values are as reported in the cited references.

Oils with a CII below 0.7 are considered stable (higher aromatics and resins or the green wedge). Those above 0.9 are considered very unstable or severely problematic, while a CII between 0.7–0.9 may be considered cases of mild problems. Table 5.2 shows the examples of some crude oils on which the asphaltene assessment methods have been tested. The reported SARA analysis data have been used to see their place-ment on the asphaltene–resin ratio chart as shown in Figure 5.3 and for calculating the CII values.

As seen in Figure 5.3, all but one oil (Boscan) falls in the stable area; however, the fairly low CII value of 0.41 solidly indicates that this oil is stable or not problematic, which is confirmed based on its trouble-free production.[13] This is precisely reversed for all other oils, except oil 7 from Lake Maracaibo in that the CII values are higher than 0.9. Consistent asphaltene assessment results, that is, stable or not problematic, only for the Lake Maracaibo oil 7, which also matches with what the authors[14] report

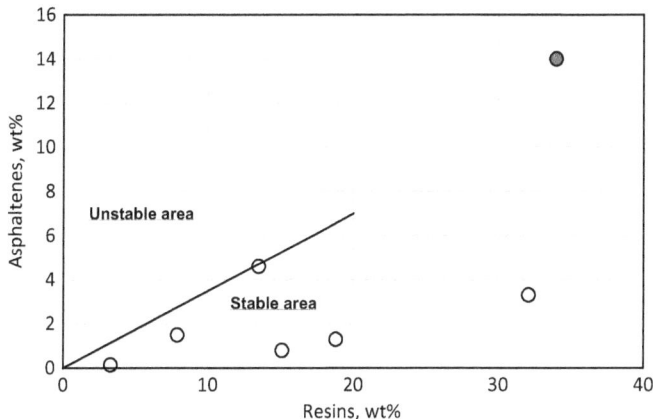

FIGURE 5.3 Placement of various crude oils on the asphaltene–resin graph[11] to determine their stability. Details of the oils are shown in Table 5.2.

as stable production. In all other cases, the CII-based assessment concurs with the field reports. If other data are available, additional assessments can be carried out via the de Boer method and the Oilphase–Schlumberger index method; however, a more rigorous confirmation can only be obtained via a detailed laboratory examination.

5.4   EXPERIMENTAL TECHNIQUES

5.4.1   FLUID STUDIES FOR WAX

Changes in the temperature condition are dominant factors affecting the separation and deposition of paraffin waxes. As the temperature is reduced, the highest temperature at which the first amount of wax crystals appears from the bulk fluid is called the WAT. In general, the WAT is classified into two categories. These are thermodynamic (or true) and experimental WAT. The thermodynamic WAT is related to the actual liquid (oil)–solid (wax) equilibrium temperature condition at a given pressure and is the maximum temperature at which both phases are in equilibrium at a given pressure. The experimental WAT is the temperature at which the first wax crystal appears in the bulk liquid phase. Experimental WAT may vary depending on the type of experimental methods. The WAT is generally measured at the atmospheric conditions as its measurement at reservoir conditions may be time-consuming and costly. Typically, the WAT measured at reservoir conditions is lower than the one at atmospheric (or stock tank) conditions. Wax deposition models based on dead oil samples cannot predict the wax deposition at the reservoir condition.

The WAT is also called wax precipitation temperature (WPT) or cloud point because below the WAT, the fluid will appear cloudy. To avoid any flow assurance issues due to wax deposition, they must be managed properly during crude transportation.[17] The WAT is an important parameter for defining crude oil properties for wax flow assurance. The precipitation of wax crystals in crude oil affects the crude oil properties such as rheology, gelling, and density. Figure 5.4 shows a schematic of the

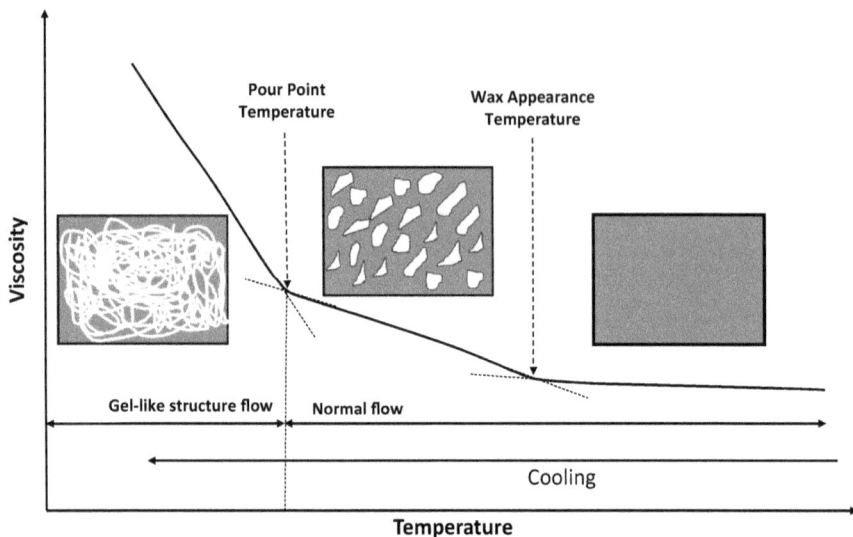

FIGURE 5.4   Schematic of the wax deposition phenomenon and the impact on the viscosity of the crude oil.

wax precipitation phenomenon and its impact on the rheology of crude oil. Until the WAT, the viscosity of the crude oil does not change much, although it depends on the composition of the crude oil. Below the WAT, the viscosity starts to increase, although the fluid is still able to flow. However, below a certain temperature (also called the pour point), the crude oil viscosity starts to increase significantly, and thus, it becomes difficult to flow. A high WAT means the crude oil is more prone to deposit wax earlier. If the temperature of the systems is above the WAT, then there is very little possibility of wax deposition in the production and transportation pipelines. The WAT is often measured at laboratory conditions using different methods. The measurement of the WAT depends on the sensitivity of the measurement methods, which might be based on cooling history and cooling rate. Therefore, measuring the WAT can be very subjective and thus defines the decision-making process.[18]

The WAT measurement depends on the composition of the crude oil, measurement method, the history of cooling or heating, and residence time during measurement. With very sensitive measurement methods, a higher WAT may be obtained. With an increase in pressure of the system, the WAT decreases as the sample solution gas is dissolved, delaying the wax crystallization.[19] The rate of cooling during measurement affects the WAT. A faster cooling rate gives a lower WAT, while a slower cooling rate will give higher WAT. Generally, at the lab scale, dead oil obtained from the separator or the stock tank oil is used to measure the WAT provided; there are no major losses of higher paraffins till the point of sampling. The following methods are used to measure the WAT of the crude oil.

5.4.1.1   Visual Method

The American Society for Testing and Materials (ASTM) standard D 2500-09 method is a visual method used to define WAT or cloud points of petroleum fluids

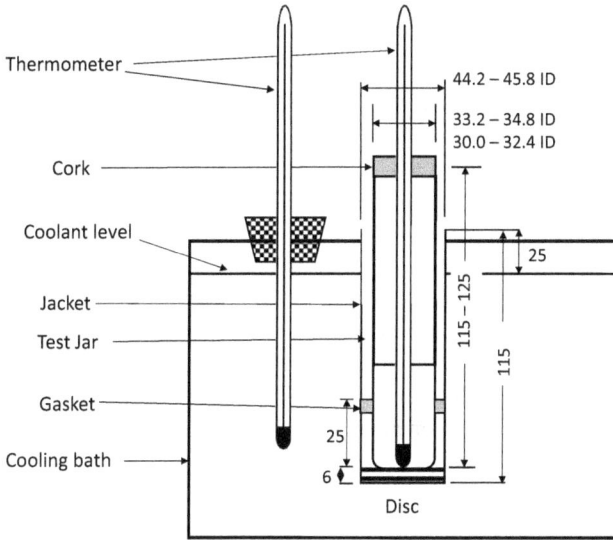

FIGURE 5.5 Schematic of the experimental setup for the ASTM D 2500 test method to determine cloud point. All dimensions are in mm.

at a specific rate. In this method, only transparent oils are preferred. In the case of nontransparent crude oil, there are chances of observation failure of wax crystals. In this method, the crude oil sample is taken in a beaker and placed in a constant temperature bath. The temperature is then slowly reduced from about 15°C above the expected WAT by 1°C. A high-precision microscope is used to find the crystal formation visually. The highest temperature at which wax crystal is observed is noted as the WAT. Figure 5.5 shows the schematic of the experimental setup used for the same.[20–23] Kruka et al.[21] altered this technique by stirring the sample. Tiwary[22] used a residence time of about 15 min at each step of 1°C and observed wax crystals.

5.4.1.2 Cold Finger

The cold-finger method is one of the simplest methods to determine the WAT. This method primarily involves a temperature-controlled rod submerged in a gently heated crude oil sample. As the temperature is reduced below the WAT, petroleum wax starts to deposit on the surface of the rod. This method needs some significant quantity of wax to be deposited before identifying the WAT. This method yields the lowest WAT.

5.4.1.3 Flow Loop Experiments

Flow loop experiments are one of the preferred methods to study wax deposition from crude oil. This method mainly consists of a pipe loop system to mimic the actual oilfield condition. The design contains a double-pipe heat exchanger in which the crude oil is circulated from the tube side while the cold fluid is circulated from the shell side.[18] Initially, the oil is heated in a reservoir tank and circulated in the flow loop. The heat exchanger in the flow loop cools the crude oil. The wax deposition is visually seen in a window attached to the flow loop, and the corresponding temperature is measured to determine the WAT. It is, however, very hard to simulate the actual oilfield conditions.

5.4.1.4 Cross-Polar Microscopy

Cross-polar microscopy (CPM) method typically provides the highest WAT values as it is a more sensitive method to the presence of the wax crystals. Thus, this method is one of the most preferred and accurate techniques to determine WAT.[24] This method requires a light source, a polarizer, a temperature controller, an infrared filter, and a microscope. A crude oil sample is placed on a glass slide on the temperature-controlled microscope stage. Initially, at a higher temperature above the WAT, the view appears black. As the temperature is reduced, at a particular temperature, wax crystals start appearing, as shown in Figure 5.6.[19, 25] It is to be noted that both samples show different WATs and different wax crystal morphologies at 0°C, although both do now show any wax formation at 60°C.

5.4.1.5 Viscometry

This method is based on the concept that the rheological behavior of the crude oil during the wax formations, particularly near the WAT, changes from Newtonian to non-Newtonian behavior.[22, 26] A rheometer or viscometer is utilized to measure the viscosity of the crude oil sample. The sample's viscosity is measured with respect to the reduction in temperature. At the WAT, there is a change in the viscosity slope or the viscosity–temperature relationship.[24, 27, 28] Figure 5.7 shows the schematic of WAT identification of a sample crude oil. The point at which the curve changes its linear behavior to nonlinear is the WAT of the crude oil.[29] This method may not be accurate all the time, as it becomes very difficult to identify the WAT from the flow curve. Often, the results of the method are compared with the differential scanning calorimetry (DSC) technique or the CPM method.

5.4.1.6 DSC

The DSC method measures heat release from the crude oil sample during the wax nucleation and crystallization process. The heat release profile at the onset of wax

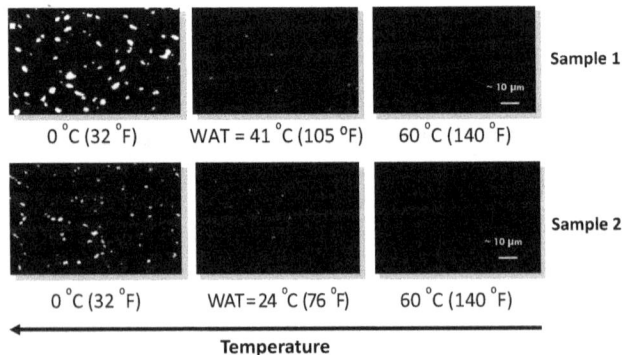

FIGURE 5.6 Cross-polar microscopy image showing the appearance of wax as the temperature is decreased for two different crude oil samples.

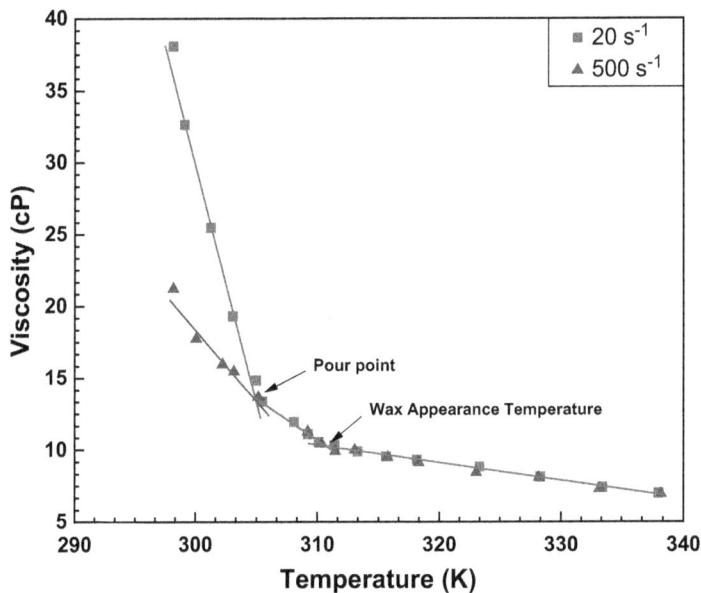

FIGURE 5.7   Identification of the WAT using viscosity measurement. Data are taken from Zhao et al.[30]

crystallization is less and shows a stable baseline. Crystallization is an exothermic process, and thus, the crystals formed will transfer heat from the crude oil sample. In Figure 5.8, the temperature at the exothermic peak indicates the WAT. The wax precipitation peak temperature (WPPT) is also shown in Figure 5.8. This method uses information on the amount of heat released during crystallization. The amount of wax can be obtained from the ratio of heat released by the heat of crystallization.[22, 24, 31] The method is more suitable for oil samples containing a small amount of wax. This method is a relatively very fast and reliable method.[32] For precise measurement of the WAT using DCS, a slow cooling or heating rate is required, along with properly calibrated equipment and erasing the previous thermal history, if any.[33] The WAT is measured using DSC method, and the viscometry method shows the difference in values due to their sensitivity for different samples.

5.4.1.7   Filter Plugging

In filter plugging (FP) method, a capillary with a filter setup is used to measure the WAT. The setup is submerged into the temperature-controlled water bath. A preheated and prefiltered waxy crude oil solution is passed through a capillary and filter while the setup is submerged in a water bath, utilizing a specific colling rate. The pressure drop across the capillary and filter is recorded as the temperature decreases. As the wax precipitates at a certain temperature, the differential pressure drops increase, indicating the WAT.[19] This method is suitable for live and less viscous crude oil.

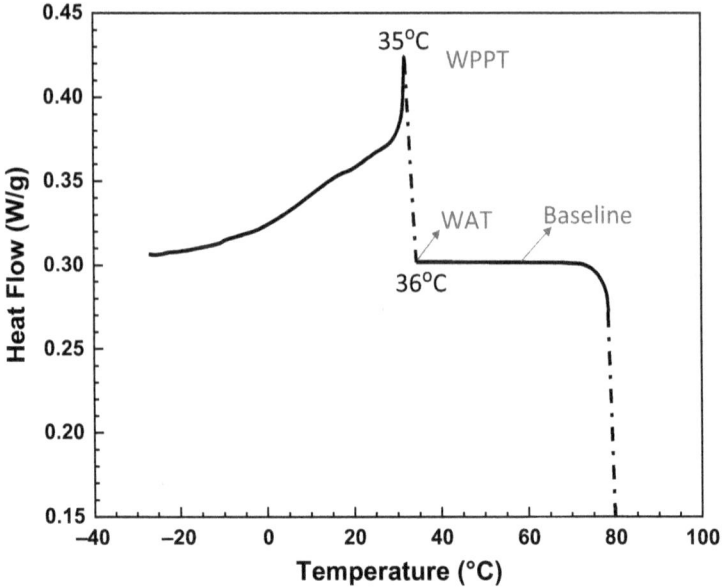

FIGURE 5.8   Determination of WAT and WPPT using differential scanning calorimetry (DSC) thermal spectrum.[34]

5.4.1.8   Gelation and Gel Strength

The pour point is defined using the standard ASTM method but is insufficient for inferring wax deposition under flow conditions. Gelation temperature provides further information on the gel-like behavior of crude oil and its potential to deposit wax under flow conditions. Rheological studies are used to understand the gelation temperature of waxy crude oil using a shear stress experiment while the oil is being cooled. The temperature at which waxy crude oil becomes gel/solid-like fluid is called gelation temperature.[35] In terms of the rheological definition, the gelation temperature or gel point is the temperature at which the solid-like behavior of fluids becomes predominant over the liquid-like behavior. During oscillatory shear stress, the fluid gives an in-phase and an out-of-phase response in the resultant motion of the fluid. Loss modulus (G'') characterizes liquid-like behavior (due to in-phase response denoted by Newtonian fluid). In contrast, the storage modulus (G') is used to characterize solid-like behavior (out-of-phase response denoted by typical Hookean solid) of the fluid. At a temperature much higher than the cloud point (or WAT), the waxy crude oil may behave like a Newtonian fluid, having the loss modulus much higher than the storage modulus. With the decrease in temperature, as the wax starts to separate from the bulk phase, the solid-like behavior of the fluid increases sharply near the pour point (the temperature at which the crude oil loses its flow behavior). The temperature at which the storage and loss moduli become equal is the gelation temperature of the waxy crude oil.[35] Another rheological method involves measuring shear rate while low constant shear stress (say 0.1 Pa) is applied on the crude oil sample while cooling. Most waxy crude oil behaves like a shear-thinning fluid; that is, the viscosity of the crude oil decreases with an increase in a shear rate below the WAT. At the gel point, the viscosity of the crude oil increases sharply. It is imperative to understand the gel strength of the crude oil to infer the flow behavior in the pipeline.

Due to the high apparent viscosity of the waxy crude oil and its gelation below the pour point, the flow of crude oil may not be possible in the pipeline due to high yield (or gel) strength. Also, after shut-in, a successful pipeline restart is very important for efficient flow assurance. During shut-in, the wax gel may form due to the cooling of the crude oil, which may plug the entire pipeline.[36] During a restart operation, to ensure the flow, a large pressure is required to break the gel and to ensure the flow. Thus, it is necessary to understand the gel strength to design a successful shut-in and restart operation. Several factors impart the gel strength to the crude oil, including temperature, shear, and fluid composition. Information on the yield stress is therefore very important to understand the restart pressure for onshore and subsea pipelines. The gel strength of the waxy crude oil can be measured using rheological studies on waxy crude oil and model pipeline test (MPT).

5.4.2 Fluid Studies for Asphaltene

The precipitation of asphaltene from the bulk crude oil is mainly characterized due to decrease in pressure. As discussed earlier, the pressure asphaltene separates from the bulk fluid is called the asphaltene onset pressure (AOP). Asphaltene studies mainly include the determination of the AOP. The following are some methods used to determine the AOP.

5.4.2.1 Gravimetric Method

In this method, the live crude oil collected in the sampler is transferred to a pressure–volume–temperature (PVT) cell as shown schematically in Figure 5.9a. Sufficient care should be taken to ensure that the temperature and pressure of the sample are maintained at reservoir conditions. Once the sample is transferred, it should be conditioned at reservoir pressure and temperature. At constant temperature, the pressure of the sample is then decreased stepwise (in the interval of 50 bars) below the bubble point pressure.[37–39] As the pressure decreases, asphaltene tends to precipitate at the bottom of the cell. Crude oil samples are then taken from the top section of the cell, where there is less or no asphaltene. The sampled oil is then flashed at the standard condition, and the amount of asphaltene in the oil is determined using n-alkane precipitation. As the pressure decreases, the asphaltene content in the crude oil increases until the saturation pressure is reached. The procedure is followed till a few pressure steps are below the bubble point. The gas phase is purged at and below the bubble point from the PVT cell. The fluid is then agitated for 24 h, and then the asphaltene content from the supernatant fluid is analyzed. The amount of asphaltenes in the crude oil *vs* pressure is shown schematically in Figure 5.9b. The trends in the curve denote the upper AOP and the lower AOP. Above the upper AOP, the asphaltene content in the crude oil phase is nearly constant. Below the upper AOP, the asphaltene content in the oil decreases as the asphaltene tends to deposit at the bottom of the cell. At the bubble point, the amount of the asphaltene in the upper oil phase is minimum, as, at this point, most of the asphaltene will have deposited at the bottom of the cell due

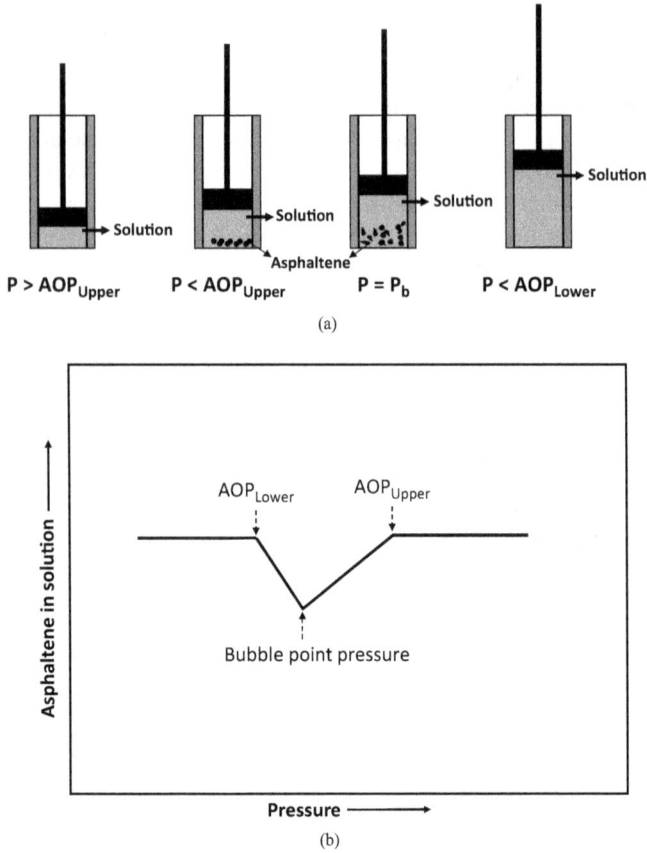

FIGURE 5.9 (a) Schematic of the asphaltene deposition in a PVT cell. (b) Asphaltene content with pressure in the crude oil phase.

to the dominance of the lighter hydrocarbons in the oil phase. A further decrease in pressure releases the gas out of the solution, which tends to redissolve the asphaltene in the oil phase, increasing its content till the lower AOP. Beyond this, the asphaltene content below the lower AOP remains constant. A significant amount of live crude oil may be required, which is one of the main drawbacks of this method.

5.4.2.2 Light Scattering Technique Using Near-Infrared

This method measures the transmittance of near-infrared (NIR) light through reservoir fluid with a change in pressure, temperature, or composition. NIR light with an 800–2,200-nm wavelength is typically used due to the black color of the crude oil. Figure 5.10a shows a schematic of the PVT cell with a visual window containing the NIR setup. In this method, a specified amount of the reservoir fluid is injected into the PVT cell at or above reservoir pressure while maintaining the reservoir temperature of the PVT cell. The fluid is mixed at high rpm for 30 min. Subsequently, a

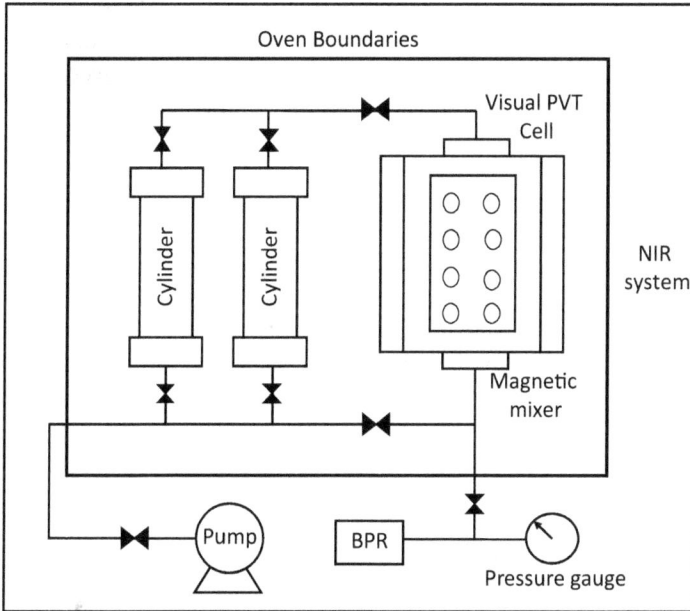

(a)

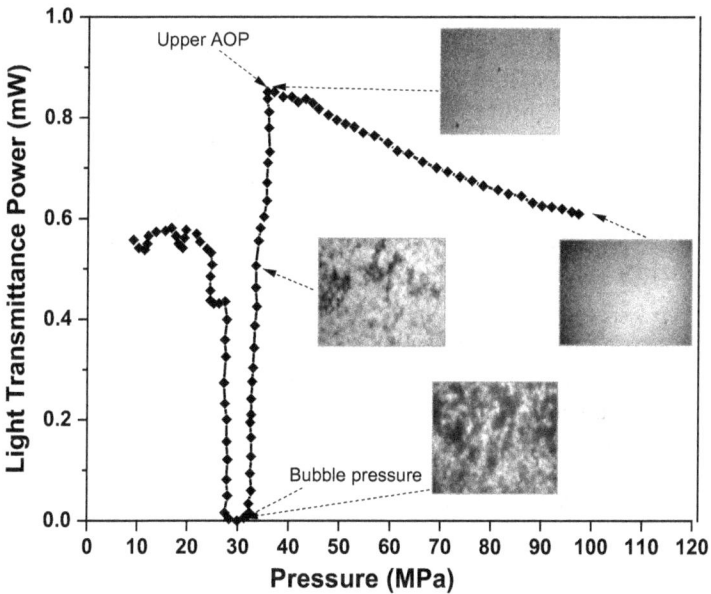

(b)

FIGURE 5.10   (a) Schematic of the PVT cell with NIR setup; (b) a plot of light transmittance with pressure.

reference baseline is obtained by scanning the fluid using light transmittance. After that, isothermal depressurization is performed, and the light transmittance intensity is measured. Pressure is reduced in steps of 30 bar. A graph of average transmitted light power with pressure is constructed, as shown in Figure 5.10b. For a homogeneous fluid with dissolved asphaltene ($P >$ upper AOP), light travels through the fluid with minimum scattering. As the pressure reduces below the upper AOP, asphaltene crystals separate from the dissolved fluid and cause partial light scattering. A gradual decrease in light transmittance is observed till the bubble point at which total scattering is observed due to asphaltene particles as the pressure nears the lower AOP. The power of the light transmittance increases due to the redissolution of asphaltene in the crude oil. At each pressure step, the system is maintained for equilibrium to be reached. One of the advantages of this method is that very little amount of live crude oil sample is required, and the test is relatively quick.[37]

5.4.2.3   Acoustic Resonance Technique

This method is mainly used to determine the upper AOP based on the principle that the speed of the sonic wave is affected due to the change in the fluid density. Thus, the phase transition can be identified by passing acoustic waves through the bulk fluid. The experimental setup, as shown in Figure 5.11a, consists of a high-pressure, high-temperature (HPHT) cylindrical resonator of 0.25-in. diameter with a wave emitter and receiver at two ends. The sample is depressurized in the cylinder, and acoustic simulations are applied using a piezoelectric element. The response is passed through a low-noise amplifier and then an analogue-to-digital converter. This method cannot detect the lower AOP because the gradual decrease in pressure does not affect the resonance characteristic below the saturation pressure. Two minima are observed in the plot: one corresponds to the upper AOP at a higher pressure, and the other is the saturation pressure (see Figure 5.11b). There is a sharp decrease in acoustic response from about 55 MPa to 43 MPa. The upper AOP is the point at which the acoustic response starts to increase. The acoustic response further increases, and it starts to decrease near the saturation pressure (bubble point).[37]

5.4.2.4   Viscosity Measurement Technique

Generally, asphaltene in the crude oil is stabilized by resins and remains colloidal. Any external disturbance (change in pressure, composition, addition of chemicals, etc.) may affect the resin layer on the asphaltene, separating it from the asphaltene. Asphaltene then tends to agglomerate in the crude oil and thus increase the viscosity of the crude oil. Thus, in this method, the crude oil viscosity is measured by adding external chemicals (precipitants), which can enhance the agglomeration of asphaltene. The onset of asphaltene precipitation can be inferred from the change in viscosity of the crude oil. This method is used for both light and heavy crude oil. The amount of precipitant required is very less. The viscosity of crude oil changes rapidly as soon as the precipitation of asphaltene in the crude oil starts. The asphaltene onset is visible when the viscosity of the crude oil change with a change in the weight fraction of solvent added to the crude oil as shown in Figure 5.12.[40, 41]

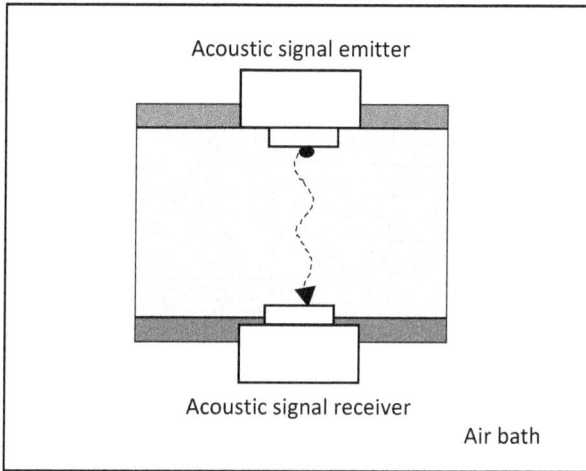

(a)

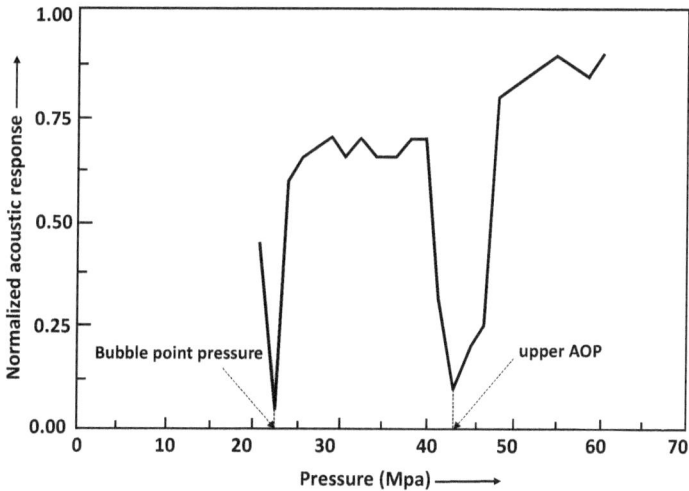

(b)

FIGURE 5.11 (a) Schematic of the acoustic resonance technique (ART) setup; (b) measurement of upper AOP using ART setup.[40]

5.4.3 FLUID STUDIES FOR GAS HYDRATE

Fluid studies on gas hydrate primary include determination of phase equilibrium conditions, kinetics, rheological, and morphological studies. Lab-scale hydrate studies for flow assurance typically include the following experiments:

a) Thermodynamic or phase stability study to determine the phase equilibrium of hydrates in the presence of various additives (promoters or inhibitors).

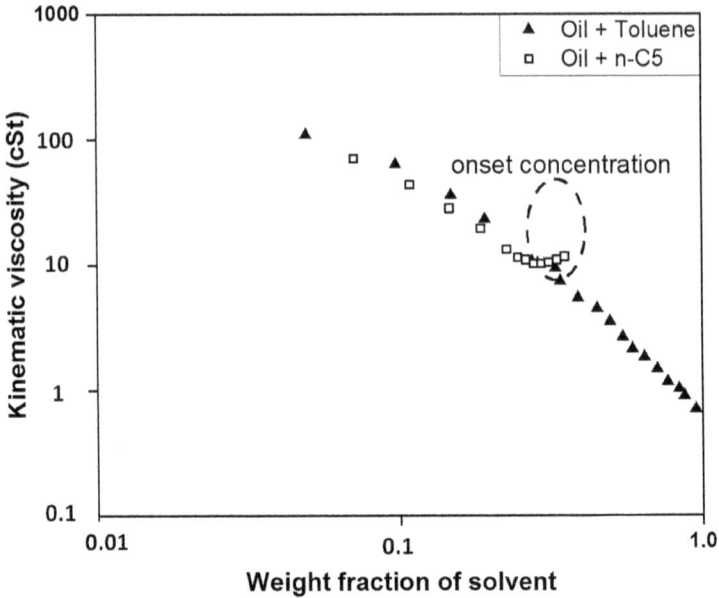

FIGURE 5.12 Change in the kinematic viscosity of crude oil with weight fraction of solvent. Onset concentration indicates the concentration of solvent (n-C$_5$ in this case) required for asphaltene precipitation. Data are taken from Escobedo & Mansoori.[41]

  b) Kinetic studies to determine the rate of hydrate formation and dissociation and related parameters such as gas-to-hydrate and water-to-hydrate conversion.
  c) Rheological studies to determine the impact of various inhibitors on the flow behavior of hydrate slurries in different systems, such as in bulk, emulsions, and dispersions.
  d) Flow loop studies to investigate the impact of hydrate formation on the multiphase flow characteristics at simulated field-scale conditions.
  e) Morphology studies of gas hydrate formation and dissociation.

Some of the methods to study gas hydrate systems have been discussed subsequently.

5.4.3.1 Phase Stability Study

A stirred reactor setup as shown schematically in Figure 5.13 is typically used for studies on the phase stability and kinetics of hydrate formation and dissociation relevant for flow assurance studies. The experimental setup contains a high-pressure reactor that can sustain typical transport conditions (high pressure and low temperature) suitable for offshore flowlines (1–20 MPa). The temperature is controlled externally by circulating a glycol–water mixture through the jacket of the reactor vessel. The pressure and temperature data are logged continuously for the entire experimental duration and are acquired using an online data acquisition system connected to the

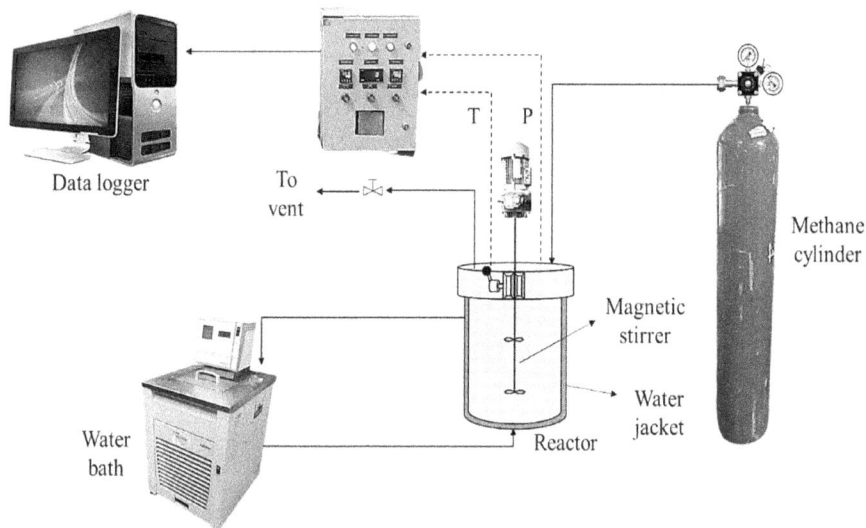

FIGURE 5.13 Schematic of the experimental setup for gas hydrate studies (thermodynamic and kinetics).

computer for storage and analysis. The typical procedure followed for phase stability studies is described in the following.

The high-pressure reactor is initially cleaned with distilled/deionized water. The required solution (water or water containing inhibitors) is injected into the reactor, which is then purged with the hydrate former gas to remove any atmospheric air. After that, the hydrate former gas is injected at desired experimental pressure, and the temperature is lowered enough to form hydrate. In this condition, the volume of gas in the system remains unchanged. Therefore, this method is also known as the isochoric pressure search method. The major steps in the isochoric pressure search method are shown schematically in Figure 5.14. After point *a*, as the temperature decreases, the pressure starts reducing due to gas contraction. At point *b*, the temperature of the system suddenly rises due to hydrate formation as it is an exothermic process. As the hydrate formation continues, the pressure in the reactor decreases from *b* to *c*. At *c*, when sufficient hydrate is formed (minimum pressure decrease of about 1 MPa from the nucleation point *b* is to be ensured), the temperature of the system is gradually increased by 1 K/h up to point *d* followed by a slow ramp heating (0.1 K/h) up to point *e*. The phase equilibrium point is determined by the intersection of the two-phase (L_w–V) contraction line (*a–b*) of the fluid mixture and the three-phase (H–L_w–V) dissociation line (*d–e*). Several such experiments are performed to generate the loci of equilibrium points, as shown in Figure 5.14.

To determine the efficacy of an additive (such as promoter or inhibitor), the additive is premixed with water and used for hydrate formation.[43, 44] A similar procedure as described earlier is followed to get an equilibrium PT curve. For example, if the equilibrium curve lies to the left of the pure methane hydrate curve (i.e., higher pressure and lower temperature), the additive acts as the thermodynamic inhibitor.

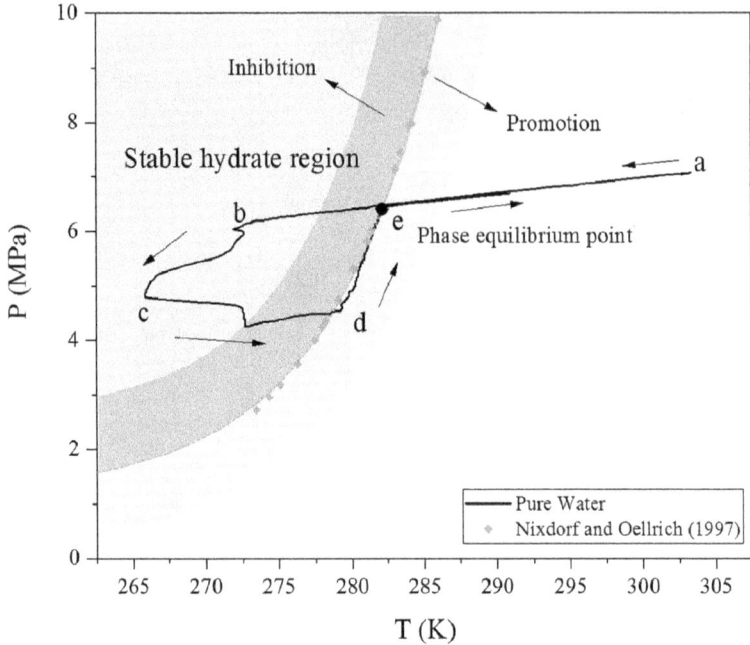

FIGURE 5.14 Isochoric P–T trace for pure methane hydrate system (black line). Methane hydrate equilibrium data are from Nixdorf and Oellrich[42] (blue diamonds).

Similarly, when the equilibrium curve shifts to the right of the pure methane hydrate curve (high temperature and low pressure), it is known as thermodynamic promotion. An additive may act as a thermodynamic promoter or an inhibitor due to the guest and host interactions, change in the activity of water in its presence in the solution, and formation of unusual water cages by the guest molecules.[45, 46]

5.4.3.2 Isothermal Kinetic Study

The experimental setup (Figure 5.13) can be used to investigate the kinetics of hydrate formation and dissociation. After cleaning the reactor with distilled water, the prepared solution is injected into the reactor and closed airtight. The reactor is then purged with the hydrate former gas to remove any atmospheric air. The stirred is switched on, and reactor temperature is set at the desired value. Hydrate former gas is injected at the desired pressure after the temperature becomes stable at the desired set point. The magnetic stirrer is kept running at a constant rpm (revolutions per minute) throughout the experiment to ensure low mass and heat transfer resistances during hydrate formation. The typical pressure and temperature profile within the reactor during the formation of the methane hydrate experiment is shown in Figure 5.15. From Figure 5.15, it is evident that during the initial phase, the gas dissolves in the liquid phase (pure water or water containing additives such as inhibitors) depending on the solubility. After some time, as the hydrate formation begins, the temperature rises, and the reactor pressure decreases suddenly. The time at which a sudden decrease in pressure is observed is called as a hydrate induction time. After some time, the pressure reaches a steady state. After a sufficient time, the formed

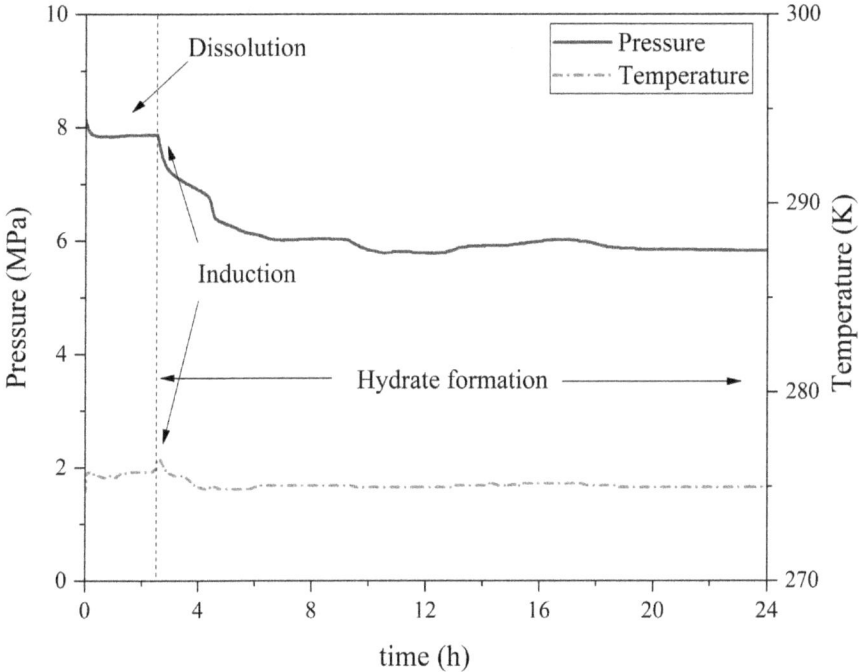

FIGURE 5.15 Typical pressure and temperature profile during methane hydrate formation experiment in pure water at 8 MPa and 275.15 K.

hydrate is dissociated at the desired rate (K/h). The gas is then vented safely. During hydrate formation and dissociation, the reactor pressure and temperature conditions are acquired continuously and used to determine various hydrate formation and dissociation parameters, as discussed in the following.[47, 48]

Moles of gas consumed in hydrate formation:
The moles of gas consumption during hydrate formation can be calculated using

$$\left(\Delta n_{g,H}\right)_{\downarrow} = V_g \left(\frac{P_i}{Z_i RT} - \frac{P_t}{Z_t RT}\right) \tag{5.3}$$

where $(\Delta n_{g,H})_{\downarrow}$ is the moles of gas consumed in hydrate at a particular time t; V_g is the volume of the gas in the reactor; T is the temperature during the hydrate formation; R is the ideal gas constant; P and Z are the pressure and compressibility factor, respectively; and i and t are the subscripts, representing the initial and any time, respectively. The compressibility factor in Equation 5.3 is calculated from Pitzer's correlation[49] given by a set of equations (Equation 5.4).

$$Z = Z^0 + \omega Z^1; \ Z^0 = 1 + B^0\left(\frac{P_r}{T_r}\right); Z^1 = B^1\left(\frac{P_r}{T_r}\right) \tag{5.4}$$

Here, ω is the acentric factor, Z^0 is the compressibility factor of fluid for nearly spherical molecules and Z^1 is the correction for nonspherical intermolecular forces. T_r is the reduced temperature.

$$B^0 = 0.083 - \left(\frac{0.422}{T_r^{1.6}}\right); B^1 = 0.139 - \left(\frac{0.172}{T_r^{4.2}}\right)$$

Typically, $(\Delta n_{g,H})_\downarrow$ is normalized by moles of water (n_w) as in Equation 5.5:

$$\left(N_t\right)_\downarrow = \frac{\left(\Delta n_{g,H}\right)_\downarrow}{n_w} \tag{5.5}$$

Water and gas-to-hydrate conversion:
The percentage conversion of gas to hydrate (G to H) and water to hydrate (W to H) are calculated according to

$$\%\mathrm{G\,to\,H} = \frac{\left(\Delta n_{g,H}\right)_\downarrow}{n_g} \times 100 \tag{5.6}$$

$$\%\mathrm{W\,to\,H} = \frac{\left(\Delta n_{g,H}\right)_\downarrow}{n_w} \times h \times 100 \tag{5.7}$$

Here, n_g is the moles of gas at the start of the experiment and h is the hydration number.

Rate of hydrate formation:
A discrete forward difference method is used to calculate the average rate of hydrate formation:

$$\frac{d\left(N_t\right)_\downarrow}{dt} = \frac{d\left(\Delta n_{g,H}\right)_\downarrow}{dt} = \frac{\left(\Delta n_{g,H}\right)_{\downarrow,\,\Delta t+t} - \left(\Delta n_{g,H}\right)_{\downarrow,t}}{\Delta t} \tag{5.8}$$

Here Δt is the time difference between the two observations.

Moles of gas released during hydrate dissociation:
The gas released during hydrate dissociation is calculated using Equation 5.9 and normalized using Equation 5.10.

$$\left(\Delta n_{g,H}\right)_\uparrow = V_g\left(\frac{P_t}{Z_tRT} - \frac{P_0}{Z_0RT}\right) \tag{5.9}$$

$$\left(N_t\right)_\uparrow = \frac{\left(\Delta n_{g,H}\right)_\uparrow}{n_w} \tag{5.10}$$

Here, subscripts 0 and *t* represent the initial (at the onset of dissociation) and any time *t* during hydrate dissociation, respectively.

Rate of hydrate dissociation:
The average rate of hydrate dissociation is calculated using the following discrete forward difference method.

$$\frac{d(N_t)_\uparrow}{dt} = \frac{d(\Delta n_{g,H})_\uparrow}{dt} = \frac{(\Delta n_{g,H})_{\uparrow,\Delta t+t} - (\Delta n_{g,H})_{\uparrow,t}}{\Delta t} \tag{5.11}$$

With the help of the preceding equations, one can determine whether an additive is a kinetic promoter or inhibitor of the hydrate in a particular system.[50, 51] It has to be kept in mind that the hydrate formation and dissociation kinetics are different for every reactor setup, so they cannot be compared directly.

5.4.3.3   Rheological Studies

Rheological studies involve investigating the viscosity, viscoelasticity, yield stress, and other flow properties of the hydrate slurries in various systems such as dispersions and emulsions. Viscosity and yield stress of hydrate slurry are critical parameters, especially during the start-up and shutdown of the system. Some of the important rheological properties of hydrate slurry include constant shear viscosity, flow curve for fluid models, storage, and loss modulus to determine solid and liquid behavior of hydrate slurry. These results are essential to determine possible locations of hydrate plugging in a flowline, given the flow conditions (e.g., pressure, temperature, and flow rate), and aid in the mitigation of the hydrate plugging. In these rheological studies, the first step is to synthesize the water/oil emulsions (with model/crude oil) and establish the emulsion stability for at least the experimental duration. Afterward, a droplet size distribution is typically determined through an optical microscope for micron-sized droplets or DLS (dynamic light scattering) equipment for the nano-meter range. A schematic of a typical setup used in rheological studies involving a high-pressure rheometer is given in Figure 5.16.

The high-pressure rheology of hydrate slurries can be studied in the rheometer, providing valuable insights for hydrate mitigation. Pure water or emulsion/dispersion of oil and gas at desired pressure are injected into the rheometer measuring cell, and then a constant shear rate measurement is employed. The temperature is set at the desired hydrate formation condition. The results of the viscosity of sample hydrate systems are shown in Figure 5.17. The pressure decreases during the cooling step, and the viscosity increases in response to the temperature change as shown in Figure 5.17a. The initial pressure and viscosity shifts are only due to temperature changes and not necessarily an indication of hydrate formation. Hydrate formation starts after cooling once the system conditions are thermodynamically favorable and is indicated by a significant viscosity increase and a sudden pressure decrease (Figure 5.17a). The viscosity profile of hydrate slurries depends on the temperature, pressure, and water volume fraction in the emulsion/dispersion system. For example, various systems in the literature have shown a rapid decrease in viscosity immediately after the initial

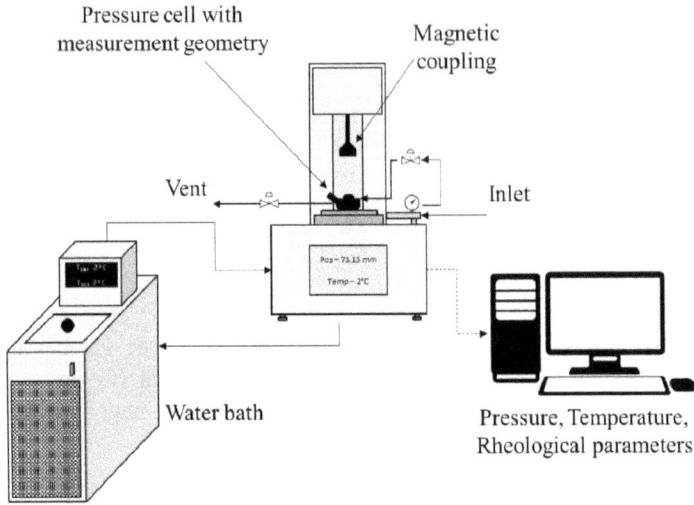

FIGURE 5.16   A schematic of a typical setup used in rheological studies including a high-pressure rheometer.

viscosity peak, indicating an alignment of newly formed aggregates under flow.[52–54] After this decrease, some experiments show decreasing viscosity, while others show increasing viscosity. After the hydrate formation, flow curve tests are carried out when viscosity is stable for enough time duration (Figure 5.17b). The hydrate slurry is heated back to a fixed temperature during the dissociation by giving a ramp profile at a constant shear rate (Figure 5.17c). As the slurry warms up, the viscosity decreases a little initially. Still, after some time, once the temperature reaches close to the hydrate dissociation temperature, a sudden spike is observed due to stronger cohesion between the hydrate particles owing to the thickening water layer on the surface of hydrate particles.[55, 56]

5.4.3.4   Flow Loop Experiments

The flow loop experiments are designed to study the hydrate-related phenomena at the simulated field-scale conditions. Often, non-hydrocarbon gases such as CO_2 and H_2S accelerate the natural gas hydrate formation during flow. Some of the parameters that are measured across a flow loop section during a multiphase flow include

- Pressure drop (ΔP) and viscosity.
- Flow rate.
- Hydrate volume fraction.
- Temperature.
- Visual observation.

These observations and results can be correlated with kinetics and thermodynamic studies to better understand the effect of various additives on hydrate formation and hydrate plugging. For example, a study has been done to investigate the kinetics of CO_2 hydrate in a flow loop.[57] The schematic of the high-pressure flow loop used in

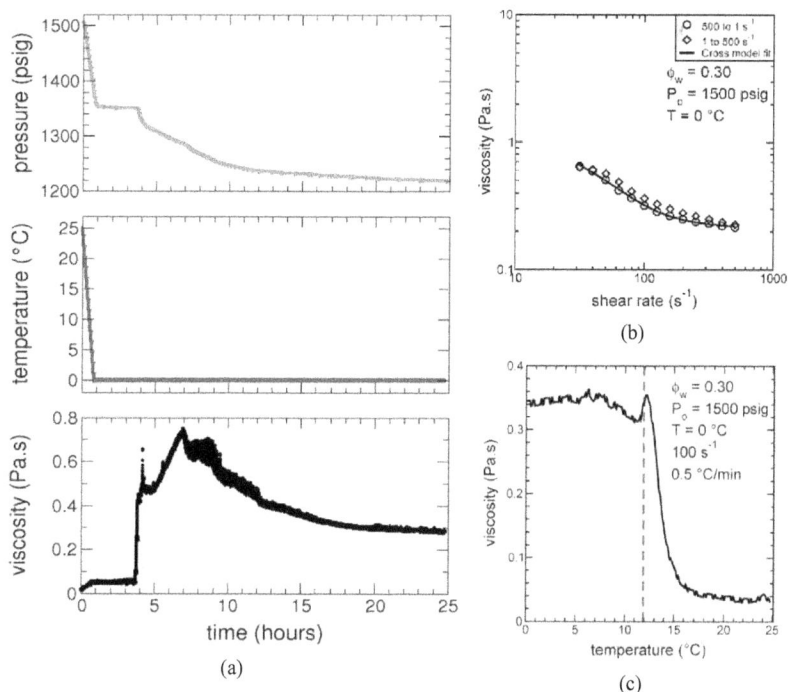

FIGURE 5.17 Graphs pertaining to water–in–mineral oil emulsion with 0.30 water volume fraction: (a) Pressure, temperature, and viscosity evolution of hydrate slurry in the methane-saturated emulsion with time at $\dot{\gamma} = 100$ s^{-1}; (b) flow curve of the hydrate slurry at 1500 psig and 0°C, (c) dissociation profile of methane hydrate slurry formed at 100 s^{-1} and 0.5°C/min. (a), (b), (c) Reproduced from Webb et al.[52] with permission from American Chemical Society (Copyright American Chemical Society, 2014).

this study is given in Figure 5.18a. The 7.5-Ltr flow loop is capable of high-pressure experiments (15 MPa). The operating temperature range is −15°C to 20°C. Pressure drop is found to increase after the hydrate formation. This is due to the increased viscosity of the fluid system during hydrate formation. Further details associated with the flow loop design can be accessed from Zhou et al.[57] Typical parameters measured during the flow loop experiments such as pressure, temperature, and pressure drop are given in Figures 5.18b and c.

5.4.3.5 Morphology Studies

Morphological studies involve observing the characteristic hydrate appearance, such as size, shape of hydrate crystals, and its growth, thus giving valuable insights into the hydrate formation process. This is of importance for optimizing the processes involving fast kinetics.[58] Morphological studies can be divided into hydrate crystal morphology and macroscopic hydrate formation. Macroscopic hydrate formation is useful for studying kinetics and hydrate formation characteristics essential for process optimization. In addition, the crystal morphology also has implications for hydrate-based gas transportation technology in which larger crystals with less

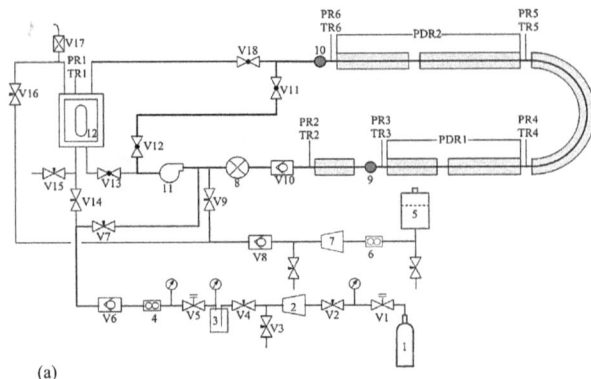

(a)

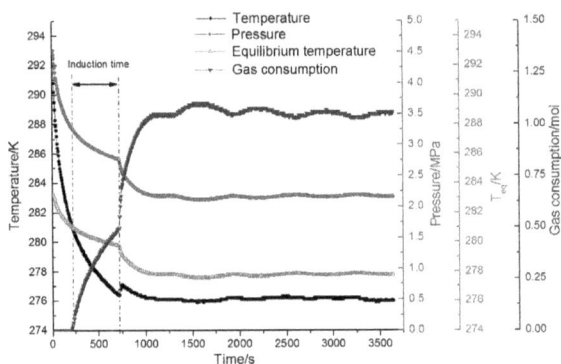

(b)

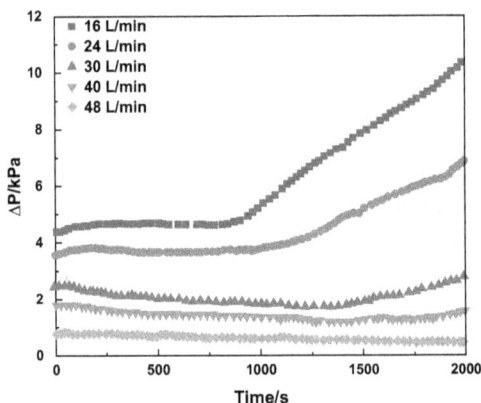

(c)

FIGURE 5.18   (a) High-pressure flow loop schematic (1: gas cylinder; 2: gas booster pump; 3: buffer tank; 4: gas mass flow meter; 5: liquid storage tank; 6: turbine flowmeter; 7: liquid booster pump; 8: liquid mass flow meter; 9, 10 – viewport; 11: circulating pump; 12: tank reactor; Symbols stand for V, valves; PR, pressure transducer; TR, temperature sensor; PDR, differential pressure transducer. (b) Equilibrium pressure, temperature, and gas consumption at operating condition (4.5 MPa, 275.15 K, 30 L/min, liquid loading: 80 % (v/v)).(c) Pressure drop with time for different flow rates. (a), (b), (c) Reproduced from Zhou et al.[57] with permission from Elsevier (Copyright Elsevier, 2018).

surface area and lower viscosity are preferred. Furthermore, the mechanical properties and permeability of hydrate-bearing sediments depend on the hydrate crystal morphology in the pores. Both macroscopic and hydrate crystal morphology are briefly discussed in the following sections.

5.4.3.5.1 Macroscopic Morphological Study

Hydrate growth during formation and the gas–liquid interface during dissociation is typically examined in a visual cell that is strong enough to withstand the pressure in the hydrate stable region (few MPa), as depicted in Figure 5.19. The morphology of hydrate growth at the interface (inside the acrylic column) has been acquired with time using a microscope. Also, the setup contains process measurements to monitor the kinetics accurately.[59] The direction of hydrate growth can be examined, which has implications for flow assurance and other industrial applications.[60] This vertical growth model of hydrate has been depicted in a schematic in Figure 5.20. Thus, various chemical inhibitors/additives can be used to study and optimize thermodynamic and kinetic performance using this method.

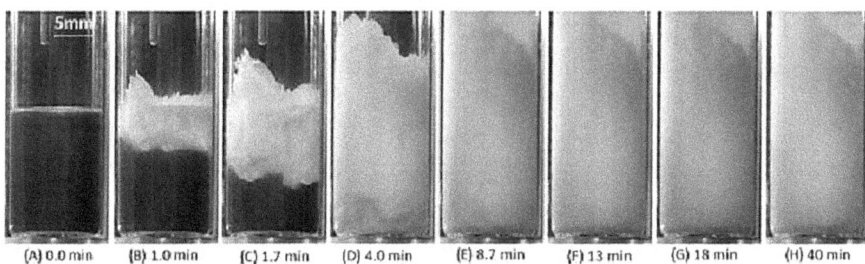

(A) 0.0 min   (B) 1.0 min   (C) 1.7 min   (D) 4.0 min   (E) 8.7 min   (F) 13 min   (G) 18 min   (H) 40 min

FIGURE 5.19   Methane hydrate formation in 2 ml of 5.6 mol% THF solution in a nonstirred visual cell at 7.2 MPa and 283.2 K. Reproduced from Veluswamy et al.[59] with permission from Elsevier (Copyright Elsevier, 2016).

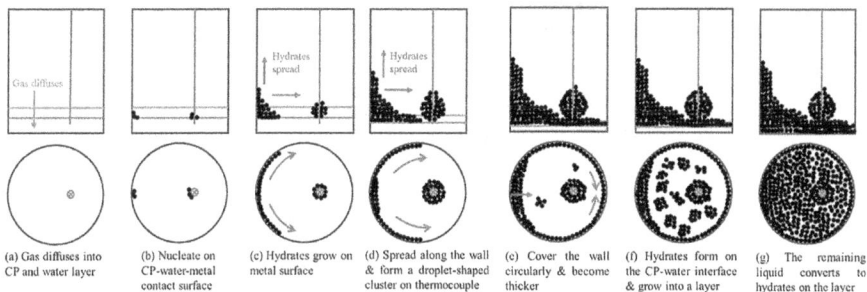

(a) Gas diffuses into CP and water layer   (b) Nucleate on CP-water-metal contact surface   (c) Hydrates grow on metal surface   (d) Spread along the wall & form a droplet-shaped cluster on thermocouple   (e) Cover the wall circularly & become thicker   (f) Hydrates form on the CP-water interface & grow into a layer   (g) The remaining liquid converts to hydrates on the layer

FIGURE 5.20   Schematic of the $CO_2/H_2/CP$ hydrate formation behavior in a nonstirred reactor (front and top view). Reproduced from Zheng et al.[60] with permission from American Chemical Society (Copyright American Chemical Society, 2018).

5.4.3.5.2   Hydrate Crystal Morphology

Crystal morphology encompasses the shape and size of hydrate crystals, affected by various parameters, including hydrate formers, additives, and subcooling. Typical guest molecules for hydrate-based technologies include CH_4, C_2H_6, C_3H_8, and CO_2. Understanding the growth of hydrate crystals is of paramount importance in industrial processes. The effect of subcooling on the hydrate formation and growth on the water droplet surface is given in Figure 5.21. Hydrate morphological changes can be studied in the presence of surfactants and inhibitors at varying concentrations, which influence the capillary action and wettability and thus, hydrate crystal shape and growth (see Figure 5.22).[58, 62] This information can then be used to optimize the surfactant/inhibitor type and its concentration for efficient

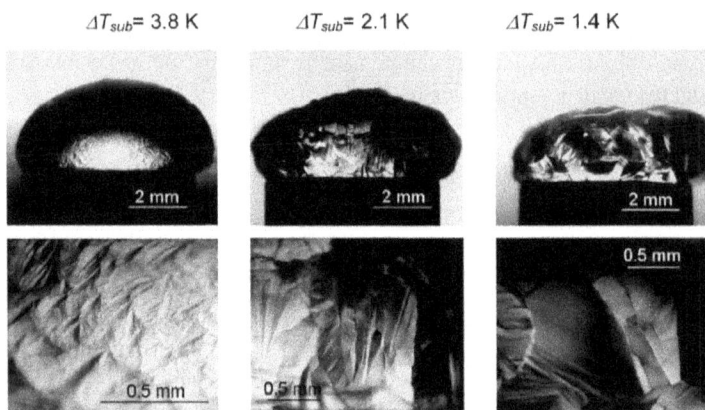

FIGURE 5.21 Effect of subcooling on the methane hydrate formation and growth on the water droplet surface at 8.15 MPa. Reproduced from Tanaka et al.[61] with permission from American Chemical Society (Copyright American Chemical Society, 2009).

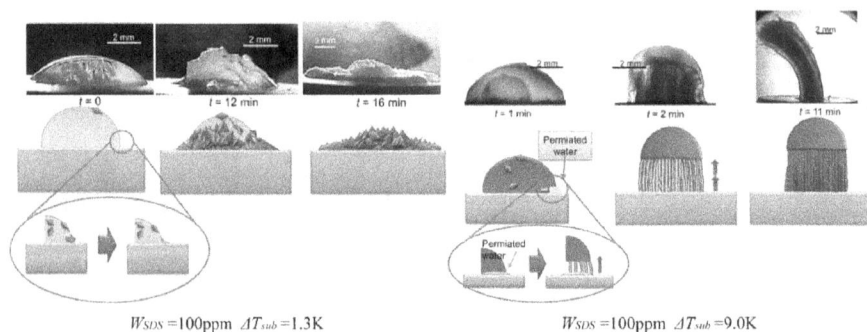

FIGURE 5.22 Methane hydrate crystal growth on the interface for sodium dodecyl sulfate (SDS) concentration of 100 ppm and subcooling of 1.3 K and 9 K. Reproduced from Hayama et al.[62] with permission from American Chemical Society (Copyright American Chemical Society, 2016).

hydrate management. Morphological studies provide a foundation to understand hydrate formation on a lab scale and have implications for industrial processes. A schematic in Figure 5.23 shows factors affecting the hydrate morphology and their significance on various processes.[58]

Although advancements have been made in morphological studies, there are certain limitations. These include unclear mechanisms responsible for morphological differences in various systems, complex interfacial phenomena (multicomponent system, heat and mass transfer), and phase transition kinetics. Some of the focus areas of the morphological studies include the following:

- Comprehensive promoter evaluation for hydrate-based applications.
- The basic control mechanism of hydrate formation, quantitative measurement, physical and mathematical modeling.
- Comprehensive assessment of hydrate formation in porous media and simulation of actual hydrate reservoir conditions.
- Combining morphological observations with analytical techniques such as X-ray diffraction, computed tomography (CT) scan, nuclear magnetic resonance (NMR) spectroscopy.

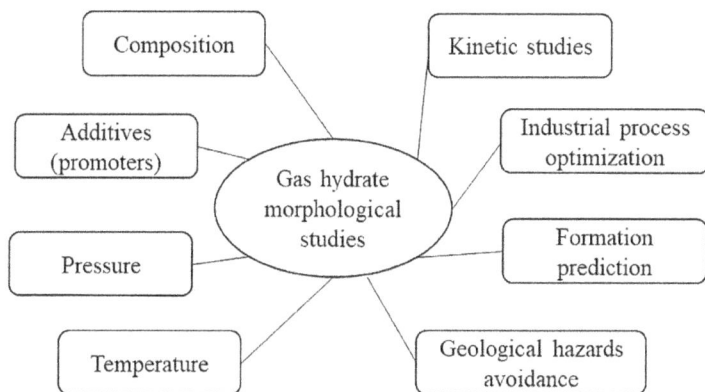

FIGURE 5.23 Factors affecting hydrate morphology and its impact on various industrial processes.

5.5 EQUATIONS OF STATE AND OTHER MODELING METHODS

The primary purpose of this section is to provide an overview of the equations-of-state (EOSs)-based or other modeling methods that are not purely empirical for solids, such as hydrates, asphaltenes, and waxes. The reader is referred to comprehensive coverage of this topic by Pedersen et al.[3]

5.5.1 HYDRATE MODELING

Unlike asphaltenes and waxes, EOS modeling of gas hydrates has significantly advanced. Hydrate equilibrium or PT conditions can be fairly accurately predicted since the chemistry is well known (typically based on the gas composition). One of the best examples of an easy-to-use method is the one by Sloan,[1] which is a combination of hydrate phase equilibria based on van der Waals and Platteeuw model and hydrocarbon phases modeled by the Soave–Redlich–Kwong (SRK) EOS with parameters regressed from experimental measurements. The method can handle non-hydrocarbon components in the gas mixture and inhibitors, such as salt, methanol, and their mixtures. The calculation results include equilibrium hydrate structure (sI or sII), equilibrium phases (e.g., liquid water–hydrate–vapor), and equilibrium pressure. Figure 5.24 compares the hydrate formation prediction from Sloan's[1] method and the empirical correlation of Towler and Mokhatab[9] with experimental data for a nine-component natural gas mixture[63] that includes 0.5mole% CO_2. As seen in Figure 5.24, overall, Sloan's method matches the experimental data quite well; the predictions without methanol are almost in line with the measured data. On the other hand, the empirical correlation is not designed for handling inhibitors. Thus, the comparison is shown only for without methanol. Although the estimated values in the lower pressure region match with the experimental data reasonably well, fairly significant deviations at higher pressures are readily apparent.

5.5.2 PARAFFIN OR WAX MODELING

WAT modeling by an EOS is basically treated similar to a bubble point calculation, as described by Ahmed[64] on a stock tank or dead oil or as suggested by Pedersen et al.[3] to perform a normal vapor–liquid–equilibria (VLE) calculation followed by a solid–liquid–equilibria (SLE) calculation on the equilibrium liquid phase. Using the

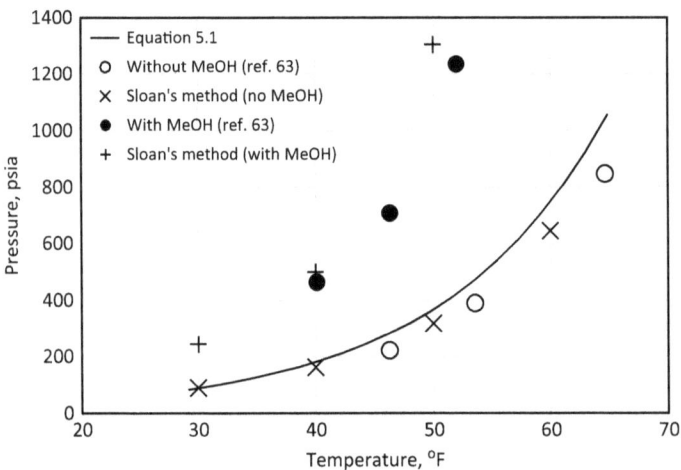

FIGURE 5.24 Comparison of experimental hydrate equilibrium conditions, for a nine-component natural gas mixture,[63] with Sloan's[1] method and the empirical correlation.[9]

basis as "n" moles of feed splitting into n_L and n_S moles of liquid and solid, respectively, $n = n_L + n_S$ or in terms of individual components $Z_i n = X_i n_L + S_i n_S$, where the left-hand side is total moles of component "i" in the feed that splits into moles of that component in the liquid and solid phases, respectively. Since all mole fractions sum up to 1; $\sum_{i=1}^{n} Z_i = 1; \sum_{i=1}^{n} X_i = 1; \sum_{i=1}^{n} S_i = 1$. If calculations are carried out on the basis of 1 mole of feed, then $1 = n_L + n_S$ or in terms of individual components $Z_i = X_i n_L + S_i n_S$. Defining the equilibrium ratio, $K_i^{SL} = S_i/X_i$ or $S_i = K_i^{SL} X_i$, $Z_i = X_i n_L + K_i^{SL} X_i n_S$, which can also be expressed as

$$X_i = \frac{Z_i}{n_L + K_i^{SL} n_S} \quad or \, S_i = \frac{K_i^{SL} Z_i}{n_L + K_i^{SL} n_S} \tag{5.12}$$

When wax forms, $n_S \cong 0$; that is, $n_L = 1$; such that $S_i = K_i^{SL} X_i$ (because feed is liquid). The summation should equal 1 at the correct/converged values of K_i^{SL};

$\sum_{i=1}^{n} K_i^{SL} X_i = 1$. K_i^{SL} can also be defined as the ratio of fugacity coefficients of individual components ($i = 1$ to n) in the liquid and solid phases, respectively; $K_i^{SL} = \frac{\Phi_i^L}{\Phi_i^S}$.

Although, in principle, iterative Rachford–Rice-type flash calculations can be carried out and updated K_i^{SL} values are determined from the fugacity coefficient ratio, the calculation is initiated by obtaining the initial Φ_i^L and Φ_i^S values from liquid and wax phase models.[3]

The fluid composition heavy end needs to be defined as wax formers and nonformers under the assumption that only C_{7+} components can form wax, and only some part of each component splits into a wax former and nonformer.[3] Pedersen et al.[3] suggested the following functional form:

$$Z_i^S = Z_i^{total} \left[1 - \left(1.074 + 0.0006584 M_i \right) \left(\frac{\rho_i - \rho_i^P}{\rho_i^P} \right)^{0.1915} \right]; Z_i^{total} = Z_i^S + Z_i^{non} \tag{5.13}$$

where Z_i^{total} is the total mole fraction of a given component, which is split into solid forming and nonforming parts, respectively, and ρ_i^P is the density of normal paraffin of the same molecular weight as carbon number fraction i (M_i in Equations 5.13 and 5.14), in g/cm³, calculated from the following empirical correlation:

$$\rho_i^P = 0.3915 + 0.0675 \ln \left(M_i \right) \tag{5.14}$$

If Z_i^S results in a negative value, the wax content is assumed to be equal to zero.[3] A numerical example shows the practical application of the preceding splitting equations.

Molar composition, molecular weight, and specific gravity of various components for a separator liquid drawn from a test separator at 600 psi and 110°F are provided in Table 5.3. A subsample of this separator liquid was transferred to a PVT recombination cell at 600 psi and ambient temperature after the sample had been sitting in the lab for a long duration. This oil is known to be waxy in nature, and it is suspected that in the transfer process, wax may have precipitated and that all of it remained in the separator liquid sample bottle while 100% of the liquid portion is transferred to the PVT cell. Equations 5.13 and 5.14 can be used to estimate (a) the composition of the wax (normalized Z_i^S) remaining in the separator liquid sample bottle following the transfer process and (b) the composition of the liquid (normalized Z_i^{non}) transferred to the PVT cell. Detailed calculations are shown in Table 5.3.

5.5.3 ASPHALTENE MODELING

Pedersen et al.[3] have shown the successful application of EOS models (Peng–Robinson (PR) as well as SRK) for predicting the AOP as a function of temperature, in conjunction with the normal VLE calculations. They, however, recommend

TABLE 5.3
Example Application of the Wax Splitting Method of Pedersen et al.[3]

Component	Z_i^{total}, mole%	M_i, lb/ lb-mole	ρ_i, g/cm³	ρ_i^P, g/cm³	Z_i^S, mole%	Z_i^{non}, mole%	Norm. Z_i^S, mole%	Norm. Z_i^{non}, mole%
N2	0.30						0.00	0.35
H2S	0.01						0.00	0.02
CO2	0.05						0.00	0.06
C1	0.69						0.00	0.81
C2	0.83						0.00	0.97
C3	2.58						0.00	3.01
i-C4	1.31						0.00	1.53
n-C4	4.07						0.00	4.76
i-C5	3.35						0.00	3.92
n-C5	2.07						0.00	2.42
C6	4.85						0.00	5.66
C7	5.53	104	0.720	0.705	2.50	3.03	17.34	3.54
C8	6.08	121	0.747	0.715	2.21	3.88	15.30	4.53
C9	4.85	129	0.768	0.719	1.49	3.36	10.35	3.92
C10	5.32	134	0.795	0.722	1.34	3.98	9.28	4.66
C11	4.76	145	0.802	0.727	1.16	3.60	8.03	4.21
C12	4.28	157	0.815	0.733	0.97	3.31	6.71	3.87
C13	4.16	172	0.826	0.739	0.88	3.28	6.12	3.83
C14	3.85	179	0.838	0.742	0.75	3.10	5.19	3.62
C15	3.68	196	0.845	0.748	0.69	3.00	4.76	3.50
C16	3.55	204	0.848	0.750	0.65	2.90	4.49	3.39
C17	2.96	214	0.846	0.754	0.56	2.40	3.85	2.81
C18	2.29	223	0.850	0.757	0.42	1.87	2.89	2.19
C19	2.38	254	0.865	0.765	0.38	2.00	2.66	2.34
C20+	26.19	399	0.956	0.796	0.44	25.76	3.03	30.10
	100.00				14.41	65.47	100.00	100.00

Perturbed Chain–Statistical Associating Fluid Theory (PC-SAFT) as a good candidate for modeling asphaltene precipitation. EOS models are typically used to determine AOPs and total asphaltene percentage as a function of pressure, temperature, and injection gas. Other suggested models that appear in the literature, such as (1) polymer solution, (2) micellization, and (3) colloidal theory-based, have been discussed by Pedersen et al.[3] However, EOS models certainly offer the convenience of linking them to a reservoir simulation model since they are an inherent component, which can be used for assessing AOPs and evaluating miscible gas injection–based enhanced oil recovery (EOR) processes in conjunction with the potential for asphaltene problems. An example of asphaltene modeling using an EOS model is shown in Figure 5.25.

Like waxes, the fluid composition heavy end needs to be defined first in terms of asphaltene former and nonformer. Kohse and Nghiem[66] suggest splitting a heavy end, such as the C_{31+}[64] fraction into C_{31+A} (nonprecipitating) and C_{31+B} (known as the precipitating [asphaltene] fraction). The proposed splitting calculation is dependent on the experimental asphaltene precipitation data, the molecular weight of the oil, and the heavy fraction and is defined as follows:[64]

$$C_{31+B} = \frac{W_{C31+B} MW_{oil}}{MWC_{31+B}} \tag{5.15}$$

where W_{C31+B} is the total mass fraction of the precipitate; MW_{oil} and MW_{C31+B} are the molecular weights of the oil and the plus fraction, respectively; and the total C_{31+} mole fraction $= C_{31+A} + C_{31+B}$. Note that the molecular weight of C_{31+A} and C_{31+B} is considered the same $(=C_{31+})$.[64] The balance of the splitting equation that reduces to the definition of mole fraction follows (note that in the balance equation,

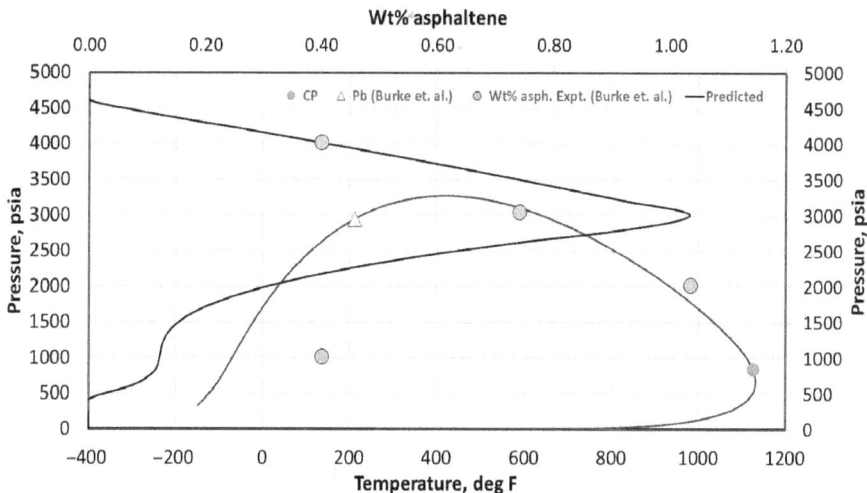

FIGURE 5.25   Example of asphaltene modeling using an EOS model. Experimental data (as shown in the legend) are from Burke et al.[65]

TABLE 5.4

Example Application of Asphaltene Splitting Method[64, 66] for Data Reported by Ramirez-Jaramillo et al.[67]

Oil	C30+, mole fraction	Asphaltene (WC30+B), Mass fraction	MWoil, g/g-mole	MWC30+, g/g-mole	C30+A, mole fraction	C30+B, mole fraction
WA	0.0756	0.0286	108.96	580	0.0702	0.0054
WB	0.1023	0.0143	111.29	580	0.0996	0.0027

mole implies g-mole; alternatively, the g can be replaced by lbm, and thus the mole becomes lbm-mole):

$$C_{31+B} = \frac{\left[\dfrac{g\ of\ (asphaltene), C_{31+B}}{g\ of\ oil}\right]\left[\dfrac{g\ of\ oil}{mole\ of\ oil}\right]}{\left[\dfrac{g\ of\ C_{31+B}}{mole\ of\ C_{31+B}}\right]} = \frac{mole\ of\ C_{31+B}}{mole\ of\ oil} \quad (5.16)$$

The splitting equation shown earlier is somewhat universal in that it can also be written for a plus fraction such as a C_{30+}. For the data reported by Ramirez-Jaramillo et al.[67] for two different oils, the heavy end C_{30+} can be split using Equation 5.15, as shown in Table 5.4.

5.6 OTHER FLOW ASSURANCE ISSUES: EMULSIONS AND FOULING

5.6.1 CRUDE OIL EMULSIONS

Emulsions are nothing but a colloidal dispersion of one liquid phase (called dispersed phase) in the other liquid (continuous phase).[68, 69] The emulsion is called as oil-in-water (o/w) emulsion when the oil is in the dispersed phase and water-in-oil (w/o) emulsion when the water is in the dispersed phase. Mixed emulsion types, such as multiple emulsions (o/w/o), are also observed. In the emulsified state, the system's free energy increases due to the increase in the interfacial area between the dispersed droplets and the bulk phase. Consequently, the emulsions are thermodynamically unstable and hence try to minimize the surface area by merging the dispersed droplets, thereby resulting in the separation of phases.

Emulsion often forms upstream and downstream in the oil and gas industry. In upstream production facilities, the crude oil may form an emulsion when it comes in contact with the formation water (brine) present in the reservoir or sourced from the injection water wells during secondary oil recovery. These fluids are immiscible with each other, but when combined, they form emulsions as the pressure drop occurs across chokes, valves, and wellheads or in the presence of turbulence.[70] The presence of surfactants and ionic compounds in the oil–water system (such as asphaltene, resin, waxes, and sediments) enhances the emulsion's stability. The stabilizers (emulsifiers) are often added in injection fluids or occur naturally and promote the kinetic stability of the emulsion. Various other factors, such as the composition of the crude oil, asphaltene, resins and wax content, properties of formation water, pH of formation water, shear rate, temperature, and pressure affect the emulsion stability. Emulsion shows non-Newtonian shear thinning behavior. Increasing temperature reduces the viscosity of the emulsion. Increasing pressures increases emulsion viscosity; however, this behavior is also expected to depend on the gas–oil ratio (GOR) and bubble point pressure. The kinetic stability of emulsions is heavily influenced by the droplet size of the water.[71, 72]

The emulsions are more viscous than pure oil and water and hence pose a formidable challenge for flow assurance downhole and in the well tubing. Severe corrosion may also arise due to the emulsified water, thus resulting in an extra cost for transportation and the processing of crude. Water–in–crude oil emulsions can also be formed during oil spill in the ocean.[73] These emulsions are difficult to break due to their high stability, thus affecting the environment. Various problems caused by stable emulsions and factors that improve or reduce the stability of emulsions have received great attention in the petroleum industry.[74] The general relationship between crude oil properties and emulsion formation is summarized in Table 5.5.[75] Due to the enhanced stability of crude oil emulsions, many researchers and scientists are concerned about developing novel economic interventions and demulsification techniques to break and separate the different phases. It is therefore required to break the emulsion with minimal cost. It is vital to understand how they are stabilized to devise an optimum treatment for w/o emulsions.

TABLE 5.5

General Relationship Between Crude Oil Properties and Emulsion Formation

S.no.	General properties	Observation
1.	Oxidation	Changes in functional group due to oxidation can engage sites for emulsion formation
2.	Viscosity	Greater viscosity often indicates large asphaltene content that enhances emulsion stability
3.	API gravity	Low gravity are also indication for high asphalt content that aids in emulsion formation
4	Aromaticity	Presence of high aromatic compound in waxy crude results in unstable emulsion formations.
5.	Resin/asphaltene ratio	high resin/asphaltene ratios decrease the emulsion stability
6.	Heteroatom content	Heteroatoms, such as nitrogen, oxygen and sulfur, causes acidic and basic traits in petroleum-based fluids, and thus stabilize water-in-oil emulsions.

In the emulsion systems, the presence of an electrical double layer or steric stabilization due to surfactants or polymer molecules prevents droplet coalescence. These surface-active molecules form steric hindrance by adsorbing on the surface of the dispersed droplets forming an elastic film that can act as a barrier against coalescence. These arrangements of molecules are referred to as micelles. The high viscosity of the bulk phase, low interfacial tension, and relatively small volumes of the dispersed phase are some of the other factors that favor the stability of the emulsion. Smaller droplets, having a narrow droplet size distribution, generally form a stable emulsion. In the case of polydisperse dispersions, the large droplets grow at the expense of smaller ones, making the emulsion unstable. This effect is termed Ostwald ripening.[76] In the case of crude oil emulsion, the emulsions are mainly formed and stabilized due to the formation of asphaltene and resins layers around the water droplets, preventing them from coalescence.[77] Asphaltene molecules are generally considered to have a complex crosslinked network, which forms micelles along with resins and sometimes waxes. These micelles encapsulate water, forming a w/o emulsion. Here, the emulsion stability is due to the physical barrier that obstructs the coalescence. As most of the oil well flows below the water cut, less than 60–80% (generally less than the water cut at the *phase inversion point*), it encounters water-in-oil type of emulsion. Indeed, an asphaltene/resin (A/R) ratio is important to understand the emulsion formation and stability mechanisms of crude oil.[78, 79] The asphaltene and resin films in emulsion show viscoelastic properties, crucial for the fluid phenomenon at various shears. Some of the recent works also indicate that wax particles also favor the stable emulsion in the presence of clays and other inorganic materials or naphthenate.[80, 81]

Although thermodynamically unstable, crude oil emulsions are simultaneously kinetically stable and can last for an extended time. According to kinetic stability, emulsions are divided into loose, medium, and high-density (tight) emulsions.[82] They differ in their separation rates, with loose emulsions separating within a few minutes and removed water separated after about 10 min, sometimes described as emulsions with free water. However, dense (or tight) emulsions take days, weeks, or even longer to separate. The general approach to reach emulsion stability is through droplet size distribution. Experiments have shown that a smaller average size of dispersed droplets increases residence time, requiring larger separation plant equipment.[82] The effect of asphaltenes solvency on the stability of water–in–crude oil emulsions shows that the solubility of asphaltenes in the crudes is crucial for emulsion stability. It is understood that the higher the asphaltene concentration favors more emulsion stability. Figure 5.26 depicts the production of asphaltene-stabilized emulsions. Surfactants in the oil–water interface aid in the formation of tiny droplets, which are important in the emulsion production process. By assuming that the surface free energy changes during emulsification, these surfactants can minimize interfacial tension. The primary role of surfactants is to enhance emulsion stability by promoting the production of emulsions, generating smaller droplets.

Crude oil emulsions are highly viscous and thus are difficult to flow in the well-tubing, pipelines, and reservoir. It is necessary to break them in situ to enhance

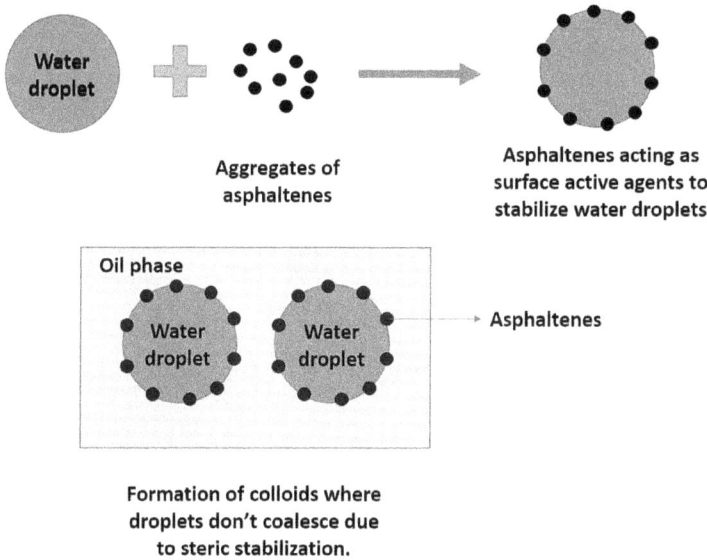

FIGURE 5.26   Role of asphaltenes as an emulsion stabilizer for crude oil–water emulsion.

crude oil production. As it is difficult to break emulsion near the well bore, emulsified crude oil is treated at the surface facilities using a gravity-settling tank, cyclone separator, pH adjustment, filtration, membrane separation, and chemical inhibitors. In the case of a deep offshore condition, installing a processing facility is challenging, as all these techniques require a high surface area. Crude oil production rate can be enhanced by injecting de-emulsifiers near the well bore and at surface facilities. It is important to check the efficacy of the de-emulsifier and an optimum quantity for the de-emulsification job. An overdose of de-emulsifier may worsen the emulsion formation tendency of crude oil along with precipitation of asphaltene.

5.6.2   FOULING

Fouling refers to the undesirable growth or production of surface deposits in technological applications. Inside pipelines, machinery, and heat exchangers, fouling occurs. Fouling substances are already dissolved in the liquid. For example, calcium carbonate and related salts are found in water, while paraffin is found in crude oil, among other things. If a system has fouling or biofouling, corrosion may occur as a secondary effect. Deposits build up in pipes and systems over time. These deposits reduces the pipe's cross-section, obstructing flow and heat transfer. Lime or similar deposits produce hard crusts in the case of mineral fouling. When it comes to organic material, biofilm comes into existence first, followed by biofouling. Because the deposits are often contained in stagnant water, corrosion occurs due to outgassing. In the case of biofouling, organisms cause microbiologically induced corrosion (MIC). There are several types of fouling observed in the oil and gas industry, which are explained briefly next.

5.6.2.1 Macrofouling

Macrofouling is caused by coarse materials that are either biological or inorganic, such as industrial waste. Such compounds can clog heat-exchange surfaces, causing the relevant heat transfer coefficient to deteriorate. They may also generate flow obstructions, redistribute flow within the components, or fretting damage.

5.6.2.2 Microfouling

Microfouling is of the following types.

5.6.2.2.1 Precipitation Fouling

Precipitation fouling, also known as scaling, occurs when solid salts, oxides, and hydroxides crystallize from fluids (e.g., calcium carbonate or calcium sulfate). Fouling from water is the most common; however, it may happen also from non-aqueous systems. Precipitation fouling is a typical problem in hard-water boilers and heat exchangers, and it commonly results in lime scale. Precipitation fouling is frequently caused by salt solubility being dependent on temperature or the evaporation. The essential distinction is between salts with a "normal" or "retrograde" temperature dependence on their solubility. Salts with a "normal" solubility increase as the temperature rises, fouling the cooling surfaces in the process. Salts foul the heating surfaces with "inverse" or "retrograde" solubility. Due to its retrograde solubility, calcium sulfate is a common precipitation foulant of heated surfaces. Pipelines or flowlines with higher saline water content are prone to precipitation fouling.

5.6.2.2.2 Particulate Fouling

Fouling is caused by particles suspended in water or gas, which buildup on a surface, usually colloidal particles, and follows a different mechanism than precipitation fouling. This technique is primarily significant for colloidal particles whose diameter is less than 1 μm in at least one dimension. Particles are carried to the surface by various processes and can attach themselves at the site, for example, by flocculation or coagulation. It is worth noting that the attachment of colloidal particles is often based on electrical forces. *Sticking probability* is a term used to describe the possibility of engagement or attachment.

5.6.2.2.3 Corrosion Fouling

Corrosion deposits are formed in situ as the substrate corrodes, such as magnetite on carbon steel surfaces. On the other hand, fouling deposits are formed from a material that has been removed from the environment. Fouling deposits formed by ex situ–generated corrosion products should not be confused with corrosion deposits. The composition of corrosion deposits is usually connected to the substrate's composition. The geometry of the metal–oxide and oxide–fluid interfaces may also make it possible to distinguish between corrosion and fouling deposits. One of the examples of corrosion fouling is the deposition of iron oxide or oxyhydroxide deposit of carbon steel.

5.6.2.2.4 Chemical Reaction Fouling

Chemical reaction fouling is caused by chemical reactions, such as the breakdown or polymerization of organic materials on heated surfaces. Chemical reactions can occur when chemical species in the process fluid contact heat transfer surfaces. In certain situations, the metallic surface can act as a catalyst. The polymerization of olefins or the deposition of heavy fractions is common in petroleum-processing systems (asphaltenes, waxes, etc.). High temperatures can cause organic materials to carbonize on the tube wall. Chemical reactions can also cause fouling in the food business, such as milk processing. Fouling caused by an ionic reaction resulting in the formation of an inorganic solid is known as precipitation fouling.

5.6.2.2.5 Solidification Fouling

Solidification fouling occurs when high-melting-point components of a flowing fluid freeze on a subcooled surface. Some of the examples of this type of fouling are the solidification and deposition of wax from crude oil and deposition of molten ash (delivered in furnace exhaust gas) on the surface of the heat exchanger. The surface must be subcooled in relation to the foulant's solidification point because it must be below a specified temperature threshold.

5.6.2.2.6 Biofouling

The undesired collection of microorganisms, plants, animals on surfaces, and algae and diatoms on offshore surface facilities, such as ship hulls or pipes and reservoirs with untreated water, is referred to as biofouling or biological fouling. This can be accompanied by corrosion influenced by microbiology (MIC). Bacteria can produce slimes or biofilms. Thus, organisms can agglomerate on surfaces due to the formation of colloidal hydrogels of water and extracellular polymeric substances (EPS) (such as polysaccharides, lipids, nucleic acids, etc.). The structure of a biofilm is frequently complex. Bacterial fouling can happen in either aerobic (water with oxygen dissolved in it) or anaerobic (water with no oxygen) environments. In reality aerobic bacteria prefer open systems with continual oxygen and nutrition delivery, commonly found in warm, and sunny conditions. When sufficient nutrients are present, anaerobic fouling develops more frequently in closed systems. Sulfate-reducing bacteria (SRB, e.g., desulfurispirillum) generate sulfide, which can cause ferrous metal corrosion (and other alloys). On the other hand, sulfide-oxidizing bacteria can create sulfuric acid and play a role in concrete corrosion (e.g., *Acidithiobacillus*).

5.6.2.2.7 Composite Fouling

Composite fouling is a regular occurrence. This sort of fouling occurs when more than one foulant or fouling mechanism are active simultaneously. Multiple foulants or methods may interact with one another, resulting in a synergistic fouling that is more complex than the sum of its parts.

REFERENCES

1. Sloan, E. D. (2000). *Hydrate Engineering*. Bloys J. B. (Ed.). Society of Petroleum Engineers, US.
2. Klinkenberg, L. J. (1941). The permeability of porous media to liquids and gases. Drilling and Production Practice, American Petroleum Institute, pp. 200–213. (API-41-200).
3. Pedersen, K. S., Christensen, P. L., & Shaikh, J. A. (2014). *Phase Behavior of Petroleum Reservoir Fluids*. CRC Press, Taylor & Francis Group, Boca Raton, FL.
4. Gudmundsson, J. S. (2018). *Flow Assurance Solids in Oil and Gas Production* (1st Edition). CRC Press, Taylor & Francis Group, London, UK.
5. Golczynski, T. S., & Kempton, E. C. (2006). Understanding wax problems leads to deepwater flow assurance solutions. *World Oil*, 227, 7–10.
6. Gilby, G. W. (1983). The Use of Ethylene-Vinyl Acetate Copolymers as Flow Improvers in Waxy Crude Oil. In *Chemicals in the Oil Industry*. In *Special Publication-RSC, Vol. 45*, Ogden, P. H., (Ed.). Royal Society of Chemistry, Manchester.
7. Notz, P. (2009). *Flow Assurance for Offshore Production*. Petroskills, Houston, TX.
8. Katz, D. (1959). *Handbook of Natural Gas Engineering*. McGraw-Hill, New York.
9. Towler, B. F., & Mokhatab, S. (2005). Quickly estimate hydrate formation conditions in natural gases: Using this simple technique to predict unfavorable conditions can save time and cost. *Hydrocarbon Processing*, 84(4), 61–63.
10. de Boer, R. B., Leerlooyer, K., Eigner, M. R. P., & van Bergen, A. R. D. (1995). Screening of crude oils for asphalt precipitation: Theory, practice, and the selection of inhibitors. *SPE Production & Facilities*, 10 (1), 55–61. https://doi.org/10.2118/24987-PA
11. Jamaluddin, A. K. M., Nighswander, J., & Joshi, N. (2001). A systematic approach in deepwater flow assurance fluid characterization. In *SPE Annual Technical Conference and Exhibition*. New Orleans, Louisiana., September 30–October 3. (SPE 71546). https://doi.org/10.2118/71546-MS
12. Yen, A., Yin, Y. R., & Asomaning, S. (2001). Evaluating asphaltene inhibitors: laboratory tests and field studies. In *SPE International Symposium on Oilfield Chemistry*. Houston, Texas, February 13–16. (SPE 65376). https://doi.org/10.2118/65376-MS
13. Minssieux, L. (1997). Core damage from crude asphaltene deposition. *In International Symposium on Oilfield Chemistry*. Houston, Texas, February 18–21. (SPE-37250-MS). https://doi.org/10.2118/37250-MS
14. del Carmen Garcia, M., & Chiaravallo, N. Asphaltenes deposition control in lake maracaibo crude oil production. In *SPE International Symposium on Oilfield Chemistry*. Houston, Texas, February 13–16. (SPE-65009-MS). https://doi.org/10.2118/65009-MS
15. Kabir, C. S., & Jamaluddin, A. K. M. (2002). Asphaltene characterization and mitigation in south Kuwait's Marrat reservoir. *SPE Production & Facilities*, 17(4), 251–258. https://doi.org/10.2118/80285-PA
16. Jamaluddin, A. K. M., Creek, J., Kabir, C. S., McFadden, J. D., D'Cruz, D., Manakalathil, J., Joshi, N., & Ross, B. (2002). Laboratory techniques to measure thermodynamic asphaltene instability. *Journal of Canadian Petroleum Technology*, 41(7), 44–52. https://doi.org/10.2118/02-07-04
17. El-Dalatony, M. M., Jeon, B. H., Salama, E. S., Eraky, M., Kim, W. B., Wang, J., & Ahn, T. (2019). Occurrence and characterization of paraffin wax formed in developing wells and pipelines. *Energies*, 12(6), 967. https://doi.org/10.3390/en12060967
18. Bidmus, H. O., & Mehrotra, A. K. (2004). Heat-transfer analogy for wax deposition from paraffinic mixtures. *Industrial & Engineering Chemistry Research*, 43(3), 791–803. https://doi.org/10.1021/ie030573v

19. Monger-McClure, T. G., Tackett, J. E., & Merrill, L. S. (1999). Comparisons of Cloud Point Measurement and Paraffin Prediction Methods. *SPE Production & Facilities*, 14(1), 4–16. https://doi.org/10.2118/54519-PA

20. Mitchell, K., & Chandler, J. (1998). The use of flow improved diesel fuel at extremely low temperatures. In *International Fall Fuels and Lubricants Meeting and Exposition*. San Francisco, California, October 19–22. (982576) https://doi.org/10.4271/982576

21. Kruka, V. R., Cadena, E. R., & Long, T. E. (1995). Cloud-point determination for crude oils. *Journal of Petroleum Technology*, 47(8), 681–687. https://doi.org/10.2118/31032-PA

22. Tiwary, D. (2002). Rheology and Phase Behavior of Highly Paraffinic "waxy" Mixtures. Master's thesis, University of Calgary, Canada. https://doi.org/10.11575/PRISM/15997

23. Leontaritis, K. J., & Leontaritis, J. D. (2003). Cloud point and wax deposition measurement techniques. In *SPE International Symposium on Oilfield Chemistry*. Houston, Texas, U.S.A., February 5–7. (SPE-80267-MS) https://doi.org/10.2118/80267-MS

24. Roenningsen, H. P., Bjoerndal, B., Hansen, A.B., & Pedersen, W. B. (1991). Wax precipitation from North sea crude oils: 1. Crystallization and dissolution temperatures, and Newtonian and non-Newtonian flow properties. *Energy & Fuels*, 5(6), 895–908. https://doi.org/10.1021/ef00030a019

25. Haj-Shafiei, S., & Mehrotra, A. K. (2019). Achieving cold flow conditions for 'waxy' mixtures with minimum solid deposition. *Fuel*, 235, 1092–1099. https://doi.org/10.1016/j.fuel.2018.08.102

26. Tiwary, D., & Mehrotra, A. K. (2008). Phase transformation and rheological behaviour of highly paraffinic "waxy" mixtures. *Canadian Journal of Chemical Engineering*, 82(1), 162–174. https://doi.org/10.1002/cjce.5450820121

27. Wavrek, D. A., & Deo, M. (1997). Wax Precipitation: Compositional Study and Cloud Point Measurements. *In Proceedings of the 2ⁿᵈ International Symposium on Colloid Chemistry in Oil Production*. Rio de Janeiro, Brazil, August 31–September 3.

28. Kök, M. V., Letoffe, J. M., & Claudy, P. (1999). DSC and rheometry investigations of crude oils. *Journal of Thermal Analysis and Calorimetry*, 56, 959–965. https://doi.org/10.1023/A:1010159718321

29. Zhu, T., Walker, J. A., & Liang, J. (2008). *Evaluation of Wax Deposition and Its Control During Production of Alaskan North Slope Oils*. University of Alaska Fairbanks, AK.

30. Zhao, J., An, X., & Qiao, S. (2019). The extraction effect of CO_2 injection on the flow properties of crude oil. *Petroleum Science and Technology*, 37(6), 710–717. https://doi.org/10.1080/10916466.2018.1564769

31. Calange, S., Ruffier-Meray, V., & Behar, E. (1997). Onset crystallization temperature and deposit amount for waxy crudes: Experimental determination and thermodynamic modelling. (SPE-37239-MS) In *International Symposium on Oilfield Chemistry*, Houston, Texas, February 16–21. https://doi.org/10.2118/37239-MS

32. Martos, C., Coto, B., Espada, J. J., Robustillo, M. D., Gómez, S., & Peña, J. L. (2008). Experimental determination and characterization of wax fractions precipitated as a function of temperature. *Energy & Fuels*, 22(2), 708–714. https://doi.org/10.1021/ef7003927

33. Webber, G. V. (2000). Wax Characterization by Instrumental Analysis. Master's thesis, University of Stellenbosch, South Africa.

34. Fan, K., Huang, Q., Li, S., & Zhao, D. (2015). Wax deposition study in a cold-finger system with model oil. In *SPE/IATMI Asia Pacific Oil & Gas Conference and Exhibition*. Nusa Dua, Bali, Indonesia, October 20-22. (SPE-176447). https://doi.org/10.2118/176447-MS

35. Venkatesan, R., Singh, P., & Fogler, H. S. (2002). Delineating the pour point and gelation temperature of waxy crude oils. *SPE Journal*, 7(4), 349–352. https://doi.org/10.2118/72237-PA

36. Zhao, Y., Kumar, L., Paso, K., Safieva, J., Sariman, M. Z. B., & Sjöblom, J. (2012). Gelation behavior of model wax–oil and crude oil systems and yield stress model development. *Energy & Fuels*, 26(10), 6323–6331. https://doi.org/10.1021/ef3012454

37. Jamaluddin, A. K. M., Creek, J., Kabir, C. S., McFadden, J. D., D'Cruz, D., Joseph, M. T., Joshi, N., & Ross, B. (2001) A comparison of various laboratory techniques to measure thermodynamic asphaltene instability. In *SPE Asia Pacific Improved Oil Recovery Conference*. Kuala Lumpur, Malaysia, October 8–9. (SPE-72154-MS) https://doi.org/10.2118/72154-MS

38. Joshi, N. B., Mullins, O. C., Jamaluddin, A., Creek, J., & McFadden, J. (2001). Asphaltene precipitation from live crude oil. *Energy & Fuels*, 15(4), 979–986. https://doi.org/10.1021/ef010047l

39. Hirschberg, A., DeJong, L. N. J., Schipper, B. A., & Meijer, J. G. (1984). Influence of temperature and pressure on asphaltene flocculation. *Society of Petroleum Engineers Journal*, 24(3), 283–293. https://doi.org/10.2118/11202-PA

40. Soleymanzadeh, A., Yousefi, M., Kord, S., & Mohammadzadeh, O. (2019). A review on methods of determining onset of asphaltene precipitation. *Journal of Petroleum Exploration and Production Technology*, 9, 1375–1396. https://doi.org/10.1007/s13202-018-0533-5

41. Escobedo, J., & Mansoori, G. A. (1995). Viscometric determination of the onset of asphaltene flocculation: A novel method. *SPE Production & Facilities*, 10(02), 115–118. https://doi.org/10.2118/28018-PA

42. Nixdorf, J., & Oellrich, L. R. (1997). Experimental determination of hydrate equilibrium conditions for pure gases, binary and ternary mixtures and natural gases. *Fluid Phase Equilibria*, 139(1–2), 325–333. https://doi.org/10.1016/S0378-3812(97)00141-6

43. Prasad, S. K., Mech, D., Nair, V. C., Gupta, P., Sangwai, J. S. (2018). Effect of high molecular weight asphaltenes on the phase stability of methane hydrates. *Proceedings of the 28th International Ocean and Polar Engineering Conference*, Sapporo, Japan, June 10-15. (ISOPE-I-18-659).

44. Bhawangirkar, D. R., & Sangwai, J. S. (2021). Phase equilibrium of methane hydrates in the presence of $MgBr_2$, $CaBr_2$, and $ZnBr_2$ aqueous solutions. *Journal of Chemical & Engineering Data*, 66(6), 2519–2530. https://doi.org/10.1021/acs.jced.1c00121

45. Shi, L., & Liang, D. Phase equilibria of double semiclathrate hydrates formed with tetraamylammonium bromide plus CH_4, CO_2, or N_2. *Journal of Chemical & Engineering Data*, 60(9), 2749–2755. https://doi.org/10.1021/acs.jced.5b00516

46. Gupta, P., Nair, V. C., & Sangwai, J. S. (2018). Phase equilibrium of methane hydrate in the presence of aqueous solutions of quaternary ammonium salts. *Journal of Chemical & Engineering Data*, 63(7), 2410–2419. https://doi.org/10.1021/acs.jced.7b00976

47. Nair, V. C., Ramesh, S., Ramadass, G. A., & Sangwai, J. S. (2016). Influence of thermal stimulation on the methane hydrate dissociation in porous media under confined reservoir. *Journal of Petroleum Science & Engineering*, 147, 547–559. https://doi.org/10.1016/j.petrol.2016.09.017

48. Mech, D., & Sangwai, J. S. (2016). Effect of molecular weight of polyethylene glycol (PEG), a hydrate inhibitive water-based drilling fluid additive, on the formation and dissociation kinetics of methane hydrate. *Journal of Natural Gas Science and Engineering*, 35(B), 1441–1452. https://doi.org/10.1016/j.jngse.2016.06.020

49. Smith, J. M., Van Ness, H. C., & Abbot, M. (2001). *Introduction to Chemical Engineering Thermodynamics* (6th Edition). McGraw-Hill, New York.

50. Sahu, C., Sircar, A., Sangwai, J. S., & Kumar, R. (2021). Kinetics of methane hydrate formation in the presence of 1-dodecyl-2-pyrrolidinone and tetrahydrofuran in pure water. *Industrial & Engineering Chemistry Research*, 60(20), 7588–7598. https://doi.org/10.1021/acs.iecr.1c00925

51. Prasad, S. K., Nair, V. C., & Sangwai, J. S. (2021). Effect of asphaltenes on the kinetics of methane hydrate formation and dissociation in oil-in-water dispersion systems containing light saturated and aromatic hydrocarbons. *Energy & Fuels*, 35(21), 17410–17423. https://doi.org/10.1021/acs.energyfuels.1c02252

52. Webb, E. B., Koh, C. A., Liberatore, M. W. (2014). High pressure rheology of hydrate slurries formed from water-in-mineral oil emulsions. *Industrial & Engineering Chemistry Research*, 53(17), 6998–7007. https://doi.org/10.1021/ie5008954

53. Pandey, G.; Linga, P.; Sangwai, J. S. (2017). High pressure rheology of gas hydrate formed from multiphase systems using modified Couette rheometer. *Review of Scientific Instruments*, 88(2), 025102. https://doi.org/10.1063/1.4974750

54. Pandey, G., & Sangwai, J. S. (2020). High pressure rheological studies of methane hydrate slurries formed from water-hexane, water-heptane, and water-decane multiphase systems. *Journal of Natural Gas Science and Engineering*, 81, 103365. https://doi.org/10.1016/J.JNGSE.2020.103365

55. Webb, E. B., Rensing, P. J., Koh, C. A., Sloan, E. D., Sum, A. K., & Liberatore, M. W. (2012). High-pressure rheology of hydrate slurries formed from water-in-oil emulsions. Energy & fuels, 26(6), 3504–3509. https://doi.org/10.1021/ef300163y

56. Dieker, L. E., Aman, Z. M., George, N. C., Sum, A. K., Sloan, E. D., & Koh, C. A. (2009). Micromechanical adhesion force measurements between hydrate particles in hydrocarbon oils and their modifications. *Energy & Fuels*, 23(12), 5966–5971. https://doi.org/10.1021/ef9006615

57. Zhou, S., Yan, H., Su, D., Navaneethakannan, S., & Chi, Y. (2018). Investigation on the kinetics of carbon dioxide hydrate formation using flow loop testing. *Journal of Natural Gas Science and Engineering*, 49, 385–392. https://doi.org/10.1016/j.jngse.2017.10.022

58. Jiang, L., Xu, N., Liu, Q., Cheng, Z., Liu, Y., & Zhao, J. (2020). Review of morphology studies on gas hydrate formation for hydrate-based technology. *Crystal Growth & Design*, 20(12), 8148–8161. https://doi.org/10.1021/acs.cgd.0c01331

59. Veluswamy, H. P., Wong, A. J. H., Babu, P., Kumar, R., Kulprathipanja, S., Rangsunvigit, P., & Linga, P. (2016). Rapid methane hydrate formation to develop a cost effective large scale energy storage system. *Chemical Engineering Journal*, 290, 161–173. https://doi.org/10.1016/j.cej.2016.01.026

60. Zheng, J., Zhang, B. Y., Wu, Q., & Linga, P. (2018). Kinetic evaluation of cyclopentane as a promoter for CO_2 capture via a clathrate process employing different contact modes. *ACS Sustainable Chemistry & Engineering*, 6(9), 11913-11921. https://doi.org/10.1021/acssuschemeng.8b02187

61. Tanaka, R., Sakemoto, R., & Ohmura, R. (2009). Crystal growth of clathrate hydrates formed at the interface of liquid water and gaseous methane, ethane, or propane: variations in crystal morphology. *Crystal Growth & Design*, 9(5), 2529–2536. https://doi.org/10.1021/cg9001048

62. Hayama, H., Mitarai, M., Mori, H., Verrett, J., Servio, P., & Ohmura, R. (2016). Surfactant effects on crystal growth dynamics and crystal morphology of methane hydrate formed at gas/liquid interface. *Crystal Growth & Design,* 16(10), 6084–6088. https://doi.org/10.1021/acs.cgd.6b01124

63. Bishnoi, P., & Dholabhai, P. D. (1999). Equilibrium conditions for hydrate formation for a ternary mixture of methane, propane and carbon dioxide, and a natural gas mixture in the presence of electrolytes and methanol. *Fluid Phase Equilibria*, 158–160, 821–827. https://doi.org/10.1016/S0378-3812(99)00103-X

64. Ahmed, T. (2007). *Equations of State and PVT Analysis*. Gulf Publishing Company, Houston, Texas.

65. Burke, N. E., Hobbs, R. E., & Kashou, S. F. (1990). Measurement and modeling of asphaltene precipitation (includes associated paper 23831). *Journal of Petroleum Technology*, 42(11), 1440–1446. https://doi.org/10.2118/18273-PA

66. Kohse, B. F., & Nghiem, L. X. (2004). Modelling Asphaltene Precipitation and Deposition in a Compositional Reservoir Simulator. In *2004 SPE/DOE Fourteenth Symposium on Improved Oil Recovery*. Tulsa, Oklahoma, U.S.A., April Tulsa, Oklahoma, U.S.A., April 17–21. (SPE 89437). https://doi.org/10.2118/89437-MS

67. Ramirez-Jaramillo, E., Lira-Galeana, C., & Manero, O. (2006). Modeling asphaltene deposition in production pipelines. *Energy & Fuels*, 20 (3), 1184–1196. https://doi.org/10.1021/ef050262s

68. Goodarzi, F., & Zendehboudi, S. A. (2019). Comprehensive review on emulsions and emulsion stability in chemical and energy industries. *Canadian Journal of Chemical Engineering*, 97(1), 281–309. https://doi.org/10.1002/cjce.23336

69. Kumar, G., Kakati, A., Mani, E., & Sangwai, J. S. (2021). Stability of nanoparticle stabilized oil-in-water Pickering emulsion under high pressure and high temperature conditions: Comparison with surfactant stabilized oil-in-water emulsion. *Journal of Dispersion Science and Technology*, 42(8), 1204–1217. https://doi.org/10.1080/01932691.2020.1730888

70. Mohyaldinn, M. E., Hassan, A. M., Ayoub, M. A. (2019). Application of Emulsions and Microemulsions in Enhanced Oil Recovery and Well Stimulation. In *Microemulsion – A Chemical Nanoreactor*, Mejuto, J. (Ed.). IntechOpen. https://doi.org/10.5772/intechopen.84538

71. De Oliveira, M. C. K., Carvalho, R. M., Carvalho, A. B., Couto, B. C., Faria, F. R. D., & Cardoso, R. L. P. Waxy crude oil emulsion gel: Impact on flow assurance. *Energy & Fuels*, 24(4), 2287–2293. https://doi.org/10.1021/ef900927g

72. Raya, S. A., Mohd Saaid, I., Abbas Ahmed, A., & Abubakar Umar, A. (2020). A critical review of development and demulsification mechanisms of crude oil emulsion in the petroleum industry. *Journal of Petroleum Exploration and Production Technology*, 10(4), 1711–1728. https://doi.org/10.1007/s13202-020-00830-7

73. Mehta, S. D. (2005). Making and Breaking of Water in Crude Oil Emulsion. Master's Thesis. Texas A&M University, US.

74. Fingas, M., & Fieldhouse, B. (2009). Studies on crude oil and petroleum product emulsions: Water resolution and rheology. *Colloids and Surfaces A: Physicochemical and Engineering Aspects*, 333(1–3), 67–81. https://doi.org/10.1016/j.colsurfa.2008.09.029

75. Auflem, I. H. (2002). Influence of Asphaltene Aggregation and Pressure on Crude Oil Emulsion Stability. Ph.D.Thesis, Norwegian University of Science and Technology, Norway.

76. Sjöblom, J. (1996). *Emulsions And Emulsion Stability* (2nd Edition). Marcel Dekker Inc, New York.

77. Kokal, S. (2005). Crude oil emulsions: A state-of-the-art review. *SPE Production & Facilities*, 20(01), 5–13. https://doi.org/10.2118/77497-PA

78. Schorling, P. C., Kessel, D. G., & Rahimian, I. (1999). Influence of the crude oil resin/asphaltene ratio on the stability of oil/water emulsions. *Colloids and Surfaces A: Physicochemical and Engineering Aspects*, 152(1–2), 95–102. https://doi.org/10.1016/S0927-7757(98)00686-4

79. Spiecker, P. M., Gawrys, K. L., Trail, C. B., & Kilpatrick, P. K. (2003). Effects of petroleum resins on asphaltene aggregation and water-in-oil emulsion formation. *Colloids Surfaces A Physicochemical and Engineering Aspects*, 220(1–3), 9–27. https://doi.org/10.1016/S0927-7757(03)00079-7

80. Ali, M., & Alqam, M. H. (2000). The role of asphaltenes, resins and other solids in the stabilization of water in oil emulsions and its effects on oil production in saudi oil fields. *Fuel*, 79(11), 1309–1316. https://doi.org/10.1016/S0016-2361(99)00268-9

81. Visintin, R. F. G., Lockhart, T. P., Lapasin, R., & D'Antona, P. (2008). Structure of waxy crude oil emulsion gels. *Journal of Non-Newtonian Fluid Mechanics*, 149(1–3), 34–39. https://doi.org/10.1016/j.jnnfm.2007.07.008

82. Zolfaghari, R., Fakhru'l-Razi, A., Abdullah, L. C., Elnashaie, S. S. E. H., Pendashteh, A. (2016). Demulsification techniques of water-in-oil and oil-in-water emulsions in petroleum industry. *Separation and Purification Technology*, 170, 377–407. https://doi.org/10.1016/j.seppur.2016.06.026

6 Deepwater Oil and Gas Environment

6.1 INTRODUCTION

Development, operation, and maintenance of oil and gas fields become more challenging in deep water. Technologies that are successful at surface conditions become less effective under harsh environmental conditions of deep water. In deep water, high-salinity conditions, marine organisms, low temperature, high-pressure conditions, and so on offer operational difficulties. Conventional drilling, well completion, and production practices at the surface conditions cannot be directly applied at the offshore condition. Hence, the deployment of oil and gas production facilities, including pipelines, needs special consideration. Deepwater oil and gas production technologies are always fraught with technological, commercial, and economic difficulties, requiring the development of radically new technologies and strategies for oil and gas production.[1]

The conditions at seabed are very dark and cold, especially below a depth of 500 m. Also, the corresponding pressure consequently increases with the water depth. The high-temperature gradient in subsea conditions is one of the challenging problems. The subsea temperature profile was often overlooked due to a lack of knowledge in the early years, and isothermal conditions were assumed during design considerations. However, with technological advancement, information on temperature profiles are nowadays available. There is a negative temperature gradient from the sea surface to the seafloor, which becomes positive below the seabed. Low-temperature conditions result in the high viscosity of crude oil and wax deposition in subsea flowlines and risers. In addition, as discussed earlier, due to high-pressure and low-temperature conditions, gas hydrate formation becomes common in the subsea, which can plug flowlines, thereby affecting the fluid flow.[2] The temperature near the seafloor is reported to be as low as 40°F and high as 150°F. Hydrate formation is very severe under low-temperature and high-pressure conditions in deep water and can result in the plugging of the tubing and affecting the operation of blowout preventers (BOPs). The presence of hydrates and low-temperature conditions can adversely influence the durability of cement, casing, and other subsea facilities. They can impact the safety of an offshore drilling operation significantly. It is difficult to reduce the pressure below hydrate formation conditions due to various factors. In addition, the saline environment offers challenges, such as corrosion and scale formation in the flowlines, which incur additional costs for operation and maintenance. Multiphase flow due to liquid dropout, water production, and sloughing are other serious issues in offshore conditions that need to be addressed. Thus, in order to develop subsea oil and gas field development plans, it is essential to know the details of all the components of subsea oil and gas production systems and the subsurface temperature profile. These information are necessary to safely exploit offshore hydrocarbon reserves.[3]

6.2 OVERVIEW OF OFFSHORE PRODUCTION SYSTEMS

As the discoveries of oil and gas reservoirs are being made offshore, the oil and gas industry has entered into a new phase. One of the first onshore production efforts began in early 1859 from the famous Col Drakes's well in Pennsylvania in the United States. However, the first-ever offshore well was drilled after almost 38 years, in 1897.[4] Until the 1980s, *deep water* meant a water depth of 800 ft. By 2008, water depth less than 1,500 ft was considered shallow. As of now, water depths of 1,500 ft and above are considered to be deepwater systems, with water depths of more than 7,000 ft are often referred to as ultradeep water.[4] The fixed oil and gas production platforms can be used for a shallow water depth, but for deep water, fixed platforms cannot be deployed, and hence, mobile or floating platforms are deployed.

While the reserves at the 300–400-ft water depth are being produced using fixed platforms, more attention is being given to alternate production systems as oil and gas production is now contemplated from deeper waters. The offshore production environment encounters extreme conditions, such as high pressure, low temperature, salinity, and microbial activities. In such conditions, various challenges concerning flow assurance emerge. A detailed understanding of the offshore production system is necessary to develop strategies for the flow assurance problem. Novel technologies are required to produce oil and gas economically from inaccessible locations and challenging conditions. Depending on the water depth, various offshore structures are used to produce oil and gas, either attached to the sea bottom and/or floating on the water.[5] Typically, water depths are classified as below:

a) Shallow water depth (<350 m)
b) Deepwater depth (350–1,500 m)
c) Ultradeep water depth (>1,500 m)

The various offshore oil and gas deepwater platform systems are shown in Figure 6.1. All the deepwater offshore production platforms are of floating type.

6.2.1 CLASSIFICATION AND TYPES OF OFFSHORE PLATFORMS

Offshore production platforms are designed to produce oil and gas underneath the seabed. These platforms have drilling equipment, oil and gas processing and storage facilities, and personal living spaces. Generally, the cost involved to deploy these platforms is huge. Thus, they are expected to have long operational life even in extreme conditions. Generally, offshore platforms are based on two major structural characteristics based on their support for structural stability: bottom-supported and floating structures. Figure 6.2 shows the details of the classification.[4] Table 6.1 shows major differences based on their specific characteristics.[4, 5]

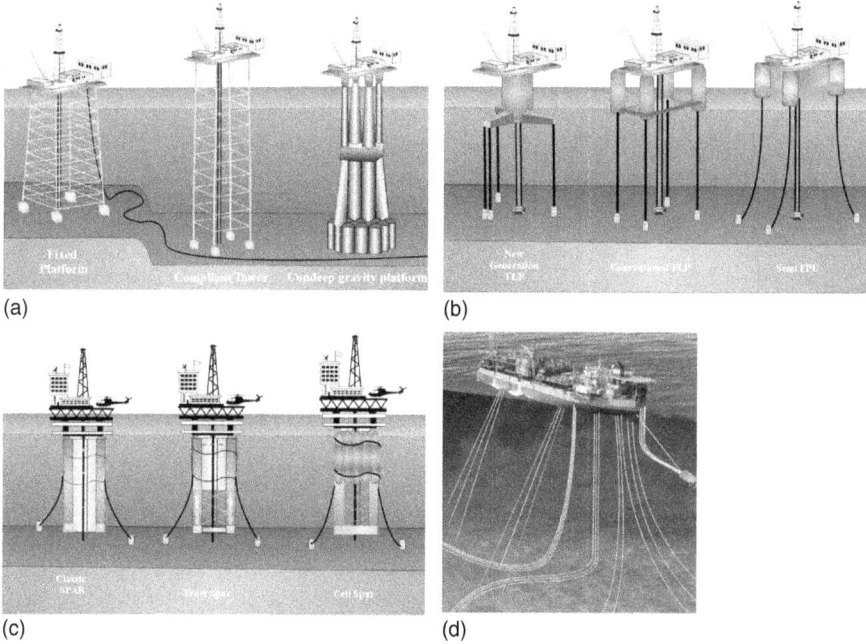

(a)

(b)

(c)

(d)

FIGURE 6.1 Schematic of commonly found deepwater oil and gas platforms.

6.2.2 Fixed Platforms

Fixed platforms, as the name suggests, are fixed to the seafloor. They are mounted on steel legs or concrete piles piled into the seafloor. If the water depth is very shallow, their weight is sometimes enough to withstand loads. Fixed platforms are relatively stable and resilient to harsh environments offshore. These can generally be used upto an offshore water depth of 150 m.

6.2.3 Compliant platforms

Compliant platforms are relatively narrow rigs having flexible towers extending from under the deck to the seafloor. They have somewhat smaller footprints than the fixed platforms. They can withstand offshore environment up to a water depth of 1,000 m and are more robust than floating platforms.

6.2.4 Mobile Platforms

A mobile offshore production unit (MOPU) is a portable platform that can be redeployed during oil and gas production and is generally used for water depths of more than 500 m. Examples are jack-up drilling rigs or floating production storage and offloading (FPSO) units. As these are portable, they are more economical and feasible oil- and gas-producing units for offshore conditions. Mobile platforms are used in deep water. It is necessary to ensure that they are located properly at the desired

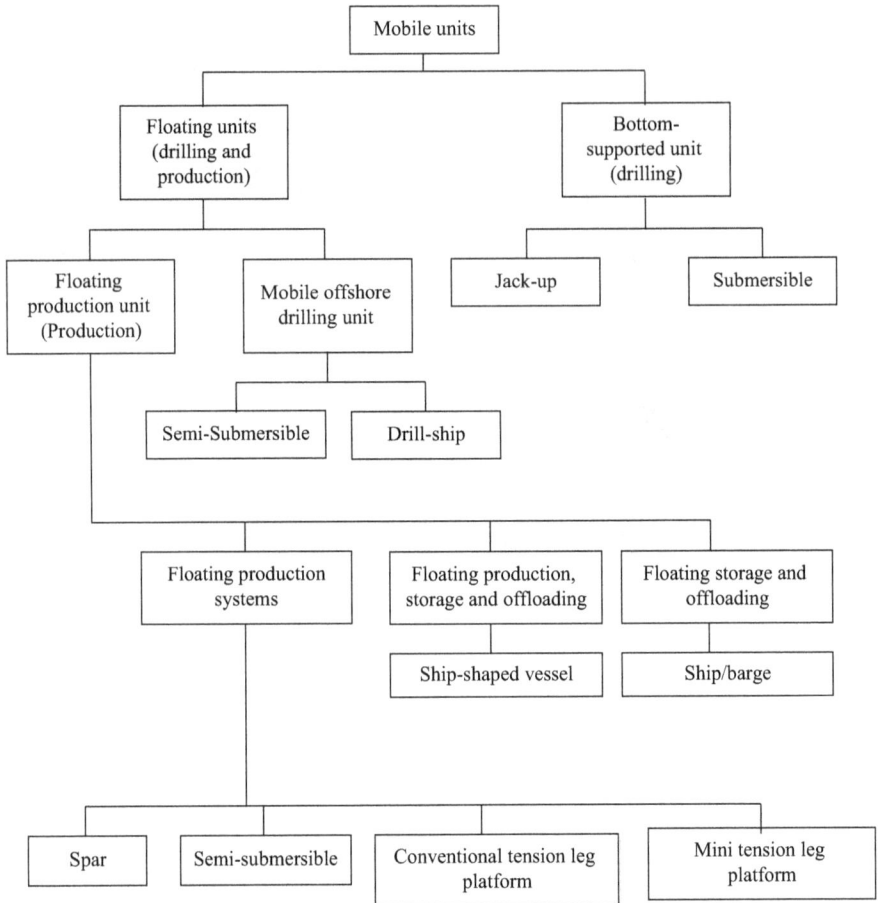

FIGURE 6.2　Different types and subtypes of mobile platform units.[4, 5]

location using a mooring system. In this case, all six degrees of freedom can destabilize the structure. The mobile platforms should be positively (buoyancy force is greater than gravity force) or neutrally (buoyancy force is equal to gravity force) buoyant. Dry tree (a surface completion system) and wet tree (a subsea completion system) systems are used with the offshore platforms. Wet and dry tree systems are used for neutrally and positively buoyant structures. Mobile units also utilize risers for connecting the wellhead to the deck. These risers are either rigid or flexible. Rigid risers are used with a dry tree system, and flexible risers are used with a wet tree system.

To ensure these mobile units are fixed at the desired location, two types of positioning systems are used:

- Dynamic positioning system.
- Mooring system.

TABLE 6.1

Major Difference between Bottom-Supported Structures versus Floating Structures[5]

Functions	Bottom-Supported Structures	Floating Structures
Payload support	The foundation of platforms is fixed to the seafloor.	These are floating types due to their buoyancy support.
Access to the wellhead	Firm conductors are employed to connect the wellhead to the deck.	The wellhead is at the seafloor, which is connected using dynamic risers.
Construction of structures	Tubular structures are made in yards and deployed in waters.	It contains hull-type structures or ship-shaped structures.
Installation to the location	The structure is dry towed and piled to the position.	It may be dry or wet towed to the place and then secured to the moorings.
External and environmental loads	It is resisted due to the heavy inertia of the compliant structure.	It is resisted due to the inertia of the structure and the moorings.
Regulations	These are governed by rules and regulations of the industry and government.	These are governed by the oil industry, rules and regulations of the government, coast guard, and international maritime.

Offshore platforms are classified into various types,[4] which are discussed later.

The selection of these systems depend on the type of floating unit, the soil condition at the seafloor, and technical feasibility at the location. Further details are provided later, along with various platforms used in offshore production methods.

6.2.5 Production Platforms

Mobile platforms can serve both drilling and offshore production operations. Following are some of the commonly used production platforms used in the industry, as shown in Figure 6.3:[6]

- Tension leg platform (TLP) and mini TLP
- Floating production storage and offloading (FPSO)
- Floating production system (FPS)
- Spar platforms

These platforms have the following main components,[4]

- *Hull*: The hull is of different shapes, ships, pontoons, cylinders, large caissons, spars, and so on, and made up of steel used for water displacement.
- *Topsides*: These decks provide space for equipment, surface oil and gas production, separation facilities, accommodations, and so on.
- *Mooring*: The mooring is a mechanism for fixing or anchoring the floating unit to a specific location. These may be done using steel wires, synthetic ropes, or a combination of both, having significant footprints at the seabed.

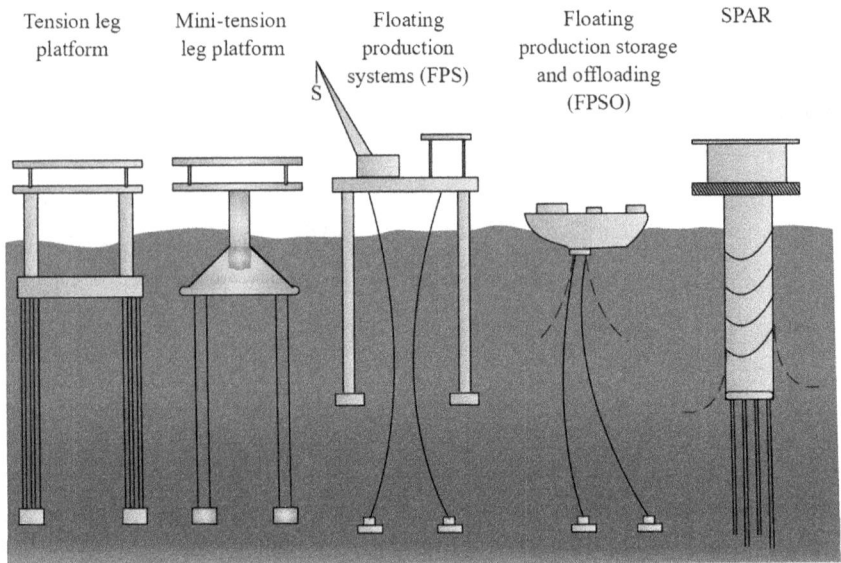

FIGURE 6.3   Schematic of various offshore floating production platforms.

- *Risers*: As the name suggests, they rise from the seabed to the surface facilities to transport produced oil and gas from the wellhead to the deck and sometimes to the shore. These are made up of steel tubes.

Following are the details on the different types of floating platforms used in the offshore industry.

6.2.5.1   TLP

TLPs (shown schematically in Figure 6.4) are floating platform systems with buoyant production services.[4] This platform is permanently moored vertically using tendons fixed to the platform and can be deployed up to a water depth of 300–1,500 m. The system of tendons is referred to as tension leg,[7] hence the name "tension leg platform". The TLP's foundation is kept stable using piling and a complex mooring design. This platform allows translational motion, such as surge and sway, but limits heave (vertical) and rotational motions, such as roll and pitch. Hence, dry trees are generally found in TLP.[8]

TLPs consist of following elements:[4]

a) *Hull structure*: These can be triangular, square, or hexagonal shapes. It entails pontoons, corner, and center nodes, and legs.
b) *Deck structure*: This provides space to accommodate and install drilling and production operations equipment.

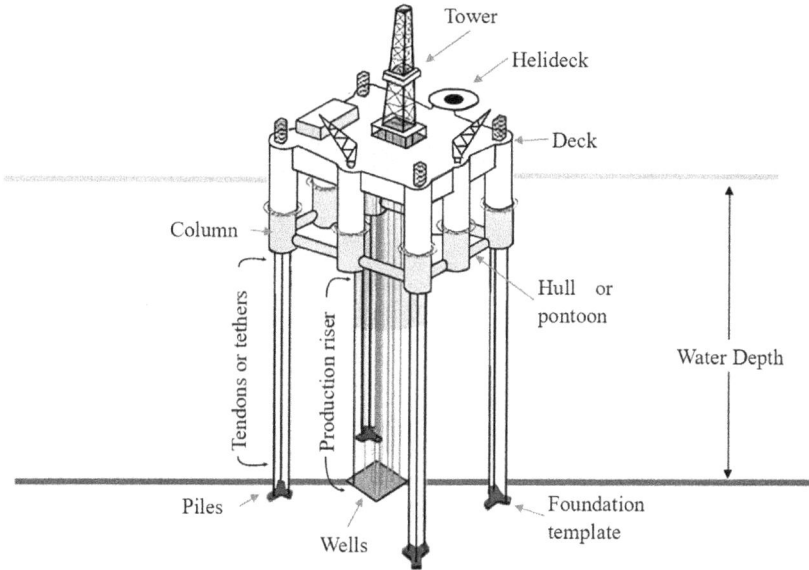

FIGURE 6.4 A schematic of the tension leg platform showing various structural elements.

c) *Foundation with piles and tendons*: Foundation contains four different steel templates in octagonal shape. Eight tubular steel pipes 72-in. in diameter are used to anchor each template to the seabed and are stabbed to the sleeves in the template with a minimum penetration of 58 m.

6.2.5.2 FPSO

FPSO unit is a ship-shaped floating structure used for offshore oil and gas production and storage of oil, shown schematically in Figure 6.5.[4] These have different types of mooring systems. FPSOs are neutrally buoyant floating systems with three degrees of freedom in motion. An FPSO vessel is designed in such a way that it can receive and store hydrocarbon from multiple wellheads, process them, and store the oil until it can be offloaded to the oil tanker. These are primarily used during production operations and generally do not provide facilities for drilling operations.[5] FPSOs are very easy to install and transfer from one location to another and hence are the preferred production platforms. FPSOs can serve the purpose of both a production platform and a storage. A slightly different floating storage and offloading (FSO) vessel is used only to store and offload oil. If a large amount of gas is produced along with oil, it is either reinjected into the well or sent to a nearby storage facility or flared into the atmosphere. As these are floating types of vessels, the wells have a subsea or wet tree completion, and the wellheads are connected to FPSOs using flexible risers.[9] FPSOs can be used in both shallow and deepwater conditions and offer resistance to harsh

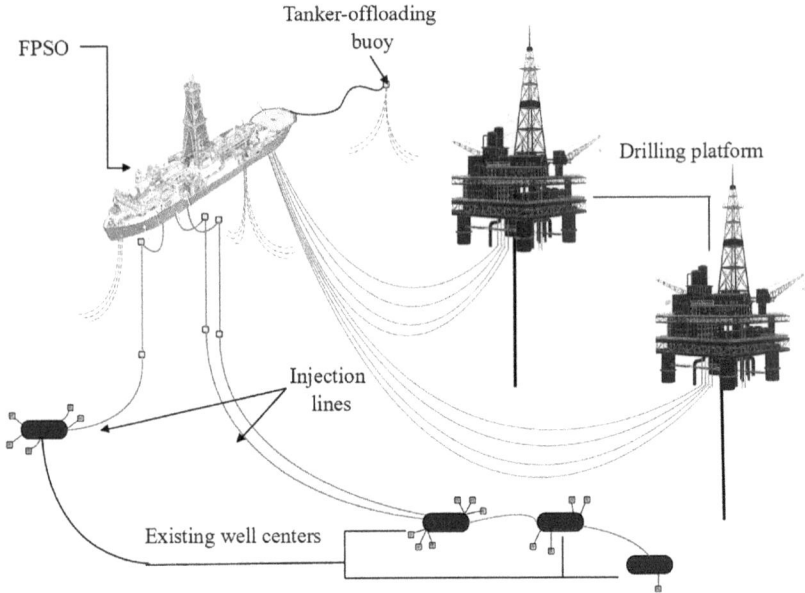

FIGURE 6.5 A schematic of the offloading operations of an FPSO unit.

environment. For positioning FPSOs, a spread mooring system is used in shallow waters, while an external and internal turret mooring system is used in deepwater conditions. FPSOs are preferred in several situations, such as the absence of pipelines infrastructure, and harsh weather conditions, where other production platforms cannot be installed economically. FPSOs need enough space on deck for oil storage, the provision for processing plant, accommodations, and enough ballast size to reduce the impact of motions and facility of subsea production turret.

Some of the significant advantages of FPSOs are that

a) Remote offshore locations can become easily accessible.
b) The need for the long-distance pipeline can be avoided, particularly for deepwater applications.
c) They can be useful for small oilfields, as an investment toward other production systems containing a series of pipelines may not be economical.
d) They are easy to move once the oilfield is depleted or other production issues.
e) Compared to other production platforms, they are cost-effective.

6.2.5.3 FPS

FPSs are ship-like semisubmersible systems or TLP, collecting and processing the oil from the subsea wells. Flexible risers transport oil from the subsea wellhead to the FPS platform, and the gas is reinjected into the reservoir, similar to FPSO. An FPS

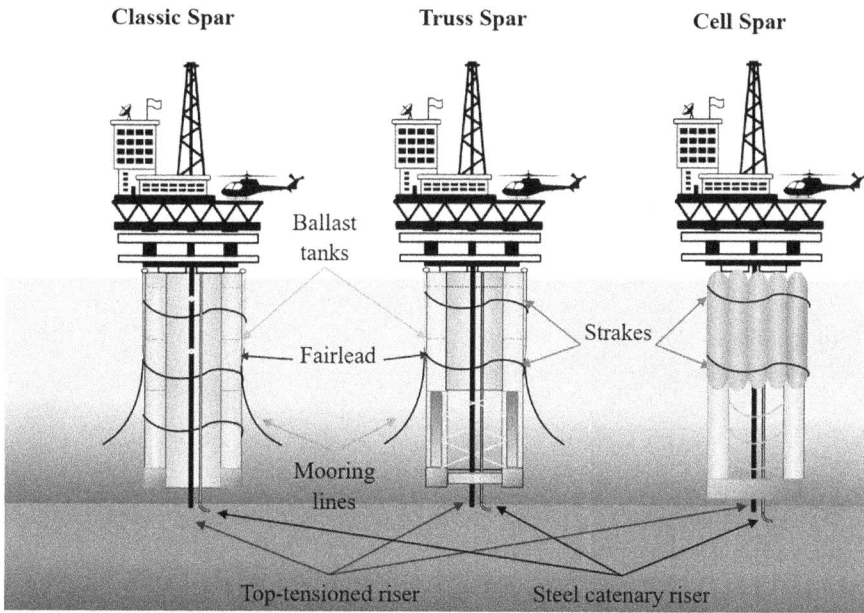

FIGURE 6.6 A schematic of various types of spar platforms.

also contains the control systems connected hydraulically or electrically to the subsea wet tree. The major difference between an FPS and an FPSO is that the FPS does not have the facility to store oil and offloading facility. The fluid is transferred via export risers to the shore.

6.2.5.4 Spar Platform

A spar is a type of floating platform used for deepwater applications.[7] Due to their deep-draft design, these platforms are more stable in harsh conditions and are not affected much due to wind, waves, and currents. These platforms allow dry and wet tree oil and gas completion systems. These platforms contain vertical buoyant cylinders of large diameter with a supporting deck and are permanently anchored to the seafloor using a spread mooring system.[4, 10] The chamber below the deck has denser material for stability. Tanks with heavy ballast are located below the bottom of the cylinders.[11] Helical strakes are used to reduce the impact of vortex-induced motions. These platforms are deployed in the Gulf of Mexico, Norway, and Malaysia.[12] Three spars type platforms are classic, truss, and cell[4] (Figure 6.6):

a) *Classic spar*: This is the oldest design of spar. Its midsection is an extension of the hard tank. The cylindrical hull having heavy ballast tanks is at the bottom of the cylinder.

b) *Truss spar*: This is an advanced version of the classic spar containing three sections: Shortened cylindrical hull with a large diameter, below that truss structure, and finally tanks for heavy ballast. Truss averts the vertical drive of the platform due to waves.

c) *Cell spar*: It is a more advanced third-generation structure. These consist of small-diameter (60–70 ft and 400–500 ft long) cylinders mainly for buoyancy and ballast. These are easy and less time-consuming to fabricate and deploy and are smaller and more economical to deploy in offshore fields.

Spar platforms can be used as drilling rigs and for the production and storage of oil and gas.[11] Spars are more attractive to the industry due to their superior stability in wide ranges of water depths and strength. One of the new variants of these platforms is called spar drilling production storage and offloading, which consists of the advantages of classic spar having a deep draft and oil storage and offloading facilities. These platforms do not depend on subsea pipelines, thus making them a cost-effective solution to produce from marginal oilfields.[13]

6.3 DEEPWATER CHALLENGES

Deepwater conditions attract several challenges to the production of oil and gas. Some of these are listed later. One of the critical factors is to select the right platform for deployment and needs careful consideration based on the ocean and environment conditions, field capacity, water depth, and flow assurance issues. The deepwater condition often faces complex weather issues, such as cyclones and tsunamis. Thus, the platforms need to be designed so that the effects of unforeseen circumstances are considered. The geotechnical issues at the seabed also offer challenges while laying subsea pipelines and production systems as there is always a chance of subsidence due to loosening soil conditions. In the absence of a pipeline, it becomes challenging to predict the production performance of offshore reservoirs. In the case of ultradeep water, deploying a suitable oil transport facility is costly, and unless there are enough economic incentives, field development plans can be suspended. The salinity and marine organism in the seawater conditions may significantly impact the structural stability due to corrosion issues. Despite several of these challenges, deepwater reservoirs attract industry due to their potential to supply oil and gas for several years. The technologies for the deepwater projects have seen maturity and became successful projects in different parts of the world.[14]

6.4 DEEPWATER STATION POSITIONING

The offshore platforms and vessels are often exposed to severe weather conditions in the atmosphere (such as cyclones) and under the ocean (such as ocean currents). It is important to ensure that the offshore platforms used for drilling, production, workover, and so on are in the desired positions. There are mainly six motions that are defined in an offshore platform, of which three are rotational motion, and the rest are translational motion.[4, 15]

The translational motions are

a) Surge (along the x-axis).
b) Sway (along the y-axis).
c) Heave (along the z-axis).

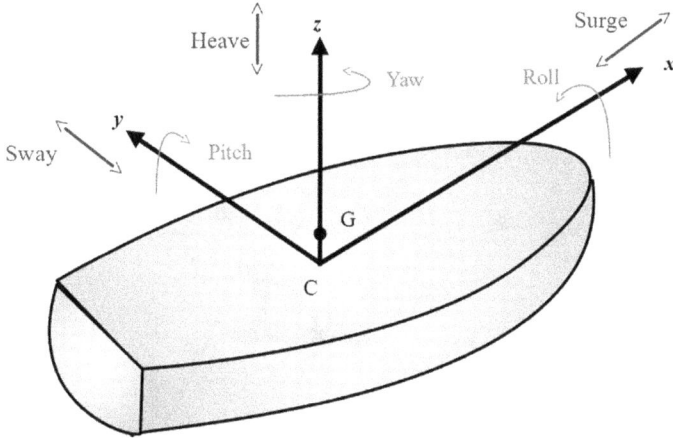

FIGURE 6.7 A schematic diagram explaining the six degrees of freedom of a floating vessel.

The rotational motions are

a) Roll (rotation around front to the back axis).
b) Pitch (rotation around the side-to-side axis).
c) Yaw (rotation around the vertical axis).

Figure 6.7 shows the details on the different types of motion that affect an offshore vessel/platform.[16] In addition to this, other forces act on a floating vessel/platform. Figure 6.8 shows the schematic of these forces impacting the offshore structure. To ensure that these platforms/vessels are kept at station, various methods are used such as conventional mooring or spread mooring, turret mooring, and dynamic positioning system.[17]

6.5 SUBSEA PRODUCTION SYSTEM AND ITS ELEMENTS

For offshore production operations, a subsea well configuration is used in which all the production systems are placed at the subsea. Typically, the production facilities underwater are referred to as "subsea". Subsea wells specifically indicate the oil- and gas-producing well drilled below the seafloor. Generally, the *subsea* prefix can also be used for other terminologies, such as fields, projects, and developments. Subsea conditions require a water-resistant and anticorrosive environment to ensure that the structures retain a longer production life. Subsea drilling operations need special types of drilling rigs, such as jack-up rigs, swamp barges, and fixed platforms. Once the well is drilled, a subsea well is completed for production. Subsea wells with

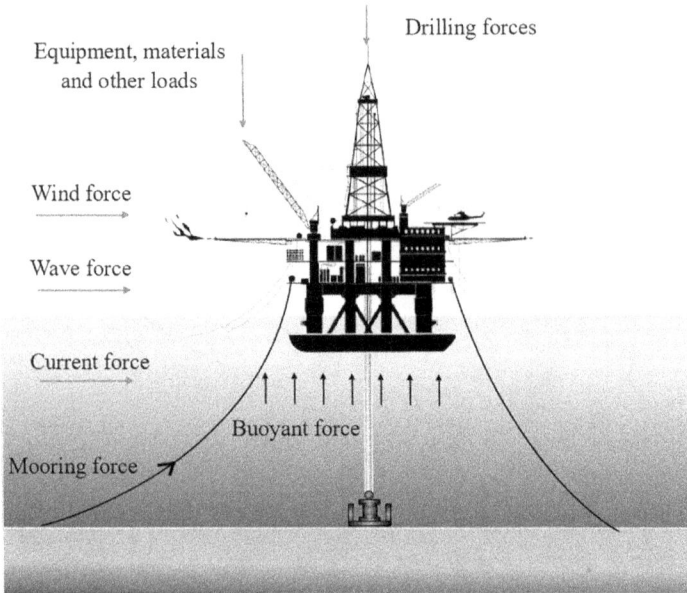

FIGURE 6.8   A schematic of various forces acting on a floating vessel.

components such as Christmas trees, valves, pressure gauges, and connected pipe-lines and risers are laid and maintained at regular intervals to mitigate environmental risks, such as oil spills, corrosions, and external damage due to oceanic conditions.

A typical subsea well production system consists of the following:[18]

- Subsea completed well
- Christmas trees
- Wellhead systems
- Subsea tie-in and flowline system
- Jumpers
- Umbilical
- Riser system
- Subsea equipment and control system to operate the well

Figure 6.9 shows the schematic of all the components and their interactions in a subsea production system.[18, 19] As part of subsea field development, a cluster of wells (or single well) can be connected to a single platform, an FPSO, or onshore produc-tion facilities. In addition, subsea tieback can connect newly developed production wells to the existing subsea production system. The subsea equipment is rated to sustain high pressure and temperature conditions, harsh salinity environment, and critical external and internal loads in the deepwater environment. Often, the sub-sea conditions, as mentioned earlier, lead to severe flow assurance issues during oil and gas transportation. Hence, the fluid produced from subsea wells (oil, gas, water)

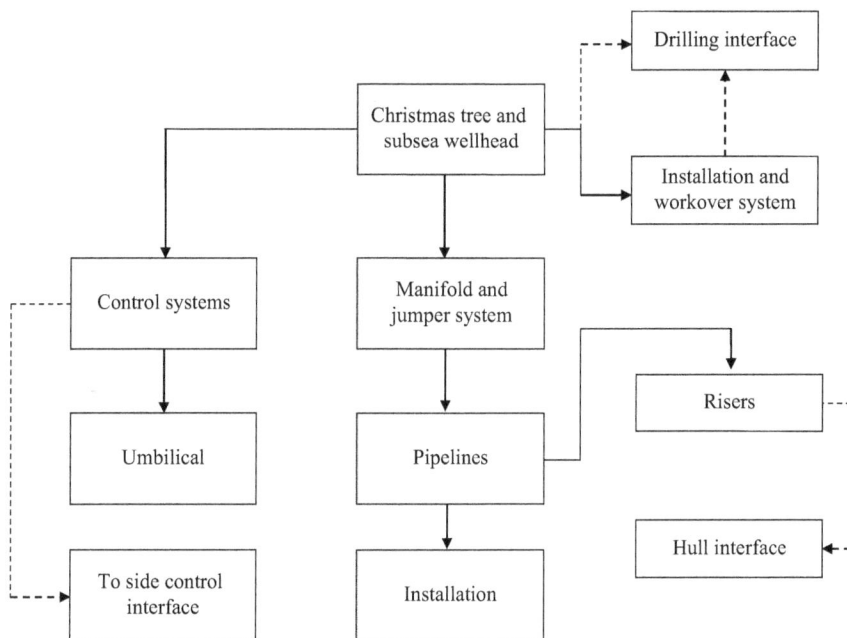

FIGURE 6.9 Schematic of all the components of subsea production system and their interaction.[18, 19]

should be carefully analyzed for their properties to avoid the risk associated with transportation.

Figure 6.10 shows the schematic representation of the subsea production system. As discussed earlier, the subsea system utilizes a dry or wet tree configuration.[18] For the shallower water depths, less than 30 m, jacketed platforms with dry trees are used. Wet tree systems, although expensive, are an attractive option for deepwater conditions due to their reliability. Subsea tiebacks are also gaining traction in the new oil and gas development projects.

6.5.1 Wet Tree and Dry Tree Systems

A schematic of the subsea wet tree and dry tree configuration is shown in Figures 6.11a and b. As the name suggests, dry trees are placed at the surface or platform. In the case of a wet tree, the Christmas tree and related components are placed at the seabed in a cluster, template, or tieback device. Dry tree configurations provide direct access to the well and are deployed with TLPs or Spars. Dry tree systems need motion-optimized hulls to house the riser systems due to constraints on water depth for which they can be used. Wet trees, which are used for deepwater production purposes, are connected and operated remotely. The wet tree configuration consists of a central moon pool for controlling marine risers and trees, manifolds, and BOP. Wet trees account for about 70% of all deepwater development projects.

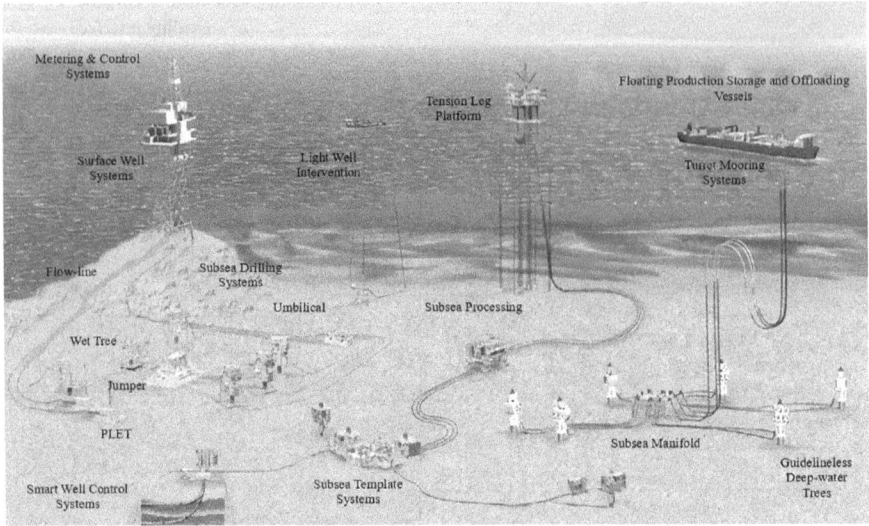

FIGURE 6.10   Schematic of the subsea production system. Modified from Bai and Bai[18] with permission from Elsevier. (Copyright Elsevier, 2010).

6.5.2   SUBSEA FLOWLINES

Subsea flowlines transport oil and gas from the subsea wellhead to the production platform. These can be connected to a single well or cluster of wells to collect the produced oil and gas streams from single/multiple wellheads directly to a manifold or process equipment.[20] Flowlines can be single or multiple bundled pipes inside a carrier pipe. The subsea flowlines can be rigid or flexible and are well insulated to maintain the temperature of the flowing stream within the pipeline under subsea conditions.[18]

Typically, flowline size varies between 1 and 24 inches. The flowlines also provide crossover spools and valves to allow maintenance and pigging operations. Various operating parameters, such as flow rates, temperature, pressure, and the like, decide the selection of flowline. High-pressure, high-temperature, and high-saline conditions in the subsea often demand technical challenges and higher material grades for construction, increasing the cost of a subsea development project. Generally, the materials used to construct flowlines are mainly carbon and stainless steel, which differ in the chromium content (higher for stainless steel). The presence of chromium forms a passive film of chromium oxide, which restricts the oxygen diffusion to the steel surface and prevents further corrosion.[21] If the production stream contains H_2S, it costs more to fabricate a suitable flowline using Inconel to address the severe corrosion issues. In addition, other challenges in subsea conditions include yield strength, reliability of coating, insulation, and maintenance/replacement of subsea flowlines in the case of production issues.

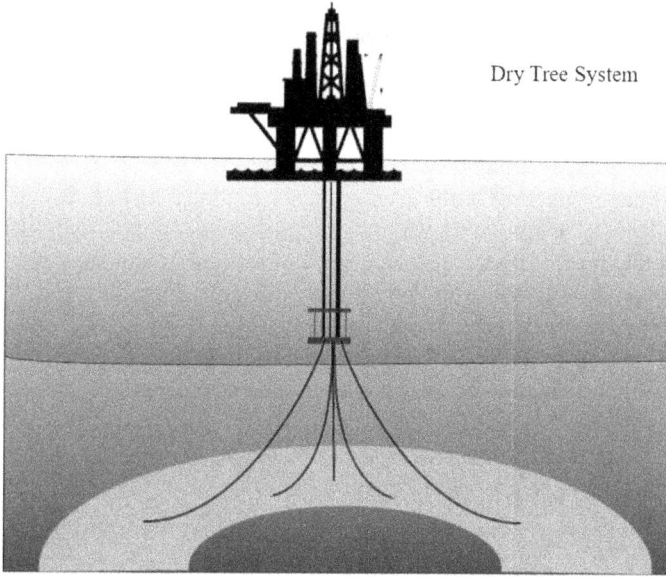

Dry Tree System

(a)

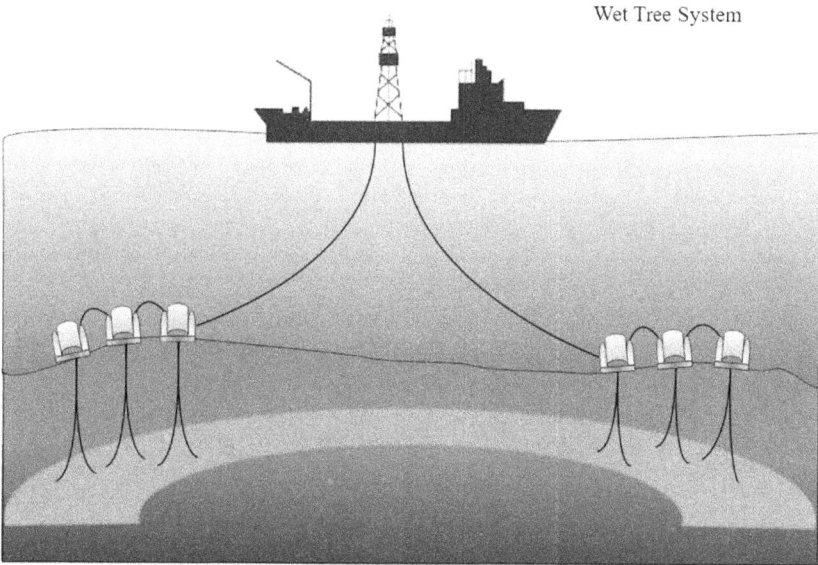

Wet Tree System

(b)

FIGURE 6.11 A schematic diagram of (a) dry tree completion and (b) wet tree completion.

6.5.2.1 Risers

As the name indicates, risers are steel pipes or tubes that rise from the seabed to the hull of the offshore platform. The main purpose of risers is to transport the production streams from the wellhead to the top of the deck.[4, 18] Risers can be used with fixed, jack-up, or mobile platforms and can be designed to be used beyond 10,000-ft water depth.[22] Figure 6.12 shows the pictorial representation of the riser system. Generally, the main function of the riser is to develop interaction between the seafloor and the host platform, and transport the production streams (oil, gas, water) to the surface/platform.[23] Risers also perform various other functions depending on the type, such as drilling and well completion operations from all types of drilling platforms. During secondary or tertiary recovery, risers can be useful for injecting water, gas, or other aids toward the wellhead. Risers can also help during completion, run down a well-completed string, downhole equipment and tools, pigging, and other workover operations.[23]

Generally, riser systems are categorized depending on their functions, as discussed above.[23] These include the following:

- Drilling risers: Used during a drilling operation
- Production risers: Used during oil and gas production operations
- Completion/workover/intervention risers: Used during completion and workover operation
- Export risers

FIGURE 6.12 A pictorial representation of multiple riser configurations.

6.5.2.1.1 Drilling Risers

Drilling risers are further categorized into two types:

- Marine drilling riser
- Tie back drilling riser

A marine drilling riser is used to transport mud to the surface during offshore drilling. It has a large-diameter pipe connected to the BOP at the seabed. A marine riser tensioner placed at the platform gives required stability to the riser during operation in harsh offshore conditions.[24] Often, buoyancy modules are attached to the joints to reduce the tension to make them neutrally buoyant. A tieback riser consists of a single or multiple concentric cylinders or high-pressure pipes extending from the well to the surface BOP.[23]

6.5.2.1.2 Production Risers

Production risers (as shown schematically in Figure 6.13) transfer produced oil, gas, and water streams from the seabed to the platform or surface facilities. These risers are connected to the wellhead and floaters on the surface. The production riser systems are categorized into the following types.

6.5.2.1.2.1 *Freestanding Flexible Risers* These consist of column towers running up to the seafloor to provide direct access to the well. Column towers provide support to the risers without the need for tensioners.[4, 23, 25]

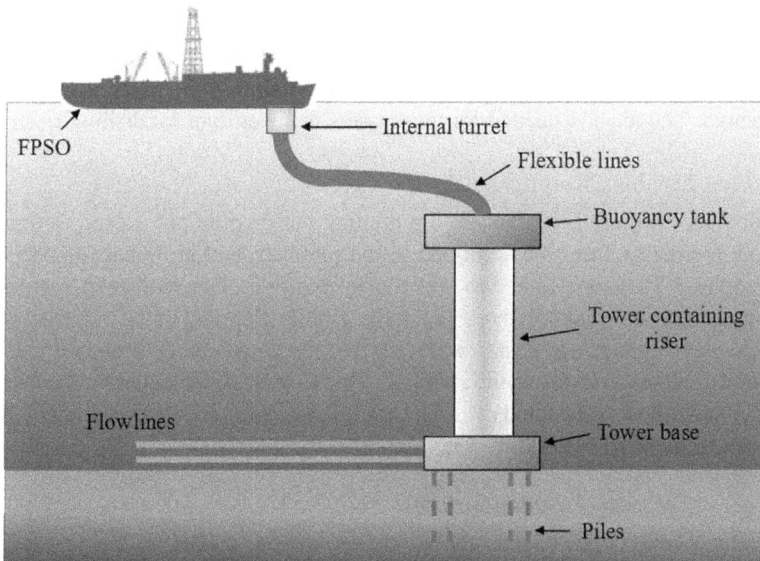

FIGURE 6.13 A schematic of the tower production riser.

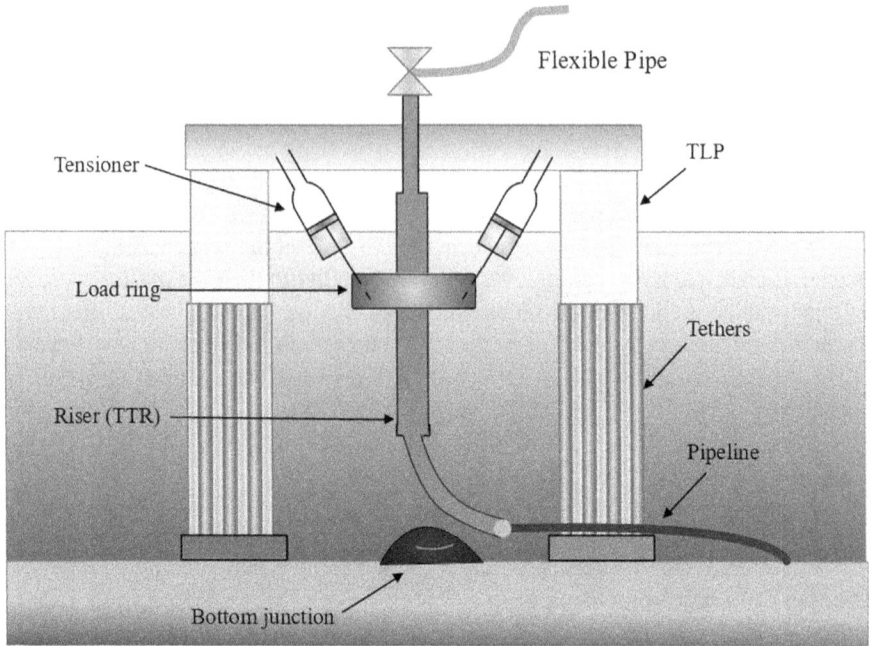

FIGURE 6.14 A schematic of the top tensioned riser.

6.5.2.1.2.2 Top Tensioned Risers Top tensioned risers (TTR) as shown in Figure 6.14 are long, flexible circular cylindrical pipes, connecting the seabed to the floating production platforms (such as TLPs or spars).[18, 26] As these risers are well submerged, hence they are subjected to continuous, steady current and oscillatory wave impacts of varying intensities. As the name indicates, these risers are associated with tensioners at the top to provide stability and maintain the angle under external load factors. The tension needed for these risers is lower than for drilling risers.[18, 23]

6.5.2.1.3 Flexible Risers

Flexible risers (Figure 6.15a and b) are nothing but flexible pipes used to transport production streams from the wellhead to the platform and are ideal for deepwater applications.[18] These are linked to wet tree systems and can withstand vertical and horizontal motions, making them ideal for mobile floating units. Desired shapes (steep-S, lazy-S, plain wavy, lazy wavy, etc.) to the risers can be imparted using the buoyancy modules to improve the stability.[4] Due to their flexible nature, they are also suitable for shallow waters and cyclonic conditions. With recent developments, these risers can be used for a 3,000-m water depth.

6.5.2.1.4 Catenary Production Risers

Steel catenary risers (SCRs), as shown in Figure 6.16, consist of a rigid steel pipe having a large diameter with enough bending stiffness. The shape of an SCR is

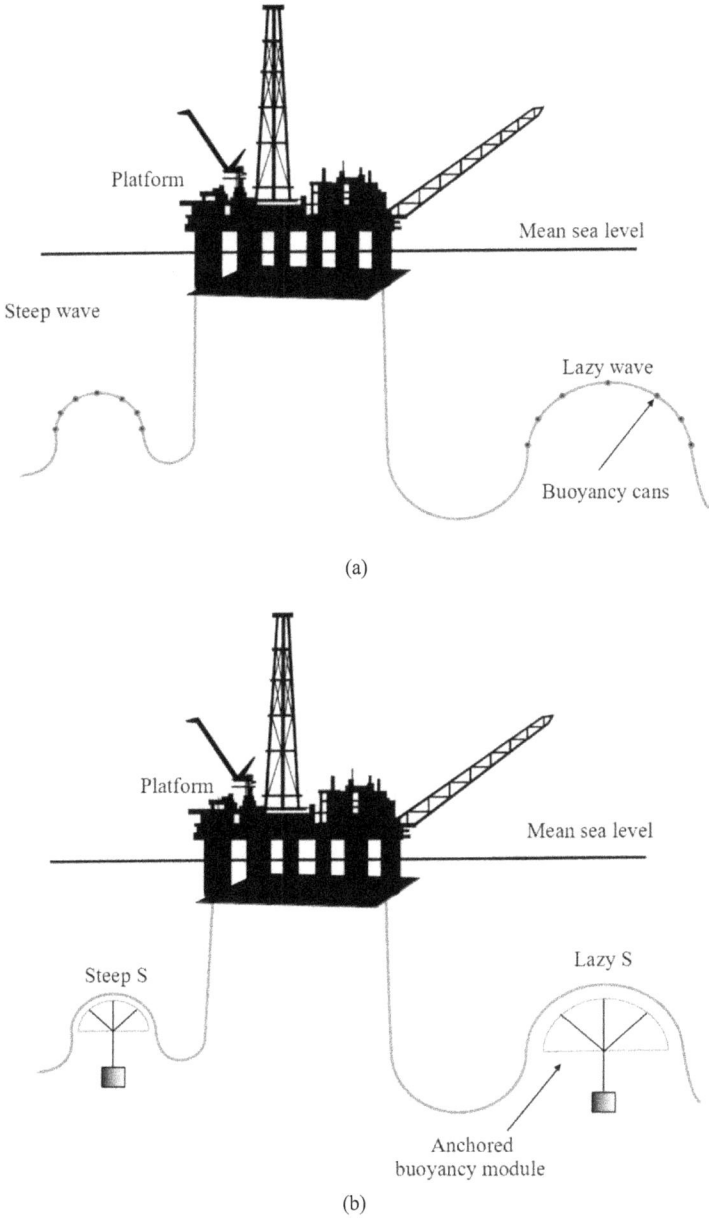

FIGURE 6.15   Types of flexible riser: (a) steep wave and lazy wave type, and (b) steep S and lazy S type.

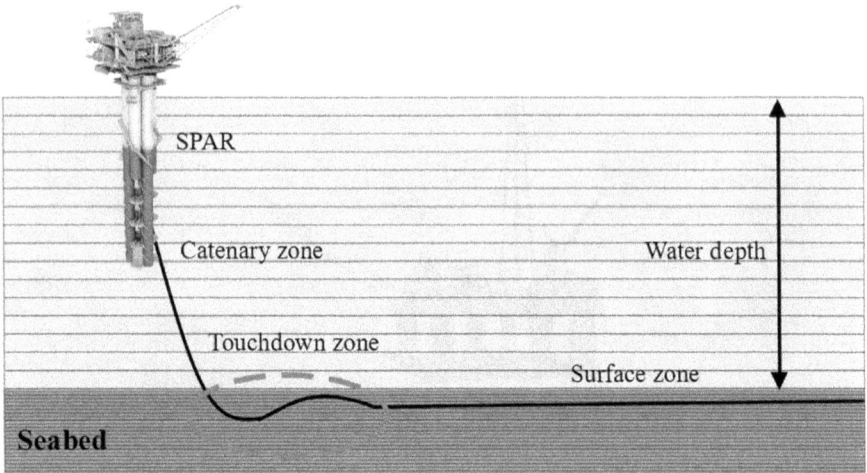

FIGURE 6.16   A schematic of the steel catenary riser.

controlled by its weight, buoyancy, and surrounding hydrodynamic forces acting on it.[27] Initially, these were used as export risers for floating and fixed production platforms.[18, 28] These are free-hanging risers without having intermediate buoys or floating devices. Ball joints allow rotation to accommodate the motion effect due to vessels, waves, and current.[18]

6.5.2.1.5   Hybrid Risers

Due to the need to improve the functions of conventional top tensioned risers and flexible or catenary risers, hybrid risers have been introduced. A hybrid riser (Figure 6.17) consists of rigid steel and flexible pipes. The part of the riser component is attached to the buoyancy modules around the central tubular section. The flexible pipes absorb all the dynamic loads, while the steel catenary risers connect to the subsurface buoy for support. The advantage of these risers is that they help reduce the decoupling effect between production units and risers. The riser load is transferred to the production unit, minimizing the fatigue. These risers can be preinstalled before the production unit is deployed at the location. These are cost-effective and provide great reach to deepwater fields.[18]

6.5.2.2   Manifolds, Jumpers, and Tie-Ins

6.5.2.2.1   Manifolds

Subsea manifolds are located at the seabed and are used in subsea oil and gas fields production operations. Manifolds are arrangements of pipes and valves that connect multiple flowlines that are used to combine, distribute, control, and monitor the flow rate. In general, a manifold is fluid-collecting and distribution equipment that connects more than one pipe to one channel or at a single place. Manifold systems can range from simple supply chambers with several outlets to multichambered flow control units consisting of many integral valves and other network interfaces. They can

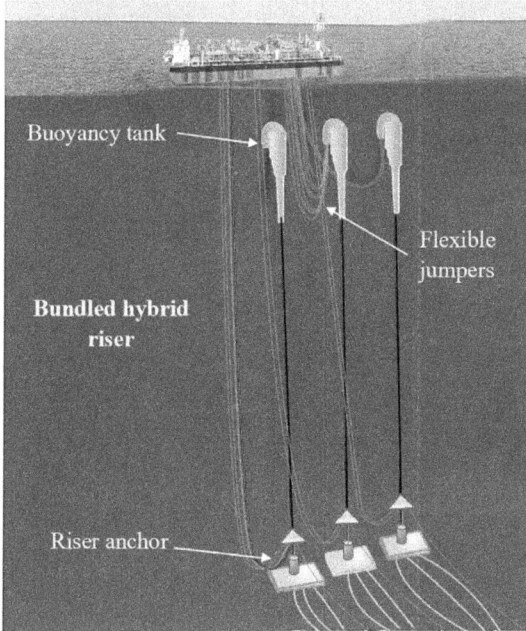

FIGURE 6.17   A pictorial representation of the bundled hybrid riser.

be installed with an array of wells to combine the production from different production wells. They can also be used for the injection well for enhanced oil recovery operation at the subsea. Several types of manifolds are practiced in the industry. Manifolds can be fixed to the seabed using piles or skirts. The manifold's size and the type depend on the number of wells and the production rate.[29]

Manifolds are typically used to fulfill the following functions:[29]

a) To blend the production streams from more than one well (production manifolds).

b) To distribute water and/or gas (injection fluids) from different producing or injection wells (injection manifolds).

c) To direct the fluid flow of the production/injection well from the manifold header.

d) To monitor and control the volumetric flow rate from different wells.

e) Transporting the fluid from a well to nearby offshore facilities during the well testing.

f) Isolate the individual wells and the flowlines using a using pipeline manifold valve.

g) Pigging operation of a subsea flowline system.

h) Develop subsea reservoir development plans by increasing the production capacity.

i) Accommodates the well control system, including sensors, flow meters, etc.

Manifolds are classified based on their primary function:[18]

a) Production and/or test manifolds: Control fluid flow rate from individual wells to the production and test headers.
b) Gas injection manifolds: Inject gas from a gas well.
c) Gas lift manifolds: Inject the gas into production tubing to improve the oil recovery (gas lift).
d) Water injection manifolds: Inject water into the injection well during the secondary or tertiary oil recovery process to increase oil recovery.
e) Choke or kill manifolds: Control the well operation during blowout control.

Manifolds are also categorized based on environmental application into two categories:

- Onshore or on-land manifolds
- Offshore or subsea manifolds

The onshore manifolds are placed at the group gathering station (GGS) or near the oil- and gas-producing well clusters and connect all the nearby well clusters with production and injection lines. Subsea manifolds are placed at the seabed within the periphery of the oil and gas production or injection wells.[18, 30] Subsea manifolds are fixed to the seabed using subsea piles foundation or mud mats with skirts.[18] Some of the offshore manifolds in use are a simple pipeline end manifold (PLEM) or pipeline end termination (PLET), and large subsea manifold structures.[18, 30]

6.5.2.2.2 Jumpers

Jumpers are a small section of pipes or connectors, used to connect two or more equipment or facilities, such as subsea Christmas tree and production manifold,[18] so that the production or injection fluids can flow between different types of equipment facilities. Jumpers can also be used to connect manifolds and the base of the riser. The length and diameter depend on the flow rate and the distance between different components. A typical jumper mainly consists of two end connectors at both sides of a spool with a connecting pipe body in between them. A spool or main pipe serves to transport the fluids while the end connectors connect the main pipe to two different types of equipment. Jumpers also account for any possible contraction or expansion during any temperature or pressure variations within or outside the system. Jumpers also help accommodate any tolerance due to fabrication or installation of existing subsea equipment or facilities.[31] Jumpers are designed and fabricated after the subsea equipments are installed at the seabed. After designing jumpers for a particular need, they are transported to the offshore field and installed at the seabed, connected to the concerned equipment, pressure tested for any leaks, and then commissioned.

Jumpers are categorized into two main types, namely, rigid and flexible.[32] As the name suggests, the main pipe body is rigid and flexible for these jumpers, respectively, due to the material used to fabricate it. Figure 6.18 shows an example of rigid jumpers installed at the seabed. Rigid jumpers are made up of solid steel pipes with connectors at both spool ends. Figures 6.19 (a–c) show different shapes of the rigid

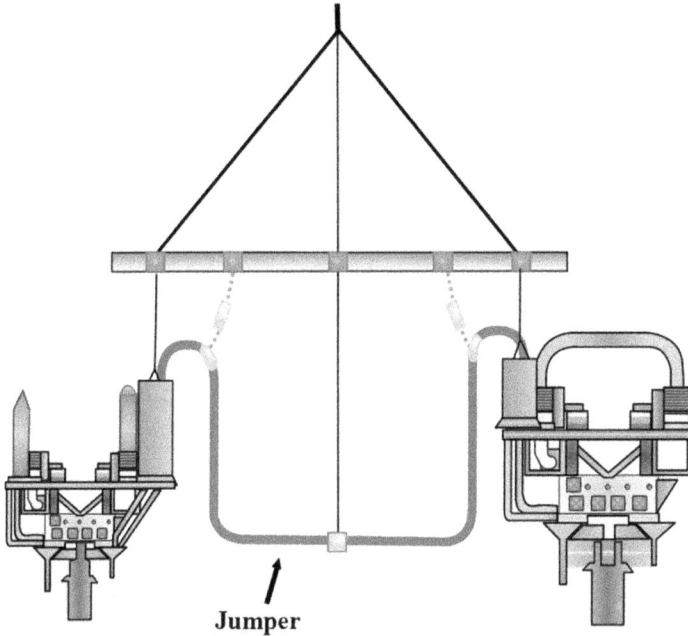

FIGURE 6.18   Subsea rigid jumper.

jumpers. Depending on the need and space constraints, rigid jumpers can also be manufactured in a variety of many other geometrical shapes. Rigid jumpers are the oldest and simplest types of jumpers used by the offshore oil and gas industry.[32] On the other hand, flexible jumpers consist of the flexible pipe due to the material used to fabricate them. Like rigid jumpers, the flexible jumpers also consist of two end connectors (one each on both the ends) and a flexible pipe body. Flexible jumpers are more versatile compared to rigid jumpers. Flexible jumpers are more suitable for static and dynamic loading environments. These are mainly used to separate rigid risers from the FPSO vessel to keep the facility safe from any fatigue failure due to motions of FPSO. Figure 6.20 shows a schematic of subsea rigid and flexible jumpers used in the freestanding hybrid riser (FSHR) system.[33]

6.5.2.2.3   Tie-Ins

Subsea tie-in systems are vital building blocks of subsea installations. They offer secure and leak-proof connections between subsea components such as flowlines, risers, umbilicals, and pipelines to safely transport production streams to the surface facilities.[32] Various tie-ins are necessary for flowline systems such as manifold-Christmas tree, interconnected Christmas trees, and flowline-manifold tie-in. For control systems, tie-ins connect the umbilicals between Christmas trees or subsea manifolds.[32] Tie-ins are also used to lock jumpers between the two subsea components and various connectors.

Based on the installation, tie-in systems are categorized into vertical and horizontal configurations. A vertical tie-line and a rigid jumper are shown in Figure 6.21.[32]

"M" Shaped (Bends)

(a)

"M " Shaped (Elbows)

(b)

Inverted "U" Shaped

(c)

FIGURE 6.19   Some of the most commonly used configurations of the subsea rigid jumpers.[32]

The vertical tie-in is directly connected to the facility during the tie-in process. These systems require relatively less operational time and do not require any pull-in requirement during their installation. Horizontal tie-in systems are generally installed between two existing subsea equipment or facilities.[32] Typical horizontal tie-in system is shown in Figure 6.22. Vertical tie-in systems typically require shorter jumpers than horizontal tie-in systems. Vertical tie-in systems are more convenient in the case of subsea space constraints such as complex subsea flowline/pipeline networks and are useful for production units in deepwater.[31, 32]

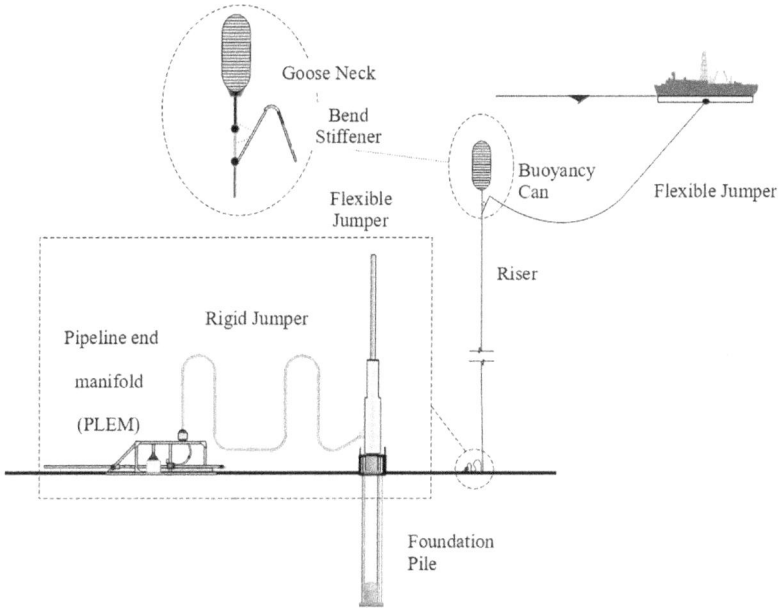

FIGURE 6.20 Rigid and flexible jumpers used in FSHR system.[34, 35]

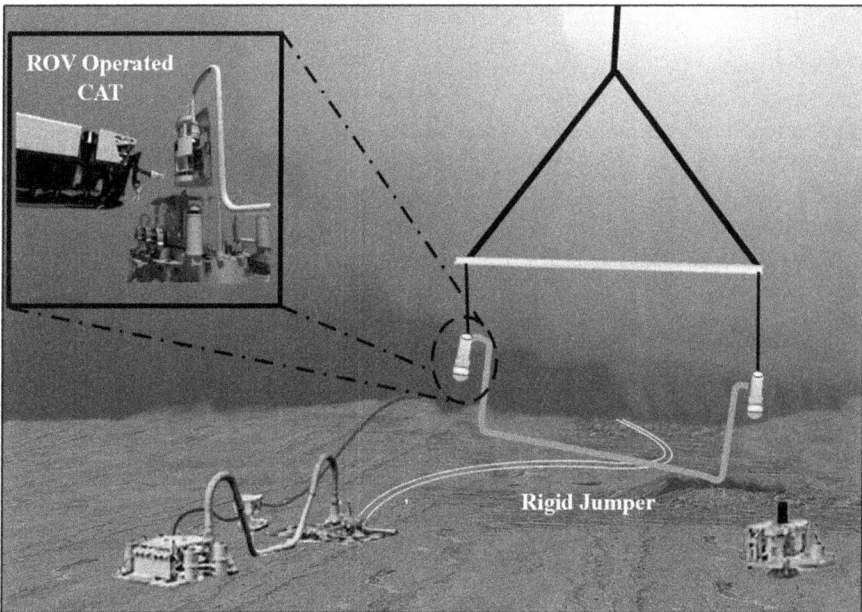

FIGURE 6.21 Vertical tie-in systems used with a rigid jumper with the remotely operated vehicle (ROV) operated connector actuation tool (CAT).[36]

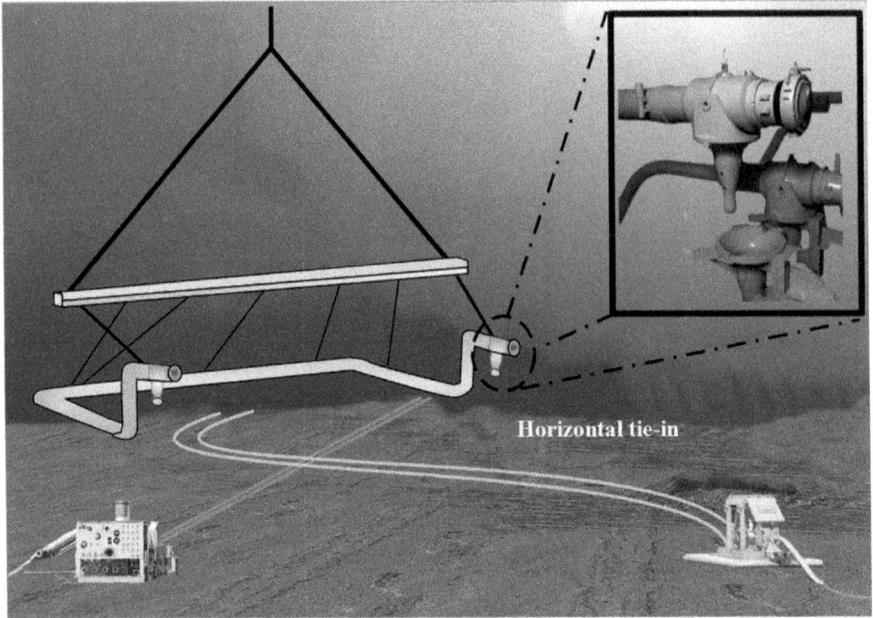

FIGURE 6.22   Pictorial representation of a horizontal tie-in system with a rigid jumper.[36]

6.6   DEEPWATER TEMPERATURE REGIMES

The temperature is one of the main operating conditions for oil and gas transport under deepwater conditions. During the initial phases of oil and gas field development, temperature profiling was not considered seriously. However, with the advent of production issues and technological challenges, information on the temperature regimes under subsea conditions attracted attention. Low-temperature conditions in deep waters often attract various flow assurance issues such as gas hydrate, wax, and asphaltene deposition. In addition, scale and corrosion of offshore facilities are other problems to be addressed. The temperature near the wellbore in the subsea could be as low as 40°F. Hydrates formation could be very severe under such low temperatures and can block pipelines and tubing and interfere with the BOP operations. Information on the subsurface temperature regimes is vital to properly plan the offshore oil and gas field development strategies.[3]

Generally, temperature conditions in the ocean range from 28–86°F. The surface water is generally warmer at low latitude regions, while it is cold near the poles (Figure 6.23).[37,38] It is interesting to observe that at the equivalent latitudes, seawater surface temperature in eastern regions is colder than in western regions due to ocean surface currents (discussed later).[37] The subsea atmosphere is typically very dark and cold below a depth of 500 m (1640 ft). As mentioned earlier, the surface water is warm, and the temperature of the water decreases as the depth increases, resulting in the average ocean seawater temperature being about 4°C, close to the refrigeration temperature.

A typical temperature profile for open and mid-latitude ocean systems is shown in Figure 6.24. The figure shows that surface water is warm due to sunrays, as they can

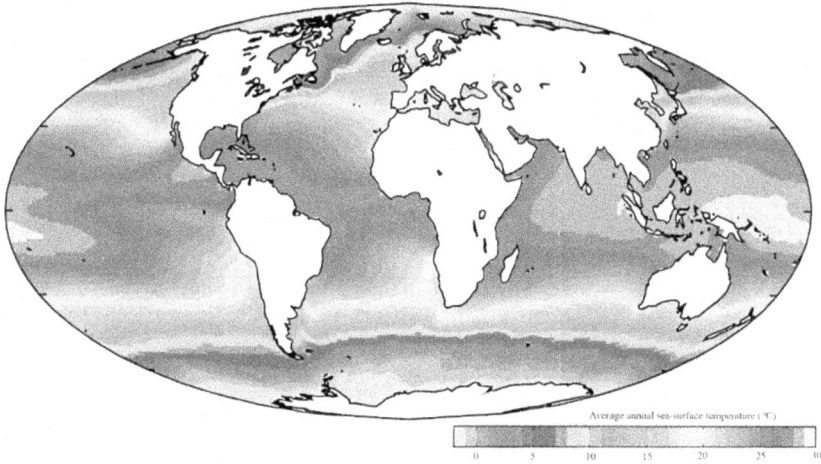

FIGURE 6.23 Global average annual sea surface temperature (Reproduced with permission under Creative Commons Attribution 4.0 International License, Earle S, "Physical Geology").[37, 38]

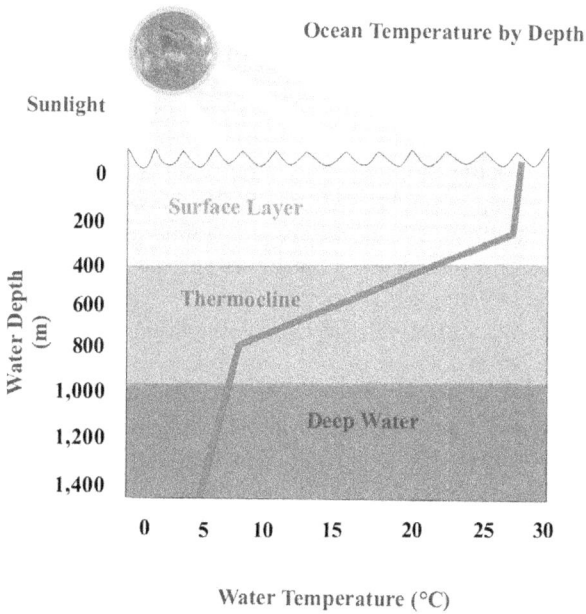

FIGURE 6.24 Schematic of a seawater temperature profile with depth.[39]

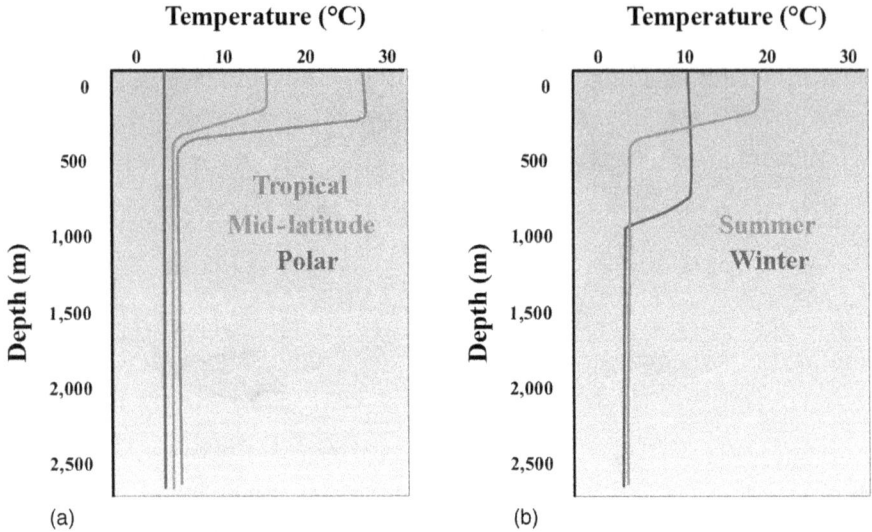

FIGURE 6.25 (a) Schematic of the temperature profiles with depth in sea for tropical, mid-latitude, and polar regions.[37] (b) Temperature variation with depth in sea during summer and winter region in seawater.[37]

only penetrate seawater up to 1000 m, resulting in a lesser density than water at deeper depths. The temperature remains relatively constant up to 100–200 m. In this region, the surface winds, ocean currents, and waves keep the water in a mixed state, and hence, this layer is referred to as a mixed layer. The temperature falls rapidly below this mixed layer over a narrow increase in depth. This profile is referred to as thermocline. A thermocline is, thus, mainly a transition layer between the warmer sea surface and cooler deep water below.[39] Below the thermocline, the temperature is relatively constant at about 2°C till the seabed. Below the thermocline, as there is hardly any heat source available, the temperature remains constant, making it one of the most thermally stable regions on earth.[39, 40]

Temperature profiles differ with the latitude. Figure 6.25a shows a schematic of the temperature variation at different locations. At tropical and mid-latitude regions, the thermocline is very distinct. In contrast, the temperature profile in the polar regions is more or less constant from the surface to the deep water. In the low- and high-latitude regions, the surface temperatures are more or less constant. In contrast, in mid-latitude regions, there is variation in surface temperatures of about 8–15°C from summer to winter due to seasonal fluctuations. The thermocline is generally more distinct in summer than in winter. In winter, the thermocline is typically stiffer than in summer in the mid-latitude region (Figure 6.25b).[37]

6.7 OCEAN WATER SALINITY AND DENSITY PROFILE

Besides thermal variation (Figure 6.26a), salinity and density also vary with water depth. It is known that pure water density is 1,000 kg/m³ at 4°C. Due to the presence

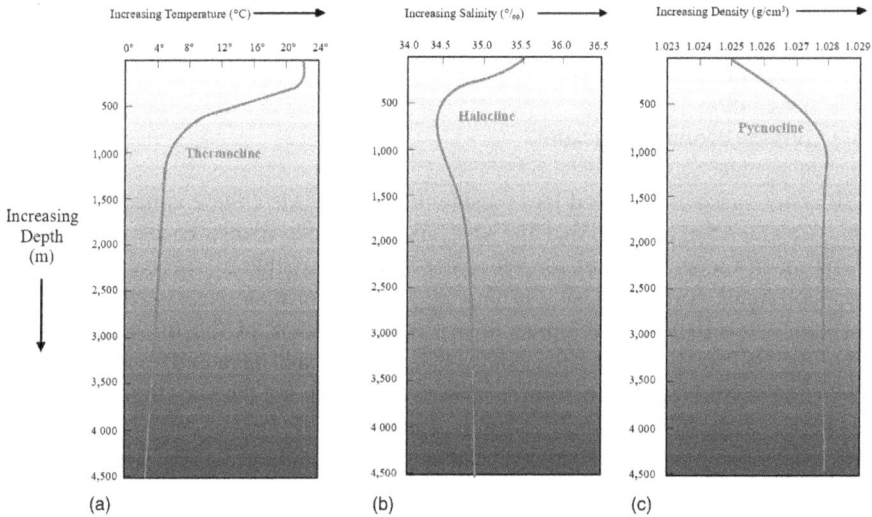

FIGURE 6.26   Schematic of the (a) thermocline, (b) halocline, and (c) pycnocline for ocean waters versus depth.

of salts, seawater is denser than pure water. The presence of salt also affects the freezing point of water. Salt decreases the freezing point of water to below 0°C. The seawater density generally varies between 1.02 to 1.03 kg/m^3.[37] The salinity of ocean water changes rapidly with depth, and this salinity profile is called halocline (Figure 6.26b). In the Atlantic Ocean, distinct haloclines are observed up to about 3300 feet. The seawater density may increase by about 0.7 kg/m^3 per increase in salinity by 1 kg/m^3.[41] Temperature also affects the density of the seawater. At low temperatures, the density of seawater increases, while with an increase in temperature the density decreases. Like the temperature profile, the density also increases as we go from the surface to the ocean bottom. The lower part of the ocean has more dense water than the surface water. The profile of density variation in ocean water is referred to as pycnocline (Figure 6.26c). Water circulation typically happens in the same layer of the density.

6.8   SUBSEA PROCESSING AND SEPARATION

The production of crude oil from a subsurface environment carries many undesirable components, such as water, sand, and chemicals. The separation of desirable components from the undesirable in the subsurface is crucial for mitigating the flow assurance issues. Subsea processing (SSP) is the collective term, encompassing the treatment and handling of the fluids produced from the reservoir to mitigate the flow assurance related issues during their transportation to the platform or onshore.[18] An aim of SSP is the production and processing hydrocarbons from a reservoir that are suitable to sell directly to the market.[42] A large number of techniques and technologies are involved in this process, either implemented or in the developing stage.[18] The common SSP techniques include

- Gas–liquid separation.
- Gas compression.
- Liquid pressure boosting.
- Desanding.
- Subsea cooling application.

SSP is not space-constrained like topside processing. Also, there is naturally available seawater for cooling, and the separation of fluids is also easier on a warm, fresh-out-of-the-well stream. Development is made possible by subsea trees, risers, and umbilicals. SSP has been a promising solution for many years and has helped revolutionize offshore developments worldwide.[43] Some of the financial benefits from the SSP include reduced capital expenditures and increased financial gains. The main driving factors for subsea processing design are the very high mean time between failure (MTBF), extreme subsea environment (pressure, temperature, salinity), mechanical and operational robustness, and redundancy, along with the spare capacity for single-component failure. Subsea processing is classified into four different categories based on separation equipment, water, and sand disposal system as shown in Table 6.2.[18, 44] Some of the important subsea processes are discussed in the following.

6.8.1 Gas–Liquid Separation

In gas–liquid separation, the gas rises to the surface in a separate pipe while the liquids and potential sand are pumped topside with a multiphase pump. This solution allows depressurizing the system to avoid forming hydrates in case of a shutdown, relevant for transporting high gas and low water level fluid over short distances to the platform for onshore separation.[45]

TABLE 6.2
Details of Various Subsea-Processing Types

Subsea Processing	Nature of Separation	Equipment Used	Sand Disposal	Water Disposal
Type 1	Direct handling of multiphase flow	Multiphase pump	None	None
Type 2	Partial separation	Multiphase pump and separator, wet gas compressor possible	None	Partial reinjection possible
Type 3	Complete separation	Single or multiphase pump, separator, and scrubber	Addressed	Reinjection of majority water stream
Type 4	Pipeline quality oil and gas export	Single-phase pump and compressor, multistage separator	Addressed	Complete reinjection

6.8.2 Gas Compression

The actual gas compression station is dependent on the well stream and the surrounding infrastructure. Gas compression at the surface is costly and consists of numerous components. Achieving the gas compression stage at the subsea level is quite challenging as many risk factors are involved. The main attention is to minimize the components without negotiation, resulting in lower cost. A typical gas compression station at a subsea level consists of:

- An inline separator,
- A cooling stage,
- A scrubber,
- Liquid pumping, and
- A compressor.

The produced fluid is cooled, and liquid is knocked out in the scrubber.[46]

6.8.3 Sand Handling

The main methods for removing sand involve using an in-line desander or removing the sand accumulated in the bottom of a separator vessel. The in-line desander separates sands and other suspended solids from a multiphase flow based on the difference in density and centrifugal forces.

6.8.4 Subsea Cooling Application

The fluid produced from a hydrocarbon well is at times very hot, sometimes over 100°C, and thus needs to be cooled to meet the pipeline design temperature. Reducing the temperature of the fluid flow decreases the costs of the pipeline installation. A lower design temperature allows for a cheaper material and standard coating. Cooling also reduces the corrosion rate due to a lower temperature and water knockout. For existing fields, coolers enable tie-in of high temperature (HT) production system in low design temperature (DT) pipeline.

6.8.5 Subsea Separation

The subsea separation gives reasonable control to the fluid flow by mitigating flow assurance issues. Unlike onshore separators, the separators used at the subsea level require robust engineering design, thicker walls, and corrosion-free materials for construction. Three important configurations of subsea separators are cylindrical, spherical, and horizontal (Figure 6.27). The spherical separators are still in the development stage, and a lot of research and development are necessary to analyze the efficiency at the subsea level. On the other hand, although vertical separators offer a low footprint level than horizontal separators do, they are prone to vortex-induced vibration.[47] Newer separation technology is the linear pipe design that saves weight and fabrication costs. Linear two-phase and three-phase separation are based

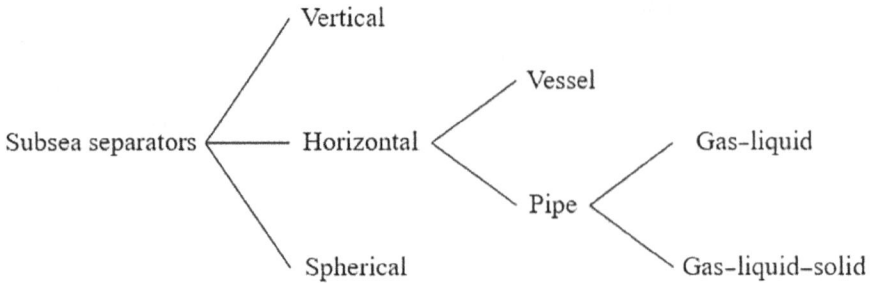

FIGURE 6.27   Flowchart of various subsea separator types.[47]

on horizontal gravity-based separators with an electrical submersible pump (ESP). The three-phase linear separator is a more compact version of the two-phase linear separator. The produced stream enters the linear separator and forms a wavy pattern due to W-A-V geometry (upward and downward branches), causing it to separate into different phases.[47] The processing and boosting technology was developed in Brazil to increase production and maximize recovery. Its lessons clearly stated the importance of coordination between the academy, university, research center, supplier, industry, and the government for achieving the fruit of SSP. Technology based on the ESP has been shown to be a better solution to maximize the well flow rate.[48] Statoil and Total are the main leading promotors of SSP, and the Norwegian government has established an OG21 ("Oil and Gas in the 21st Century") strategy to enable longer distance transport.[42]

6.8.6   Deployment Window and Success of SSP

Close collaboration between suppliers and operators is vital for any subsea project to be successful. Suppliers have to deal with the feasibility and market potential of new and expensive technology. In contrast, operators are concerned about the performance potential of subsea technologies in existing and new developments. There is an opportunity window in the development and deployment of SSP to be profitable. Identifying this window is vital and can be accomplished via technology development–driven and long-term collaboration between suppliers and field operators. Suppliers work on the inputs provided by the operators but understanding the process themselves is important. On the other hand, operators gain knowledge of possible deployment scenarios with existing technology. Apart from collaboration for a successful subsea processing deployment, some additional factors are clear goals, cost management, and visible benefits. The operators and industry need to set clear goals for their technology requirements. Suppliers need to have a clear vision of the subsea processing value addition and asset enhancement. When evaluating an asset, there should also be a benchmark case for cost management. The overall cost is influenced by many factors, including future expectations of oil and gas prices, thus motivating the development of low-cost solutions.[42] With the revolution of technology and advancement in science, everything has been made possible in recent years.

REFERENCES

1. Oil & Gas Portal. Subsea Technology and Equipments. http://www.oil-gasportal.com/subsea-technology-and-equipments/ (accessed August 5, 2021).
2. Valenchon, C. P., Anrès, S. J., Baudouin, B. F., & Biolley, F. M. (2000). Early production systems (EPS) in ultra deep water, a way to improve reservoir management and field economics. In *SPE European Petroleum Conference*, Paris, France, October 24–25. (SPE-65167-MS). https://doi.org/10.2118/65167-MS
3. Feng, M. (2011). *The Temperature Prediction in Deepwater Drilling of Vertical Well.* PhD. Thesis. Texas A&M University, Texas, USA.
4. Laik, S. (2018). *Offshore Petroleum Drilling and Production.* CRC Press, New York. https://doi.org/10.1201/9781315157177
5. Chakrabarti, S. K. (2005). *Handbook of Offshore Engineering* (Vol. 1). Elsevier, Amsterdam, The Netherlands.
6. Singh, B. different types of offshore oil and gas production structures https://www.marineinsight.com/offshore/different-types-of-offshore-oil-and-gas-production-structures/ (accessed August 5, 2021).
7. Speight, J. G. (2015). Offshore Platforms. In *Subsea and Deepwater Oil and Gas Science and Technology.* Speight, J. G. (Ed.). Gulf Professional Publishing, Elsevier, Oxford, UK, pp. 71–106. https://doi.org/10.1016/B978-1-85617-558-6.00003-9
8. Rigzone. How Does a Tension Leg Platform (TLP) Work?. https://www.rigzone.com/training/insight.asp?insight_id=305&c_id= (accessed August 5, 2021).
9. Modec. FSO Ta'Kuntah. https://www.modec.com/project/detail/cantarell.html (accessed August 5, 2021).
10. Offshore. (2002) State-of-the-art of spread moored systems for deepwater floating production platforms. https://www.offshore-mag.com/production/article/16759723/stateoftheart-of-spread-moored-systems-for-deepwater-floating-production-platforms (accessed August 6, 2021).
11. Rigzone. How do spars work? https://www.rigzone.com/training/insight.asp?insight_id=307&c_id=12 (accessed August 6, 2021).
12. Global Security. Spar platform. https://www.globalsecurity.org/military/systems/ship/platform-spar.htm (accessed August 6, 2021).
13. Liu, W., Wang, J., Huang, J., Liu, Y., & Li, Y (2017). Spar platform oil storage and offloading system design. In *27th International Ocean and Polar Engineering Conference*, San Francisco, California, USA, June 25–30. (ISOPE-I-17-013).
14. Guzman R., Carvajal P., & Thuriaux-Alemán, B. (2013). Opportunities and challenges for global deepwater players. https://www.adlittle.com/sites/default/files/prism/Global_deepwater.pdf (accessed August 6, 2021).
15. Patel, M. H. (1994). Offshore Engineering. In *Mechanical Engineer's Reference Book* (12th Edition). Smith, E. H. (Ed.). Elsevier, Oxford, UK, pp. 14-1–14-34. https://doi.org/10.1016/B978-0-7506-1195-4.50018-X
16. Ibrahim, R. A., & Grace, I. M. (2010). Modeling of ship roll dynamics and its coupling with heave and pitch. *Mathematical Problems in Engineering*, 2010, 934714. https://doi.org/10.1155/2010/934714
17. Wilson, J. F. (2003). Offshore Structures (Marine Engineering). In *Encyclopedia of Physical Science and Technology* (3rd Edition). Meyers, R. A. (Ed.). Academic Press, Elsevier, pp. 161–168. https://doi.org/10.1016/B0-12-227410-5/00512-3
18. Bai, Y., & Bai, Q. (2010). *Subsea Engineering Handbook.* Gulf Professional Publishing, Elsevier, Burlington, MA, USA.

19. Woo, J. H., Nam, J. H., & Ko, K. H. (2014). Development of a simulation method for the subsea production system. *Journal of Computational Design and Engineering*, 1(3), 173–186. https://doi.org/10.7315/JCDE.2014.017

20. IPIECA. Flowlines. https://www.ipieca.org/resources/energy-efficiency-solutions/units-and-plants-practices/flowlines/ (accessed August 7, 2021).

21. Markforged. Carbon Steel vs Stainless Steel. https://markforged.com/resources/blog/carbon-steel-vs-stainless-steel (accessed August 7, 2021).

22. Rigzone. How Do Risers Work? https://www.rigzone.com/training/insight.asp?insight_id=308&c_id (accessed August 7, 2021).

23. Miller, C. A. (2017). Risers Introduction. In *Encyclopedia of Maritime and Offshore Engineering*. Carlton, J., Jukes, P., Choo, Y. S. (Eds.). John Wiley & Sons, Ltd, Chichester, UK, 3307. https://doi.org/10.1002/9781118476406.emoe485

24. Sparks, C. P. (1976). *Fundamentals of Marine Riser Mechanics*. PennWell Corporation, Oklahoma, USA.

25. Luppi, A., Cousin, G., & O'Sullivan, R. (2014). Deepwater hybrid riser systems. In *Offshore Technology Conference-Asia*. Kuala Lumpur, Malaysia, March 25–28. (OTC-24802-MS). https://doi.org/10.4043/24802-MS

26. American Petroleum Institute. (2020). *API Standards: International Usage*.

27. Langner, C. G. (1984). Relationships for deep-water suspended pipe spans. In *Proceedings of The Offshore Mechanics and Artic Engineering Symposium (OMAE 84)*, New Orleans, Luisiana, pp. 552–558.

28. Phifer, E. H., Kopp, F., Swanson, R. C., Allen, D. W., & Langner, C. G. Design and installation of auger steel catenary risers. In *26th Annual Offshore Technology Conference*. Houston, Texas, USA, May 2–5. (OTC-7620-MS). https://doi.org/10.4043/7620-MS

29. Americal Petroleum Institute. (2019). Recommended practices for offshore / subsea structure & manifold (API RP 17P). In *A.P.I. Recommended Practices*. American Petroleum Institute, p. 17.

30. Havard, D. (2013). *Oil and Gas Production Handbook - An Introduction to Oil and Gas Production, Transport, Refining and Petrochemical Industry*. ABB Oil and Gas.

31. Toleman, B., Huang, C., Sonawane, M., Shankaran, R., & Foster, D. (2019). Subsea rigid jumper design optimization for sour service application. In *38th International Conference on Ocean, Offshore and Arctic Engineering (OMAE 2019)*. Glasgow, Scotland, June 9–14.

32. Koto, J. (2017). *Subsea Connection and Jumper* (2nd Edition). Ocean and Aerospace Research Institute, Indonesia.

33. Horn, C. (2008). Flowline Tie – Ins. In *SUT Subsea Seminar of Year*.

34. Kim, K. S., Choi, H.S., & Kim, K. S. (2018). Preliminary optimal configuration on free standing hybrid riser. *International Journal of Naval Architecture and Ocean Engineering*, 10(3), 250–258. https://doi.org/10.1016/j.ijnaoe.2017.10.012

35. Roveri, F. E., Velten F., A. G., Mello, V. C., & Marques, L. F. The Roncador P52 oil export system – hybrid riser at a 1800m water depth. In *2008 Offshore Technology Conference*. Houston, Texas, USA, May 5–8. (OTC-19336-MS). https://doi.org/10.4043/19336-MS

36. FMC Technologies. Subsea tie-in systems. https://www.technipfmc.com/en/what-we-do/subsea/subsea-systems/subsea-infrastructure/connection-and-tie-ins/ (accessed August 8, 2021).

37. Webb, P. Introduction to Oceanography. https://rwu.pressbooks.pub/webboceanography/chapter/6-2-temperature/ (accessed August 8, 2021).

38. Earle, S. (2019). *Physical Geology* (2nd Edition). BCcampus, Victoria, Canada.

39. National Ocean Services. https://oceanservice.noaa.gov/facts/thermocline.html (accessed on December, 2021).

40. Kolade, O. (2009). *Deepwater Drilling Problems*. University of Stavenger, Norway.
41. International Marine Consultancy. Halocline Effect. https://www.imcbrokers.com/halocline-effect/ (accessed December 14, 2021).
42. Ruud, T., Idrac, A., McKenzie, L. J., & Høy, S. H. (2015). All subsea: A vision for the future of subsea processing. In *Offshore Technology Conference*. Houston, Texas, USA, May 4–7. (OTC-25735-MS). https://doi.org/10.4043/25735-MS
43. John, S. M., Liu, M.L., Wang, Z., & Lin, J. (2018). Qualification of new technologies in subsea processing. In *Offshore Technology Conference*. Houston, Texas, USA, April 30–May 3. (OTC-29054-MS). https://doi.org/10.4043/29054-MS
44. Devegowda, D., & Scott, S. L. (2003). An assessment of subsea production systems. In *SPE Annual Technical Conference and Exhibition*. Denver, Colorado, October 5–8. (SPE-84045-MS). https://doi.org/10.2118/84045-ms
45. Speight, J. G. (1984). *Handbook of Offshore Oil and Gas Operations*. Gulf Professional Publishing, Elsevier, Oxford, UK.
46. Fantoft, R. (2005). Subsea gas compression - challenges and solutions. In *Offshore Technology Conference*, Houston, Texas, May 2–5. (OTC-17399-MS). https://doi.org/10.4043/17399-MS.
47. Prescott, N., Mantha, A., Kundu, T., & Swenson, J. (2016). Subsea separation - Advanced subsea processing with linear pipe separators. In *Offshore Technology Conference*, Houston, Texas, USA, May 2–5. (OTC-27136-MS). https://doi.org/10.4043/27136-MS
48. Roberto, M. A. R., Koelln, H. P., & De Rezende, R. I. (2021). Brazil subsea processing & boosting technologies yard: More than 20 years of lessons learned. In *Offshore Technology Conference*, Virtual and Houston, Texas, August 16–19. (OTC-31032-MS). https://doi.org/10.4043/31032-MS

7 Management and Control of Solids in Flow Assurance

7.1 INTRODUCTION

Long-chain hydrocarbons, including wax and asphaltene, typically remain dissolved in the crude oil at reservoir conditions. As the crude oil flows from the reservoir to the surface, heavier hydrocarbons tend to separate from the bulk fluid and deposit in the production tubing and surface facilities. Gas hydrates, wax, asphaltene, and scale formation and deposition are some of the leading issues of pipeline blockage in the petroleum industry and often lead to pressure abnormalities, resulting in production interruptions. The hydrocarbon deposits induce fouling in the oil reservoir, pipelines, wells, oil processing and production facilities. The transportation of oil and gas from the wellhead to production facilities and sale locations are often disturbed due to these deposits. In order to develop mitigation methods, it is important to know the mechanism by which these deposits form. It is also necessary to precisely locate these deposits in offshore and onshore oil and gas production facilities. After that, flow assurance needs to be ensured by various means so that the oil and gas stream can flow without any issues to the point of sale. To ensure flow assurance of the production streams, oilfield industries and service companies have introduced and commercialized a diverse range of reasonably efficient flowline and pipeline remediation methods, treatment systems, and maintenance services for various deposits. This chapter provides details on various methods used to address flow assurance issues.

7.2 MECHANISM OF ORGANIC AND INORGANIC SOLID DEPOSITION

7.2.1 Wax Deposition

It is important to understand the mechanism of solid deposition within the pipeline and surface facilities and analyze these blockages physically and chemically to choose the best remediation method for a specific flow assurance problem. Wax deposition involves the separation of long-chain paraffins from crude oil, followed by their eventual growth and deposition on the surface of pipelines when the temperature of the system falls below the wax appearance temperature (WAT). The wax initially nucleates at the WAT, agglomerates, and subsequently precipitates to form solid deposits, rendering oil and gas production difficult and making the oil transport capacity fragile.[1] Typically, as the reservoir reaches a mature stage, the percentage of long-chain paraffins increases (up to 10–70%), attracting the potential threat of wax deposition. High-paraffin content also results in a higher viscosity and pour point of the crude oil, resulting in pressure drop, reduced flowability, and high cost of pumping.[2] The wax molecules also start to develop gel at the cold surface. The wax crystals trap oil molecules within, making them more viscous.[3] Wax crystallization, precipitation, and deposition can increase due to factors such as shear dispersion, gravity settling, and Brownian motion. Wax deposition happens in two stages: wax gel formation and aging. The severity of the wax deposition issues is mainly due to the type of crude oil and concentration of wax crystals. The wax deposits from crude oil are more difficult to manage than condensate because the paraffins in crude oil are typically longer in carbon number than those present in the condensate. Various wax deposition mechanisms have been proposed by Burger et al.[4] These include molecular diffusion, shear dispersion, Brownian diffusion, and gravity settling. The molecular diffusion considers the dissolved wax molecules in the oil, while the other three mechanisms deal with the suspended wax particles precipitated from the oil. The Brownian diffusion is unlikely to be a major cause of wax deposition as the wall temperature is lower than the bulk, leading to enhanced precipitation at the wall. Also, gravity settling and shear dispersions are insignificant as there are no thicker wax deposits at the bottom. The wax deposition rates remain unchanged even with the increasing shear rate.[5, 6] This leaves molecular diffusion as the only accurate way to describe the wax deposition, as shown in Figure 7.1.

There are four major stages in the wax deposition in the flowline:[5]

  i. Precipitation.
 ii. Dissolved wax radial concentration gradient generation.
iii. Wax deposition on the existing deposits.
 iv. Internal diffusion and precipitation of wax in the deposits.

The waxy components nucleate and precipitate out from the crude oil and form wax crystals as it nears the WAT in the first stage. This can also happen at the wall, provided the temperature is below WAT at the location. In the second stage, wax precipitation is more significant on the wall than in crude oil, leading to a concentration

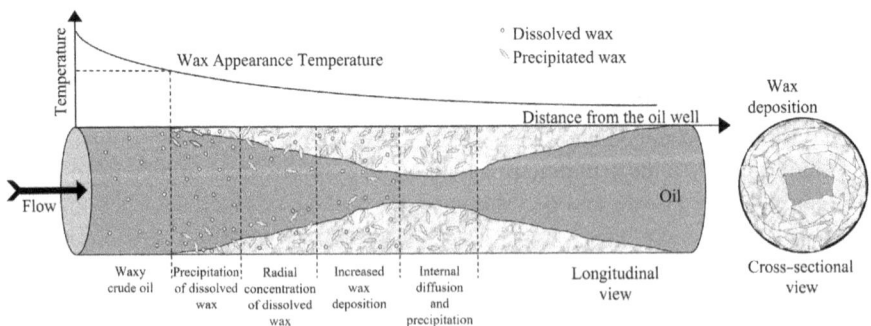

FIGURE 7.1 Schematic of the wax deposition mechanism inside flowline during transportation of waxy crude oil.

gradient and attracting further wax deposition. In the third stage, this process continues and leads to an enhanced wax deposition. In the final stage, the waxy components diffuse into the existing deposits instead of forming a new layer, increasing the wax content in the deposits. This phenomenon is also known as "deposit aging".

7.2.2 ASPHALTENE DEPOSITION

Pressure, temperature, and compositional variation during crude oil transportation can change the equilibrium of asphaltene in the crude oil. The effect of composition and pressure variation on asphaltene precipitation is generally higher than temperature. During the transport of crude oils in a wellbore/pipeline, the pressure depletes, and, on a pressure-temperature diagram, it follows the path indicated by arrows (Figure 7.2). At point 1, the pressure is high enough so that asphaltenes remain stable (dissolved in oil). As the pressure decreases, point 2 is reached, which lies on the upper asphaltene onset pressure curve. Below this curve, asphaltenes begin to precipitate, and point 3 is reached, where the maximum asphaltene precipitation occurs. Further reduction in pressure makes lighter fractions come out of the crude oil, and the asphaltene solubility increases.[7, 8] For the plugging to occur, it makes sense that the transport condition should be around the point of maximum asphaltene precipitation (i.e., just above the bubble pressure curve). The asphaltene plugging results from a few stages, such as precipitation, flocculation, and deposition.[7] Since asphaltenes have a tendency to aggregate, they flocculate after the precipitation and form larger clusters. They can still be transported until the size is too large to be carried by the crude oil. After that, they begin to settle down on the flowline wall, and sufficient deposition takes place with time, reducing the cross section of the flow.

7.2.3 HYDRATE DEPOSITION

Hydrate formation occurs once the temperature falls inside the hydrate phase stability region (high pressure and low temperature). In an offshore system, hydrate formation and deposition occur downstream of water accumulations for many reasons, including a change in the flow geometry or some nucleation site, such as sand, weld slag, among others. The hydrate formation and plugging consists of four major steps:

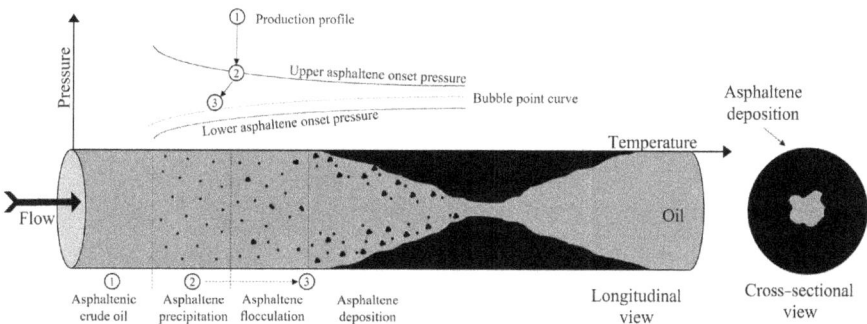

FIGURE 7.2 Schematic of the asphaltene deposition mechanism inside flowline during transportation of heavy crude oil.

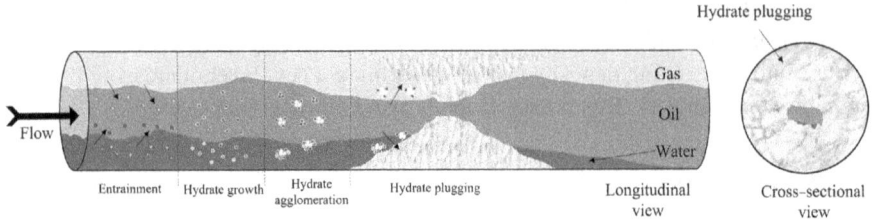

FIGURE 7.3 Schematic of the hydrate deposition mechanism inside flowline during multi-phase (oil, gas, water) flow in the flowline.

entrainment, hydrate growth, hydrate agglomeration, and hydrate plugging, as shown in Figure 7.3, modified from Turner's[9] work. The first step is the entrainment of droplets. Due to turbulent flow, the water droplets get entrained in the oil phase. If the water phase is sufficient enough, oil droplets can also be entrained in the water phase. Also, the gas is dissolved in the oil and water phases as per the solubility (shown by small drops). In the second step, hydrate formation occurs at the water–oil interface due to the mass transfer of gas through the interface.[10] After the hydrate shell forms at the oil–water interface, it grows inward for the water droplets entrained initially in the oil phase. In the third stage, hydrate agglomeration occurs where the hydrate crystals form a bigger agglomerate. Finally, the large hydrate aggregates deposition occurs, and the deposit or plug anneals with time, leading to a solid structure.[11]

7.2.4 SCALE DEPOSITION

Although scale deposition is a common problem in the petroleum industry, it is detrimental to flow assurance as it hinders the flow of produced fluids, leading to increased mitigation costs, thus making marginal wells uneconomical. Scale deposition occurs due to the crystallization and precipitation of minerals from the produced water in a laminar flow regime. Different types of oil field scales include carbonates and sulfates. Scale can also form due to corrosion, for example, sulfide, hydrous oxides, and carbonate scales on the flowline wall. Two major causes of the oil field scale deposition include the pressure and temperature changes during production or using an injection brine incompatible with formation water (e.g., mixing calcium-, barium-, strontium-rich formation water with sulfate-rich seawater).[12] Basic steps in a scale deposition process are given in the schematic diagram (Figure 7.4).

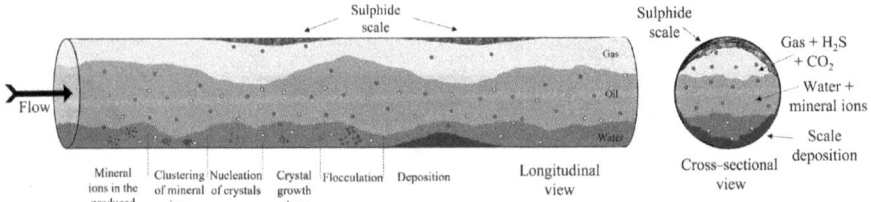

FIGURE 7.4 Schematic of the scale deposition mechanism inside flowline during transportation of produced fluids.

The scale formation involves several steps such as the formation of mineral ions cluster, nucleation, crystal growth, and finally, flocculation, and deposition. The mineral ions initially present in the produced water form clusters. Afterward, inorganic salt crystals nucleation occurs due to a chemical reaction between the ions. In the next major step, the formed crystals flocculate. Finally, when the clusters become so big that they cannot be transported any farther, they begin to deposit on the flowline wall, and with time, they harden into scales.

7.3 LOCATING SOLID BLOCKAGE IN FLOWLINE INFRASTRUCTURE

Detecting solid blockage in the pipeline and flowline is an important aspect of flow assurance management practices. Solid blockages such as hydrates, wax, and asphaltene, often deposit in flowline systems at various locations, including valves, bends, and internal surfaces of the pipeline. Blockage does not always show any visible external indicators like leakage, thus identifying their location is always challenging. Precise information on the blockage and/or leakage location inside the pipeline systems would significantly leverage their mitigation strategies. Various techniques identified in the literature and industrial practices are used to identify the location of the solid blockages in the flowlines. These methods are categorized on different parameters, namely local and global, physical (visual), model-based, intrusive and nonintrusive, and others[13] and are discussed briefly.

7.3.1 Intrusive Methods and Nonintrusive Methods

Flowline blockages are typically identified via intrusive methods such as a closed-circuit camera and intelligent or robotic pigging. The use of a robotic pig or camera does have some limitations and uncertainties, including traveling distance, speed, bulk fluid viscosities, and the severity of deposition. The pig or camera may often get stuck due to external factors. Nonintrusive methods mainly include applying external sensors to detect the blockage within the pipeline. These include methods based on ultrasound techniques to detect the blockage within the pipelines.

7.3.2 Nondestructive Testing Methods

Several nondestructive testing (NDT) methods are used to detect the blockage in flowlines, categorized as either local or global. The local methods (mainly for onshore facilities) may contain visual checks of the systems, radiography, dye injection, ultrasound, electromagnetic methods, magnetic particle inspection (MPI), and pigging while global methods include pressure transient profile and acoustic wave methods.

7.3.2.1 Visual Method

The visual method is time-consuming as the service engineer has to visually scan the pipeline to detect potential leakage and blockage around the valve and manifold. The service engineer can use a handheld device to record the pressure differential across a different section of the pipeline and then analyze the data for the potential location of the blockage. This method is suitable for short-distance onshore pipelines and the pipeline under maintenance.[14]

7.3.2.2 Radiographic Method

Radiography is an NDT method used locally to detect any leakage, potential blockage in the pipeline, and structural integrity of the material. Often, the blockage within the pipeline (such as wax, asphaltene, and hydrate and corrosion/scale) may impact the structural integrity of the pipelines in the forms of cracks and the like. In this method, X-rays or gamma rays are passed through the surface to generate information about the internal structure of the material.[15] Due to the use of radiation, this method is not suitable for online applications and thus is mainly recommended for offline applications. Safety concerns should be considered as the radiation may pose a threat to the health and safety of personnel.[13]

7.3.2.3 Pig-Assisted Monitoring System

A pig-assisted flowlines monitoring system is one of the global methods used to detect blockage locations within the flowline. Pigs are used during flowline clearing or remediating solid deposits. Initially, a hydraulic test is carried out, after which a pig injection is done to remove the solids. With the advancement in pigging operations, intelligent pigs are used for monitoring and surveillance of flowlines.[16] The intelligent pigs are attached with ultrasound sensors, which can pass data to the operator to generate information on the internal structure of the flowlines. The pigs can also monitor different parameters in situ. The information collected over time can be analyzed for the location and severity of blockage within the pipeline.

7.3.2.4 Acoustic Reflectometry Method

This method introduces a sound pulse in the flowline using an acoustic pulse generator. As the acoustic wave passes through the pipe, it gets reflected when it encounters a blockage. Acoustic sensors placed outside the pipe measure the pulse signal, and the information is then used to generate a noise profile in the pipe. The deviation from the baseline profile is measured to get the location of the blockage inside the flowlines. Parker[17] is one of the early researchers to propose the use of acoustic reflectometry. He proposed a correlation between the acoustic pulse and the noise, thus providing an approach to detect any small acoustic signal-to-noise ratio even for long-distance flowlines. Koyama et al.[18] and Wang et al.[19] used this method to locate the pipeline blockage and its severity.

7.3.2.5 Ultrasonic Testing Method

Ultrasonic testing (UT) is one of the advanced and local methods used to detect a solid blockage. The method uses high-frequency acoustic waves to gauge the thickness of the flowlines wall with precision and reliability.[13] It provides information to the indicator on any changes in the wall thickness or defects in the surface. The principle of the UT method is based on the travel time required for an acoustic wave to pass through the material.[20] The UT setup consists of an ultrasonic probe, a receiver, a transducer, and a display screen. The sensor is placed at a certain angle on the flowline wall. The high-frequency acoustic wave is reflected upon passing through the material. The thickness is then measured based on the reflection time. The information on the wave velocity, energy reflected, and time is annualized to precisely determine the location of the blockage. This method can be categorized into contact and noncontact methods (Figure 7.5). In the contact method, an ultrasonic transducer is directly in contact with the fluid system. The direct contact method has limitations while detecting the blockage contour for natural gas pipelines. The hydrate deposits within the pipeline result in shielding of the pipeline, making it inconvenient for direct contact measurement. Also, the direct contact method may generate a strong reflection of the wave from the pipe surface, resulting in signal attenuation. Thus, the noncontact method is preferred over the contact method for natural gas pipelines.[21]

Figure 7.5 shows the ultrasonic transducer, which is the heart of the ultrasonic detection system.[21] Figure 7.5a shows the working principle of the system, while Figure 7.5b shows the transducer and the waterproof connector, respectively.[21] The precise detection of blockage location and measurement of blockage counters within flowlines requires accurate placement of the ultrasonic transducer. Figure 7.5 shows the schematic of the related component of the UT method applied on the flowline surface. The systems can be moved across the pipeline surface using robotic arms and a virtually guided vehicle for deepwater applications.

7.3.2.6   Mass and Fluid Composition Measurement

The mass balance method is traditional and global, involving a basic principle of conservation of mass. Heavy hydrocarbons, such as asphaltenes, and wax precipitate out while lighter hydrocarbons, such as gas, escape from the crude oil during the solid organic or inorganic deposition in the subsea flowlines and surface facilities, leading to a change in the fluid composition at the exit. Any discontinuity in the fluid composition and mass will indicate the possibility of deposition and leakage in the flowlines. This online sensor for mass flow rate, composition, and density can be employed at different flowline infrastructure locations. This process mainly depends on flow rates and pressure information from flow meters or pressure gauges at periodic intervals and needs constant and accurate flow rate monitoring across the section of flowlines.[21]

7.3.2.7   Densitometry Method

The density of the solid deposits formed by hydrate, wax, asphaltene, or scales will vary from the bulk fluid based on the in situ conditions. In this method, a densitometer and a scanning device are used to detect the fluid phase density at specific pipeline internals. The variation in density along the flow path provides information on the location of the blockage. However, this method has limitations to accurately predict the blockages due to the low-density difference between hydrate and water. The density difference between the blockage and fluid within the pipeline affects the accuracy.[22] However, this method can provide a quick estimate on the possible blockage locations based on engineering field experience along with the details on the fluid properties and operating conditions.[23]

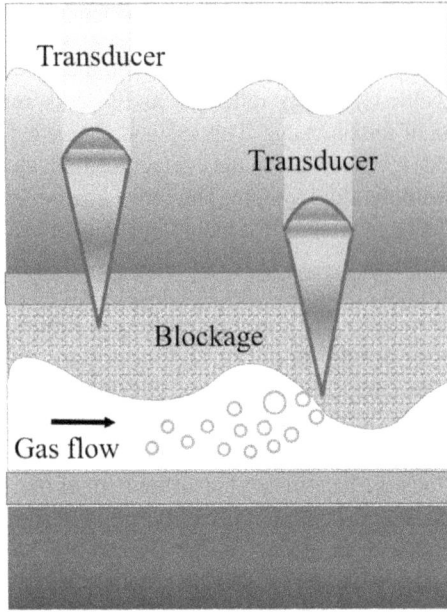

(a)

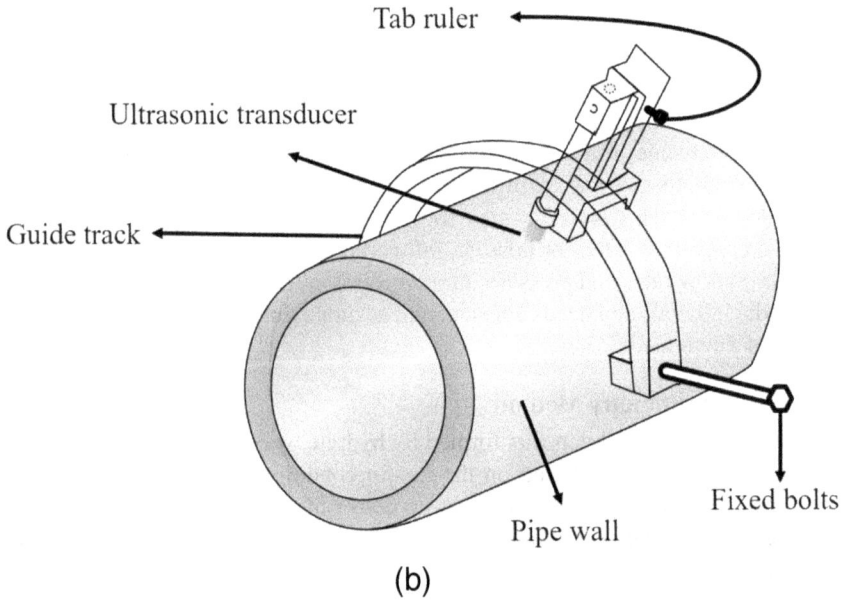

(b)

FIGURE 7.5 (a) Working principle of the ultrasonic transducer; (b) transducer and the waterproof connector.

7.3.2.8   Isotope Tracking Method

In this method, higher velocity at a particular section indicates a restriction based on fluid continuity. In this method, the isotope is injected into the flowline, and its movement is tracked using tracking devices. The online data reported using the tracking devices are used to analyze the velocity of the fluid. The velocity profile thus indicates the possible location of the blockage in the flowline.[23–25]

7.3.2.9   Model-Based Method

The real-time transient model (RTTM) is one of the advanced global methods used to detect leakage and potential solid blockage in the pipelines. This method utilizes the basic principle of conservation of mass, momentum, and energy-balance equations to mathematically model the entire flowline system. This measurement makes it convenient to measure the flow parameters and model the steady-state and unsteady-state fluid flow behavior.[26, 22]

The supervisory control and data acquisition system (SCADA) is one of the advanced global methods used to detect leakage and sloid blockage in the pipeline. In this method, the SCADA sensor collects process data over time and the length across the flowlines. The information is recorded at a remote terminal unit (RTU) and the central recording terminal in real time. The data are then processed and used alongside mathematic models, describing the fluid behavior to understand the possible location of the solid blockage in the pipeline based on the discrepancies in process parameters. The online data provide an operator with real-time details of the flowline condition.

The SCADA method uses a mathematical model based on the hydraulic equation of the fluid in the pipe. These equations are primarily based on the laws of conservation of mass, momentum, and energy along with the equation of state model of the fluid phase, correlating with physical properties of fluid such as density, viscosity with pressure, and temperature conditions in situ. The equations governing the conservation laws in the SCADA method are nonsteady partial differential equations. The hydraulic parameters and fluid phase properties are a direct function of the time, in situ conditions, and the distance along the flowline. The model prediction provides information on pressure differential, flow rate, and various properties of fluid phase behavior inside the pipe. The model predictions are compared with the measured values to detect any variation that indicates the possibility of blockage at a given location. The SCADA-based method is advantageous for detecting discrepancies in the flow behavior inside the flowlines.[23, 27]

7.3.2.10   Pressure Measurement–Based Method

7.3.2.10.1   Backpressure Technology

In this method, a pressure drop profile is established for various flow rates for a flowline.[23] Any deviation from the established profile during operation indicates the existence of the blockage. Scott and Satterwhite[28] used this method for monitoring the development of hydrate growth in a gas pipeline system. The backpressure method has also been proposed for detecting the blockage location.[28] Scott and Yi[29] used this method for liquid flowline to get a rough estimate of blockage. Liu and Scott[30] modified the backpressure method to locate the partial blockage in the pipeline based on three different tests. These tests include a steady-state backpressure test to estimate the blockage factor, a simultaneous shut-in test to get the average pressure, and a bleed-off test to estimate the volume factor related to the pipeline blockage. The backpressure method may not always be accurate and sometimes time-consuming. The pressure difference can also be correlated with the strain difference in the pipe wall to determine the blockage location.[31-33] For this, it is important to provide strain measurement instrumentation on the outer wall of the flowline. This method can be employed for various fluids flowing through the pipeline.

7.3.2.10.2 Pressure Transient Pulse Technology

Several studies have been reported using transient-based methods for detecting the location of pipeline blockage. These methods have advantages in response time, accuracy, and low cost.[34-38] In this method, pressure pulse wave is generated using quick-acting valves. Based on the reflection time, amplitude, and frequency of the reflected wave, accurate detection of the blockage can be made. As shown in Figures 7.6 and 7.7, the location of the blockage can be detected based on pressure information. This method is primarily categorized into time-domain and frequency-domain methods based on the processing technique.[37] Flow analysis–based methods have been explored to detect blockages. This method is mainly based on fluid transient profile during fluid flow, which is obtained from the response of the systems to injected transient profile for identification of location, size, and intensity of the blockage. The flow transients are analyzed in frequency or time domain.[28-36] Chu et al.[32] have investigated the suitability of the pressure pulse wave method for detecting the location of the blockage in the flowlines. This method uses three high-frequency dynamic pressure sensors to understand the blockage characteristic from reflected and transmitted wave patterns. It has been observed that the pressure pulse wave method is accurate, fast, and reliable and can address operational complexity. The method is useful for providing an early warning for potential flowline/pipeline blockages ensuring a low cost and safer operation.

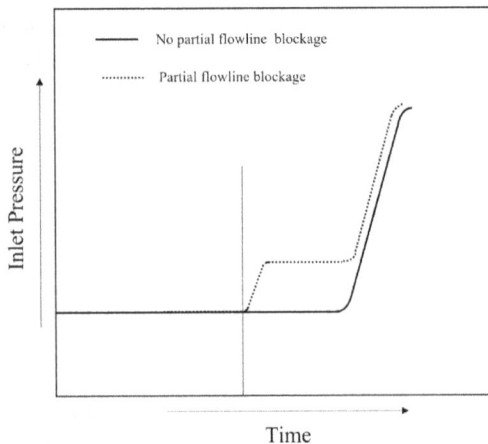

FIGURE 7.6 Impact of partial blockage on the inlet pressure.

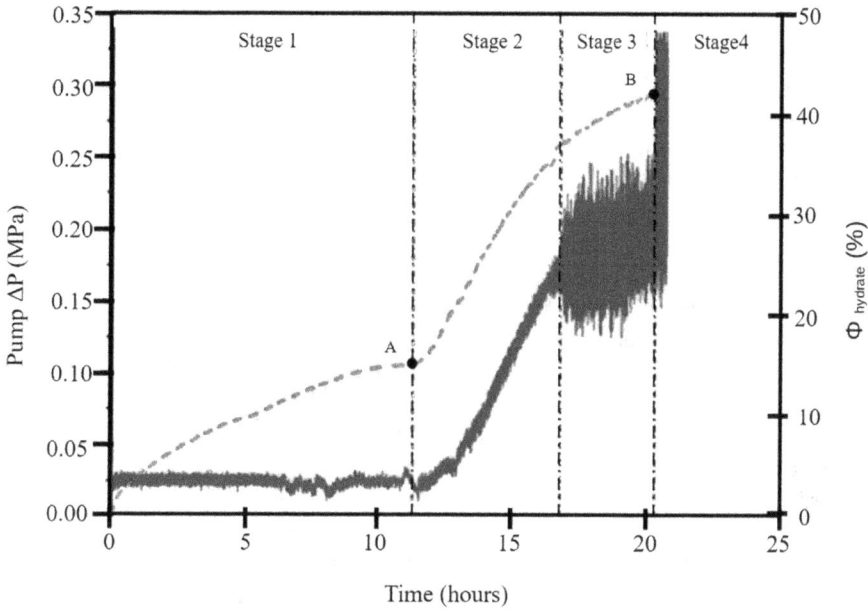

FIGURE 7.7 Time-variant association between differential pressure and volume percentage of hydrate. Reproduced from Liu et al.[39] with permission from the American Institute of Physics (Copyright American Institute of Physics, 2019).[39]

In the frequency response method, pressure pulse information is generated for varying frequencies by opening and closing a valve periodically.[23] Using different frequencies, the amplitude of the fluctuation of the valve is analyzed using a transfer function matrix. The relation between peak pressure frequency provides information on the location and size of the blockage inside the flowline. Mohapatra et al.[40] used a frictionless model to describe the location and size of the blockage in the pipeline. Some studies have reported the use of frequency response analysis to detect pipeline blockage. Antonopoulos-Domis,[41] and Qunli and Fricke[42] used the axial resonance frequency of the duct to infer the blockage using the frequency-domain technique. Lee et al.[43] developed a frequency response diagram for a flowline and inferred that various partial blockages are affected due to the peak in the frequencies of an oscillatory pattern.

7.4 SOLID MITIGATION METHODS

7.4.1 WAX REMEDIATION TECHNIQUES

Reservoir crude oil wax composition and crude oil type determine the success of any treatment methods. In several instances, two or more methods are simultaneously utilized to minimize and/or remove wax depositions from the facilities. Wax remediation techniques are broadly classified into three types: thermal, chemical, and mechanical.

7.4.1.1 Thermal Techniques

As the deposition of wax is primarily governed due to reduction in temperature and is highly temperature-dependent, therefore, thermal methods are one of the highly effective methods to control and remove wax deposition. Hot oiling is typically used to remove the deposited wax from the flowlines, pipelines, and wellbores. If the wellbore temperature falls below the WAT, wax deposition occurs in the wellbore and blocks it. Hot oiling is usually used for wellbore application. In this method, the oil is heated above the melting temperature wax and then injected and circulated in the wellbore, generally through the annular space dissolving the wax, which flows up with the crude oil to the surface production system. In this method, the wax dispersant can also be added to improve the flowability and avoid further precipitation and deposition of wax. The details on various possible dispersant systems are discussed subsequently. It is to be noted here that typically high-molecular-weight waxes deposit at the bottom of the wellbore. In contrast, lower molecular-weight waxes deposit at the upper section of the wellbore. In hot oiling, as the oil passes toward the bottom of the wellbore, the temperature may decrease. This affects the wax dissolution and carrying capacity of hot oiling. Hence, sufficient oil should be used during the process.[44] Hot oiling is not recommended when the flashpoint of the crude oil is low. In situations, where the wax deposition and WAT are low, hot oil can also be substituted with steam or hot water to melt and dissolve the wax. However, in such a situation, the possibility of emulsion formation is very high. The wax is known to stabilize the oil–water emulsion, thus may aggravate the flow assurance issue. Due to the significant cost to heat the oil, this method is not used for subsea flowlines. Hot water and steam give the required solvency effect of wax, similar to the hot oiling method. Surfactants are often used along with hot water to improve the wax dispersion during the fluid flow. The combination of hot water with a surfactant system helps improve the suspension of wax in the bulk phase due to the bipolar interaction of surfactants at the water and oil interface. An alternate way of addressing wax issues is direct heating of the surface on which wax is deposited by maintaining the surface temperature above the WAT. In this method, an electric current is passed around the pipeline wall to generate heat. It can also be used for deepwater pipeline systems effectively. One of the first applications of successful electric heating of offshore pipeline was done in 1996 on various fields in the North Sea and was used for seven flowlines having 6–16 km in length and flexible risers using hot-water heating. Electrical heating is used to maintain the temperature above the WAT and hydrate formation temperature through thermally insulated flowline and riser system.[45] Using the cable strapped (piggybacked) to the surface of the thermal insulation, the maximum amount of heat for a given system of current can be obtained. Each riser is supplied with a piggyback cable. Figure 7.8 shows a schematic representation of the electric heating system. The direct electrical heating method is typically designed to keep the temperature of the production stream above the WAT of the crude oil, using the shutdown. Temperature can be increased in a stepwise manner so that once the steady-state flow is achieved, continuous heating may not be necessary.

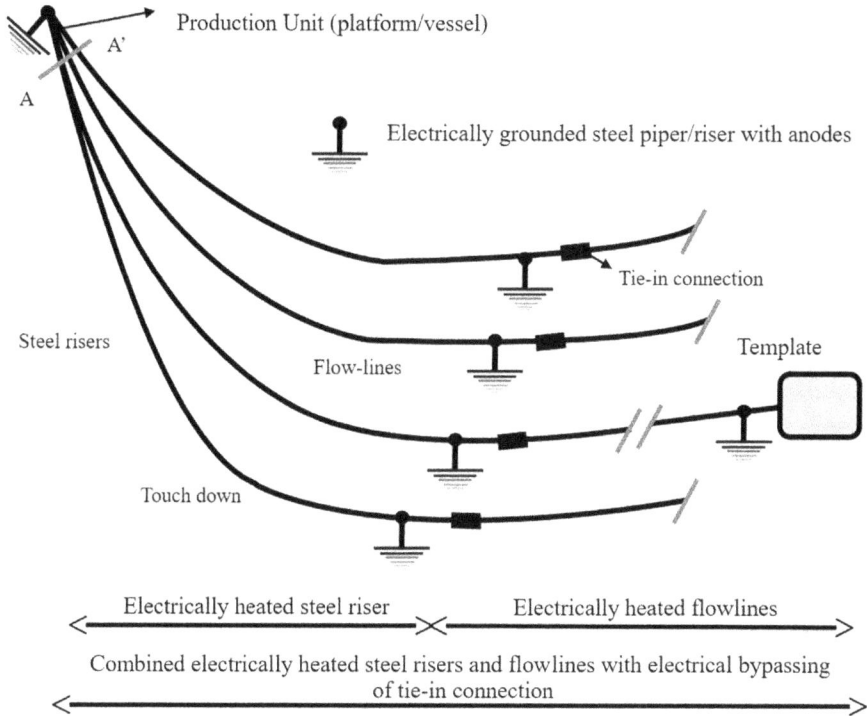

Production Unit (platform/vessel)

Electrically grounded steel piper/riser with anodes

Tie-in connection

Steel risers

Flow-lines

Template

Touch down

Electrically heated steel riser   Electrically heated flowlines

Combined electrically heated steel risers and flowlines with electrical bypassing of tie-in connection

FIGURE 7.8   Schematic representation of the electric heating system.[45]

7.4.1.2   Mechanical Techniques

Mechanical methods are simple, primitive, and frequently used for wax, hydrate, and asphaltene deposit removal and are not used to avoid or stop the deposition process. Once a sufficient amount of wax is deposited into the pipeline, which can be predicted from the flow rate and pressure drop across the pipelines, the mechanic methods are employed. In the wellbore, cutters and scrapers are generally used to remove the wax. Two types of tools are used. One cuts the deposits and takes them to the surface, and the other scratches and removes the deposits from the well tubing (see Figures 7.9 and 7.10).[46, 47] In this method, the scraper or the cutting tool attaches to the tubing wall due to metal-to-metal contact. These tools are a part of the physical method for removing wax deposits from the well tubing and pipelines without melting them. Some of the operational issues using these mechanical tools are that if the wax deposits are significant and hard enough, there is a strong chance of wax plug formation. Once the plug covers the flowline's cross-sectional area, mechanical removal is very difficult, and it may increase the load, resulting in the snapping away of the tools. Furthermore, due to the occurrence of hard wax deposits, pulling out tools may become difficult. The mechanical method is often combined with other methods (thermal or chemical methods). A scraper with a wireline is prevalent during the wax removal operation. There are, however, chances that if the wax is hard, then the tool may be stuck in the well, and the wireline may break during operations.[46]

Tool for casing

Section A – A'

Split allow cutter to expand or contract with the irregularities in the pipe

Section B – B'

A ———————— A'

B ———————— B'

Cutter

Outside diameter of the cutter

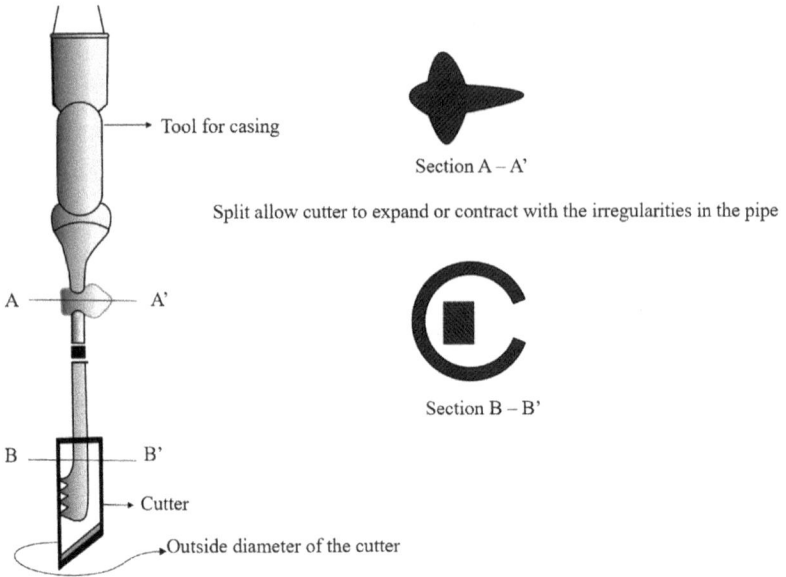

FIGURE 7.9 Schematic representation of a wax cutting tool.[47]

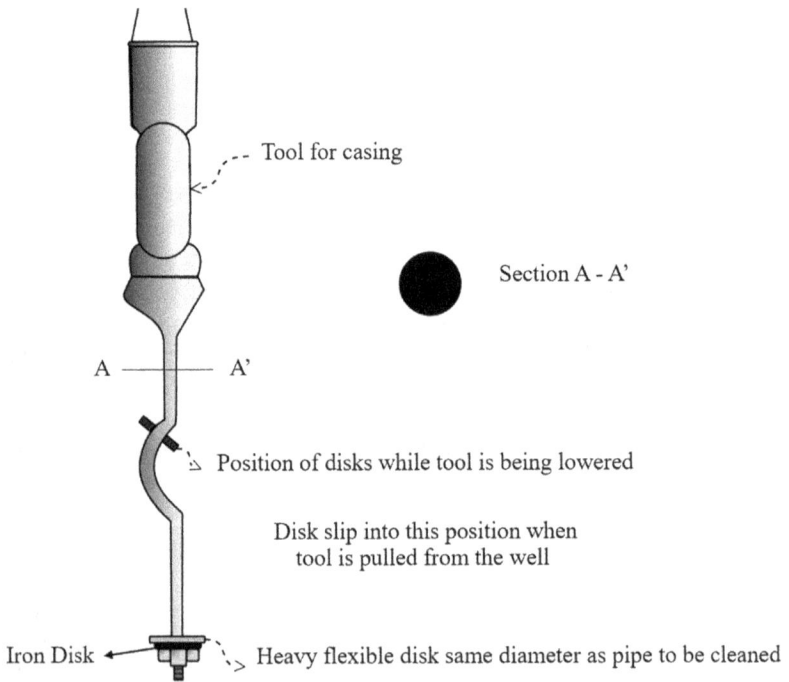

Tool for casing

Section A - A'

A ———————— A'

Position of disks while tool is being lowered

Disk slip into this position when
tool is pulled from the well

Iron Disk ←— Heavy flexible disk same diameter as pipe to be cleaned

FIGURE 7.10 Schematic representation of a wax hook tool.[47]

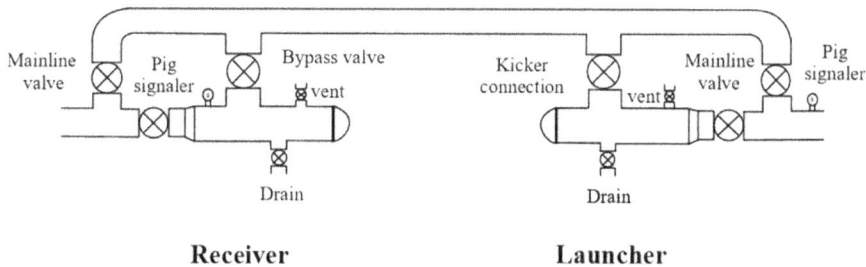

FIGURE 7.11 Schematic of the pig launcher and pig receiver station.

7.4.1.2.1 Pigging

Pigging is one of the oldest and traditional methods used to clear the wax deposits in pipelines and is now a well-established method. A pig is a piston-type unit propelled inside a pipe due to the pressure differential in a process known as pigging.[48–50] A pipeline inspection gauge (known as pig or scraper) is launched and received in/from the pipeline using pig launcher and receiver mechanism, which is typically designed at a designated section or sections of the pipeline where wax deposition is a serious issue. A schematic diagram of a pig launcher and receiver station is given in Figure 7.11. A pig launcher is a section of pipe relatively larger in dimension that connects to the regular pipe. Once the pig is pushed from the launcher, it is closed. Due to the pressure of the fluid flow, the pig moves within the pipe until it reaches the receiver (or receiving station, also called as pig catcher). Pigs are typically considered to drive away any debris and loose deposits inside the pipeline using a mechanical force. Pigging is done while the fluid is flowing within the pipeline. With the increased pig length, the amount of wax removed may increase but at the cost of a higher pressure drop. Pigging is widely used across various oilfields depending on the number of wax deposits. Pigs can be of various types, made from metal to synthetic polymeric materials such as polyurethane foam. A pig can be bullet-shaped or can be stiff aggressive tools. Very hard wax deposits require aggressive pigs made up of metal having brushes, ploughs, scrapers, and pinwheels to increase the wax removal efficiency. Pigging can be performed in intervals of days to months. As the pig moves inside the pipeline, the liquid is bypassed from a small hole inside a pig. A proper pigging operation must define the type and number of pigs used during remedial operation. If the pigging operation is not performed properly, it is most likely that the pig may be stuck in the pipeline and may lead to complete blockage.[51]

7.4.1.3 Chemical Techniques

Chemical methods are considered very good options for deepwater wax issues. Various types of chemicals are used in the industry, namely, solvents, pour point depressants (PPDs), wax crystal modifiers, and wax dispersants. PPDs inhibit the growth of wax crystals due to interaction with wax crystals edges. PPDs may reduce the viscosity, yield stress, and pour point of the oil but do not reduce the wax deposition rate.[52–54] Wax inhibitors include wax crystal modifiers, dispersants, and detergents. They keep the wax crystals dispersed and reduce their deposition on the surfaces. Detergent and surfactants, being surface-active agents, partially modify the wettability of the surface of the pipeline towards water wet, thus affecting the deposition of wax molecules. Some surfactants molecules help solubilize the wax molecules and prevent agglomeration.[55, 56] The chemicals inhibitors also weaken the wax deposits. Thus, a reduced shear can remove the wax deposits from the surface, particularly when combined with other methods (mechanical methods such as pigging). The frequency of mechanical methods may be reduced while increasing the ease of operation when used along with chemical methods, thus reducing the potential issues of pipeline blockage due to stuck pigs during operations. The chemical methods need to be used carefully for high-temperature application, deep water, and the compatibility with crude oil and other chemicals (corrosion inhibitors, de-emulsifiers, etc.) need to be performed. Various types of chemicals used for wax mitigation are discussed in the following points.

7.4.1.3.1 Chemical Solvents

Chemical methods involving solvents for wax (and for asphaltene) deposits are used frequently in the oil and gas industry successfully. The chemical methods may sometimes be expensive and hence are preferred when thermal and mechanical methods are not successful. Solvents help to dissolve wax into crude oil, thus making oil flowable in the pipeline. The dispersant helps keep the wax molecules dispersed in the bulk oil, avoiding their agglomeration and precipitation. Commonly used chemical solvents include xylene, toluene, benzene, carbon tetrachloride, trichloroethylene, perchloroethylene, carbon disulfide, and a mixture of lighter hydrocarbons such as gasoline, kerosene, or benzol.[52, 56, 57] Gasoline is considered better than kerosene for removing wax deposits. Mixing aromatic solvents such as xylene or toluene with an aliphatic solvent was observed to improve the wax removal efficiency.[58, 59] Chlorinated hydrocarbons are inexpensive and have good solubility for wax due to their high specific gravity. However, they may affect the poisoning of catalysts in the downstream processes. Aromatic solvents, such as xylene and toluene, have low specific gravity and thus are difficult to apply at the bottom of the well. Carbon disulfide is highly flammable, restricting its use.[56, 60] Paraffins solvent can be used to remove wax when asphaltene deposition is not an issue, as asphaltene tends to separate if the lighter alkane content increases in the oil. They are also preferred when surfactants and other dispersants cannot be used due to emulsion formation, which is difficult to break down in surface facilities.[57] Solvents may be effective only for specific wax molecules depending on their molecular weight and in situ pressure and temperature conditions, but are less effective for large wax plugs.[57, 61, 62]

7.4.1.3.2 Pour Point Depressants

Pour Point Depressants (PPDs) are wax inhibitors, also referred to as wax crystal modifiers. Other components of wax inhibitors include detergents and dispersants. Wax crystal modifiers interact with the wax (paraffin) molecules at their edges, hindering their formation and growth. Several studies have reported the use of PPDs.[52, 55, 63–66] PPDs reduce the pour point temperature of the crude oil, thus reducing the chances of wax crystal deposition at the operating conditions. The PPD also helps to destabilize the wax deposition, thus helping to quickly remove deposited wax from the surface of the pipeline during mechanical operation. PPDs do not impact the WAT significantly, but impact (reduces) the pour point, as PPDs only affect crystal growth, not crystallization. Some examples of PPDs include polyethylene vinyl acetate (EVA), methyl methacrylate (MMA), olefin-maleic anhydride copolymer (MAC), and diethanolamine (DEA). EVA copolymer is made from ethylene and vinyl acetate (VA) and is the most widely used PPDs in the industry.[67, 68] The main mechanism by which EVA works as PPDs is as follows. VA, a polar compound with methyl and methylene groups, exhibits strong interaction with wax molecules due to van der Waals forces, while the nonpolar long alkyl moieties of EVA interact with the paraffin, altering the wax crystallization process. This interaction results in the dispersion of wax molecules in bulk, thus avoiding their deposition.[67, 69]

Comb polymers are also widely used as PPDs to mitigate wax deposition issues. There are two types of comb polymer in use. These are MAC and poly-acrylate (PA)/polymethacrylate (PMA) ester polymers. These polymers with longer alkyl chain lengths were found to be more efficient as PPDs than those with shorter alkyl chain lengths. The pendant chain in these polymers and the size of wax molecules should be similar in size for having better efficacy to avoid a wax deposition. The backbone of these polymers does not impact the wax inhibition efficiency.[70, 71] MAC is not only observed to be efficient PPD but also helps to reduce the rate of wax deposition and yield stress.[72] Combability issues of comb polymers with wax crystals may be the limiting factor for their efficient usage for various crude oils.[69] High-molecular-weight polymers have been recommended for high-molecular-weight paraffin, while low-molecular-weight polymers are recommended for low–carbon number paraffinic crude oil systems.[69, 73, 74]

Due to the waxy nature of crude oil, PPDs may show limiting wax inhibition efficacy. Recently, it has also been shown that nanotechnology can aid in developing novel wax inhibitors. Polyoctadecyl acrylate, EVA copolymers, and methacrylate are examples of potential polymeric nanohybrids.[75] The nanoparticles can interact and embed into the nucleating wax crystal sites, decreasing their growth and agglomeration. Thus, they can suppress the pour point better than only EVA.[74, 76] The use of an organic solvent and PPDs has shown a positive effect on crude oil flowability.[77] The use of organic solvents, such as benzene, xylene, toluene, carbon disulfide, and chlorinated hydrocarbons, when used along with PPDs, can reduce the wax deposition significantly,[78] but one needs to be careful when handling this organic solvent due to its toxicity and highly flammable nature.[60, 79]

Natural surfactants produced from microorganisms are also being explored as PPDs. These natural surfactants are less toxic and more environmentally friendly than conventional surfactants.[80, 81] Some examples of natural surfactants are biosurfactants produced from a microorganism such as *Bacillus subtilis* and *Pseudomonas aeruginosa*. They can reduce crude oil viscosity significantly, improving their flowability.[82–88] The polyamine amide derived from canola oil was also effective in reducing pour points.[89] Jatropha seed and castor seed oil showed a reduction in pour point from 24 to 8°C along with a reduction in viscosity for Nigerian waxy crude oil when used in low dosage (0.1–0.3 vol %).[89]

7.4.1.3.3 Wax Dispersant

Dispersant molecules closely resemble the surfactant molecules. Dispersant molecules contain two ends. One end is attached to the oil or water phase, while the other is attached to wax molecules, making the wax molecules remain dispersed in the system, thus avoiding their deposition.[52] Dispersants can also attach to the pipe wall surface and change the wettability to water-wet by forming a thin layer on the pipe wall surface, thus rejecting the wax molecules and their deposits. In crude oil, dispersant molecules adsorb on the wax crystal surface, making a crystal lattice structure that leads to wax crystal growth and the development of larger wax crystals. This modified spherical-like large crystal with dispersant help improve the flowability of the crude oil. When used along with a polymeric flow improver, a wax dispersant works exceptionally well to avoid wax deposition while simultaneously improving the flowability of crude oil. Examples of surfactants as wax dispersants are alkyl sulfonates and fatty amine ethoxylates. These surfactants work well with longer chain esters, resulting in reduction of the pour point, surface tension, and co-crystallization.[56]

7.4.1.3.4 Wax Crystal Modifiers

Wax crystal modifiers work similarly to PPDs by interacting with wax molecules and limiting their growth. Due to hydrocarbon chains, wax crystal modifiers coprecipitate with wax and participate with wax crystals by residing on the crystal lattice. They provide steric hindrance on the wax crystals, thus interfering with the agglomeration of new wax molecules and inhibiting the growth.[60] Wax crystal modifiers act like PPDs and reduce the pour point. They interact with wax molecules during the nucleation process and affect the growth by forming micelle-like aggregates, resulting in smaller wax crystals that remain stable in the bulk oil (see Figure 7.12a and b).[90–92] Exact mechanism by which these modifiers work is still unclear.[93] Examples of wax crystal modifiers include polyalkyl methacrylate, vinyl acetate copolymers, methacrylic acid ester, polymeric fatty ester, and crystalline-amorphous copolymers such as polyethylene–polyethylene propylene (PE-PEP), olefin/ester copolymers and polyethylene butene (PEB), polyacrylates, polymethacrylates, and alkylphenol resins.[52, 94, 95] A review on different wax crystal modifiers developed and used in industry is provided by Wei.[93] These are mainly ethylene-vinyl acetate, poly(ethylene-propylene), poly(ethylene-butene), and poly(maleic anhydride amide co-α-olefin), among others. It has also been observed that these modifiers show higher efficiency when used with solvents.[93]

7.4.1.4 Advanced Methods

7.4.1.4.1 Cold-Flow Technology

Cold-flow technology is one of the emerging methods used to mitigate wax precipitation issues.[97] The main principle of this method is to eliminate the temperature gradient between the fluid and ambient conditions, resulting in the formation of wax slurry in a controlled manner such that they do not agglomerate and deposit inside the pipeline.[97, 98] The flow profile of waxy crude oil in the pipelines can be divided into hot and cold flow regimes [97, 99–101] as shown schematically in Figure 7.13. Between

Sand, silt iron oxide particles, entrapped water

Cold Surface

Agglomeration (below cloud point)

Wax crystal (above cloud point)

Depositing to the cold surface

(a)

Crystal modifier

Crystal modifier interfering with agglomeration

Crystal Modifier interfering with deposition

(b)

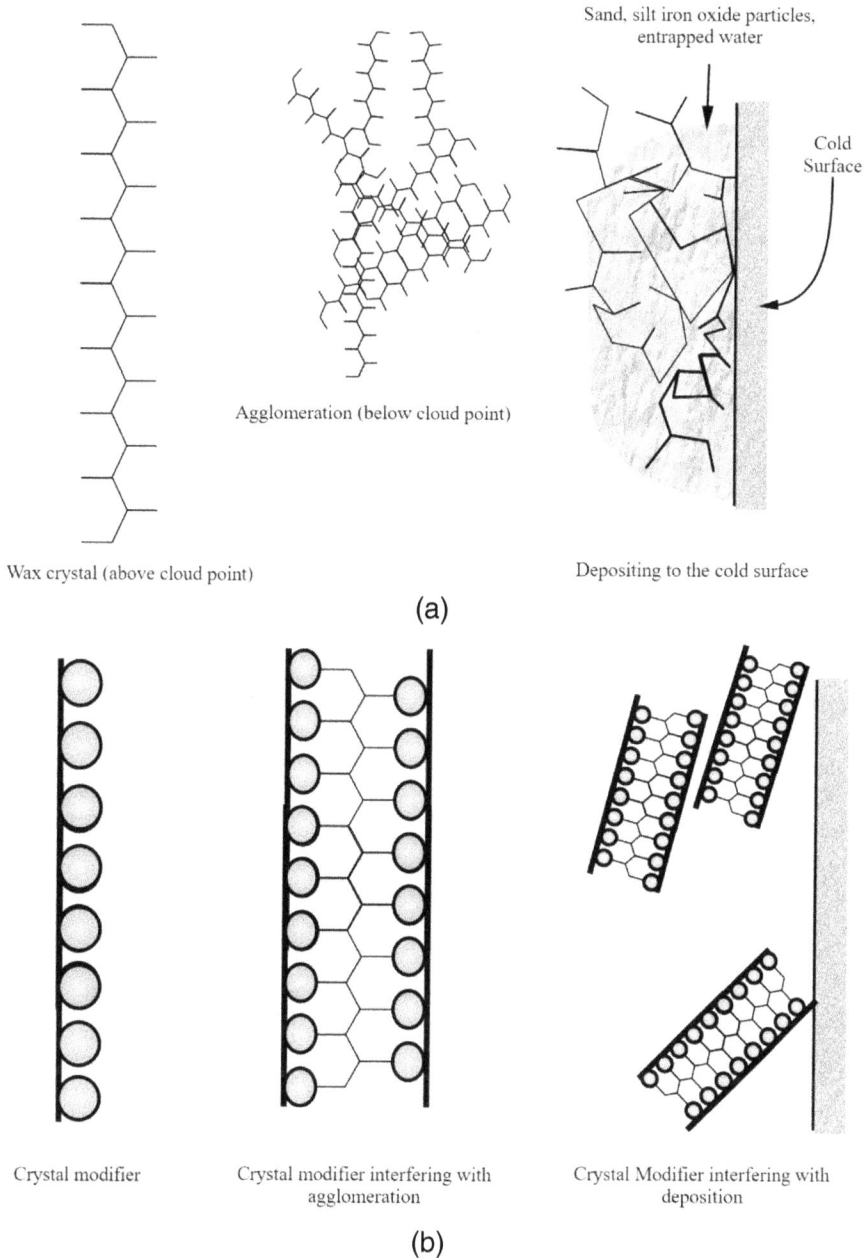

FIGURE 7.12 Schematic representation of the mechanism of wax crystal modifier.[56, 96]

points A and B, the average temperature of the crude oil (T_h) is in hot flow conditions, and above the WAT, the wax deposition occurs. The wax deposition increases axially as the fluid flow temperature approaches the WAT (B). When the fluid temperature further drops below the WAT, the flow enters the cold-flow regime, in which the wax

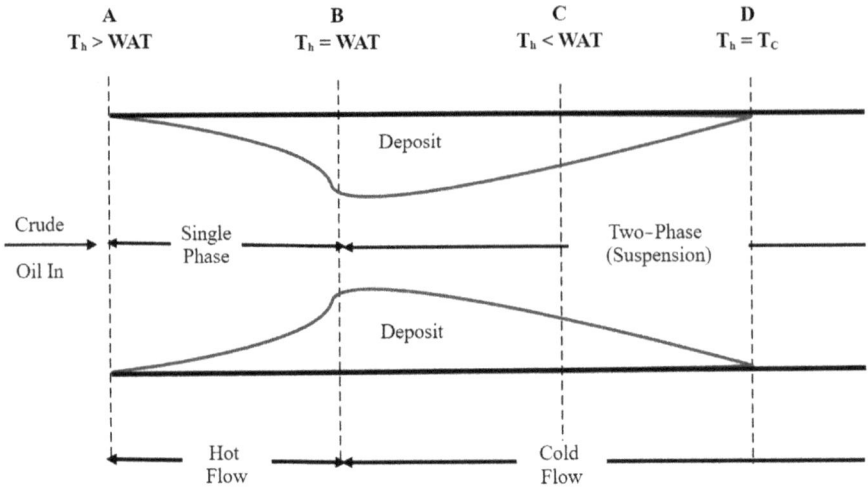

FIGURE 7.13 Illustration of flow profile of waxy crude oil in pipelines

crystals diffuse back into the bulk fluid, leading to the wax slurry fluid flow behavior (after point B). Subsequently, as the fluid flow temperature decreases, the thickness of the wax deposition continues to decrease until it reaches zero as the fluid temperature nears the temperature of the surrounding (between point "B" and "D"). After point D, and the crude oil is in thermal equilibrium with ambient, no further wax deposition occurs as there is no temperature gradient.

Figure 7.14 shows the schematic representation of the cold flow scheme used in the real field operation.[62, 97] In this method, the temperature of the fluid is reduced below the WAT by externally circulated fluids or adding the hydrocarbon fluid at a

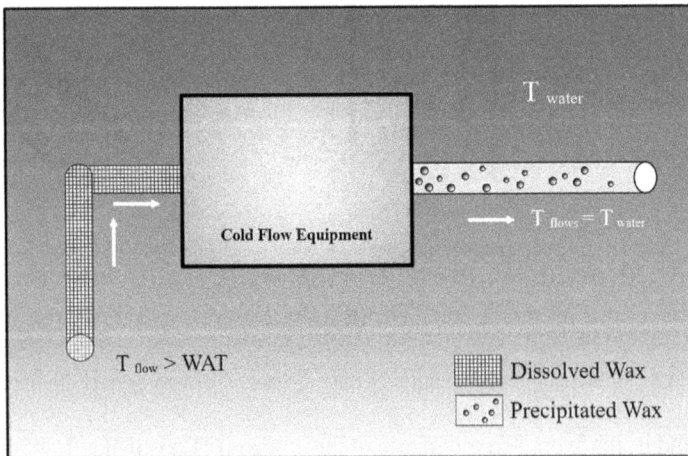

FIGURE 7.14 Schematic representation of the cold-flow scheme used in the real field operation.

lower temperature so that the average fluid temperature falls below the WAT. Argo et al.[102] provided similar methods to address wax and asphaltene deposition issues in the pipelines. A modified method of wet-cold-flow technology has been proposed by Haghighi et al.[103] and Azarinezhad et al.[104] for wax mitigation.

7.4.1.4.2 Thermal Insulations and Coating Materials

For subsea flowlines and risers, it is important to ensure that the heat loss to the surrounding should be prevented so that the fluid flow temperature is maintained above the WAT. It has also been reported that the system pressure does impact the WAT for live and dead oil systems.[105] To avoid heat loss, the insulation of pipelines and risers is practiced. Plastic insulation done on the pipelines has shown a reduction in wax deposition by about 30%.[106] The understanding of the detailed mechanism of how wax attaches and deposits on the surface of the pipeline catalyzed further research in the development of anti-wax coating materials. Such novel materials are wax repellent and can help to reduce the frequency of mechanical cleaning. The coating materials should be chemically inert, anti-abrasive, anti-corrosive, and resistant to all other possible deposits (such as asphaltene, scales, etc.). Some of the coating materials used for this purpose are ethylene-tetrafluoroethylene (ETFE), polyvinyl chloride, glass-reinforced epoxy resin (GRE) for internal and external pipe coating.[107]

7.4.1.4.3 Microbial Treatment Method

Naturally occurring marine organisms can help mitigate wax deposition issues in pipelines. These organisms can absorb the paraffin, use it as a carbon source for energy, and thus, remove these deposits over a certain period.[56, 57] In this method, organisms are injected with crude oil and nutrients to support the initial growth phase. Microorganisms remove the paraffin deposits by using them as a carbon source for energy. They degrade heavier hydrocarbons into smaller molecules and release by-products such as acids, biosurfactants, and gases, further dissolving these waxes in the oil phase and resulting in a decrease in viscosity, pour point, and surface tension of the crude oil system. In the field, organisms are fed periodically to maintain the microbial colony size to avoid the wax deposition. The major advantages of this method are that it is environment friendly, nonflammable, noncarcinogenic, and economical.[55-57] Some major disadvantages could be that the organisms may not survive under extreme pressure and temperature conditions. Also, the bio-sludge may plug the pipeline if not controlled properly.

This method is successfully used in Mehsana asset of Oil and Natural Gas Corporation Limited.[108] Biosurfactant-producing microorganisms such as *Pseudomonas* and *Actinomyces* species have been used to treat wax issues in pipelines. These bacteria have been shown to reduce heavier hydrocarbon percentage from the crude oil, thus making it flowable. These organisms help lower the WAT, cloud point, and viscosity, thus making crude oil less prone to deposit wax inside the pipelines in about a few days.[109, 110] Studies show that bacterial strains *Geobacillus* TERI NSM can degrade paraffin under high temperatures.[111] Bacterial consortium has also been investigated for the mitigation of wax issues. A bacterial consortium (SBC1) has shown to be effective in addressing the solid and semisolid paraffin deposition issues.[112] For good application, these methods should be used with oil well–producing water and below 200°F so that the organisms can survive down-hole.[62, 113]

The microbial treatment method, when used with nutrient medium and growth enhancer, showed that it reduces the frequency of mechanical scrapping and pigging operation.[114] *Pseudomonas aeruginosa* (N2) and *Bacillus licheniformis* (KB18) are some other strains of *Pseudomonas* and *Bacillus* found to be effective to remove paraffin deposits.[115] They also observed that the organism could change the wettability of the stainless steel to water wet and decrease the adhesion of wax molecules by forming a small layer on the steel surface. It has been observed that the well treatment done using bacteria showed no wax issues in the wellbore and pipelines for 6 months.[115] Field tests have shown *Bacillus* and *Pseudomonas* species to have a good tendency in addressing wax deposition issues.[116]

7.4.2 ASPHALTENE REMEDIATION TECHNIQUES

The deposition of asphaltene attracts many issues during oil and gas production. As mentioned earlier, asphaltene may deposit near the wellbore, contributing to skin damage. Their deposition on the rock surface can change the wettability to oil-wet conditions, reducing oil recovery. They can deposit in a wellbore, pipelines, and surface separators and processing facilities, thereby attracting flow assurance issues and decreasing oil and gas well productivity. Several methods have been employed in the industry to address asphaltene deposition issues. The mitigation of asphaltene deposition can be done by controlling the pressure and temperature conditions, changing the composition, or through an external aid. These methods are mainly categorized into chemical, mechanical, and thermal methods. These methods are discussed earlier in the context of wax deposition issues, and the same with or without modifications can be used for asphaltene deposition issues.

7.4.2.1 Thermal Methods

Thermal methods to address asphaltene issues mainly include hot oiling or injecting hot oil into the well to dissolve asphaltene deposits. The other methods include downhole thermal heaters. A downhole heater is a heating element placed in the wellbore for a short time to melt and dissolve the asphaltene in the wellbore or the tubing. Once the asphaltene is dissolved, it can be produced along with crude oil. Electrical heating is costly and depends on the availability of an electrical source around the well. Other methods include a heat-releasing mixture of chemicals, such as a solution containing an equimolar mixture of sodium nitrate and ammonium chloride. Once these chemicals are released in the wellbore, they react and generate exothermic heat, melting and dissolving the asphaltene. A buffer chemical is used to decelerate the exothermic reaction downhole. However, this method is costly and can only be used in specific reservoir conditions where other methods may not be suitable.[117]

The plasma heating method has also been used as a sidewall-melting process to melt and dissolve asphaltene, resin, and paraffin (wax) deposits from the pipeline and tubing. The plasma torch is used externally to heat the pipe wall surface, while from inside, a mechanical method such as scrubbing, pigging, and so on is used. The plasma heating ensures that the pipe wall temperature exceeds the melting temperature for the deposits, which helps soften the deposit. Mechanical methods aid in removing the softened or melted deposit and fluid from inside the pipe. The plasma heating process can be repeated several times until all the deposits are removed.[117]

7.4.2.2 Mechanical Techniques

The mechanical method consists of scrapping of deposits inside the well and pipelines.[118–120] The other methods include pigging and mechanical vibrations. Pigging is often used to remove wax deposits. However, its use to specifically asphaltene deposits is not well known. Pigging operation removes wax and asphaltenes deposits together as the pig move through the pipelines.[121] Solvents and some specific additives are injected in pipelines during pigging operations to disintegrate the solid deposits, reduce surface tension, and improve the efficiency of the pigging operation. The addition of solvent also helps overcome technical issues during the cleaning of pigs. Recently, there has been advancement in intelligent pigs for pigging operations that mainly works on remote visualization, advanced control operation, and localized heating. For wellbore applications, wireline operation is used to scrap the solids. Another method uses a hydro-blasting tool to cut the deposits, which are done using a coiled tubing operation.[122] In hydro-blasting, a high-pressure jet fluid through the nozzle removes the deposits. Methods like brushing and scrapping require shutting down the well for an extensive time,[123] thus impacting production with significant expenses. Coiled tubing operation is also suited to remove deposits in a wellbore and pipeline.[123] These methods should be used where the wellbore or pipeline is easily accessible. Some of these methods may have a limitation on working pressure downhole to cut the deposits.[123] Due to time-consuming mechanical operations, other methods to address asphaltene deposition are considered.

7.4.2.3 Chemical Method

Chemical methods are widely used to address the asphaltene deposition issue, particularly where mechanical methods fail. Chemicals primarily involve using solvents, polymers, and anti-agglomerate for asphaltenes. One of the main objectives of chemicals is to dissolve the asphaltene in the bulk fluid. Unlike waxes, in which solvents primarily require lighter hydrocarbons, asphaltene needs aromatic solvents. Asphaltene is known to be soluble in aromatics (such as toluene and xylene) and insoluble in light paraffin.[124] Allenson and Walsh[122] have suggested using asphaltene inhibitor squeezes near the wellbore to avoid asphaltene deposition. It has also been suggested to use deasphalted oil containing high aromatic content to dissolve asphaltene into the bulk fluid. The idea behind this approach is to break the intermolecular interactions between asphaltene molecules and keep them suspended in the hydrocarbon fluid. Al-Sahhaf et al.[125] investigated the impact of deasphalted oil, resin, toluene, and other surfactants such as nonylphenol, dodecylbenzene sulfonic acid, and dodecyl resorcinol to reduce asphaltene deposition. It has been observed that surfactants are more effective in reducing asphaltene precipitation than toluene and deasphalted oil. Asphaltene deposits can also be dissolved using hot mineral oil along with surfactants solutions. The most common aromatic solvents used are xylene and toluene. These aromatic hydrocarbons may show less dissolving efficiency for high-molecular-weight asphaltenes. The efficiency can be increased by the use imidazolines or polyamine condensation product lines of fatty acids. However, using these solvents is discouraged due to their high volatility, flammability, and corrosive and toxic nature. A patent by Wolfgang and Herold[126] proposes the use of fatty acid esters and esters for safe and environmentally friendly alternate solvent for asphaltene deposition mitigation.

Asphaltenes contain a low H/C ratio, indicating that they are highly unstable molecules in the oil.[127] It has been observed that increasing the amount of resin in the oil increases asphaltene stability.[128] It is known that during the oil flow from the reservoir to the surface, associated resins and aromatics with asphaltene molecule tends to separate, making asphaltene unstable and prone to agglomeration and deposition in the wellbore and surface facilities. Gas injection methods in which hydrocarbon gas is injected for oil recovery or artificial lift operation could destabilize the asphaltene. On the contrary, it has been proposed that reinjecting the dead oil from the stock tank/surface into the wellbore or reservoir could help reduce asphaltene deposition issues, as the dead oil primarily contains resins and aromatics help stabilize asphaltene.[129] It has been observed that using Perturbed Chain Statistical Associating Fluid Theory (PC-SAFT) EoS, the unstable asphaltene phase envelop (APE) region becomes smaller. The use of commercial and non-commercial inhibitors has been investigated by Ghloum et al.[127] They proposed to use a mixture of high doses of non-commercial inhibitor along with a low dose of commercial inhibitors. Some researchers indicated that the mixture (50:50) of toluene with diesel has proved to be a good solvent to dissolve asphaltene.[130] Indeed, extensive field studies are required to confirm this strategy.

The use of polymeric dispersant has been studied by Schantz and Stephenson[131] to reduce asphaltene deposits. Alkylbenzene-derived amphiphiles reduce the asphaltene deposition by improving the asphaltene stability.[132] The polar headgroup of the amphiphile helps to stabilize the asphaltene molecule. Ionic liquids and alkylbenzene-derived amphiphiles improve the stability of asphaltene, reducing its deposition.[132, 133] Dodecylbenzene sulfonic acid (DBSA) in low doses was found to dissolve asphaltenes.[134] It is observed that resins work synergistically with DBSA to improve the stability of asphaltene and reduce its precipitation. The use of xylenes was also found to dissolve asphaltene and improve well productivity.[134]

7.4.2.4   Surface Coating Method

Internal surface coating can help prevent the deposition of heavier hydrocarbons such as asphaltene. The use of epoxy resin, polyurethane, phosphate, and two different fluoropolymers, polytetrafluoroethylene and perfluoroalkyl has been studied for their potential application as a coating material to mitigate asphaltene deposition in a flow loop experimental setup.[119, 135] Hydrophobic polyurethane and epoxy resin coatings have shown weak performance on asphaltene deposition, though the reduction in a deposition is in the order of 10 and 50%, respectively. The super-hydrophilic phosphate coating helped form a water film on the surface, reducing the asphaltene deposition by 90%. Finally, fluoropolymer internal surface coatings showed higher performance in reducing the asphaltene deposition by 99.89% and 99.94%. The field-scale studies of such coating could provide an economic advantage as this may reduce frequent maintenance using mechanical operation. The interaction between electroless nickel–phosphorus and oil–water emulsion containing asphaltene has been studied by Gong et al.[136] It has been observed that the electroless nickel–phosphorus coating could reduce asphaltene deposition. However, field tests are necessary to confirm the efficacy.

Rigorous studies are required to understand the performance of various coating materials for anti-asphaltene deposition use. Although some lab studies show these various materials to be useful for the application due to low cost, issues such as their durability as coating materials, adhesion, environmental compatibility, and others need to be investigated.[137, 138] Unforeseen circumstances such as cracks and coating defects may attract other issues such as corrosion.[138] Although surface coating can be an advantageous and economical method for asphaltene mitigation, the development of durable coating material is the need of the hour. The risk associated with coating failure can be huge. Using nanotechnology, there is scope for developing advanced materials to address flow assurance issues due to wax and asphaltene.

7.4.3 HYDRATE REMEDIATION TECHNIQUES

As the oil and gas industry is venturing into the offshore deepwater environments, the low temperature and high-pressure conditions attract the potential of hydrate formation in the offshore pipeline and risers, which has been one of the deepwater flow assurance issues. The continuous research and development in hydrate flow assurance mitigation have resulted in the development of new technologies and even proposed altering the hydrate paradigm. Deepwater oil and gas operations are common nowadays at several offshore locations, including the Gulf of Mexico (GOM), West Africa (WA), and the North Sea (NS). Gas hydrate issues become evident commonly during drilling operations and offshore production. Hydrates form in the presence of free water (produced with natural gas and oil), at high-pressure and low-temperature conditions, and over a period, resulting in flowline blockage. Hydrate formation and blockage are prevalent at a section where the flow changes direction, particularly at the junction of well, pipeline, and risers. Hydrate plugs, once formed, are difficult to remove because hydrate dissociation is endothermic, and the heat transfer is very gradual during dissociation. In addition, a sudden dissociation of the hydrate plug can be disastrous, and 1 m³ of hydrate can result in the release of gas of about 160–180 m³, thus increasing the in situ pressure within the pipelines. Therefore, it is important to ensure that hydrate mitigation and prevention strategies should be applied carefully. This section provides the methods used to mitigate hydrate flow assurance issues.

7.4.3.1 Thermal Techniques

Direct and active heating is one way to prevent and dissociate hydrate plugs. In this method, active heating is applied to the section to increase the temperature in situ. As mentioned earlier, safety issues should be considered as sudden dissociation may release a large amount of gas, and may damage the flowline. Heat application should be uniform for a safe and easy hydrate remediation process. The advantage of this method is that it can be quick against depressurization methods, which may take a long time to dissociate hydrate. The active heating method can be combined with thermal insulation for flowlines and risers to reduce power consumption.[139]

Active heating methods are of different types: (a) hot fluid circulation and (b) direct and indirect electric heating (DEH/IDEH).[140–144]

7.4.3.1.1   Hot-Fluid Circulation

Hot-fluid circulation provides the same benefit as like electrical heating approach. In this method, dead oil or hot-fluid circulation containing hot water is performed. The hot circulation fluid approach uses hot dead oil to heat the dual pipelines, flow-lines, or riser systems. This method is often used during a restart after a prolonged shutdown and is more suited for the bundled pipeline. The looped pipeline may not be suitable for hot-fluid circulation due to its longer length. The hot-fluid circulation with water is suitable for continuous operation and during shutdown and restart of production. The hot water provides heat to produce fluid, maintaining the temperature above the hydrate formation condition. Thermal and mechanical design, configuration, production, life cycle, and risk considerations must all be addressed during the construction of these packages. Active heating techniques provide adequate protection. Hydrate control can become easier with a combination of insulation and heating. Active heating can provide the operational flexibility for a subsea production system by addressing the production issues, water cut, start-up, operating flow rate, and depressurization times during hydrate dissociation.[144]

7.4.3.1.2   Electrical Heating

Electrical heating (EH) is a rapidly evolving technique that has to gain acceptance in offshore fields such as Nakika, Oregano, Asgard, Huldra, and Sleipner in the NS and Habanero in the GOM.[145] The benefits of EH include standardized restart operations, the elimination of flowline depressurization, and the capability to effectively remediate the hydrate blockages. The EH method is categorized mainly into the following types:

i) Direct heating
    This method uses the streamline as an electrical conductor for resistance heating.[145] In this method, the current is passed axially from the pipe wall to heat the surface directly. Four main configurations are used:
    a) *Pipe-in-pipe (PIP) systems:* The current is passed from one pipe and returned from the second pipe.
    b) *Fully insulated single pipe (closed system):* The current is passed using a cable and returns from the flowlines.
    c) *Earthed current single pipe (open system):* This method is similar to a closed system, except that a part of the current can go into the surroundings, helping avoid the electric insulation of the flowline.
    d) *Direct heating for the pipe bundle:* In this method, the current is passed from one flowline and returned from another flowline.[144]

ii) Indirect heating
    In indirect heating, the EH element is installed on the surface of the pipeline or flowlines. The flowline is heated due to the thermal conduction of heat from the EH element. The main types include the following:
    a) *Induction heating:* The tube conducts the current from a cable, and the flowline gets heated due to thermal conduction from the tube.

b) *EH trace method:* This method generates the heat using a resistant element attached to the pipelines or flowlines. A self-regulating polymer is used in which when the temperature falls, the polymer creates electrical conduction pathways to allow the flow of current for heating purposes. Once the temperature increases, the electrical conduction reduces or stop. This method is highly energy-efficient and is used for a bundle of pipe or PIP systems, tiebacks, and its use is limited to the onshore piping system.

c) *Electric gut trace heating*: This method uses a heating element installed inside the pipeline to generate heat for the need. The heating element is an electrical heater (or hot fluid). This method is safe as the heating element is inside the pipeline surface. However, this method limits the application of mechanical methods, such as pigging and pipeline with multiple bends and valves.[144]

7.4.3.1.3 Thermal insulation

Thermal insulation is one of the best preventive strategies to mitigate hydrate formation issues in flowlines and risers. The thermal insulation method has the following types.[142, 144, 146, 147]

a) *Cast in place (CIP)*: In this method, an insulation material (single or multiple layers of the same or different material) is cast directly on the surface of the pipeline.

b) *Pipe in Pipe (PIP)*: In this method, one pipeline is fixed onto another pipeline to provide better insulation than CIP. However, this method is expensive and can be employed for the short-distance pipeline.

c) *Sandwich pipe (SP)*: Concerning the PIP system, the SP pipeline is a potential option for deepwater applications due to its lower weight while providing good thermal insulation for flow assurance issues.[148, 149] One pipe is sandwiched between another pipe using a thick and flexible core with high strength and thermal insulation.

d) *Burial pipelines*: This method provides better thermal insulation than open pipelines. In this method, the pipeline is buried under a mudline by covering it with sand, rock, gravels, or soil exhumed from the seabed while trenching, which provides additional safety from shipping and marine life. This method is useful for shallow water depths. Additional insulation can be provided to the pipeline.

7.4.3.2 Mechanical Methods

Pipeline inspection gauge (pig) is the most commonly used mechanical method to remove the hydrate deposition in the pipeline. The details of pigging operation have already been discussed earlier for wax and asphaltene remediation. As the pig moves from the entire pipeline, they remove deposits.[150] Pigging should be done frequently to avoid plug formation. For deep water, pigging is more complex. For such conditions, the pigs are passed from the flowline loop and recover from the topside.[141, 144] Once the plug is formed (fully or partially), pigging is not recommended, as it will be impossible to pass the pig through the pipeline and may be stuck inside. Drilling the plug is also not a good idea as it may suddenly release significant gas, causing harm to the pipeline. Coiled tubing is a potential alternative when pigging cannot be performed.

7.4.3.3   Hydraulic or Depressurization Method

The depressurization process is also known as blowdown. In this method, the pressure of the system is reduced from both the end to below the hydrate dissociation pressure at a given temperature. Depressurization from only one side should be avoided as a large pressure difference may generate hydrate plug projectile, which may be catastrophic. As the pressure is reduced below hydrate phase equilibrium conditions, the hydrate plug temperature may further reduce below its surroundings, helping heat flow from the surrounding environment (or ocean) to hydrate plug, thus enhancing the dissociation process at the plug boundaries near the pipeline surface. The hydrate formation temperature decreases due to reduced system pressure, preventing further hydrate formation. This method is recommended for flowlines that are not perfectly insulated as the hydrate dissociation could be faster in such situations than perfectly insulated and buried pipelines systems.[144, 151]

This method is quite popular when the hydrate plug is completely formed and blocks the pipelines. The method helps in fast restart after a complete shutdown. This method is difficult to be applied for offshore conditions and uneven pipelines due to operational issues. For select subsea flowlines, the method may be difficult to be employed. Also, the complicated flowline system geometry and possible liquid head in the flowline or riser can hinder the application of the blowdown method. The other main issue is that the dissociation of the hydrate plug can be time-consuming. This method is unsuitable for pipelines containing lighter liquid hydrocarbon due to vaporization during depressurization.[144, 151] Rapid depressurization should be avoided as it may worsen the hydrate problem due to Joule Thomson cooling effect. This process may attract many operation challenges due to handling a large amount of gas and liquid during blowdown. During this process, it is required to stop the production, thus incurring additional revenue loss.

7.4.3.4   Hydraflow

Hydraflow is a relatively advanced method used to address hydrate deposition issues. In this method, hydrate formation is allowed, but the hydrate agglomeration and deposition into the pipeline are prevented. Most of the gas is allowed to convert into hydrate crystals forming the hydrate slurry in the pipeline. The pressure and temperature conditions in the pipeline are carefully monitored. Once the hydrate slurry is formed during flow, it is allowed to flow in the pipeline until the surface facilities. The dosage of anti-agglomerate can be added into the pipeline based on rigorous calculation and understanding of the hydrate phase envelop. The anti-agglomerate can be separated at surface facilities and recycled, reducing the operation cost. In addition to avoiding hydrate plug formation, the main advantage of this method is that it increases the fluid carrying capacity of the pipeline as most of the gas is converted into hydrate slurry. This method is called cold flow because no heating or insulation is employed, hydrate formation is not prevented, but their deposition and the flow conditions are relatively cold.[144, 152]

7.4.3.5   Operational Methods

Operational methods mainly include hydrate monitoring and water removal techniques. The hydrate monitoring technique involves mainly three main mechanisms: (1) optimizing the inhibitor dosage to control the safety margin of hydrate formation, (2) sensing early-warning signs of the hydrate formation, and (3) low-pressure operations. The monitoring system needs dynamic information on the pressure and temperature condition to understand hydrate formation behavior in addition to helping to reduce the inhibitor dosing amount. Thus, this method helps to reduce the cost toward remedial, maintenance, revenue loss due to possible deferred production, and potential environmental impact.[153]

The second method involves removing water from the production streams wherever and as early as possible to minimize hydrate formation. The dehydration process is commonly used to remove water from the production streams. For subsea operations, subsea separation systems can remove water from the production streams. The separation of water at subsea prevents hydrate formation, increases the recovery of hydrocarbon fluids, and reduces the energy required to lift water from flowlines. Other advantages include reduced challenges associated with handling formation water at surfaces, treatment, and disposal. Efficient subsea separation and disposal systems are being developed to separate the water from production streams at the seafloor. The basic component of the subsea separation unit includes a separator, pump, valves, chemical dosing system, and water injection well. Although subsea separation units remove significant water from the streams, they do not completely remove it. Thus, the hydrate formation possibility may be reduced but cannot be completely avoided. For this, continuous injection of chemical inhibitor is injected into subsea flowlines.[144]

The third process to prevent hydrate formation uses low-pressure operations. The pressure of the flowlines can be managed below the hydrate formation pressure at a given temperature. This method is suitable for onshore pipelines. In deep water, the ambient temperature is quite low. Hence, a significantly low pressure is required (maybe less than 30 bars) to avoid hydrate formation. Such low-pressure operations for subsea pipelines and risers are not practical. An alternate approach is to lower the operating pressure by employing a subsea choke to reduce the driving force for hydrate formation, which will also help reduce the inhibitor dosing rate.[150]

7.4.3.6   Chemical Techniques

The chemical method is used widely to prevent hydrate formation from onshore and offshore subsea flowlines. This method either aids in removing hydrate plug formation or helps in removing it. The various chemicals used for hydrate mitigation are categorized into different types. These are thermodynamic hydrate inhibitors (THIs), kinetic hydrate inhibitors (KHIs), anti-agglomerates (AAs), and cationic starch.

7.4.3.6.1   THIs

These inhibitors are the most commonly used chemicals as a hydrate prevention strategy for deepwater applications. These chemicals mainly affect the phase equilibrium of hydrate systems by increasing the hydrate formation pressure or reducing the hydrate formation temperature, thus making the operating conditions lie outside the hydrate phase stability conditions. The most commonly used THIs are methanol, ethanol, MEG, and salts, as they have low costs and are available easily. Methanol is quickly vaporized to the gas phase and is also water-soluble, thus affecting the chemical potential to prevent hydrate formation. As hydrate formation mainly happens in the water phase, the inhibitors are generally water-soluble. As methanol is highly volatile, it vaporizes into the gas phase. Thus, there could be a major loss of methanol while using them as chemical inhibitors. However, they are one of the highly effective THIs. One of the main issues is that they need to be used continuously and in large quantities. Thus, they need high storage volumes, and the cost for onshore or offshore pumping could be more, thus inviting additional capital expenditure; they are also flammable.[141, 142, 151, 153–157] These are mainly used as hydrate prevention techniques, although they can also be used for hydrate remediation. In such cases, the inhibitor's contact with the potential location of the hydrate plug could be very tricky due to the long distance from the injection point. In such situations, they are injected into the pipeline using a coiled tubing operation.

Some of the challenges using THIs are selecting proper and precise dosing rates, as underdosage may result in poor hydrate inhibition performance. Often, the injection of inhibitor location is not properly identified. Thus, they may lose their efficacy by the time they reach the potential hydrate zones in the flowlines. There could be environmental concerns in case of leakage to the ocean affecting marine life. The compatibility with materials and corrosion issues, particularly using glycols, should be carefully considered, and they should be handled carefully during transportation and at the injection site to avoid environmental impact.[150]

7.4.3.6.2   KHIs

KHIs help delay or slow hydrate formation kinetics by delaying the hydrate nucleation and growth process.[142, 158] However, they don't affect the phase equilibrium conditions of hydrate. KHIs are typically low-molecular-weight water-soluble polymers or copolymers that bind to the hydrate surface, forming a hindrance for further growth kinetics and thus reducing hydrate plug formation. They are generally dissolved in some carrier solvent, including produced stream and water, and pumped into the flowlines or pipelines. They are required in low quantities (typically less than 1 wt%) and often trend as low-dosage hydrate inhibitors (LDHIs). These are mainly water-soluble long-chain compounds, showing a high efficiency in delaying hydrate formation in flowlines, and are environmentally friendly. They are effective up to 15–23°F subcooling conditions and thus have limitations for ultra-deepwater conditions in which subcooling can be of the order of 250°F and more. Thus, THIs is preferred for ultra-deepwater application.[142, 156] These alone may not be effective at certain applications, and thus, it has been suggested to use THIs along with KHIs to enhance the performance of hydrate prevention.[158, 159] KHIs use less energy than THIs and thus are very commonly used for hydrate prevention, providing lower operating expenses (OPEX)/capital expenditures (CAPEX) with increased performance. The well-known KHIs is polyvinyl pyrrolidone (PVP) and polyvinyl caprolactam (PVCap). Other examples of copolymers as KHIs include vinyl methyl acetamide (VIMA) and vinyl caprolactam (VCAP) or poly(VIMA/VCAP). KHIs are being used in offshore operations in GOM and the NS. These are safe to use as compared to THIs. One of the main disadvantages is that KHIs have time-limited inhibition efficiency, thus restricting their use during the shutdown. Depressurization is one way to address the long-term shutdown to ease the restart operation.[160] Simultaneous use of thermal method consisting of electrically heated flowlines and chemical method is recommended to address hydrate prevention in the more complex subsea system. Electrical heating is highly effective compared to blowdown methods during hydrate plug removal.[145] The use of electrical heating can further reduce the amount of LDHIs required for injection. However, hydrate inhibitors must be continuously injected to prevent hydrate formation in subsea flowlines.

7.4.3.6.3 AAs

AAs are a type of polymers and surfactants that stick on the surfaces of hydrate while avoiding their agglomeration. AAs do not delay the hydrate formation significantly like LDHIs but form tiny hydrate crystals droplets while keeping them suspended in the hydrocarbon or water fluid phase without affecting flow behavior. Thus, they do not allow hydrate to form large agglomerates, resulting in plug formation. Thus, AAs forms hydrate crystals that are easily transportable.[144] AAs can work with higher subcooling conditions and can be effective as high as 400°F subcooling.[142, 151, 156, 159] They can be used effectively for deepwater applications due to their high subcooling properties. AAs have been used in field trials successfully, such as in the deepwater of the GOM for a relatively high subcooling of 40°F. Various factors may affect their efficiency: water-cut, operating pressure and temperature conditions, salinity, and hydrocarbon fluid composition. Typically, water should be less than 50% so that the hydrate crystals can remain suspended in the hydrocarbon fluid phase, thus avoiding further hydrate growth. AAs can be simultaneously used along with THIs for specific applications during shutdown and restart of the production facilities. AAs are more toxic than KHIs. AAs can also show dual action of corrosion inhibition and hydrate mitigations, thus reducing the cost of operation in some situations.[161] AAs are effective in restricting the hydrate plug formation, and thus once injected, they continue to be effective during the flow regime.

7.4.4 SCALE AND CORROSION INHIBITORS

7.4.4.1 Scale Inhibitor

The best technique to prevent scale from forming in the first place is to use a scale inhibitor. Scale inhibitor is a chemical treatment used to prevent or delay scale deposition in production and completion systems. Scale prevention is preferable to scale removal because it is less expensive, more efficient, and effective to keep production running. The most commonly used scale inhibitor in oilfields is phosphonates and polymers. They are extensively used for scale treatment of water systems because of their high solubility, thermal stability, and dosage efficiency. Following are some of the methods used for scale management.

7.4.4.1.1 Continuous Treatment

One strategy for preventing scale formation in petroleum-producing wells is continuously injecting scale inhibitors into injection wells. An aqueous solution of scale inhibitors is continually injected into the completion through a downhole injection point. The scale inhibitor is injected at the point of turbulence to achieve a homogeneous mixture. A steady supply of the scale inhibitor is maintained at a controlled rate during the addition. The exact inhibitor concentration required to prevent scale formation is carefully monitored, resulting in efficient scale inhibitor utilization. Although this method is considered more effective, its implementation is hampered by a lack of availability and the high cost of the necessary facilities.[162]

7.4.4.1.2 Squeeze Treatment

A scale squeezing is performed by injecting the scale inhibitor solution into a producing formation and anchoring the inhibitor in the formation. Scale squeezing is the best form of scale prevention. The squeeze treatment method comprises three stages. The first stage is called the preflush stage, which includes injecting a small volume of fluids containing chemicals (acids, chelating agents, surfactants) to clean the wellbore and production tubing. A scale inhibitor mixed with makeup water is injected into the formation following the preflush in the next stage. Finally, the overflush stage or displacement stage includes injecting another volume of fluid that pushes the scale inhibitor farther into the formation.[162] As the production of oil/gas starts, the scale inhibitor is produced with the formation water. The scale inhibitor flows back at a small concentration after the squeeze. The return scale inhibitor concentration rises quickly, peaks at a certain level, and then drops to a low plateau concentration within a few days, which lasts throughout the squeeze. The plateau flowback inhibitor concentration is usually enough to keep the scale at bay.

7.4.4.1.3 Batch Treatment

In this treatment, a considerable quantity of scale inhibitors is injected regularly and utilized for a long time. Large volumes of scale inhibitors are injected into the tubing at the top, then moved to the bottom of the tubing by the fluids in the oil well. Before production resumes after batch treatment, the well is closed for some time. The main disadvantage of this treatment is that the production comes to a halt during this treatment.[162]

7.4.4.1.4 Scale Inhibitors Extracted from Natural Sources

Green scale inhibitors (derived from plant extracts) are gaining research interest in the prevention of groundwater contamination and protecting the environment from the hazardous chemical used as scale inhibitors.[163] Green scale inhibitors provide several advantages, including biodegradability, high efficiency, and nontoxicity. Abdel-Gaber et al.[164] reported the plant-derived extracts with an economic benefit on a broader scale in the Mediterranean coastal zone, growing well under calcareous soil conditions. The potential of these plants to accumulate a higher percentage of calcium and minerals is exceptional.[164–166] They also studied the scale inhibition from the *Punica granatum* leaf and hull extract for the $CaCO_3$ scale.[166] They observed that the scale inhibitor derived from hull extract showed better scale inhibition than the leaf extract.

7.4.4.1.5 Biodegradable Polymers

As environmental restrictions are becoming stricter, biodegradable, non-toxic, and ecologically friendly polymers are getting attention in the research and development of the scale inhibitors.[163] The dehydration of maleic acid yields the synthetic polymer based on maleic anhydride, which is frequently employed in scale inhibition. Poly(aspartic acid) (PASP), a biodegradable polymer with no phosphorus atom is very effective on sulfate and carbonate scale. PASP can be made in various ways, with or without a catalyst. The scale inhibitor (PASP) inhibits the $CaCO_3$ scale precipitation, where PASP and Ca^{2+} ion are 4 mg/L and 14,225 mg/L, respectively. Euvrard et al.[167] had observed that when PASP was used as an antiscalant, it reduces the growth rate of $CaCO_3$ crystals and reported meager scale deposition on the surface of stainless steel. Nowadays, modified natural polymers are also used as antiscalants in the oil and gas industry.

7.4.4.2   Corrosion Inhibitor

Oil and gas fields are facing major problems due to the corrosion of pipelines and pumps, which eventually needs maintenance and replacement of some of the major equipments. Maintenance and replacement of the equipment increases the production cost. Hence, it is essential to control corrosion by various methods, among which the application of corrosion inhibitors is one of the well-known techniques. Corrosion inhibitors are the chemical substance used to reduce the effect of the corrosion-forming corrosive components present in the production fluids. Chemical inhibitors are used in small concentrations, which get adsorbed on the metal surface and form a thin protective layer that prevents the interaction of the corrosive components present in the oil, gas, and aqueous phases with the metal surface. The oil and gas industry uses corrosion inhibitors to control corrosion, some of which are discussed henceforth. Figure 7.15 shows the classification of the various inhibitors based on the mode of corrosion.

7.4.4.2.1   Interface Inhibitors

Interfacial inhibitors are also called as a barrier inhibitors. Interfacial inhibitors develop a protective layer on the metal surface due to the strong chemisorption, leading to a relative reduction in penetration of the corrosive components. This kind of inhibitors also provides a protective layer against the metal ions in the aqueous solution that is ready to interact with acidic and basic elements of the

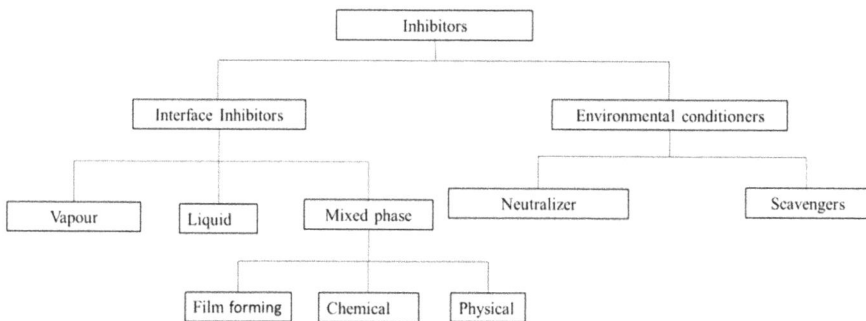

FIGURE 7.15   Classification of the corrosion inhibitors used in the oil and gas field.

crude oil. The interface inhibitors are the highest class of inhibitors, and they do not interact with corrosive components effectively.[168] These inhibitors are classified into vapor, liquid, and mixed-phase inhibitors. Vapor phase inhibitors provide temporary protection against corrosion in a closed environment. However, vapor phase inhibitors perform two things, that is, self-transportation of the inhibitors and protecting the metal surface from the corrosion causing elements. Liquid phase inhibitors mostly contain oxide, hydroxide ions that passivate the film and inhibit metal dissolution. Some of the liquid phase inhibitors control corrosion by reducing the reduction rate. Most effective inhibitors are those inhibitors that form a protective layer by transfer of the charged particles from inhibitors to metal or vice versa.[169]

Mixed inhibitors are generally formed when furfuryl alcohol is mixed with liquid inhibitors.[170] These inhibitors work by parallel adsorption on the metal surface due to more active substances. The inhibition mechanism combines surface protection due to adsorption and electrostatic force of repulsion between adsorbed species and chloride ions of the aqueous medium. Mixed inhibitors are further divided into three types, that is, film-forming, chemical, and physical. Film-forming inhibitors develop a protective film over the metal surface and restricted the interaction of corrosive species with metal. Chemical mixed inhibitors are the most effective one, as the protective layer on the metal surface is formed by charge transfer, and it is irreversible. Physical inhibitors adsorb on the metal surface due to the electrostatic attraction force among inhibitors and metal surface molecules. These kinds of inhibitors interact rapidly but get removed easily.[171]

7.4.4.2.2 Environmental Conditioners Corrosion Inhibitor

Inhibitors that reduce the corrosivity of the medium by deactivating the corrosion-causing ions of the fluid medium are called environmental conditioner inhibitors. These include the following types:

a) *Neutralizing inhibitors*
   The acid present in the oil and gas mixture or acid added during acid fracturing causes corrosion of the equipment involved in production. The acidization can be managed with the help of neutralizing inhibitors that reduce the concentration of the hydrogen ions in the aqueous solution. These kinds of a neutralizer are also used in the boiler in water treatment and the flowlines.[172] Some of the frequently used neutralizers are alkylamines, polyamines, morpholine, sodium hydroxide, and ammonia.

b) *Scavenging inhibitors*
   Scavenging inhibitors are mostly used to deactivate the corrosion-causing ions present in reservoir fluids. Deactivation of corrosion-causing ions stops the formation of impurities and unwanted reactions. For oilfield applications, sulfite and hydrazine are the most commonly used scavenging inhibitors. However, due to the toxic nature of hydrazine and the nonvolatile nature of sodium sulfite, these inhibitors fail to protect against oxygen-based corrosion. Ascorbic acid is used as an oxygen scavenger corrosion

inhibitor. Similarly, organotin compounds are generally used in polymer manufacturing industries as hydrochloric acid scavengers.

7.5 CHEMICAL DOSING RATE FOR BLOCKAGE PREVENTION

The choice of a chemical dosing solution to mitigate hydrate, scale, wax, and asphaltene problems is fundamental to the flow assurance procedure and directly influences capital and operational expenditures. The inhibitor injection strategy should include a comprehensive review of deposits, an evaluation of crude oil properties, and a quantification of stock tank liquid tests, which forms an integral part of the formulation of the remediation action plan.

7.5.1 WAX AND ASPHALTENE INHIBITOR

Although there are no universal direct empirical correlations in the literature for the determination of the amount of dosing required for mitigation of asphaltene and wax, yet field-scale case studies can be referred to for gaining knowledge about its use and dosage optimization. The sequence of events that should be followed before selecting any dosing concentration includes but is not limited to crude oil characterization of that particular field, PPD performance testing, rheological tests for PPD determination, dynamic cold finger tests for wax deposition, and flow loop tests for combined wax deposition and PPD determination. The flow loop experiments feature prominently in simulating a pipeline restart pressure (pressure needed to overcome the yield stress of the gelled crude and return to a flowing liquid state). A simplified form of the equation is[173]

$$p = \tau \frac{2L}{R} \tag{7.1}$$

where p is the minimum restart pressure, τ is the yield stress, and L and R are the pipeline length and radius, respectively.

However, during a field trial, the checklist of parameters that needs to be inspected includes, but is not limited to, pour point (twice daily), yield stress (every dosage reduction), kinematic viscosity (weekly), wax inhibition (per dosage reduction), restart pressure (during urgent shutdown), and pigging (biweekly). Using inhibitor doses in various major deepwater oil pipelines or field tubulars can help understand the concentration of doses that might be optimum for mitigation of wax, asphaltene, and scales. For instance, in a deepwater oilfield in Malaysia of depth about 1300 m having light-waxy crude oil is tied back to a host facility through a 203-km long, 18-in. internal diameter (ID) oil export pipeline.[174] The oil production rate from the well was approximately 13×10^4 bbls/day. The seafloor temperature is around 4°C, the pour point temperature is 18°C, and the WAT is around 25°C. Since the minimum subsea ambient temperature is around 4°C, this condition would have led to the risk of fluid gelling, especially during emergency shutdown operations and wax deposition during low flow rate conditions. Thus, a conservative approach of chemical inhibitor dosing was adopted, initially at a high dosing rate and gradually being optimized downward in a stepwise manner. Subsequently, following each reduction in the dosing rate, samples were collected from the receiving end of the pipeline and sent to the laboratory for detailed testing. The most promising inhibitors from the pour point tests were then subjected to rheological investigation to simulate their effectiveness for field conditions. Initially, steady-state shear viscosity screening was done to ensure correlation with the pour point results.[174]

One of the studies showed that using 2000 ppm of *Sapindus mukorossi* as a wax dispersant improved the flow of crude oil through pipelines. Also, it significantly reduced the yield stress, viscosity, and interfacial tension. Moreover, it was also responsible for reducing the pour point by decreasing the size of wax crystals and their structure reorientation.[175] In another study on Indian crude oil and wax dispersant, 2 wt.% dosages of a surfactant tri-triethanolamine monosunflower ester showed a significant decrease in the pour point. The pour point of the wax before the treatment was reported to be 42°C, and the resulting pour point of the wax was 1°C after the wax dispersant treatment.[176] Al-Sabagh et al.[177], in his study, showed that lower dosages (250 and 500 ppm) are not as effective as higher dosages (1000 and 1,500 ppm) of polymeric derivatives of styrene–maleic anhydrides (SMAs), such as hexadecyl-amine SMACA, octadecyl alcohol SMASA, benzyl-alcohol SMABA, and aniline SMAAn. The change in pour point using a higher concentration of the derivatives ranged from 12–21°C. The maximum efficacy was observed in the case of SMACA, an acetyl side chain, and an amine group derivative of SMA, where the change in pour point was observed to be 21°C at 1,500 ppm. Another study on Gemini surfactants to reduce pour point suggested that as the concentration of the additives increased (to 2,000 ppm), PPD is observed as well.[178] Another ester group additive, ethylene vinyl acetate co-diethanolamine (EVA co-DEA), was studied, and it was found that the optimum dosage of 28.42 wt% EVA displayed the most effective results in terms of inhibition of wax.[179] EVA alone cannot influence the depression in pour point when mixed with 20 wt% VA; 50 ppm of the polymer was able to reduce the pour point by 26°C.[180] Jatropha seeds and castor seed oil were able to reduce the viscosity of wax deposits and depress the pour point from 24 to 8°C on the usage of low dosage, such as 0.1–0.3 vol.%.[181]

In another instance, in an onshore Abu Dhabi oilfield located about 185 km south of Abu Dhabi city, few wells have production problems due to obstructions in the wellbore due to asphaltene precipitation.[182] At the onset, the action plan for softening was focused on using solvent-soak treatments with an additional dispersant. However, this action plan had minimal effect and can be detrimental to the well due to fluid disparity. Therefore, it was decided to use commercially available asphaltene inhibitors through a chemical injection line from the surface to the wellbore. Although the inhibitors assessed will not prevent asphaltene precipitation completely but can substantially prolong the process of deposition. Multiple asphaltene deposit samples were collected from wells and were characterized to ascertain the intrinsic properties, such as water content, organic content, inorganic content, and occluded oil.[182]

The effect of unplanned inhibitor dosing expenses can significantly increase the operational expenditures and abbreviate the economic life of an asset. Therefore, the need for an effective production chemical management program is of utmost importance.[183] An inclusive approach encompassing system design, selection of materials, flow assurance, operability strategy, process parameter development, and injection of chemicals can result in huge cost savings, environmental burden reduction, and, in general, reduce associated flow assurance issues.

7.5.2 GAS HYDRATE INHIBITOR

Commonly used techniques for hydrate mitigation include dehydration, injection of thermodynamic inhibitors, insulation, or heating to maintain the temperature above the hydrate stability temperature, usage of AAs or KHIs as LDHIs. The use of LDHIs is considered a relatively new technique compared to the others.[184–187] The main advantage of using kinetic inhibitors as LDHIs is that the chemical dosage required to operate these chemicals effectively is very low, about 0.5–1%. Poly(N-vinyl pyrrolidone) (PVP), poly(N-vinyl valerolactam), poly(N-methyl-N-vinyl acetamide) (VIMA), poly(acryloyl morpholine), and poly(acryoyl pyrrolidine) are examples of such inhibitors. Similarly, AA also has a low chemical dosage.[188–190] A second-generation KHI N-vinylpyrrolidone, N-vinyl caprolactam, and dimethyl amino–ethyl–methacrylate, a terpolymer, commercially known as Gaffix VC-713 (a terpolymer of vinyl caprolactam/vinylpyrrolidone/dimethylaminoethyl methacrylate) displays good hydrate inhibition at moderate pressure. 0.5 wt.% VC-713 delayed the formation of hydrates by 20 h. However, 0.75 wt% VC-713 delayed the hydrate formation by 10 h. A second-generation KHI, PVCap, shows optimum hydrate inhibition at a concentration of 0.5 wt%. Without salt, PVCap inhibited the formation of hydrates for about 16 h at 1 wt%, while at 0.75 wt%, hydrate formation was negligible. The copolymer N-vinylpyrrolidone-co-N-vinyl caprolactam, also known as VP/VC; if used at a ratio equal to or less than 25/75, then it can be as effective as VC-713 or PVCap.[191]

Alcoholic THIs like methanol is used in deepwater operations at 10–20 wt.% to mitigate the formation of hydrates in production wells. In some cases, MEG is used at about 20–60% to have a maximum effect on gas hydrate inhibition. Methanol provides a better outcome in hydrate inhibition than MEG. A dosage of 10% sodium chloride (NaCl) as an electrolytic THI lowers the temperature by 2.5–4°C. With the addition of 3.5 wt.% of NaCl, the hydrate formation temperature reduces by 2–3°C between 20–100 bar pressure range. Other salts can be used as THIs such as potassium chloride (KCl), potassium carbonate (K_2CO_3), sodium bromide (NaBr), calcium chloride ($CaCl_2$), and so on. A dosage of 20% of any THI proved to be the optimal dosing required to lower the temperature and shift the hydrate equilibrium curve.[188]

Thermo-kinetic inhibitors are those gas hydrate inhibitors that alter the hydrate-liquid–vapor-equilibrium (HL_wVE) curve and delay the process of hydrate formation. Some ionic liquids are specially crafted to act as both kinetic and thermodynamic inhibitors. A study was conducted to compare five such thermo-kinetic inhibitors at −12°C and 114 bars: 1-ethyl-3-methylimidazolium tetrafluoroborate (EMIM-BF_4), 1-ethyl-3-methyl-imidazolium tetrafluoroborate (BMIM–BF_4), 1-ethyl-3-methyl-imidazolium trifluoro methane sulfonate (EMIM–CF_3SO_3), 1-ethyl-3-methyl-imidazolium dicyanamide (EMIM–$N(CN)_2$), and 1-ethyl-3-methyl-imidazolium ethyl sulfate (EMIM–$EtSO_4$). For 10 wt.% of each of these ionic fluids, EMIM-BF_4 showed the best results with the highest mean induction time of 6.48 h. Although the alkyl group of EMIM-BF$_4$ and BMIM-BF$_4$ differs, EMIM-BF$_4$ displayed better performance than BMIM-BF$_4$. More experiments were conducted with EMIM-BF$_4$ of concentrations of 0.1, 0.5, and 1 wt%. Moreover, 1 wt% EMIM-BF$_4$ was further compared with 1 wt% of Luvicap and 1 wt% of PVCap. In the preceding experiment, 1 wt.% EMIM-BF$_4$ showed better results. Also, an insignificant change in induction time of hydrate formation was seen as the concentration of EMIM-BF$_4$ increased from 1–10 wt.%, indicating that higher than 1 wt.% concentration of ionic liquid does not offer much advantage.[192]

7.5.2.1 Empirical Correlations for Hydrates

Several comparatively simple and widely used empirical methods to approximate the effect of chemicals on the hydrate depression temperature are as follow.

7.5.2.1.1 Hammerschmidt Equation

The Hammerschmidt equation provides information on the hydrate depression temperature with respect to the concentration of inhibitors in the water phase and the molecular weight of the inhibitor.[193]

$$\Delta T = \frac{K \times w}{1.8 \times \text{MW}_{\text{inb}} \times (1 - w)} \tag{7.2}$$

where ΔT is the depression of the gas hydrate freezing point in °C, K is the dimensionless constant (e.g., methanol = 2335; MEG = 2700); w = mass fraction of inhibitor in the final water phase; MW_{inb} is the molecular weight of the inhibitor, (e.g., methanol = 32, MEG = 62).

7.5.2.1.2 Nielsen–Bucklin Equation

For ΔT values greater than 13.5 when using methanol, the Nielsen–Bucklin equation is more valid when the mass fraction of methanol in the water phase increases beyond 0.25.[194] The equation is given as

$$w = x_{\text{m}} \times 32.04 / (18.015 + x_{\text{m}} \times 14.025) \tag{7.3}$$

where w is the mass fraction of methanol in the final water phase; $x_{\text{m}} = 1 - \exp(-\Delta T/72)$, where ΔT is the depression temperature of the gas hydrate freezing point in °C.

7.5.2.1.3 The Carroll Method

A modified version of the Nielsen–Bucklin equation is the Carroll equation. It includes an activity coefficient to account for the concentration of the inhibitor. The Carroll equation can be used for a wide range of inhibitor concentrations.[195]

$$\Delta T = -72 \ln(\gamma_{\text{w}} x_{\text{w}}) \tag{7.4}$$

where γ_{w} is the activity coefficient of water and x_{w} is the mole fraction of water.

Margules equation can be used to determine the activity coefficient:

$$\ln \gamma_w = \frac{a}{RT} x_1^2 \tag{7.5}$$

where γ_w is the activity coefficient of water, x_1 is the mole fraction of solvent (water), and a/RT is the Margules coefficient.

7.5.2.1.4 McCain Method

The McCain correlation for computing the effect of brine on the hydrate formation temperature is given by[195]

$$\Delta T = AS + BS^2 + CS^3 \tag{7.6}$$

where ΔT is the hydrate temperature depression (°F) and S is the salinity of water (wt%).

$$A = 2.20919 - 10.5746\gamma + 12.1601\gamma^2 \tag{7.7}$$

$$B = -0.106056 + 0.722692\gamma - 0.85093\gamma^2 \tag{7.8}$$

$$C = 0.00347221 - 0.0165564\gamma + 0.049764\gamma^2 \tag{7.9}$$

McCain's method is used till the 20 wt% salt concentrations and for gas gravities (γ) in the range 0.55–0.68.

7.5.2.1.5 Ostergaard Method

Ostergaard presented a correlation applicable to both inorganic salts and organic compounds considering the effect of pressure:[196]

$$\Delta T = \left(c_1 W + c_2 W^2 + c_3 W^3\right)\left(c_4 \ln P + c_5\right)\left[c_6\left(P_0 - 1000\right) + 1\right] \tag{7.10}$$

where ΔT is the hydrate temperature depression (°C), P is the pressure of the system (kPa), W is the concentration of the inhibitor (mass%), P_0 is the dissociation pressure of hydrocarbon fluid at 0°C (in kPa) and C_i are constants for inhibitor.

7.5.3 SCALE INHIBITOR

In the early 20th century, several significant contributions were made to asserting chemical inhibitor dosing for scale inhibition. One such substantial contribution by Tomson et al.[197] states the concentration of calcium carbonate solution at 25°C in equilibrium with calcite or its intrinsic solubility to be $10^{-5.1}$ M. An effective inhibitor concentration of 0.1–0.4 mg/L of Aminotri (methylenephosphonic acid) at 25°C is

TABLE 7.1
Inhibitor Dosage (mg/L) to Control Calcium Carbonate Scales

Inhibitors	Tomson and Matly[197]	Vanderpool[198]	Cognetti et al.[199]	Re and Gill[200]
Aminotri (methylenephosphonic acid)	0.10	0.40	0.40	0.45
1-hydroxyethylidene 1,1-diphosphonic acid	0.09	0.25	0.25	0.25
2-phosphonobutane-1,2,4-tricarboxylic acid			0.75	0.60

obtained if the nucleus is positively charged, resulting in the acceleration of Aminotri (methylenephosphonic acid) toward the nucleus as compared to $CaCO_3$ by a factor of 4–6. Similarly, the observations of Vanderpool, Tomson, and other investigators[198–200] are reported in Table 7.1.

Reactions generally do not occur abruptly. A propagation delay is seen when all reactants are added together since they must react in parallel. This time delay is called the induction time. Scale inhibitors do not usually prevent precipitation, but they retard the inescapable by stretching the induction time. The induction time can be expressed as[201]

$$\text{Induction time} = \frac{1}{K(\text{Saturation ratio} - 1)^{P-1}} \tag{7.11}$$

where K is the temperature-dependent constant, the saturation ratio is the degree of supersaturation, and P is the critical number of molecules in a cluster before phase change.

The guidepost generally followed is that reaction rates roughly double for every 10°C rise in temperature. The aforementioned temperature constant was observed to equate well with the Arrhenius relationship, as mentioned in the following expression[201]

$$K = A.e^{-E_a/RT} \tag{7.12}$$

where K is the temperature-dependent constant, E_a is the activation energy, R is a gas constant, and T is the absolute temperature. Models in literature for optimizing dosing explain the impact of dosage on prolonging induction time. Factors influencing the antiscalant dosage requirement to delay precipitation are time, temperature, pH, and active sites. Figure 7.16 illustrates the effect of phosphonate dosing on extending induction time. The goal of the inhibitor dosage is to extend the time before precipitation. The induction time can be modified as[201]

$$\text{Induction time} = \frac{\text{Inhibitor}^M}{K(\text{Saturation ratio} - 1)^{P-1}} \tag{7.13}$$

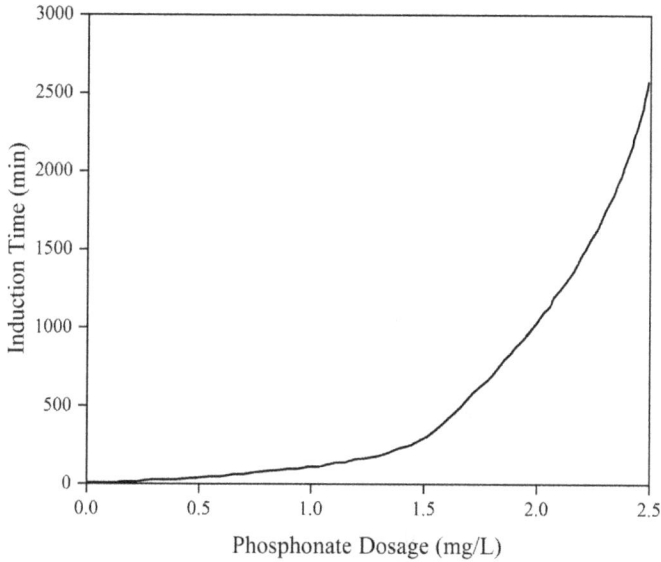

FIGURE 7.16 Effect of phosphonate dosage on induction time.

On further rearranging Equation (7.13), the dosage rate can be calculated. The coefficients in the equation were estimated using regression analysis. The efficiency of ultra-low scale inhibitor dosages was calculated for $CaCO_3$ scales using the preceding model.[202] Threshold inhibitors are kinetic scale inhibitors that delay the precipitation process and can be used typically in 2–20 ppm concentration.[202] The dosage for $CaCO_3$ scale inhibitor for a particular system ranged from 0.01–0.02 ppm active phosphonate. Similarly, the dosages were calculated for calcium phosphate scale control using the earlier model. The model predicted the final dosage of 1,1 hydroxyethylidene diphosphonic acid to mitigate calcium carbonate deposition to be approximately 10% of the final optimized value, that is, 0.2 mg/l. The optimized dosage calculated for polyacrylic acid, a copolymer, as a calcium phosphate inhibitor was 9.53 mg/l.[202] Earlier studies on optimizing dosages of certain calcium carbonate inhibitors were conducted to test the efficacy of ultralow dosages, as little as 0.001–0.2 mg/l. The dosages calculated using the earlier model are accurate or reliable if the field parameters and the chemistry of water and other constituents are similar to the data considered while developing the model.[202]

Certain factors are responsible for having a significant impact on dosage. Additionally, as the residence time increases, the dosage for chemicals also increases. Similar proportionality was also seen in the case of the degree of supersaturation and temperature. Both of the entities have an increasing effect on dosage as they increase. With the increase in temperature, the reaction rate increases, and hence dosage also increases. pH has little contribution to the dosage as the pH range considered in the experiment had an insignificant impact. Although, factors like suspended solid particles in the water share an inverse relationship with dosage.

In another study, the effect of different dosages of maleic acid–acrylic acid (MA-AA), with changing temperatures and pH values, were studied to mitigate the calcium carbonate scaling. The inhibition efficiency was calculated using the following formula:

$$\%\text{Inhibition efficiency} = \left(C_a - C_b / C_c - C_b \right)100 \tag{7.14}$$

where C_a is the concentration of C_a^{2+} ions in the treated solution after precipitation, C_b is the concentration of C_a^{2+} ions in the blank solution after precipitation, and C_c is the concentration of C_a^{2+} ions in the blank solution before precipitation.[202]

The optimal dosage of MA-AA was evaluated using Dynamic Tube Blocking Test. Using the test, a different dosage of MA-AA for different pH values at 90°C was evaluated. Moreover, the dosages were also calculated for different temperatures at 8.5 pH value. The maximum and lowest inhibition efficiency displayed for 1 ppm of MA-AA was 41% at a pH value of 7 and 30% at a pH value of 8.5, respectively, at a constant temperature of 90°C. Also, as the dosage of MA-AA gradually increased from 1–20 ppm, the inhibition efficiency increased. The test results showed that 100% inhibition efficiency was achieved at pH values of 7 and 7.5 when the dosage of MA-AA was 20 ppm. As the temperature increased from 100°C to 150°C, the inhibition efficiency decreased for a particular dosage. Although the dosage of MA-AA increased, the inhibition efficiency also increased. The maximum inhibition efficacy of 98% was displayed by 60 ppm of MA-AA at 100°C at a constant pH value of 8.5. The copolymer achieved good results at temperatures as high as 150°C. A scale inhibition of 71% was achieved with a 60-ppm MA-AA dosage at 150°C. The study showed the dependency of dosage on temperature and pH values.[202]

Suspended solids in the water can also severely impact dosing requirements. They can act as active sites and can minimize inhibitor effectiveness by adsorption. Interestingly scale inhibitors have upper limits and are inefficient above the saturation level driving force, irrespective of dosing concentrations.[201]

Recently, a new method has been proposed to measure a polymeric scale inhibitor concentration calculation in water by high-performance liquid chromatography with an ultraviolet detector. This formula is useful to calculate scale inhibitor concentration at any point in time using input parameters like the volume of scale inhibitor squeezed, the volume of water squeezed, and the flow rate. This method, used on a production well of the Piltun–Astokh oilfield (Sakhalin shelf), which was treated with inhibitor SCW85370, gave good results.[203] The following expression has been proposed to calculate the scale inhibitor concentration, C:[203]

$$C = A.\frac{V_{SI}}{V_{SW}}.e^{-\left(B.\frac{Q_w}{V_{SW}+V_{SI}}.t \right)} \tag{7.15}$$

where A and B are coefficients, C is the concentration of scale inhibitor in water (mg/dm³), V_{SI} is the volume of inhibitor solution squeezed (m³), V_{sw} is the volume of water squeezed (m³), Q_w is the well water production rate (m³/day), and t is the time from well start-up after squeeze (h).

7.6 PRODUCTION CHEMICALS INVENTORY MANAGEMENT

Maintaining an accurate inventory is vital to monitor and manage production chemical handling and storage as it reduces the nonproductive time on the offshore platforms and other facilities. A chemical inventory is an essential element of the chemical management plan, necessary for a safe working environment, greater productivity, and less costly remedial action. A schematic given in Figure 7.17 shows the flowchart of the chemical management plan.[204] Production chemicals are vital for the mitigation of flow assurance–related issues. These chemicals are utilized through various methods, including a topside or a downhole injection system. A proper injection system design and inventory can maximize production. In addition, chemical injection is important for new fields due to the CPEX–OPEX trade-off.[205]

One of the major concerns of the field operator is the performance of the production chemical as per the standards at an affordable price. Production chemicals should be thoroughly checked for performance verification and should be stable for the intended usage duration. These chemicals are classified based on risk and hazards to health and safety and are accompanied by a safety data sheet. There are some points to be kept in mind while selecting chemicals for oil and gas production:[206]

i) *Qualification of chemicals*: Production chemical qualification depends on the conditions expected throughout the product delivery cycle. Ensuring product integrity throughout the manufacturing and delivery process is also essential.

ii) *Stability*: Transport conditions, such as pressure and temperature, are variable. Therefore, the chemicals are "torture-tested" for several days in a flow loop according to the usage scenario and are selected once they pass this test.

iii) *Compatibility*: Production chemicals are required to have chemical compatibility with the construction and production fluids and other chemicals at reservoir pressure and temperature conditions.

Apart from the preceding criteria, other factors affecting the selection of production chemicals are appropriate grades of raw materials and manufacturing processes,

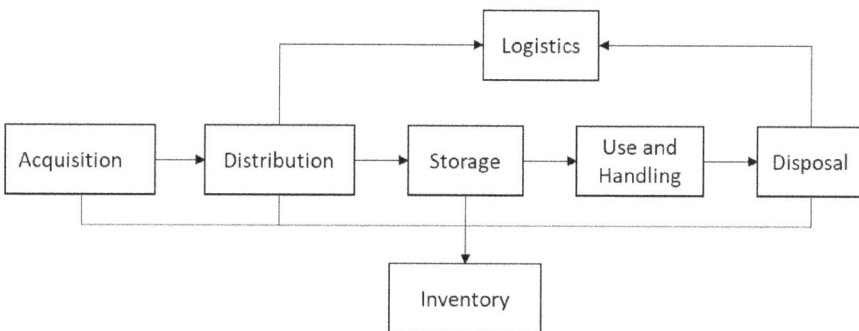

FIGURE 7.17 Flowchart of the chemical management plan.

handling and transportation, and appropriate chemical maintenance and injection practices. In addition, the flow assurance effectiveness depends on accurate data acquisition and analysis, along with production chemistry, engineering, and flow assurance considerations.[207] Early lab analyses and frequent upgradation provide a basis for an effective flow assurance strategy. Real-time data can be helpful as a tool to monitor the effectiveness of production chemicals.

7.6.1 STRATEGY FOR MANAGING CHEMICALS AND INVENTORY

Cost optimization is crucial because of the higher OPEX, especially offshore conditions. A strategy for chemical management is therefore useful with the following key focus points[207]

1. Technical data-driven decisions: These are important for some assets whose technical quality is more important than cost savings.
2. Process and workload prioritization by the engineering team: This eliminates field-based decisions for normal operations, thus saving time.
3. Reduced dependence on the chemical service provider: Operators should have the freedom to create roles and responsibilities independent of the service provider.
4. Engineered and controlled metrics: These enable data-driven decision-making.

7.6.2 REAL-TIME CHEMICAL INVENTORY AND USAGE PLATFORM

Automation in the oilfield industry improves productivity and simultaneous inventory and usage monitoring. Traditionally, real-time chemical usage and production volume data have not been readily accessible mainly due to the manual nature of sensor data aggregation, which hinders the insights and timely/useful data from reaching the production engineers and analysts. A real-time system utilizes commercial software (off the shelf) and related network equipment conforming to industry standards. Data from diverse sources are accumulated and brought to a single point by utilizing a secure local area network (LAN). Equipment-related performance details, alarms, and notifications against key indicators are then delivered to the operator via a graphical interface.[208] The various application areas of real-time chemical monitoring systems include the following

1. Tank-level monitoring
2. Chemical injection monitoring and control
3. Chemical inventory and field data
4. Simulation tools and logistics management

7.6.3 ADVANTAGES OF REAL-TIME CHEMICAL MONITORING SYSTEM

There are obvious short- and medium-term benefits to implementing a real-time chemical inventory monitoring system. In the short term, immediate results can be seen due to time savings in shipment and application. Service-level agreements can be defined for the medium term, enabling upset and monitoring changes. But all these benefits can be realized only when there are accurate data and correct workflow and interface setup. Over time, the data gathered through the system can be used to optimize chemical dosage and related applications.

REFERENCES

1. Chouparova, E., Lanzirotti, A., Feng, H., Jones, K. W., Marinkovic, N., Whitson, C., & Philp, P. (2004). Characterization of petroleum deposits formed in a producing well by synchrotron radiation-based microanalyses. *Energy & Fuels*, 18(4), 1199–1212. doi: 10.1021/ef030108a

2. Ajayi, E. O. (2018). Modelling of controlled wax deposition and loosening in oil and gas production systems. Ph.D. Thesis, Norwegian University of Science and Technology. Norway.

3. Theyab, M. A. (2018). Wax deposition process: Mechanisms, affecting factors and mitigation methods. *Open Access Journal of Science*, 2(2), 109–115. https://doi.org/10.15406/oajs.2018.02.00054

4. Burger, E. D., Perkins, T. K., & Striegler, J. H. (1981). Studies of wax deposition in the trans alaska pipeline. *Journal of Petroleum Technology*, 33(06), 1075–1086. https://doi.org/10.2118/8788-PA

5. Huang, Z., Zheng, S., & Fogler, H. S. (2015). Wax Deposition: Experimental Characterizations, Theoretical Modeling, and Field Practices (1st Edition). In *Emerging Trends and Technologies in Petroleum Engineering*, Dandekar, A. Y. (Ed.). CRC Press, Boca Raton, Florida.

6. Bern, P. A., Withers, V. R., & Cairns, R. J. R. (1980). Wax deposition in crude oil pipe-lines. In *European Offshore Technology Conference and Exhibition*, London, United Kingdom, October 21–24. (SPE-206-1980-MS). https://doi.org/10.2118/206-1980-MS

7. Khanifar, A., Demiral, B., & Darman, N. B. (2011). Modelling of asphaltene precipita-tion and deposition during WAG application. In *International Petroleum Technology Conference*. Bangkok, Thailand, November 15–17. (IPTC-14147-MS). https://doi.org/10.2523/IPTC-14147-MS

8. Panuganti, S. R., Vargas, F. M., Gonzalez, D. L., Kurup, A. S., & Chapman, W. G. (2012). PC-SAFT characterization of crude oils and modeling of asphaltene phase behavior. *Fuel*, 93, 658–669. https://doi.org/10.1016/j.fuel.2011.09.028

9. Turner, D. J. (2005). Clathrate hydrate formation in water-in-oil dispersions. *Chemical Engineering Department*. Ph.D Thesis, Colorado School of Mines, USA.

10. Turner, D. J., Miller, K. T., & Sloan, E. D. (2009). Methane hydrate formation and an inward growing shell model in water-in-oil dispersions. *Chemical Engineering Science*, 64(18), 3996–4004. https://doi.org/10.1016/j.ces.2009.05.051

11. Ballard, A., Creek, J., Eaton, M., Lachance, J., McMullen, N., Palermo, T., Shoup, G., & Talley, L. (2010). *Natural Gas Hydrates in Flow Assurance*, E. D. Sloan, C. A. Koh, & A. K. Sum (Eds.). (1st Edition). Gulf Professional Publishing.

12. Bai, Y., & Bai, Q. (2010). *Subsea Structural Engineering* (2nd Edition). Gulf Professional Publishing, Burlington, USA.

13. Abdullahi, M. (2019). *Detection of Leakage and Blockage in Pipeline Systems*. Ph.D. thesis, The University of Manchester, United Kingdom.

14. PRCI. (2011). Pipeline leak detection operational improvements. In *6th Pipeline Technology Conference*. Hannover, Germany, April 4–5.

15. ESAB Knowledge Centre. Radiographic and ultrasonic testing of welds. Welding Inspection. https://www.esabna.com/us/en/education/blog/radiographic-and-ultrasonic-testing-of-welds.cfm (accessed July 27, 2021)

16. Chis, T. (2007). Pipeline Leak Detection Techniques. *Anale Seria Informatica*, Volume V fasc I, 5th Tome 1st Fasc.

17. Parker, J. G. (1981). Acoustic detection and location of leaks in underground natural gas distribution lines. *Johns Hopkins APL Technical Digest (Applied Physics Laboratory)*, 2(2), 90–101.

18. Koyama, H., Watanabe, K., & Himmelblau, D. M. (1993). Identification of partial plugging in a gas-transport pipeline by an acoustic method. *Applied Acoustics*, 40(1), 1–19. doi: 10.1016/0003-682X(93)90018-2

19. Wang, X., Lennox, B., Turner, J., Lewis, K., Ding, Z., Short, G., & Dawson, K. (2009). Blockage detection in long lengths of pipeline using a new acoustic method. In *16th International Congress on Sound and Vibration 2009 (ICSV16)*. Krakow, Poland. July 5–9. (Vol. 1, pp 140–147).

20. Beller, M., & Schneider, U. (2007). Pipeline inspection utilizing ultrasound technology: On the issue of resolution. Pigging products & services association. In *1st International Conference on Pipeline Integrity*, Stutensee, Germany, 139–148.

21. Li, X., Liu, Y., Liu, Z., Chu, J., Song, Y., Yu, T., & Zhao, J. (2020). A hydrate blockage detection apparatus for gas pipeline using ultrasonic focused transducer and its application on a flow loop. *Energy Science & Engineering*, 8(5), 1770–1780. doi: 10.1002/ese3.631

22. Datta, S., & Sarkar, S. (2016). A review on different pipeline fault detection methods. *Journal of Loss Prevention in the Process Industries*, 41, 97–106. https://doi.org/10.1016/j.jlp.2016.03.010

23. Yang, L. (2018). Analysis of pressure distribution along pipeline blockage based on the CFD simulation. Ph.D thesis, Northeast Petroleum University, China.

24. Charlton, J. S., Heslop, J. A., & Johnson, P. (1981). Radioisotope techniques for the investigation of process problems in the chemical industry. In *International Conference for Industrial Application of Radioisotopes and Radiation Technology*. Grenoble, France, September 28–October 2. (IAEA-CN-40/110).

25. Kasban, H., Zahran, O., & El-Samie, F. E. A. (2010). New trends for on-line troubleshooting in industrial problems using radioisotopes. *The Online Journal on Electronics and Electrical Engineering*, 2(3), 284–292.

26. Geiger, G., Werner, T., & Matko, D. (2003). Leak detection and locating - A Survey. In *PSIG Annual Meeting*, Bern, Switzerland, October 15–17. (PSIG-0301).

27. Jolly, W. D., Morrow, B. T., O'Brien, F. J., Spence, H. F., & Svedeman, J. S. (1992). New methods for rapid leak detection in offshore pipelines. In *Minerals management service, U.S. department of the interior*, Southwest Research Institute, Houston, USA.

28. Scott, S. L., & Satterwhite, L. A. (1998). Evaluation of the backpressure technique for blockage detection in gas flowlines. *Journal of Energy Resources Technology*, 120(1), 27–31. https://doi.org/10.1115/1.2795005

29. Scott, S. L., & Yi, J. (1999). Flow testing methods to detect and characterize partial blockages in looped subsea flowlines. *Journal of Energy Resources Technology*, 121(3), 154–160. https://doi.org/10.1115/1.2795975

30. Liu, L. J., & Scott, S. L. (2001). A new method to locate partial blockages in subsea flowlines. In *SPE Annual Technical Conference and Exhibition*. New Orleans, Louisiana, September 30–October 3. (SPE 71548). https://doi.org/10.2118/71548-MS

31. Rogers, L. M. (1995). Pipeline blockage location by strain measurement using an ROV. In *Offshore Technology Conference*. Houston, Texas, May 1–4. (OTC-7862-MS). https://doi.org/10.4043/7862-MS

32. Chu, J., Liu, Y., Lv, X., Li, Q., Dong, H., Song, Y., & Zhao, J. (2021). Experimental investigation on blockage predictions in gas pipelines using the pressure pulse wave method. *Energy*, 230, 120897. https://doi.org/10.1016/j.energy.2021.120897

33. Heibert, L. (2003). Non-intrusive locating of a blockage in a pipeline. US Patent No: US 2003 / 0124222 A1.

34. Chen, X., Tsang, Y., Zhang, H.-Q., & Chen, T. X. (2007). Pressure-wave propagation technique for blockage detection in subsea flowlines. In *Proceedings of SPE Annual Technical Conference and Exhibition*, California, USA, November 11–14. (SPE 110570). https://doi.org/10.2523/110570-MS

35. Duan, H. F., Lee, P. J., Ghidaoui, M. S., & Tuck, J. (2014). Transient wave-blockage interaction and extended blockage detection in elastic water pipelines. *Journal of Fluids and Structures*, 46, 2–16. https://doi.org/10.1016/j.jfluidstructs.2013.12.002

36. Duan, H.-F., Pan, B., Wang, M., Chen, L., Zheng, F., & Zhang, Y. (2020). State-of-the-art review on the transient flow modeling and utilization for urban water supply system (UWSS) management. *Journal of Water Supply: Research and Technology-Aqua*, 69(8), 858–893. https://doi.org/10.2166/aqua.2020.048

37. Meniconi, S., Duan, H. F., Lee, P. J., Brunone, B., Ghidaoui, M. S., & Ferrante, M. (2013). Experimental investigation of coupled frequency and time-domain transient test–based techniques for partial blockage detection in pipelines. *Journal of Hydraulic Engineering*, 139(10), 1033–1040. https://doi.org/10.1061/(ASCE)HY.1943-7900.0000768

38. Adeleke, N., Ityokumbul, M. T., & Adewumi, M. (2013). Blockage detection and characterization in natural gas pipelines by transient pressure-wave reflection analysis. *SPE Journal*, 18(02), 355–365. https://doi.org/10.2118/160926-PA

39. Liu, Z., Yang, M., Zhang, H., Xiao, B., Yang, L., & Zhao, J. (2019). A high-pressure visual flow loop for hydrate blockage detection and observation. *Review of Scientific Instruments*, 90(7), 074102. https://doi.org/10.1063/1.5088993

40. Mohapatra, P. K., Chaudhry, M. H., Kassem, A. A., & Moloo, J. (2006). Detection of partial blockage in single pipelines. *Journal of Hydraulic Engineering*, 132(2), 200–206. https://doi.org/10.1061/(ASCE)0733-9429(2006)132:2(200)

41. Antonopoulos-Domis, M. (1980). Frequency dependence of acoustic resonances on blockage position in a fast reactor subassembly wrapper. *Journal of Sound and Vibration*, 72(4), 443–450. https://doi.org/10.1016/0022-460X(80)90356-9

42. Qunli, W., & Fricke, F. (1990). Determination of blocking locations and cross-sectional area in a duct by eigenfrequency shifts. *The Journal of the Acoustical Society of America*, 87(1), 67–75. https://doi.org/10.1121/1.398914

43. Lee, P. J., Vítkovský, J. P., Lambert, M. F., Simpson, A. R., & Liggett, J. A. (2008). Discrete blockage detection in pipelines using the frequency response diagram: Numerical study. *Journal of Hydraulic Engineering*, 134(5), 658–663. https://doi.org/10.1061/(ASCE)0733-9429(2008)134:5(658)

44. Becker, J. R. (2000). Oilfield paraffin treatments: Hot oil and hot water compared to crystal modifiers. In *SPE Annual Technical Conference and Exhibition*. Dallas, Texas, October 1–4 (SPE-63123-MS). https://doi.org/10.2118/63123-MS

45. Chun, S. J. (2000). In *Proceedings of the Tenth International Offshore and Polar Engineering Conference*, Seattle, USA, May 28–June 2.

46. Ashton, J. P., Kirspel, L. J., Nguyen, H. T., & Credeur, D. J. (1989). In-situ heat system stimulates paraffinic-crude producers in gulf of mexico. *SPE Production Engineering*, 4(2), 157–160. https://doi.org/10.2118/15660-PA

47. Reistle, C. E. (1932). *Paraffin and congealing-oil problems*. Report, University of North Texas Libraries. https://digital.library.unt.edu/ark:/67531/metadc12519/ (accessed December 8, 2021)

48. Schlumberger. How it works: Pipeline pigging. https://www.slb.com/resource-library/article/valve-insights/how-it-works-pipeline-pigging (accessed July 26, 2021)

49. Engineered Power. Pipeline Inspection Gauge (PIG). https://engineeredpower.com/pig/ (accessed July 26, 2021)

50. Chowdhury, M. S., & Abdel-Hafez, M. F. (2016). Pipeline inspection gauge position estimation using inertial measurement unit, odometer, and a set of reference stations. *ASCE-ASME Journal of Risk and Uncertainity in Engineering Systems, Part B: Mechanical Engineering*, 2(2), 021001. https://doi.org/10.1115/1.4030945

51. Tordal, A., & Statoil, A. S. A. (2006). Pigging of pipelines with high wax content. *Publication of Pigging Products and Services Association,* Statoil ASA, Stavanger, Norway.

52. Kang, P. S., Lee, D. G., & Lim, J. S. (2014). Status of wax mitigation technologies in offshore oil production. In *The Twenty-fourth International Ocean and Polar Engineering Conference.* Busan, Korea, June 15–20. (ISOPE-I-14-285).

53. Adeyanju, O. A., & Oyekunle, L. O. (2013). Experimental study of wax deposition in a single phase sub-cooled oil pipelines. In *SPE Nigeria Annual International Conference and Exhibition.* Lagos, Nigeria, July 30 July–August 1. (SPE-167515-MS). https://doi.org/10.2118/167515-MS

54. Jennings, D. W., & Newberry, M. E. (2008). Paraffin inhibitor applications in deepwater offshore developments. In *International Petroleum Technology Conference.* Kuala Lumpur, Malaysia, December 3–5. (IPTC-12127-MS). https://doi.org/10.2523/IPTC-12127-MS

55. Theyab, M. A. (2020). A review of wax mitigation methods through hydrocarbon production. *Journal of Petroleum and Enviromental Biotechnology*, 11(5), 412.

56. Al-Yaari, M. (2011). Paraffin wax deposition: Mitigation & removal techniques. In *SPE Saudi Arabia Section Young Professionals Technical Symposium.* Dhahran, Saudi Arabia, March 14–16. (SPE-155412-MS). https://doi.org/10.2118/155412-MS

57. Bimuratkyzy, K., & Sagindykov, B. (2016). The review of flow assurance solutions with respect to wax and asphaltene. *Brazilian Journal of Petroleum and Gas*, 10(2), 119–134. https://doi.org/10.5419/bjpg2016-0010

58. Thierheimer, C. L. (1990). *Solvent for paraffin removal from oilfield equipment.* United State Patent. (Patent No. 4925497).

59. Armacanqui, J. S., de Fátima Eyzaguirre, L., Flores, M. G., Zavaleta, D. E., Camacho, F. E., Grajeda, A. W., Alfaro, A. D., & Viera, M. R. (2016). Testing of environmental friendly paraffin removal products. In *SPE Latin America and Caribbean Heavy and Extra Heavy Oil Conference.* Lima, Peru, October 19–20. (SPE-181162-MS). https://doi.org/10.2118/181162-MS

60. Woo, G. T., Garbis, S. J., & Gray, T. C. (1984). Long-term control of paraffin deposition. In *SPE Annual Technical Conference and Exhibition.* Houston, Texas, September 16–19. (SPE-13126-MS). https://doi.org/10.2118/13126-MS

61. Newberry, M. E., & Barker, K. M. (1985). Formation damage prevention through the control of paraffin and asphaltene deposition. In *SPE Production Operations Symposium.* Oklahoma, USA, March 10–12. (SPE-13796-MS). https://doi.org/10.2118/13796-MS

62. White, M., Pierce, K., & Acharya, T. (2018). A review of wax-formation/mitigation technologies in the petroleum industry. *SPE Production & Operations*, 33(03), 476–485. https://doi.org/10.2118/189447-PA

63. Koshel, K. C., & Kapoor, S. (1999, October). Effect of flow improver on Indian western offshore crude-carrying pipe line: A case study. In *1999 SPE Annual Technical Conference and Exhibition.* Houston, Texas, October 3–6. SPE-56570. https://doi.org/10.2118/56570-MS

64. Craddock, H. A., Mutch, K., Sowerby, K., McGregor, S. W., Cook, J., & Strachan, C. (2007). A case study in the removal of deposited wax from a major subsea flowline system in the Gannet field. In *2007 SPE International Symposium on Oilfield Chemistry*, Houston, Texas, USA, February 28–March 2. (SPE-105048). https://doi.org/10.2118/105048-MS

65. Goodman, N. T., & Joshi, N. (2013, September). A tale of two flowlines—paraffin plugging and remediation. *In SPE Annual Technical Conference and Exhibition.* New Orleans, Louisiana, USA, September 30–October 2. SPE-166196. https://doi.org/10.2118/166196-MS

66. Sousa, A. L., Matos, H. A., & Guerreiro, L. P. (2019). Preventing and removing wax deposition inside vertical wells: a review. *Journal of Petroleum Exploration and Production Technology*, 9, 2091–2107. https://doi.org/10.1007/s13202-019-0609-x

67. Ridzuan, N., Adam, F., & Yaacob, Z. (2014). Molecular recognition of wax inhibitor through pour point depressant type inhibitor. In *International Petroleum Technology Conference.* Kuala Lumpur, Malaysia, December 10–12. (IPTC-17883-MS). https://doi.org/10.2523/IPTC-17883-MS

68. Pedersen, K. S., & Rønningsen, H. P. (2003). Influence of wax inhibitors on wax appearance temperature, pour point, and viscosity of waxy crude oils. *Energy & Fuels*, 17(2), 321–328. https://doi.org/10.1021/ef020142+

69. Al-Sabagh, A. M., El-Kafrawy, A. F., Khidr, T. T., El-Ghazawy, R. A., & Mishrif, M. R. (2007). Synthesis and evaluation of some novel polymeric surfactants based on aromatic amines used as wax dispersant for waxy gas oil. *Journal of Dispersion Science and Technology*, 28(6), 976–983. https://doi.org/10.1080/01932690701463167

70. Zhang, C., Gao, C., Gao, F., Wang, J., Zhang, D., Wang, Y., & Xu, D. (2014). Synthesis of comb bipolymers and their pour point depressing properties. *Petroleum Science*, 11, 155–160. https://doi.org/10.1007/s12182-014-0327-4

71. Xu, J., Zhang, X., Sun, J., Li, L., & Guo, X. (2009). How comb-type poly(maleic acid alkylamide-co-α-olefin) assemble in waxy oils and improve flowing ability. *Asia-Pacific Journal of Chemical Engineering*, 4(5), 551–556. https://doi.org/10.1002/apj.281

72. Li, L., Xu, J., Tinsley, J., Adamson, D. H., Pethica, B. A., Huang, J. S., Prud'homme, R. K., & Guo, X. (2012). Improvement of oil flowability by assembly of comb-type copolymers with paraffin and asphaltene. *AIChE Journal*, 58(7), 2254–2261. https://doi.org/10.1002/aic.12729

73. Borthakur, A., Chanda, D., Dutta Choudhury, S. R., Rao, K. V., & Subrahmanyam, B. (1996). Alkyl fumarate–vinyl acetate copolymer as flow improver for high waxy indian crude oils. *Energy & Fuels*, 10(3), 844–848. https://doi.org/10.1021/ef950237u

74. Yang, F., Yao, B., Li, C., Shi, X., Sun, G., & Ma, X. (2017). Performance improvement of the ethylene-vinyl acetate copolymer (EVA) pour point depressant by small dosages of the polymethylsilsesquioxane (PMSQ) microsphere: An experimental study. *Fuel*, 207, 204–213. https://doi.org/10.1016/j.fuel.2017.06.083

75. Sivakumar, P., Sircar, A., Deka, B., Silviya Anumegalai, A., Suresh Moorthi, P., & Yasvanthrajan, N. (2018). Flow improvers for assured flow of crude oil in midstream pipeline -A review. *Journal of Petroleum Science and Engineering*, 164, 24–30. https://doi.org/10.1016/j.petrol.2018.01.022

76. Wang, F., Zhang, D. M., Ding, Y. F., Zhang, L. X., Yang, M. S., Jiang, B. L., Zhang, S. M., Ai, M. Y., Liu, G. W., Zhi, S. J., Huo, L. F., Ouyang, X., & Li, L. (2011). The effect of nanohybrid materials on the pour-point and viscosity depressing of waxy crude oil. *Chinese Science Bulletin*, 56(1), 14–17. https://doi.org/10.1007/s11434-010-4174-4

77. Gateau, P., Hénaut, I., Barré, L., & Argillier, J. F. (2004). Heavy oil dilution. *Oil & Gas Science and Technology*, 59(5), 503–509. https://doi.org/10.2516/ogst:2004035

78. Bello, O. O., Fasesan, S. O., Teodoriu, C., & Reinicke, K. M. (2006). An evaluation of the performance of selected wax inhibitors on paraffin deposition of nigerian crude oils. *Petroleum Science and Technology*, 24(2), 195–206. https://doi.org/10.1081/LFT-200044504

79. Straub, T. J., Autry, S. W., & King, G. E. (1989, April 4). An investigation into practical removal of downhole paraffin by thermal methods and chemical solvents. In *SPE Production Operations Symposium*. Oklahoma, USA, March 13–14. (SPE-18889-MS). https://doi.org/10.2118/18889-MS

80. Okoliegbe, I. N., & Agarry, O. O. (2012). Application of microbial surfactant (a review). *Scholarly Journals of Biotechnology*, 1, 15–23.

81. Oguntimein, G. B., Erdmann, H., & Schmid, R. D. (1993). Lipase catalysed synthesis of sugar ester in organic solvents. *Biotechnology letters*, 15(2), 175–180. https://doi.org/10.1007/BF00133019

82. El-Sheshtawy, H. S., & Khidr, T. T. (2016). Some biosurfactants used as pour point depressant for waxy egyptian crude oil. *Petroleum Science and Technology*, 34(16), 1475–1482. https://doi.org/10.1080/10916466.2016.1204317

83. Sakthipriya, N., Doble, M., & Sangwai, J. S. (2017). Enhanced microbial degradation of waxy crude oil: A review on current status and future perspective. *International Journal of Oil, Gas and Coal Technology*, 16(2), 130. https://doi.org/10.1504/IJOGCT.2017.086315

84. Sakthipriya, N., Doble, M., & Sangwai, J. S. (2016). Efficacy of bacillus subtilis for the biodegradation and viscosity reduction of waxy crude oil for enhanced oil recovery from mature reservoirs. *Energy Sources, Part A: Recovery, Utilization, and Environmental Effects*, 38(16), 2327–2335. https://doi.org/10.1080/15567036.2015.1044624

85. Sakthipriya, N., Doble, M., & Sangwai, J. S. (2015b). Action of biosurfactant producing thermophilic bacillus subtilis on waxy crude oil and long chain paraffins. *International Biodeterioration & Biodegradation*, 105, 168–177. https://doi.org/10.1016/j.ibiod.2015.09.004

86. Sakthipriya, N., Doble, M., & Sangwai, J. S. (2018). Kinetic and thermodynamic behavior of the biodegradation of waxy crude oil using bacillus subtilis. *Journal of Petroleum Science and Engineering*, 160, 412–421. https://doi.org/10.1016/j.petrol.2017.10.056

87. Sakthipriya, N., Doble, M., & Sangwai, J. S. (2015a). Fast degradation and viscosity reduction of waxy crude oil and model waxy crude oil using Bacillus subtilis. *Journal of Petroleum Science and Engineering*, 134, 158–166. https://doi.org/10.1016/j.petrol.2015.08.002

88. Sakthipriya, N., Kumar, G., Agrawal, A., Doble, M., & Sangwai, J. S. (2021). Impact of biosurfactants, surfactin, and rhamnolipid produced from bacillus subtilis and pseudomonas aeruginosa , on the enhanced recovery of crude oil and its comparison with commercial surfactants. *Energy & Fuels*, 35(12), 9883–9893. https://doi.org/10.1021/acs.energyfuels.1c00679

89. Akinyemi, O. P., Udonne, J. D., & Oyedeko, K. F. (2018). Study of effects of blend of plant seed oils on wax deposition tendencies of nigerian waxy crude oil. *Journal of Petroleum Science and Engineering*, 161, 551–558. https://doi.org/10.1016/j.petrol.2017.12.017

90. Atta, A. M., El-Ghazawy, R. A., Morsy, F. A., Hebishy, A. M. S., & Elmorsy, A. (2015). Adsorption of polymeric additives based on vinyl acetate copolymers as wax dispersant and its relevance to polymer crystallization mechanisms. *Journal of Chemistry*, 2015, 1–8. https://doi.org/10.1155/2015/683109

91. Marie, E., Chevalier, Y., Eydoux, F., Germanaud, L., & Flores, P. (2005). Control of n-alkanes crystallization by ethylene–vinyl acetate copolymers. *Journal of Colloid and Interface Science*, 290(2), 406–418. https://doi.org/10.1016/j.jcis.2005.04.054

92. Naiya, T. K., Banerjee, S., Kumar, R., & Mandal, A. (2015). Heavy crude oil rheology improvement using naturally extracted surfactant. In *SPE Oil and Gas India Conference and Exhibition*. Mumbai, India, November 24–26. (SPE-178133-MS). https://doi.org/10.2118/178133-MS

93. Wei, B. (2015). Recent advances on mitigating wax problem using polymeric wax crystal modifier. *Journal of Petroleum Exploration and Production Technology*, 5, 391–401. https://doi.org/10.1007/s13202-014-0146-6

94. Ashbaugh, H. S., Fetters, L. J., Adamson, D. H., & Prud'homme, R. K. (2002). Flow improvement of waxy oils mediated by self-aggregating partially crystallizable diblock copolymers. *Journal of Rheology*, 46(4), 763. https://doi.org/10.1122/1.1485280

95. Leube, W., Monkenbusch, M., Schneiders, D., Richter, D., Adamson, D., Fetters, L., Dounis, P., & Lovegrove, R. (2000). Wax-crystal modification for fuel oils by self-aggregating partially crystallizable hydrocarbon block copolymers. *Energy & Fuels*, 14(2), 419–430. https://doi.org/10.1021/ef9901544

96. Allen, T. O., & Roberts, A. P. (1978). *Production operations: Well completions, workover, and stimulation*. (Vol 1, 3rd Edition). Oil & Gas Consultants International. Tulsa, USA.

97. Merino-Garcia, D., & Correra, S. (2008). Cold flow: A review of a technology to avoid wax deposition. *Petroleum Science and Technology*, 26(4), 446–459. https://doi.org/10.1080/10916460600809741

98. Coberly, C. J. (1942). Methods of preventing wax depositions in tubing (Patent No. 2303823). In *United States Patent Office* (Patent No. 2303823).

99. Arumugam, S., Kasumu, A. S., & Mehrotra, A. K. (2013). Modeling of solids deposition from "waxy" mixtures in "hot flow" and "cold flow" regimes in a pipeline operating under turbulent flow. *Energy & Fuels*, 27(11), 6477–6490. https://doi.org/10.1021/ef401315m

100. Ehsani, S., Haj-Shafiei, S., & Mehrotra, A. K. (2019). Experiments and modeling for investigating the effect of suspended wax crystals on deposition from "waxy" mixtures under cold flow conditions. *Fuel*, 243, 610–621. https://doi.org/10.1016/j.fuel.2019.01.089

101. Bidmus, H. O., & Mehrotra, A. K. (2012). Comments on "The effect of operating temperatures on wax deposition" by Huang et al. *Energy & Fuels*, 26(6), 3963–3966. https://doi.org/10.1021/ef300486y

102. Argo, C. B., Bollavaram, P., Hjarbo, K. W., Makogon, T. Y., Oza, N., Wolden, M., Lund, A., Larsen, R. (2007). *Method and system for transporting flows of fluid hyfrocarbons containing wax, asphaltenes, and/or other precipitating solids*. United States Patent. Patent No. US 7261810 B2.

103. Haghighi, H., Azarinezhad, R., Chapoy, A., Anderson, R., & Tohidi, B. (2007). Hydraflow: Avoiding gas hydrate problems. In *Proceedings of EUROPEC/EAGE Conference and Exhibition*. London, United Kingdom, June 11–14. (SPE-107335-MS). https://doi.org/10.2523/107335-MS

104. Azarinezhad, R., Chapoy, A., Anderson, R., & Tohidi, B. (2010). A wet cold-flow technology for tackling offshore flow-assurance problems. *SPE Projects, Facilities & Construction*, 5 (2): 58–64. (SPE-132349-PA). https://doi.org/10.2118/132349-PA

105. Jiang, B., Qiu, L., Li, X., Yang, S., Ke, L., & Chen, H. (2014). Measurement of the wax appearance temperature of waxy oil under the reservoir condition with ultrasonic method. *Petroleum Exploration and Development*, 41(4), 509–512. https://doi.org/10.1016/S1876-3804(14)60059-8

106. Patton, C. C., & Casad, B. M. (1970). Paraffin deposition from refined wax-solvent systems. *SPE Journal*, 10(01), 17–24. https://doi.org/10.2118/2503-PA

107. Szilas, A. P. (1975). *Production and Transport of Oil and Gas*. Elsevier Scientific Publishing Company, Amsterdam.

108. Biswas, S. K., Kukreti, V., Rana, D. P., Sarbhai, M. P., Bateja, S., & Misra, T. R. (2012). Application of microbial treatment for mitigating the paraffin deposition in down hole tubulars and surface flow lines of wells - A success story. In *SPE Oil and Gas India Conference and Exhibition*. Mumbai, India, March 28-30. (SPE-154662-MS). https://doi.org/10.2118/154662-MS

109. Etoumi, A. (2007). Microbial treatment of waxy crude oils for mitigation of wax precipitation. *Journal of Petroleum Science and Engineering*, 55(1–2), 111–121. https://doi.org/10.1016/j.petrol.2006.04.015

110. Etoumi, A., El Musrati, I., El Gammoudi, B., & El Behlil, M. (2008). The reduction of wax precipitation in waxy crude oils by Pseudomonas species. *Journal of Industrial Microbiology & Biotechnology*, 35(11), 1241–1245. https://doi.org/10.1007/s10295-008-0420-z

111. Sood, N., & Lal, B. (2008). Isolation and characterization of a potential paraffin-wax degrading thermophilic bacterial strain Geobacillus kaustophilus TERI NSM for application in oil wells with paraffin deposition problems. *Chemosphere*, 70(8), 1445–1451. https://doi.org/10.1016/j.chemosphere.2007.08.071

112. Lazar, I., Voicu, A., Nicolescu, C., Mucenica, D., Dobrota, S., Petrisor, I., Stefanescu, M., & Sandulescu, L. (1999). The use of naturally occurring selectively isolated bacteria for inhibiting paraffin deposition. *Journal of Petroleum Science and Engineering*, 22(1–3), 161–169. https://doi.org/10.1016/S0920-4105(98)00065-5

113. Towler, B. F., & Rebbapragada, S. (2004). Mitigation of paraffin wax deposition in cretaceous crude oils of Wyoming. *Journal of Petroleum Science and Engineering*, 45(1–2), 11–19. https://doi.org/10.1016/j.petrol.2004.05.006

114. Rana, D. P., Bateja, S., Biswas, S. K., Kumar, A., Misra, T. R., & Lal, B. (2010). Novel microbial process for mitigating wax deposition in down hole tubular and surface flow lines. In *SPE Oil and Gas India Conference and Exhibition*. Mumbai, India, January 20–22. (SPE-129002-MS). https://doi.org/10.2118/129002-MS

115. Xiao, M., Li, W.H., Lu, M., Zhang, Z.Z., Luo, Y.J., Qiao, W., Sun, S.S., Zhong, W.Z., & Zhang, M. (2012). Effect of microbial treatment on the prevention and removal of paraffin deposits on stainless steel surfaces. *Bioresource Technology*, 124, 227–232. https://doi.org/10.1016/j.biortech.2012.07.063

116. He, Z., Mei, B., Wang, W., Sheng, J., Zhu, S., Wang, L., & Yen, T. F. (2003). A pilot test using microbial paraffin-removal technology in Liaohe oilfield. *Petroleum Science and Technology*, 21(1–2), 201–210. https://doi.org/10.1081/LFT-120016942

117. Samigullin, A. D., Galiakbarov, A. T., & Galiakbarov, R. T. (2016). Study of the petroleum schedules thermal cleaning process from asphalt, ressin and paraffin deposits using low- temperature plasma. *Journal of Physics: Conference Series*, 669, 012017. https://doi.org/10.1088/1742-6596/669/1/012017

118. Gharbi, K., Benyounes, K., & Khodja, M. (2017). Removal and prevention of asphaltene deposition during oil production: A literature review. *Journal of Petroleum Science and Engineering*, 158, 351–360. https://doi.org/10.1016/j.petrol.2017.08.062

119. Al-Hosani, A., Ravichandran, S., & Daraboina, N. (2021). review of asphaltene deposition modeling in oil and gas Production. *Energy & Fuels*, 35(2), 965–986. https://doi.org/10.1021/acs.energyfuels.0c02981

120. Akbar, S. H., & Saleh, A. A. (1989). A comprehensive approach to solve asphaltene deposition problem in some deep wells. In *Proceedings of Middle East Oil Show*. Bahrain, March 11–14. (SPE-17965-MS). https://doi.org/10.2523/17965-MS

121. Schaefer, E. F. (1991). Pigging of subsea pipelines. In *Proceedings of the 23rd Annual Offshore Technology Conference.* Houston, Texas, May 6–9. (OTC-6769-MS). https://doi.org/10.4043/6769-ms

122. Allenson, S. J., & Walsh, M. A. (1997). A novel way to treat asphaltene deposition problems found in oil production. In *SPE International Symposium on Oilfield Chemistry.* Houston, Texas, February 18–21. (SPE-37286-MS). https://doi.org/10.2118/37286-MS

123. Li, J., Misselbrook, J., & Sach, M. (2010). Sand cleanouts with coiled tubing: choice of process, tools and fluids. *Journal of Canadian Petroleum Technology*, 49(08), 69–82. https://doi.org/10.2118/113267-PA

124. Mitchell, D. L. & Speight, J. G. (1973). Preparation of Mineral Free Asphaltenes. United State Patent. (Patent No. 3779902).

125. Al-Sahhaf, T. A., Fahim, M. A., & Elkilani, A. S. (2002). Retardation of asphaltene precipitation by addition of toluene, resins, deasphalted oil and surfactants. *Fluid Phase Equilibria*, 194–197, 1045–1057. https://doi.org/10.1016/S0378-3812(01)00702-6

126. Wolfgang, B., & Herold, C. P. (2002). Process for removing solid asphalt residues produced in the petroleum industry (Patent No. US 6368422 B1).

127. Ghloum, E. F., Al-Qahtani, M., & Al-Rashid, A. (2010). Effect of inhibitors on asphaltene precipitation for Marrat Kuwaiti reservoirs. *Journal of Petroleum Science and Engineering*, 70(1–2), 99–106. https://doi.org/10.1016/j.petrol.2009.10.003

128. Al-Kafeef, S. F., Al-Medhadi, F., & Al-Shammari, A. D. (2005). A simplified method to predict and prevent asphaltene deposition in oilwell tubings: Field case. *SPE Production & Facilities*, 20(02), 126–132. https://doi.org/10.2118/84609-PA

129. Khaleel, A. T., Abutaqiya, M. I. L., Sisco, C. J., & Vargas, F. M. (2020). Mitigation of asphaltene deposition by re-injection of dead oil. *Fluid Phase Equilibria*, 514, 112552. https://doi.org/10.1016/j.fluid.2020.112552

130. Rashed, A. M., Ghloum, E. F., Al-Matrook, M. F., Oskui, G. R., Mali, P., Telang, M., & Al-Jasmi, A. (2012). Continuous solvent flush approach for asphaltene precipitations in a Kuwaiti reservoir: Phase I - An experimental solvent screening. In *Society of Petroleum Engineers -SPE Kuwait International Petroleum Conference and Exhibition.* Kuwait City, Kuwait, December 10–12. (SPE-163316-MS). https://doi.org/10.2118/163316-ms

131. Schantz, S. S., & Stephenson, W. K. (1991). Asphaltene deposition : Development and application of polymeric asphaltene dispersants. In *Proceedings - SPE Annual Technical Conference and Exhibition.* Dallas, Texas, October 6–9. (SPE-22783-MS). https://doi.org/10.2523/22783-ms

132. Chang, C. L., & Fogler, H. S. (1994). Stabilization of asphaltenes in aliphatic solvents using alkylbenzene-derived amphiphiles. Effect of the chemical structure of amphiphiles on asphaltene Stabilization. *Langmuir*, 10(6), 1749–1757. https://doi.org/10.1021/la00018a022

133. Hu, Y. F., & Guo, T. M. (2005). Effect of the structures of ionic liquids and alkylbenzene-derived amphiphiles on the inhibition of asphaltene precipitation from CO_2 -injected reservoir oils. *Langmuir*, 21(18), 8168–8174. https://doi.org/10.1021/la050212f

134. Hashmi, S. M., & Firoozabadi, A. (2013). Self-assembly of resins and asphaltenes facilitates asphaltene dissolution by an organic acid. *Journal of Colloid and Interface Science*, 394, 115–123. https://doi.org/10.1016/j.jcis.2012.11.069

135. Moradi, S., Amirjahadi, S., Danaee, I., & Soltani, B. (2019). Experimental investigation on application of industrial coatings for prevention of asphaltene deposition in the well-string. *Journal of Petroleum Science and Engineering*, 181, 106095. https://doi.org/10.1016/j.petrol.2019.05.046

136. Gong, L., Qiu, X., Zhang, L., Huang, J., Hu, W., Xiang, L., Zhu, D., Sabbagh, R., Mahmoudi, M., Fattahpour, V., Luo, J.L., & Zeng, H. (2019). Probing the interaction mechanism between oil-in-water emulsions and electroless nickel–phosphorus coating with implications for antifouling in oil production. *Energy & Fuels*, 33(5), 3764–3775. https://doi.org/10.1021/acs.energyfuels.8b03132

137. Rodriguez, D. L. G. (2008). *Modeling of asphaltene precipitation and deposition tendency using the PC-SAFT equation of state*. Ph.D Thesis. Rice University. Texas, U.S.A. https://hdl.handle.net/1911/22140.

138. Britton, J. N. (2004). Early coating failures on offshore platforms. In *Corrosion 2004*, New Orleans, Louisiana, March 28–April 1. (NACE-04003).

139. Cochran, S. (2003). Hydrate control and remediation best practices in deepwater oil developments. In *Proceedings of the Annual Offshore Technology Conference*. Houston, Texas, May 5–8. (OTC-15255-MS). https://doi.org/10.4043/15255-MS

140. Zain, Z. M., Yang, J., Tohidi, B., Cripps, A., & Hunt, A. (2005). Hydrate monitoring and warning system: A new approach for reducing gas hydrate risks. In *SPE Europec/EAGE Annual Conference*. Madrid, Spain, June 13–16. (SPE-94340-MS) https://doi.org/10.2523/94340-MS

141. Denniel, S., Perrin, J., & Felix-Henry, A. (2004). Review of flow assurance solutions for deepwater fields. In *Proceedings of the Annual Offshore Technology Conference*, Houston, Texas, May 3–6. (OTC-16686-MS). https://doi.org/10.4043/16686-MS

142. Guo, B., Song, S., Ghalambor, A., Lin, T. R., & Chacko, J. (2005). *Offshore Pipelines*. Gulf Professional Publishing, Burlington.

143. Chakkalakal, F., Sweigart, R., Pulter, J., & Saldarriaga, G. (2008). Electric gut tracing - An innovative approach to pipeline heating. In *55th IEEE Petroleum and Chemical Industry Conference*. Cincinnati, OH, USA, September 22–24. https://doi.org/10.1109/PCICON.2008.4663994

144. Haniffa, M. A. M., & Hashim, F. M. (2011). Review of hydrate prevention methods for deepwater pipelines. In *2011 IEEE Conference on Open Systems*. Langkawi, Malaysia, September 25–28. https://doi.org/10.1109/ICOS.2011.6079275

145. Pattee, F. M., & Kopp, F. (2000). Impact of electrically-heated systems on the operation of deep water subsea oil flowlines. In *Proceedings of the Annual Offshore Technology Conference*. Houston, Texas, May 1–4. (OTC-11894-MS). https://doi.org/10.4043/11894-MS

146. Castello, X., & Estefen, S. F. (2008). Sandwich pipes for ultra deepwater applications. In *Proceedings of the Annual Offshore Technology Conference*, Houston, Texas, USA, May 5–8. (OTC-19704-MS). https://doi.org/10.4043/19704-MS

147. Ruschau, G. R., Rogers, R. H., Woodley, S. A., & Wright, E. J. (2010). Evaluation and qualification of materials for subsea insulation systems. In *Society of Petroleum Engineers - International Oil and Gas Conference and Exhibition in China*. Beijing, China, June 8–10. (SPE-131618-MS). https://doi.org/10.2118/131618-MS

148. Estefen, S. F., Netto, T. A., & Pasqualino, I. P. (2005). Strength analyses of sandwich pipes for ultra deepwaters. *Journal of Applied Mechanics, Transactions ASME*, 72(4), 599–608. https://doi.org/10.1115/1.1940667

149. Estefen, S. F., Lourenço, M. I., Feng, J., Paz, C. M., & Bessa de Lima, J. D. (2016). Sandwich pipe for long distance pipelines: flow assurance and costs. In *Proceedings of the ASME 2016 35th International Conference on Ocean, Offshore and Arctic Engineering*. Busan, South Korea, June 19–24. https://doi.org/10.1115/OMAE2016-54950

150. Bai, Y., & Bai, Q. (2018) *Subsea Engineering Handbook* (2nd Edition). Gulf Professional Publishing, Amsterdam, Netherlands.

151. Chatti, I., Delahaye, A., Fournaison, L., & Petitet, J. P. (2005). Benefits and drawbacks of clathrate hydrates: A review of their areas of interest. *Energy Conversion and Management*, 46(9–10), 1333–1343. https://doi.org/10.1016/j.enconman.2004.06.032

152. Azarinezhad, R., Chapoy, A., Anderson, R., & Tohidi, B. (2008). Hydraflow: A multiphase cold flow technology for offshore flow assurance challenges. In *Offshore Technology Conference*. Houston, Texas, USA, May 5–8. (OTC-19485-MS). https://doi.org/10.4043/19485-ms

153. Tohidi, B., Chapoy, A., & Yang, J. (2009). Developing a hydrate-monitoring system. *SPE Projects, Facilities & Construction*, 4(1), 1–6. (SPE-125130-PA). https://doi.org/10.2118/125130-PA

154. Mokhatab, S., Wilkens, R. J., & Leontaritis, K. J. (2007a). A review of strategies for solving gas-hydrate problems in subsea pipelines. *Energy Sources, Part A: Recovery, Utilization and Environmental Effects*, 29(1), 39–45. https://doi.org/10.1080/009083190933988

155. Paez, J. E., & Blok, R., (2001). Problems in hydrates: mechanisms and elimination methods. In *SPE Production Operations Symposium*, Oklahoma City, Oklahoma, March 24–27. (SPE-67322-MS). https://doi.org/10.2523/67322-ms

156. Fu, B., Neff, S., Mathur, A., & Bakeev, K. (2001). Novel low dosage hydrate inhibitors for deepwater operations. In *SPE Annual Technical Conference and Exhibition*. New Orleans, Louisiana, September 30–October 3. (SPE-71472-MS). https://doi.org/10.2118/71472-ms

157. Budd, D., Hurd, D., Pakulski, M., & Schaffer, T. D. (2004). Enhanced hydrate inhibition in Alberta gas field. In *SPE Annual Technical Conference and Exhibition*. Houston, Texas, September 26–29. (SPE-90422-MS). https://doi.org/10.2118/90422-ms

158. Sloan, E. D. (2000). *Hydrate Engineering*. Bloys J. B. (Ed.). Richardson, Tex Henry L. Doherty Fund of AIME, Society of Petroleum Engineers, USA.

159. Wu, M., Wang, S., & Liu, H. (2007). A study on inhibitors for the prevention of hydrate formation in gas transmission pipeline. *Journal of Natural Gas Chemistry*, 16(1), 81–85. https://doi.org/10.1016/S1003-9953(07)60031-0

160. Cochran, S., & Gudimetla, R. (2004). Hydrate management: Its importance to deepwater gas development success. *World Oil*, 225 (9), 55–60.

161. Hoppe, R., Martin, R. L., Pakulski, M. K., & Schaffer, T. D. (2006). Corrosion mitigation with gas hydrate inhibitors. In *SPE Gas Technology Symposium*, Calgary, Alberta, Canada, May 15–17. (SPE-100474-MS). https://doi.org/10.2118/100474-ms

162. Mpelwa, M., & Tang, S. F. (2019). State of the art of synthetic threshold scale inhibitors for mineral scaling in the petroleum industry: A review. *Petroleum Science*, 16(4), 830–849. https://doi.org/10.1007/s12182-019-0299-5

163. Mazumder, M. A. J. (2020). A review of green scale inhibitors: process, types, mechanism and properties. *Coatings*, 10(10), 928. https://doi.org/10.3390/coatings10100928

164. Abdel-Gaber, A. M., Abd-El-Nabey, B. A., Khamis, E., & Abd-El-Khalek, D. E. (2008). Investigation of fig leaf extract as a novel environmentally friendly antiscalent for $CaCO_3$ calcareous deposits. *Desalination*, 230(1–3), 314–328. https://doi.org/10.1016/j.desal.2007.12.005

165. Abdel-Gaber, A. M., Abd-El-Nabey, B. A., Khamis, E., & Abd-El-Khalek, D. E. (2011). A natural extract as scale and corrosion inhibitor for steel surface in brine solution. *Desalination*, 278(1–3), 337–342. https://doi.org/10.1016/j.desal.2011.05.048

166. Abdel-Gaber, A. M., Abd-El-Nabey, B. A., Khamis, E., Abd-El-Rhmann, H., Aglan, H., & Ludwick, A. (2012). Green anti-scalent for cooling water systems. *International Journal of Electrochemical Science*, 7(12), 11930–11940.

167. Euvrard, M., Martinod, A., & Neville, A. (2011). Effects of carboxylic polyelectrolytes on the growth of calcium carbonate. *Journal of Crystal Growth*, 317(1), 70–78. https://doi.org/10.1016/j.jcrysgro.2011.01.006

168. Larsen, A. (1973). Film forming hydrazine-containing corrosion inhibitor. United States Patent (U.S. Patent No. 3770055).

169. Olajire, A. A. (2017). Corrosion inhibition of offshore oil and gas production facilities using organic compound inhibitors - A review. *Journal of Molecular Liquids*, 248, 775–808. https://doi.org/10.1016/j.molliq.2017.10.097

170. Nam, N. D., Hien, P. V., Hoai, N. T., & Thu, V. T. H. (2018). A study on the mixed corrosion inhibitor with a dominant cathodic inhibitor for mild steel in aqueous chloride solution. *Journal of the Taiwan Institute of Chemical Engineers*, 91, 556–569. https://doi.org/10.1016/j.jtice.2018.06.007

171. Taghavikish, M., Dutta, N. K., & Choudhury, N. R. (2017). Emerging corrosion inhibitors for interfacial coating. *Coatings*, 7(12), 217. https://doi.org/10.3390/coatings7120217

172. Furman, A., & Kharshan, M. (1998). Incorporating vapor corrosion inhibitors (VCIs) in oil and gas pipeline additive formulations. In *Corrosion 98*. San Diego, California, March 22–27. (NACE-98236).

173. de Souza Mendes, P. R., de Abreu Soares, F. S. M., Ziglio, C. M., & Gonçalves, M. (2012). Startup flow of gelled crudes in pipelines. *Journal of Non-Newtonian Fluid Mechanics*, 179–180, 23–31. https://doi.org/10.1016/j.jnnfm.2012.05.003

174. Marsidi, N. A., Bak, A. S. A., Nawi, M. S. M., Lajawai, M. T., Istiyarso, I., Norhamidi, M., Kelly, C., & Gusting, K. (2019). Re-evaluating the wax flow assurance chemical mitigation strategy for a major deepwater oil pipeline in malaysia yields significant benefits. In *SPE/IATMI Asia Pacific Oil & Gas Conference and Exhibition*, Bali, Indonesia, October 29–31. (SPE-196472-MS). https://doi.org/10.2118/196472-MS

175. Kumar, R., Banerjee, S., Mandal, A., & Naiya, T. K. (2017). Flow improvement of heavy crude oil through pipelines using surfactant extracted from soapnuts. *Journal of Petroleum Science and Engineering*, 152, 353–360. https://doi.org/10.1016/j.petrol.2017.02.010

176. Kumar, S., & Mahto, V. (2017). Emulsification of Indian heavy crude oil using a novel surfactant for pipeline transportation. *Petroleum Science*, 14(2), 372–382. https://doi.org/10.1007/s12182-017-0153-6

177. Al-Sabagh, A. M., Khidr, T. T., Moustafa, H. M., Mishrif, M. R., & Al-Damasy, M. H. (2017). Investigating the synergistic effect between oil soluble surfactants and styrene–maleic anhydride copolymers to enhance the flow properties of waxy crude oil. *Petroleum Science and Technology*, 35(13), 1381–1388. https://doi.org/10.1080/10916466.2017.1331243

178. Ahmed, S. M., Khidr, T. T., & Ismail, D. A. (2018). Effect of gemini surfactant additives on pour point depressant of crude oil. *Journal of Dispersion Science and Technology*, 39(8), 1160–1164. https://doi.org/10.1080/01932691.2017.1385483

179. Anisuzzaman, S. M., Abang, S., Bono, A., Krishnaiah, D., Karali, R., & Safuan, M. K. (2017). Wax inhibitor based on ethylene vinyl acetate with methyl methacrylate and diethanolamine for crude oil pipeline. In *IOP Conference Series: Materials Science and Engineering*. *In 29th Symposium of Malaysian Chemical Engineers (SOMChE)*. Miri, Sarawak, Malaysia, December 1–3. https://doi.org/10.1088/1757-899X/206/1/012074

180. Machado, A. L. C., Lucas, E. F., & González, G. (2001). Poly(ethylene-co-vinyl acetate) (EVA) as wax inhibitor of a Brazilian crude oil: oil viscosity, pour point and phase behavior of organic solutions. *Journal of Petroleum Science and Engineering*, 32(2–4), 159–165. https://doi.org/10.1016/S0920-4105(01)00158-9

181. Akinyemi, O. P., Udonne, J. D., & Oyedeko, K. F. (2018a). Study of effects of blend of plant seed oils on wax deposition tendencies of Nigerian waxy crude oil. *Journal of Petroleum Science and Engineering*, 161, 551–558. https://doi.org/10.1016/j. petrol.2017.12.017

182. Abdallah, D., Bazuhair, M., Zwolle, S., Grutters, M., Ramanathan, K., & Artur, S. (2010). Asphaltene studies in on-shore Abu Dhabi oil fields, Part III: Optimization of field chemicals for remediation and inhibition of asphaltene deposition. In *Abu Dhabi International Petroleum Exhibition and Conference*, Abu Dhabi, UAE, November 1–4. (SPE-138040-MS). https://doi.org/10.2118/138040-MS

183. Osorio, N. F., Chipuk, T. A., & Anderson, L. (2017). Chemical injection optimization: How leverage of flow assurance, materials science and production chemistry knowledge can reduce uncontrolled expenses in production chemicals utilization. In *Offshore Technology Conference*, Houston, Texas, USA, May 1–4. (OTC-27875-MS). https://doi. org/10.4043/27875-MS

184. Frostman, L. M. (2000). Anti-agglomerant hydrate inhibitors for prevention of hydrate plugs in deepwater systems. In *Proceedings of SPE Annual Technical Conference and Exhibition*. Dallas, Texas, October 1–4. (SPE-63122-MS). https://doi. org/10.2523/63122-MS

185. Alapati, R. R., Lee, J., & Beard, D. (2008). Two field studies demonstrate that new LDHI chemistry is effective at high water cuts without impacting oil/water quality. In *Offshore Technology Conference*, Houston, Texas, USA, May 5–8. (OTC-19505-MS). https://doi.org/10.4043/19505-MS

186. Kelland, M. A. (2006). History of the development of low dosage hydrate inhibitors. *Energy & Fuels*, 20(3), 825–847. https://doi.org/10.1021/ef050427x

187. Mokhatab, S., Wilkens, R. J., & Leontaritis, K. J. (2007a). A review of strategies for solving gas-hydrate problems in subsea pipelines. *Energy Sources, Part A: Recovery, Utilization and Environmental Effects*, 29(1), 39–45. https://doi.org/10.1080/009083190933988

188. Nasir, Q., Suleman, H., & Elsheikh, Y. A. (2020). A review on the role and impact of various additives as promoters/ inhibitors for gas hydrate formation. *Journal of Natural Gas Science and Engineering*, 76, 103211. https://doi.org/10.1016/j.jngse.2020.103211

189. Arjmandi, M., Tohidi, B., Danesh, A., & Todd, A. C. (2005). Is subcooling the right driving force for testing low-dosage hydrate inhibitors? *Chemical Engineering Science*, 60(5), 1313–1321. https://doi.org/10.1016/j.ces.2004.10.005

190. Freer, E. M., & Sloan Jr, E. D. (2000). An engineering approach to kinetic inhibitor design using molecular dynamics simulations. *Annals of the New York Academy of Sciences*, 912(1), 651–657. https://doi.org/10.1111/j.1749-6632.2000.tb06820.x

191. Lederhos, J. P., Long, J. P., Sum, A., Christiansen, R. L., & Sloan, E. D. (1996). Effective kinetic inhibitors for natural gas hydrates. *Chemical Engineering Science*, 51(8), 1221–1229. https://doi.org/10.1016/0009-2509(95)00370-3

192. Xiao, C., & Adidharma, H. (2009). Dual function inhibitors for methane hydrate. *Chemical Engineering Science*, 64(7), 1522–1527. https://doi.org/10.1016/j.ces.2008. 12.031

193. Lal, B., & Nashed, O. (2020). *Chemical Additives for Gas Hydrates* (1st Edition). Springer International Publishing, Cham, Switzerland.

194. Nielsen, R. B., & Bucklin, R. W. (1983). Why not use methanol for hydrate control. *Hydrocarbon Processing*, 62(4), 71.

195. Carroll, J. (2014). Inhibiting Hydrate Formation with Chemicals. In *Natural Gas Hydrates* (3rd Edition). Gulf Professional Publishing. https://doi.org/10.1016/ B978-0-12-800074-8.00005-3

196. Østergaard, K. K., Masoudi, R., Tohidi, B., Danesh, A., & Todd, A. C. (2005). A general correlation for predicting the suppression of hydrate dissociation temperature in the presence of thermodynamic inhibitors. *Journal of Petroleum Science and Engineering*, 48(1–2), 70–80. https://doi.org/10.1016/j.petrol.2005.04.002

197. Tomson, M. B., & Matly, J. (1989). A mechanism of inhibition of nucleation of sparingly soluble salts. In *Corrosion 89*, New Orleans, Houston.

198. Vanderpoo, D. P. (1997). New Calcium Carbonate Scale Inhibitors: Understanding Complexation Constants as a Tool for Finding Improved Performance. In *58th Annual Meeting*. International Water Conference Engineers' Society of Western Pennsylvania, November. (IWC-97-40).

199. Cognetti, A., Geraghty, M., Parsons, J. R., & Vanderpool, D. (1996). Operation of cooling towers at high cycles of concentration: Corrosion and scale control. In *Corrosion 96*, Denver, Colorado, March 24–29. (NACE-96575).

200. Re, M. L., & Gill, J. S. (1996). The effect of polyamino polyether methylene phosphonate on the crystallization kinetics of calcium carbonate: A constant composition study. In *Corrosion 96*, Denver, Colorado, March 24–29. (NACE-96157).

201. Ferguson, R. J., Baron, R. F., & Stancavage, F. (2011). Modeling scale formation and optimizing scale inhibitor dosages in membrane systems. In *AWWA Membrane Technology Conference*. Long Beach, CA, USA, March 30.

202. Senthilmurugan, B., Radhakrishnan, J. S., Arana, V., & Al-Foudari, M. (2019). High temperature kinetic scale inhibitor for flow assurance application. *International Journal of Petroleum Science and Technology*, 13(1), 21–38.

203. Markin, A. N., Sukhoverkhov, S. V., Zadorozhny, P. A., Polyakova, N. V., & Brikov, A. V. (2018). Measurement and calculation of polymeric scale inhibitor concentration in water. *International Journal of Corrosion and Scale Inhibition*, 7(2), 250–259. https://doi.org/10.17675/2305-6894-2018-7-2-10

204. Xu, X. Chemical management plan in oil and gas industry. https://www.corrosionguru.com/wp-content/uploads/2016/09/chemical-management-plan-in-oil-and-gas-industry (accessed December 7, 2021).

205. Willmon, J. G., & Edwards, M. A. (2005). Precommissioning to startup: getting chemical injection right. In *SPE Annual Technical Conference and Exhibition*, Dallas, Texas, October 9–12. (SPE-96144-MS). https://doi.org/10.2118/96144-MS

206. Shirah, A. D., Place, M. C., & Edwards, M. A. (2003). Reliable sub-sea umbilical and down-hole injection systems are an integral component of successful flow assurance programs. In *SPE Annual Technical Conference and Exhibition*, Denver, Colorado, October 5–8. (SPE-84046-MS). https://doi.org/10.2523/84046-MS

207. Arciero, B. (2017) Optimization of flow assurance applications through life-of-field: an integrated approach to production engineering and chemical management across Gulf of Mexico deepwater assets. In *Offshore Technology Conference*, Houston, Texas, USA, May 1–4. (OTC-27560-MS). https://doi.org/10.4043/27560-MS

208. Piovesan, C. M. (2015). A real time chemical monitoring platform for inventory and usage. In *SPE Annual Technical Conference and Exhibition*, Houston, Texas, USA, September 28–30. (SPE-174780-MS). https://doi.org/10.2118/174780-MS

8 The Role of Solids in Subsurface and Surface Flow

8.1 INTRODUCTION

Precipitation of any type of solid ([predominantly organic in this chapter] or inorganic) in the flow path, be it subsurface in the reservoir or in the tubular, flowline–riser system or long-distance pipelines will always lead to choking the flow stream. Essentially, resulting in increased pressure drops. Precipitation of solids subsurface is typically an area dealt with by reservoir engineers; however, we believe it is rather a domain on the interface of reservoir and flow assurance engineers because the solids precipitation problem, notwithstanding its origin in the subsurface, may actually have the potential to carry over and become a flow assurance issue. Among the solids, asphaltenes are perhaps the most common that may be potentially detrimental subsurface as well as in other flow infrastructure. The negative impact of asphaltenes, especially in conjunction with miscible gas injection, was recognized by Stalkup[1] back in 1983. He stated reduction in permeability, well injectivity, productivity, and plugging of producing wells. Direct experimental evidence of solid (asphaltene) precipitation during swelling tests conducted on a black oil from Middle East using a rich injection gas was presented by Dandekar et al.[2] Visual evidence of asphaltene precipitation (see Figure 8.1) was corroborated by other analyses, such as flocculation onset titration, elemental analysis, the molecular weight of the deposit, and the densities of liquid phases recovered from the pressure–volume–temperature (PVT) cell. Their mass balance indicated 600 ppm of asphaltene precipitation. As far as inorganic solids are concerned, fines migration can play a negative role subsurface as well as in non-reservoir flow infrastructure, although producers fitted with sand screens will help hinder[3] sand production. The interplay of solids in subsurface and surface flow are topics covered in this chapter.

8.2 FORMATION DAMAGE CONCEPTS, CAUSES, AND SKIN EFFECTS

Formation damage is a somewhat self-explanatory term that is basically used to define the permeability impairment of the reservoir rock, typically during drilling operations, and, in particular, the invasion of the drilling mud filtrate into the near-wellbore region. A conceptual illustration is shown by Dandekar.[4] The near-wellbore region in which the invasion of the mud filtrate occurs is known as the skin or the damaged zone, given the reduced formation permeability. This particular zone can extend from a few inches to several feet from the wellbore.[5] A well-known dimensionless parameter, called the skin factor, is used to quantify the degree of formation damage, as shown by the following equation:[5]

DOI: 10.1201/9781003091301-8

FIGURE 8.1 Visual evidence of "sticky" oil phase on the piston of the PVT cell (left) and asphaltenic material precipitated on the magnetic stirrer (right) that mounts on the piston.

$$S = \left[\frac{k_{\text{formation}}}{k_{\text{skin}}} - 1\right] \text{Ln}\left[\frac{r_{\text{skin}}}{r_{\text{w}}}\right] \qquad (8.1)$$

where $k_{\text{formation}}$ is the original formation permeability in mD or D, k_{skin} is the permeability of the skin zone in mD or D, r_{skin} is the skin radius in ft or m, and r_{w} is the wellbore radius in ft or m.

Equation 8.1 is consistent in that the skin factor will be zero ($S = 0$) if $k_{\text{formation}} = k_{\text{skin}}$, vis-à-vis no damage, whereas $S > 0$ means damage or lowering of the near-wellbore zone permeability or the skin zone and $S < 0$ indicates permeability improvement. The positive S caused due to drilling operations can be reduced or even turned negative by acid treatment.[6] From the standpoint of completeness, a derivation of Darcy's law that reconciles with the mathematical definition of the skin factor is presented in the following.

Basically, two expressions of Darcy equation in field units for steady-state radial flow in series can be written as

$$Q = 1.1271 \frac{2\pi k_{\text{formation}} h \left(P_{\text{e}} - P_{\text{wf}}\right)}{\mu \left[\text{Ln}\left(\frac{r_{\text{e}}}{r_{\text{w}}}\right) + \left[\frac{k_{\text{formation}}}{k_{\text{skin}}} - 1\right]\text{Ln}\left[\frac{r_{\text{skin}}}{r_{\text{w}}}\right]\right]}$$

$$= 1.1271 \frac{2\pi k_{\text{avg}} h \left(P_{\text{e}} - P_{\text{wf}}\right)}{\mu \left[\text{Ln}\left(\frac{r_{\text{e}}}{r_{\text{w}}}\right)\right]} \qquad (8.2)$$

where Q is the flow rate in stb/day, k_{avg} is the average absolute permeability in D, h is the formation thickness in ft, P_e is the pressure at drainage radius in psi, P_{wf} is the well flowing pressure in psi, μ is the fluid viscosity in cP, r_e is the drainage radius in ft, r_w is the wellbore radius in ft, and S is the dimensionless skin factor.

Equation 8.2 simplifies to

$$k_{avg} = \frac{k_{formation} Ln\left(\dfrac{r_e}{r_w}\right)}{Ln\left(\dfrac{r_e}{r_w}\right) + \dfrac{k_{formation}}{k_{skin}} Ln\left(\dfrac{r_{skin}}{r_w}\right) - Ln\left(\dfrac{r_{skin}}{r_w}\right)} \tag{8.3}$$

$$k_{avg} = \frac{Ln\left(\dfrac{r_e}{r_w}\right)}{\dfrac{1}{k_{formation}} Ln\left(\dfrac{r_e}{r_{skin}}\right) + \dfrac{1}{k_{skin}} Ln\left(\dfrac{r_{skin}}{r_w}\right)} \tag{8.4}$$

Equation 8.4 is basically the mathematical definition of average permeability in series (one being the formation and the other being the skin zone). Alternatively, k_{avg} can be substituted in the second right-hand side of Equation 8.2 to arrive at the Darcy equation in terms of formation permeability and the skin factor.

As far as flow assurance is concerned, we have coined a new term for formation damage and that is "*pore scale flow assurance*", which is particularly linked to the (organic) solids, such as asphaltenes. A schematic representation of this type of (solid organic phase) formation damage is depicted in Figure 8.2. Leontaritis[7] states that in the case of asphaltenes, the formation damage may extend to many feet inside the reservoir depending on the drawdown, and in miscible recovery, the damage may even be much deeper inside the reservoir and is closely related to the miscible oil bank. As stated before, this particular reservoir problem will potentially carry over and manifest itself as a flow assurance problem. However, it may be relatively easier to handle the precipitation and deposition in the tubing or other flow systems than in the reservoir pore spaces, given the better accessibility and reach.

8.2.1 ASPHALTENE FORMATION DAMAGE OR TUBING-RELATED PROBLEMS

Somewhat similar to the acid treatment[6] for mud filtrate invasion to reduce the skin, theoretically a similar treatment such as xylene or any aromatic solvent (any aromatic is a good solvent to dissolve the asphaltenes, much like acid dissolves a mud filtrate) soak could be considered for eliminating the organic solid induced skin. However, solvents such as xylene and toluene are expensive and pose health, safety, and environmental (HSE) constraints. For this particular reason, alternative treatments such as the ones that use "deasphalted oil" (DAO) have been patented.[8] The DAO can be produced from any oil from which asphaltene molecules have been precipitated by using n-alkane as a precipitant.[8] Kabir and Jamaluddin[9] reported perhaps the most robust study on asphaltene characterization and mitigation in the Marrat reservoir in Kuwait, which was mainly a problem in the production tubing. They suggested a

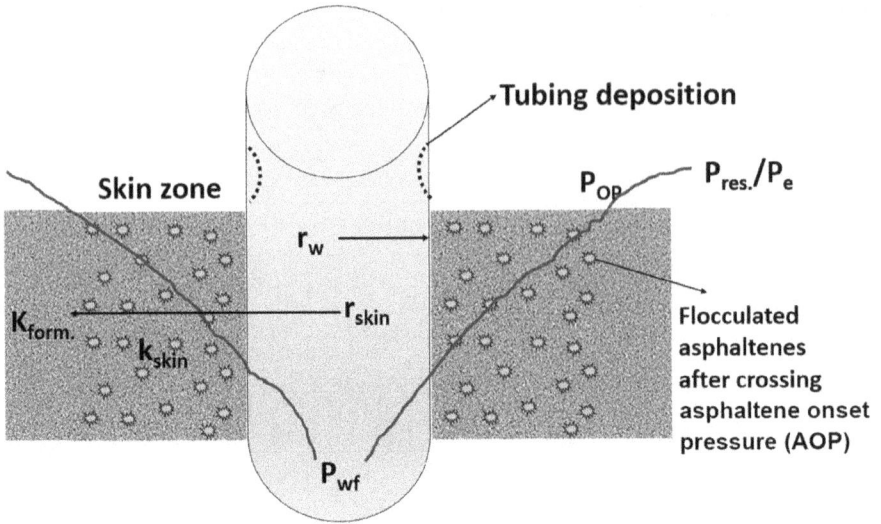

FIGURE 8.2   Formation damage and resulting "pore scale flow assurance".

rather practical approach for producing the DAO on field using a readily available condensate (much like a relatively light blend of lighter n-alkanes). Their final recipe was 1% (commercial) dispersant and 20% DAO in the tubular, which a dissolved considerable amount of asphaltenes after a 24-h soak period. They found DAO to be 50% lower in cost compared to toluene and, of course, environmentally friendly since regulations are forcing operators to seek other solutions. The effectiveness of deasphalting of the oil by both n-pentane and condensate is evident from the SARA (saturates, aromatics, resins, asphaltenes) comparison shown in Figure 8.3 (note the significant reduction in asphaltene wt%).

Ortiz et al.[10] presented a specific asphaltene-related formation damage study in a fractured carbonate reservoir in Mexico, where skin effects most likely caused by asphaltene precipitation were attributed to the severe reduction in the well productivity. Their (squeeze) treatment basically consisted of bullheading an asphaltene inhibitor (AI) into the formation penetrating a radius of 3 ft and a shut-in period of 12–24 h followed by opening the well to production. The shut-in period is intended for the AI adsorption process to be complete such that the AI remains in the formation providing protection against developing a positive skin due to asphaltenes. We use a modified version of Figure 8.1 as shown in Figure 8.4 to conceptually depict the AI treatment, which may be customized to other cases as needed.

8.2.2   WAX OR PARAFFIN FORMATION DAMAGE OR TUBING-RELATED PROBLEMS

Theoretically, wax deposition also can result in formation damage; however, waxes and asphaltenes have one major difference in that the former is much more sensitive to (relatively cold) temperatures and a loss of light ends (which possess the solvent effect). Generally, under reservoir conditions (or near the wellbore), both

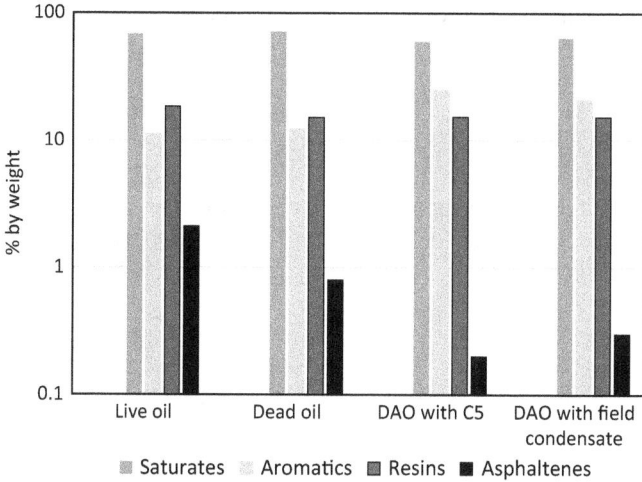

FIGURE 8.3 Comparison of SARA content in dead oil, live oil, and DAO obtained from n-pentane asphaltene precipitation and DAO obtained from field condensate asphaltene precipitation. Plot constructed based on data tabulated in Kabir and Jamaluddin.[9] Note that as stated by the authors, the live oil is basically generated from flashing the single-phase sample whereas the dead oil is stock tank oil from the separators.

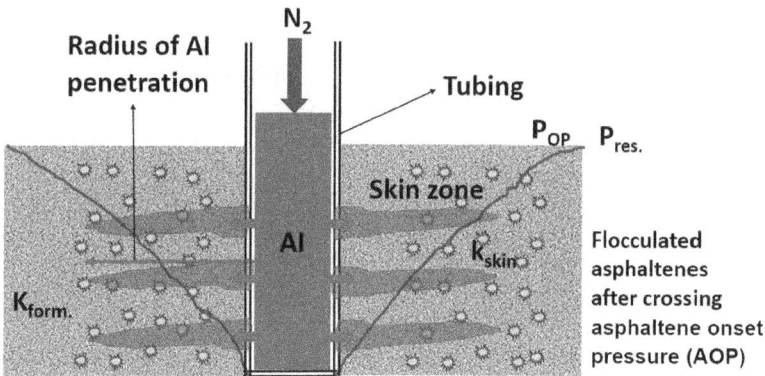

FIGURE 8.4 Asphaltene inhibitor squeeze treatment via bullheading to mitigate asphaltene-induced formation damage in a fractured carbonate reservoir in Mexico. Figure 8.1 is used to conceptually depict the treatment design in Ortiz et al.[10]

are favorable in that temperatures are higher and much of the lighter components are still in the liquid phase (unless the fluid is highly saturated); these conditions are thus conducive to the heavier alkanes remaining in the liquid phase. Therefore, cases that report wax caused formation damage are rather rare. Sutton and Roberts[11] reported paraffin deposition near the wellbore, resulting in formation damage due to the use of relatively large volumes of cold fracturing fluids injected into low-temperature wells. Of course, if this cooling of the formation is below the cloud point of the reservoir fluid, there is a potential for formation damage to occur. Compared to asphaltenes, the severity of formation damage due to wax precipitation is rather low in that the damaged area is very close to the wellbore (0–1 ft).[6]

8.3 PRODUCTIVITY INDEX AND SKIN

The effect of skin can be understood by comparing the pre and post, or before and after, well productivity index. The general expression for well productivity index, PI, denoted by J is given by the following equation, which is essentially Darcy's law rearranged:

$$\text{PI} = J = \frac{Q}{\left(P_e - P_{wf}\right)} = \frac{2\pi kh}{\mu B_o \left[\text{LN}\left(\dfrac{r_e}{r_w}\right) + S \right]} \tag{8.5}$$

Equation 8.5 can be rewritten as $J_O = \dfrac{Q}{\left(P_e - P_{wf}\right)} = \dfrac{2\pi kh}{\mu B_o \left[\text{LN}\left(\dfrac{r_e}{r_w}\right) \right]}$, where J_O is

the original or before or pre well PI with no skin or $S = 0$. The same equation can be

expressed as $J_S = \dfrac{Q}{\left(P_e - P_{wf}\right)} = \dfrac{2\pi kh}{\mu B_o \left[\text{LN}\left(\dfrac{r_e}{r_w}\right) + S \right]}$, where J_S is the well PI with skin,

which could be positive (due to damage) or negative (with acidizing, cleanup of the near-wellbore region). The ratio of two PIs will be

$$\frac{J_O}{J_S} = \frac{\left[\text{LN}\left(\dfrac{r_e}{r_w}\right) + S \right]}{\text{LN}\left(\dfrac{r_e}{r_w}\right)} \tag{8.6}$$

In the case of zero skin, or in other words, formation and skin zone permeability being equal, $J_O/J_S = 1$, whereas J_O/J_S will be greater than 1 and less than 1, when the near-wellbore region is damaged and acidized or cleaned, respectively.

Equation 8.6 can be used to calculate the PI ratio for various skin values to get an idea of PI impairment as well as improvement. As an example, if the damage is due to asphaltene precipitation near the wellbore ($r_w = 0.375$ ft) with a drainage radius of ($r_e = 750$ ft) and the asphaltene deposition confined to within 3 ft of the wellbore, gradually reducing the formation permeability by a factor of 0.05–0.9, then various (positive) skin factors can be calculated using Equation 8.1. Similarly, if the formation permeability is now restored with xylene or other treatment and even improved by acidizing then the same equation can be used to compute the (negative) skin factors. For example, if the (original) formation permeability is reduced due to damage by a factor of 0.05, then $k_{skin} = 0.05 \times k_{formation}$ and with $r_{skin} = 3.375$ ft,

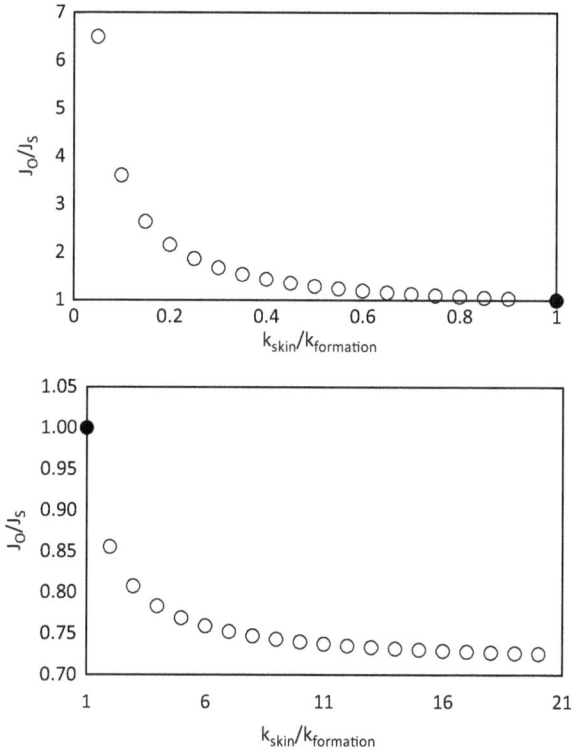

FIGURE 8.5 Plot of PI ratio vs. $k_{skin}/k_{formation}$ ratio for positive (top figure) and negative (bottom figure) skin factors. Note that the PI ratio is greater than 1 for positive skin, less than 1 for negative skin, and 1 when there is no change in the permeability (solid markers).

$$S = \left(\frac{k_{formation}}{0.05 \times k_{formation}} - 1 \right) \times LN \left(\frac{3.375}{0.375} \right) = 41.75 \text{ , whereas if the original permeability}$$

is increased by a factor of 20 then $k_{skin} = 20 \times k_{formation}$ and other data being the same,

$$S = \left(\frac{k_{formation}}{20 \times k_{formation}} - 1 \right) \times LN \left(\frac{3.375}{0.375} \right) = -2.09 \text{ . Accordingly, a plot of } J_o/J_s \text{ for vari-}$$

ous S values versus $k_{skin}/k_{formation}$ can be constructed (as shown in Figure 8.5) to get an idea of the PI ratio.

8.4 DEPOSITS IN WELLBORE/PRODUCTION TUBING AND SURFACE FACILITIES

Given the (somewhat) omnipresence of asphaltenes from the reservoir to the surface, we consider some peculiar cases of asphaltene deposition in this section. Perhaps one of the unique reported cases of asphaltene deposition everywhere (wellbore, pipelines, separators, pumps, etc.), right from day 1, is that of the Prinos field in the northern Aegean Sea.[12, 13] This severe problem almost killed further development of the

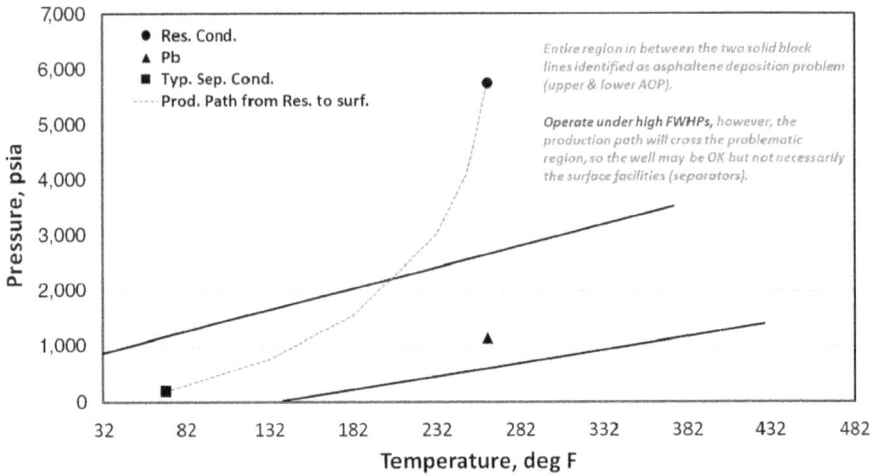

FIGURE 8.6 Interpretation of Prinos field PT conditions with respect to asphaltenes. The composite plot is prepared based on the data in Leontaritis and Mansoori.[13]

field. With the exception of the unusually high H_2S in excess of 35 mole%, the crude in this field is a typical 28° American Petroleum Institute (API) textbook-type black oil. The fluid samples were extensively studied in the laboratory from which the asphaltene deposition envelope (very broad) was bracketed, from which the company derived the rule (based on laboratory testing) of producing the wells at high-flowing wellhead pressures (FWHPs). However, as the interpreted plot in Figure 8.6 shows, the curved PT reduction path will cross the bracketed region and the surface facilities may suffer from precipitation and deposition, at least the well will not be affected. Additionally, it is easier to (mechanically for the most part) clean large-volume separator facilities because they are easier to reach in terms of access.

Haskett and Tartera[14] proposed a practical solution for the asphaltene deposition problem in the tubing for the Hassi Messaoud field in Algeria. Asphaltene deposition problems were encountered in the well tubings from the first production. The oil in Hassi Messaoud is of excellent quality and perhaps fits the profile of a volatile oil, having a light green color, an API gravity of 42.3°, a high gas–oil ratio (GOR), and an almost negligible total asphaltene content of 0.06% by weight (based on 500 mg/liter of stock tank oil[14]). The difference between the original reservoir pressure and the bubble point pressure varies between about 4,000–4,700 psi on an east–west traverse of the field. Within a time span of merely 15–20 days, 20–25% loss in FWHPs was observed, resulting in a considerable loss in production. At Hassi Messaoud, it was found that the majority of the deposition is in the single-phase fluid and that the point in the tubing most conducive to deposition is just below the bubble point.[14] However, asphaltene deposition in the tubing does not occur after the pressure falls relatively well below the bubble point, and previously deposited asphaltenes are re-peptized by the two-phase crude.[13, 14] Based on these observations, the deposition problem was largely controlled by flowing the wells at low wellhead pressures.

It should also be noted that the low total asphaltene content, such as in Hassi Messaoud, may be deceptive in that it does not necessarily mean no flow assurance problems, because for example, in the case of the Boscan crude in Venezuela, despite having 17% total asphaltenes, no problems have been reported.[13]

Somewhat similar is the case of the Ventura Avenue field in California[15] in which significant problems were encountered during the early producing life of some of the wells, which was attributed to asphaltene deposition that plugged the liners and the tubing. Additionally, produced sands and solids from the mud also contributed to the asphaltene deposition problems. Like the Hassi Messaoud oil, this particular oil also was undersaturated in the range of 4,000–5,000 psi. The severity of asphaltene deposition problems in the Ventura Field led to redrilling of many wells during the early history of the field. Apparently, like Hassi Messaoud, the asphaltene deposition problem was overcome or diminished after the flowing bottomhole pressure fell below the bubble point of the oil.[15] These wells produced without any asphaltene troubles since the early 1970s, which means about 10 years, given the fact that Tuttle's paper was published in 1983. From the comparison of the Prinos, Hassi Messaoud, and Ventura Avenue fields, it is rather interesting to note the differing recommendations and experiences of basically operating at a high-enough bottomhole pressure in the case of the former and vice versa in the case of the latter two. Tuttle[15] also reported on asphaltene deposits that were found in the production tubing in the Little Creek CO_2 injection Enhanced Oil Recovery (EOR) pilot in Mississippi. Since no asphaltene problems were reported previously during primary depletion and waterflood (which is immiscible anyway), the problem was obviously attributed to CO_2, producing an effect similar to propane deasphalting crude in refinery processes. So, this particular case indicates that not every oil intrinsically will exhibit asphaltene deposition problems, and they may be induced by miscible-type EOR processes such as CO_2 and/or rich-gas injection. Therefore, specialized PVT experiments that test the compatibility between the miscible injectant and the target oil should be a prerequisite for these types of pilot projects so that the propensity of asphaltene deposition can be ascertained a priori.

Thawer et al.[16] extensively described the problems associated with asphaltenes in the offshore (~11,000-ft depth below seabed) Ula field in Norway. The oil is high API (~39°) and has only 0.57 wt% asphaltenes. Reservoir pressure, as well as temperature, is high: 7,115 psi and 289.4°F. For a total oil production of approximately 98,000 bbl/day (converted from API and mass rates) from the Ula field, the asphaltene deposition rate in the separators was measured at about 160 kg/day (total). Big mounds of asphaltene in the separator were found. Over a period of 9 months, 25,000 kg of aphaltenes were found in two of the separators. Clearly, the problem was **so severe** that the only way of removal was by shoveling them out manually. Six different asphaltene dissolvers were tested on asphaltene pellets, however, the best results were obtained by using pure toluene and xylene.

How did such large amounts of the asphaltenes reach the separator? Thawar et al.[16] state that they originate from the continual erosion of previously deposited particles in the production tubing. Since the residence time of the oil in the separators is sufficient, it allows settling of these particles, which, in the long term, build up, resulting in self-compaction. The Kinley caliper surveys run on one of the wells indicated a deposition of up to an inch in the production tubing. Despite the deposition,

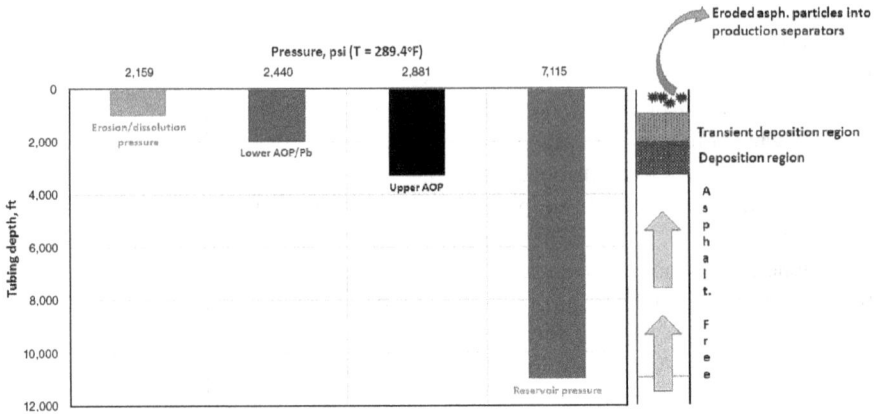

FIGURE 8.7 Graphical interpretation/analysis of Ula field asphaltene problems. Plot constructed based on numerical data and observations of Thawer et al.[16] Note AOP = asphaltene onset pressure, and P_b is bubble point.

given the large enough FBHPs, the authors' multiphase modeling indicated only a small increase in frictional pressure drop along the tubing attributable to asphaltenes and having no significant effect on the oil production. A self-explanatory graphical interpretation of the Ula field asphaltene experiences documented by Thawer et al.[16] are depicted in Figure 8.7, and relevant key points follow:

- Turbulence through two-phase flow is high enough to not allow additional deposition.
- Solvency of the oil for asphaltene increasing below the bubble point (lighter ends into the gas phase).
- Deposition below the bubble point does not cease immediately because there is a transient phase or a delay or a time lag between the saturation pressure and the erosion/dissolution pressure.
- Interestingly, Ula field observations share similarities with Hassi Messaoud as well as the Ventura Avenue field in California.

Al-Kafeef et al.[17] reported an experience, quite similar to that of the Ula field, for the Marrat wells in Kuwait. The asphaltene deposition region was found to be within the asphaltene onset pressure (AOP) and the saturation pressure depths. However, no distinction is made between the upper and lower AOPs, and the authors came up with a somewhat different interpretation of their observations. With time, the well flowing pressure as well as the saturation pressure will change, which basically means the difference between AOP and bubble point will continuously shift in the 2.488-in.-diameter tubing, that is, a variable deposit depth. Their plot of well depth versus time at least for the first year indicates a decrease in the asphaltene deposition depth from 9,000 ft to 6,000 ft.

Based on the FWHP data collected over a period of about one year, the authors observed a drawdown slope of –2 psig/day, which was representative of normal well flow and reservoir fluid properties, for 130 days and, thereafter, a rapid decline in the drawdown slope of −7 psig/day for 25 days attributed to the onset of deposition. Using the variable drawdown slope information, the authors developed an equation to estimate the asphaltene deposition thickness, expressed by the following equation:

$$\frac{h_{\text{asphaltene}}}{r_{\text{tubing}}} = 1 - \left[\frac{\left(\frac{dp}{dt}\right)_{\text{DD}}}{\left(\frac{dp}{dt}\right)_{\text{AD}}} \right]^{1/4} \tag{8.7}$$

where $h_{\text{asphaltene}}$ is deposit thickness, r_{tubing} is the radius of tubing, and $(dp/dt)_{\text{DD}}$ and $(dp/dt)_{\text{AD}}$ are the normal and deposition-related drawdown slopes, respectively. Therefore, with a tubing radius of 1.244 in., −2 psig/day, and −7 psig/day drawdown slopes, the thickness of deposit is estimated as $1.244 \times \left[1 - \left(\frac{-2}{-7}\right)^{1/4} \right] = 0.334$ inch which the authors indicate matches very well with the caliper measurements.

Finally, based on the reported data for Ula, Marrat, and Hassi Messaoud, the ratio of difference between the deposit depths or interval and the difference between the AOP and bubble point is shown in Figure 8.8 for all three cases. As seen in the figure, the ratios are somewhat similar and the oil API gravities are quite comparable, which perhaps can be used as an analog or guideline for asphaltene-related deposit prediction.

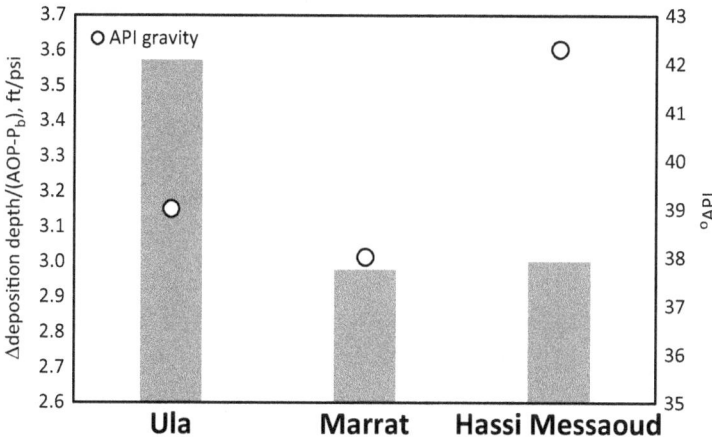

FIGURE 8.8 Comparison of Ula, Marrat, and Hassi Messaoud field asphaltene experiences. Kargarpour and Dandekar[18] have also reinterpreted the Marrat well observations and have come up with calculations that show the mass of asphaltenes that leave the tubing and deposit the tubing, which is somewhat (qualitatively) similar to the Ula field case.

8.5 DEPOSITS IN LONG-DISTANCE PIPELINES

For the most part, in this discussion, *long-distance pipelines* refer to those that typically transport oil over large distances, such as the famous Trans-Alaska Pipeline System (TAPS). Generally, asphaltenes are not a major concern in these pipelines; however, paraffin waxes are because of the loss of light ends and relatively colder temperatures, which may also promote ice crystal formation if free water is present. In the case of long-distance gas pipelines, usually the gas that would enter the pipeline would be treated (dehydrated), as well in terms of water; additionally, from a compositional standpoint, typically the gas would be lean (mostly containing methane) so that way the potential hydrate formation is virtually eliminated or minimized for operating conditions.

Zettlitzer[19] documented the successful application of chemical flow improvers in a 60-km (37.5-mile)-long feeder pipeline transporting a highly paraffinic Akshabulak oil in Kazakhstan. This particular oil is fairly light with a 42° API gravity and a gas chromatograph of the dead oil that shows peaks in carbon numbers between C_{11} and C_{18}, gradually tapering off to C_{35} and then leveling to about C_{43}, indicating the paraffinicity of the oil. Interestingly, a relatively high reservoir temperature of 81°C or 178°F means no paraffin deposition problems in the tubing; however, a high wax appearance temperature (WAT) of 28°C (based on oil viscosity vs. inverse temp), a pour point of +17°C, and a high viscosity at low temperatures (common in winter months) definitely pose transport challenges. Clearly, such oil characteristics also impart high gel strengths, vis-à-vis high restart pressures, potentially leading to even loss of a pipeline. An increase in the pressure drop in the pipeline by a factor of about 15–20 times the normal gives an idea of the negative impact of the paraffinicity of the Akshabulak oil. As part of the workflow, the operator considered and then eliminated most methods, such as heating and insulation (no electricity available due to rural location), microbial methods, and others, due to long-distance incompatibility. They considered 18 inhibitors from nine different suppliers that were tested, of which only eight were found to be eligible, nine showed no effect at all, and one had a low efficiency. Performance indicators, such as long-term effectiveness, a significant reduction in gel strength, viscosity, low toxicity, and low concentration, that is, cost-effectiveness, were considered in the extensive laboratory testing that resulted in the selection of one inhibitor at a concentration of 100 ppm. Long term efficacy of the inhibitor was apparent from the –4 to –6°C (a fairly stable) pour point over a period of 38 days.

The 800-mile-long, 48-in.-diameter TAPS is basically the lifeline of Alaska's oil industry as well as the state's economy. The TAPS is not heated but is insulated, and nearly half of the pipeline below ground is the only protection against extreme cold Alaska climate. Initially designed as a "hot oil" line in an Arctic environment, primarily transporting relatively much warmer Prudhoe Bay oil, reached a maximum throughput of 2.1 million barrels per day in 1988 after the start-up on June 20, 1977. Since achieving that peak rate, dwindling oil production on Alaska North Slope and changing characteristics of the oil blend (addition of heavier oils) has basically resulted in a gradual reduction in the throughput, which currently stands at less than 0.5 million barrels per day. So, TAPS is becoming somewhat of a "cold oil" line, which manifests some unique flow assurance challenges not usually encountered elsewhere. Obviously, the low flow through TAPS has a snowball effect in that it means slower transport rate and transit time (oil residing for a much longer duration in the TAPS), being conducive to oil and water separation, additional cooling of oil, and finally cooling leading to potential ice and (additional) wax crystal formation and accumulation. Additionally, the separated water may also result in the creation of a corrosive environment.[20] According to TAPS operator, Alyeska Pipeline Service Company, data analysis with additional investments, the pipeline can be operated with throughputs as low as 0.2 million barrels per day; however, from a long-term standpoint, the economic sustainability may be questionable, and thus, a better solution is producing more oil on the North Slope to enhance the TAPS throughput. This is currently ongoing, albeit at a slower pace, and that is the recovery of heavier viscous oils. Several parallel mitigation strategies are considered, which include slip-stream heat addition, adding corrosion inhibitors in the mainline, and wax (and water) accumulation management via regular runs of scrapper pigs.[20] Burger et al.[21] in their 1981 paper stated that the Sadlerochit crude (also known as Prudhoe Bay oil) could potentially precipitate between 14–20% high-molecular-weight material (paraffinic) at temperatures below 40 and 0°C respectively, with only a small fraction actually depositing. It should be noted that current conditions are far removed from those in 1981, that is, a much lower TAPS temperature gradient as well as changing oil blend characteristics, thus requiring frequent pigging. A collage of TAPS photographs shown in Figure 8.9 gives an idea of the extent of the unique flow assurance challenges in the long-distance oil pipeline in Alaska's Arctic environment.

An interesting wax management reassessment for a brownfield was reported by Lavenson et al.[22] that resulted in the reduction in operating expenses exceeding US\$5 million per year. Basically, via this particular example, they emphasized the continual monitoring of flow assurance strategies, which, in this case, pertains to the original oil itself and paraffinic solids potentially changing character over time. This West African offshore brownfield is operated by Chevron for decades, which means

(a) TAPS in typical summer time, snaking through Alaska   (b) TAPS in winter time   (c) Exhibit of TAPS pig

FIGURE 8.9   Collage of the TAPS photos and the "exhibit" flow through pig. (a) Photographer is Dave Houseknecht, United States Geological Survey (USGS), source/usage is Public Domain; (b) photographer is Tim Dawson, USGS, source/usage is Public Domain; and (c) photo taken by Abhijit Dandekar at the University of Alaska Museum of the North.

conditions significantly altered from "first oil".[22] Two subsea export pipelines, A and B, transport medium API gravity waxy oil to an intermediate platform and then onshore. The authors state that the export pipelines carry commingled production from multiple wells; thus, the blend contains varying amounts of wax and the WAT having a range between 70–85°F. Although no reasoning is given, it is quite likely given the age of the field and the long lengths, the export pipelines are not insulated. Surrounding seawater temperatures below 50°F create a temperature gradient, and thus, the bare lines and long lengths result in favorable conditions for cooling the oil below the WAT well before reaching the intermediate platform. Since this was known right from first oil, the initial or original wax management plan was already in place that entailed the combination of pigging as well as a paraffin inhibitor (crystal modifier – inhibits wax crystal growth and thus reduces deposition rates[23]). The paraffin inhibitor was requalified in 2006, and it is assumed that the use continued until about the fall of 2015 (based on the timeline[22]), following which the new wax management strategy was implemented and the paper published in 2017. Although there was no increase in lost production events due to wax, the amount or dosage of the inhibitor and the pig return characteristics posed important questions or intrigues and subsequent action items such as the following:

- The flow assurance strategy that was deployed at the time of "first oil" may not be valid at the present time, which needs to be reassessed; otherwise, it is business as usual and thus the high cost of inhibitors (keep injecting by default?).
- Over time, the amount of inhibitor required to minimize pig returns slowly increased – thus the motivation to reassess other options.
- The inhibitor used showed no efficacy at dosages at or above field levels, and yet there was no significant loss in production. In that case, **is the inhibitor even needed** because weekly pigging was already providing the needed protection?
- The third point/question was, in a way, addressed by a inhibition comparison of the incumbent and a new inhibitor (aimed at identifying a new chemical to potentially replace the incumbent), as shown in Figure 8.10, which basically shows negligible inhibition effect of the incumbent at 2–3 times higher concentration used in the field (~150–200 ppm). The (% inhibition) performance of the new inhibitor at relatively lower concentrations appears to be comparable with the incumbent; however, at a concentration of 1,500 ppm it does show a superior performance (~70% inhibition).

The aforementioned analysis is what the authors name as "identifying the opportunity", which guided them to conduct a field trial with "no inhibition" and continued use of pigging and low-cost dispersant combination. Note that as the name suggests, a dispersant does not prevent wax from precipitating, but disperses the deposit after it is already formed.[22, 23] After implementing the field trial with no inhibition, no significant changes were observed in average pressure drops in both lines, when monitored over a period of about 9–12 months (range for lines A and B); that is, a steady state is achieved. As stated by the authors in 2017, as of now, the field continues to deploy

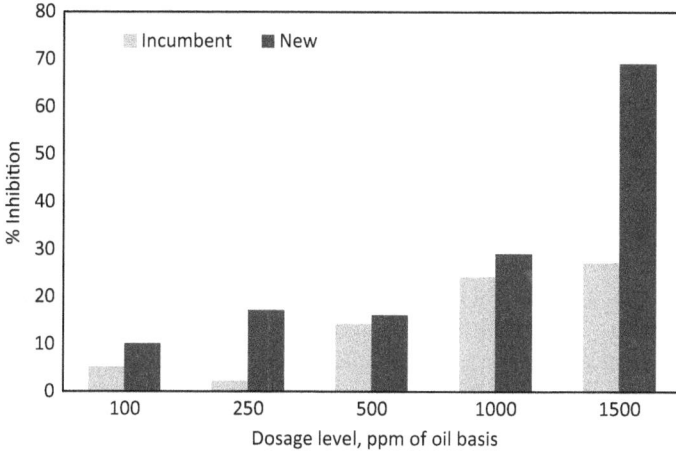

FIGURE 8.10 Cold finger–based inhibition comparison of the incumbent and new inhibitor at various dosage rates for wax management of export pipelines A and B. The plot is constructed based on tabular data reported in Lavenson et al.[22]

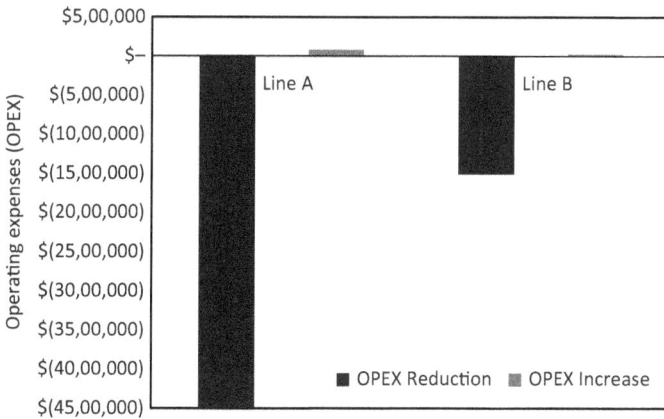

FIGURE 8.11 Reduction in the annual operating expenses realized via a "no inhibitor" wax management strategy in export pipelines A and B. The slight increase in OPEX is attributed to the paraffin dispersant costs, while the pigging cost is not tied to any dollar value and is reported as minimal. The plot is constructed based on tabular data reported in Lavenson et al.[22]

the "no inhibitor" strategy to manage wax risk in these two export pipelines, bringing the total savings of $5.907 million in operating expenses per year (due to high cost of inhibitor that was not even necessary or effective) and minimal cost incurred in frequent pigging and dispersants (see Figure 8.11).

In summary, the aforementioned cases serve as good examples of lessons learned.

8.6 TOPICS RELATED TO FLOW ASSURANCE IN LONG-DISTANCE PIPELINES

Finally, we cover two areas that are quasi-related to solids in long-distance pipelines and thus have obvious significance from a flow assurance standpoint. These are gel strength and the addition of diluents for transporting heavier crude oils. By definition, the gel of crude oil is sometimes referred to as a *soft solid*[3] or semisolid. Similarly, when diluents (relatively lighter hydrocarbons) are used in blending in long-distance pipeline operations, the primary purpose is to reduce the viscosity (of a heavier oil); however, adding a lighter fraction may promote the precipitation of apshaltenes and defeat the dilution purpose. Although drag-reducing agents, or DRAs, are not particularly related to any solids but are employed to improve the pumpability; vis-à-vis ensuring smooth flow assurance, they are included in this section given their significance from the context of long-distance pipelines.

8.6.1 GEL STRENGTH

As far as the temperature hierarchy in a waxy crude oil is concerned, it can be defined as follows: WAT > pour point (typically 5–15°C below WAT) > *gelling point*. In other words, when a waxy crude oil is cooled below the pour point, the network of wax crystals continues to strengthen the interlocking structure[24] and forms a gel, which basically consists of wax crystals in a viscous matrix.[25] This is perhaps the reason why a gel may be defined as a soft solid. In long-distance pipelines, this particular sub–pour point drop in the temperatures can occur in two different circumstances, one is due to a planned or operational reason and the second one may be due to unplanned or emergency shutdown of the pipeline operations. In both cases, the surrounding or ambient temperature plays a crucial role; for example, the TAPS traversing the arctic or potentially bare offshore export lines in deep water, albeit relatively warmer compared to the arctic, but still colder in the range of 40–50°F and thus conducive for gel formation. It should, however, be noted that both the shutdown time and the ambient temperature are two of the main variables that play an important role in the formation of the gel and in imparting particular characteristics to the gel.

In order to resume flow in a gelled pipeline, sufficiently higher pressure is required such that it overcomes the strength of the formed gel. This particular gel strength is alternatively referred to as yield stress, which is defined as the minimum stress necessary to cause a shear flow.[24] As flow in a gelled pipeline partially resumes, shearing causes a gradual breakdown in the gel structure leading to a progressive increase in the flow rate.[25] The resumption of flow in a gelled oil is accomplished by pumping in an incompressible fluid (e.g., un-gelled crude oil) to displace the gelled oil.[25] Although water also is mentioned as a displacing fluid by Davidson et al.,[25] it may not be practical in a case such as the TAPS. However, the degree of stagnation of the crude oil in the pipeline will impart a certain gel strength, and if that happens to be prolonged or if there is a delay in attempting to resume pipeline operations, then there may be a serious HSE risk in that the restart may not be possible owing to the (developed) gel strength being too great for the available applied pressure.[25] Note, however, that this type of restart would be categorized as "cold" because this is carried out at the prevailing temperature. Of course, other alternatives, such as thermal, can theoretically be partially or fully applied to reduce the gel strength and/or melt the gel and revert to liquid conditions.

Based on the foregoing discussion, it should be apparent that operating a long-distance crude oil pipeline, such as TAPS, is susceptible to gel strength issues given the harsh arctic environment. Therefore, in order to accurately assess the "upper limits" of gel strength, that is, in winter, representative oil samples are slowly cooled down from the normal flowing temperature of circa 90°F to a simulated temperature of −20°F over a 21-day period. What this ramp-down means is it would take 21 days for the oil temperature to drop down to −20°F if flow ceases and the oil is stagnant. Note that this is somewhat of a worst-case scenario for winter operations when ambient temperatures can range from −30 to −50°F; however, the insulation on TAPS would prevent the precipitous drop in the oil temperature to such low values. Theoretically, gel strength values could be measured at various different low temperatures, following the 21-day ramp-down, but in the case of TAPS studies, these are carried out at 20°F, 0°F, and −20°F to get a trend.[26, 27] Basically, it is of significance to determine how the gel strength progressively increases as the oil sample is cooled down. Since the standard slow cold ramp procedure is time-consuming and expensive, Timmcke[26] developed an alternative "fast cold ramp" method that uses a constant ramp-down rate of 2°C/h to rapidly determine gel strength values. For details of the experimental procedure for gel strength measurement and its application in pipeline (cold) restart operations, the reader is referred to Chapters 5 and 9, respectively.

Given the high gel strengths encountered in drilling operations, Timmcke[26] used the following equation from drilling applications, as a guide, to estimate the limiting gel strength, along the section of the TAPS between pump stations. For an 80-mile length, 48-in. diameter, and the maximum achievable pressure drop of 1,200 psi between pump stations,

$$\Delta P = \tau_{gel} \frac{L}{300D} \tag{8.8}$$

$$\tau_{gel} = \frac{1200 \times 300 \times 48}{80 \times 5280} = 40.9 \ lbf \, / \, 100 \ ft^2 = 196 \ dyne \, / \, cm^2$$

In Equation 8.8, ΔP is in psi, τ_{gel} is gel strength in lbf/100 ft², L is pipe section length in ft and D is pipe diameter in in. (note that 1 mile = 5,280 ft; 1 lbf/ft² = 478.8 dyne/cm²; and the TAPS ID is not 48 in. but does vary, and hence, a rounded-off value is used in these estimates). The laboratory equipment and gel strength measurements were based on the limiting gel strength values. Gel strengths reported by Timmcke[26] were carried out using the fast cold ramp to speed up the testing of multiple samples. However, note that the measured values are not directly representative of oil in TAPS conditions because the samples are chilled much more rapidly compared to the more representative slow cold ramp. All fast cold ramp gel strength values were compared with those that were measured in a commercial laboratory using the slow cold ramp method from which Timmcke[26] developed a preliminary correlation to predict the latter from the former. As noted,[26] it is a preliminary correlation, which certainly paves the way for a more rigorous, data-rich correlation. The restated or a more generic form of the statistically best (highest R^2 of 0.9098) correlation developed by Timmcke[26] is as follows:

$$\tau_{gel}(\text{SlowColdRamp}) = \tau_{gel}(\text{FastColdRamp}) \times 293.765 \\ \times \gamma^{28.63} \times \exp\left[0.01642(T+40)\right] \tag{8.9}$$

where γ is the oil specific gravity, T is temperature in °F, and the correlation is valid for specific gravities and temperatures between 0.72–0.88, and −30°F and +30°F, respectively.

An example of the application of Equation 8.9 is illustrated in Figure 8.12 which compares the measured fast and predicted slow cold ramp gel strengths, respectively, for three different temperatures for a 25%–75% blend of Alaska Gas to Liquid (AKGTL) syncrude and Alaska North Slope (ANS) crude oil. Note that Timmcke[26] developed the fast and slow cold ramp correlation based on much earlier data; that is, the gel strengths in Figure 8.12 were not part of the regression analysis. As seen in Figure 8.12, obviously, gel strengths increase by at least an order of magnitude with each temperature, and the smallest *measurable* value at 20°F (~1.3 dyne/cm²) can be construed as gelling point based on the earlier definition or characterization.[24] Let us now consider a hypothetical, albeit practical, example (modified from Timmcke[26]) of the use of gel strength in management/restart of a long-distance pipeline such as TAPS transporting a 25–75% blend of AKGTL and ANS crude. The scenario

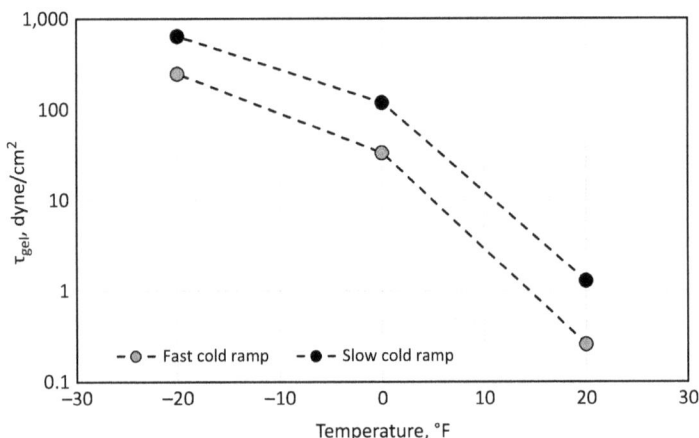

FIGURE 8.12 Measured[27] fast cold ramp gel strengths and their equivalent slow cold ramp values predicted from Equation 8.9. The respective data shown versus temperature are for a blend of 25% AKGTL (Alaska Gas to Liquids Syncrude) and 75% Alaska North Slope crude oil. Note that these values are valid for the crude characteristics at the time (ca. 15–20 years ago) the sample was collected and gel strengths measured, which has changed in the present time.

considered is a seismic event in winter that necessitates the shutdown of TAPS due to damage to the vertical support members, or VSMs. The key question is the time window available to repair the VSMs before the gel strength becomes too high to restart the pipeline. Now, assuming a limiting gel strength of 196 dyne/cm^2 (Equation 8.8) but the operational goal of restarting before the blend achieves a gel strength of 150 dyne/cm^2, slow cold ramp data such as the one in Figure 8.12 and the 21-day cool-down period curve can be combined to predict the desired time window. Based on a linear interpolation between 0°F and –20°F and the corresponding gel strengths, the blend will have a gel strength of 150 dyne/cm^2 and 196 dyne/cm^2 at about –1°F and –3°F, respectively. Finally, as shown in Figure 8.13, the time difference between the limiting and desired (goal) gel strengths amounts to about half a day, which is basically the safety margin in casualty control planning for TAPS restart. Note that these computations are dependent on the prevailing conditions; for example, the result is a 2.5-day safety margin in Timmcke's[26] scenario.

8.6.2 DILUENTS FOR FLOW IMPROVEMENT

Transporting high-density (low API) and -viscosity oils in long-distance pipelines usually requires an a priori reduction in both. This can be accomplished thermally, that is, using heat or by blending with lighter hydrocarbons known as diluents that improve the mobility of heavy oils. The resulting blend density and viscosity is a function of the applied heat, the diluent used, and the fraction in the total mixture. The application of thermal methods for long-distance transport, however, may not be feasible or practical. Typical diluents used can be naphtha, light or high API crudes, and degassed condensates (liquid hydrocarbon fraction produced from either wet or retrograde gas reservoirs). Some important considerations in using the diluents is obviously (1) the availability, (2) the possibility of "recapturing" part of the diluents

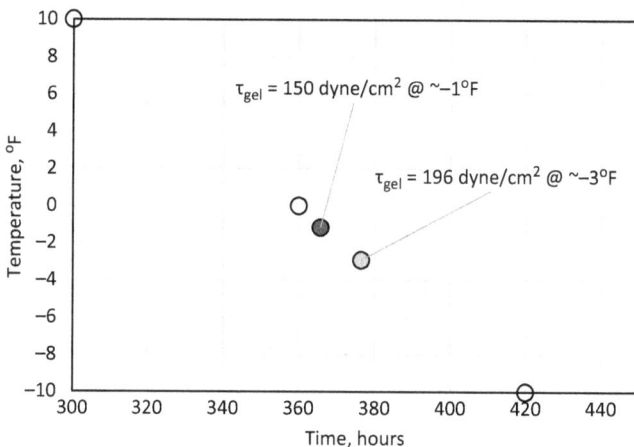

FIGURE 8.13 Zoomed in portion of the 21-day (expressed in hours) cooldown curve showing the limiting 196 dyne/cm^2 and operational (goal) 150 dyne/cm^2 gel strengths for the TAPS restart scenario.

given its premium quality or grade, and, most important, (3) the chemical compatibility of the heavy oil and the diluent from the standpoint of asphaltene precipitation and deposition in the pipeline and/or settling in the receiving infrastructure, that is, storage tanks and oil tankers.

A classic case of dilution for flow improvement and its sustainability is TAPS, which was originally designed and built for transporting 28° API Prudhoe Bay–type crude oil from ANS to the marine terminal in Valdez. Combined oil production from ANS steadily declined since the peak output of 2.1 million barrels per day in 1988. The dwindling ANS oil production is, however, exerting an increased burden on the economic operation of TAPS, raising an important question related to its future sustainability. The proven and recoverable reserves of conventional natural gas in the developed and undeveloped fields on ANS are estimated to be as much as 38 trillion standard cubic feet.[27] Additionally, estimates of unconventional gas resources in the form of hydrates and coalbed methane are also significant. This, in fact, creates an interesting scenario in that these vast gas resources are stranded because the domestic gas market is far away from the ANS. Similarly, the in-place resources of heavy oil on ANS are also huge, ranging between 20—25 billion barrels. However, heavy oil resources on ANS have received little attention, mainly because these oils are viscous and are difficult to transport. A feasible solution, in fact, lies in the combination of the foregoing problems and scenarios. The vast natural gas resources on the ANS can be chemically converted to produce GTL products on ANS which will provide additional liquid capacity to fill up TAPS. Additionally, ANS' heavy oil resources also have the potential to become future TAPS throughput. In this manner all three issues can be addressed, that is, increase the economic life of TAPS and monetize the huge gas resources on the ANS by bringing it to market in the form of GTL and bringing Alaska's heavy oil to market. Although, transportation of commingled or blended GTL products, conventional and unconventional (heavy) oils is theoretically feasible; it is this diverse throughput and Alaska's arctic environment that poses several operational challenges, such as pumpability, gel strength (cold restart problem), effect on density and viscosity, and solid organic deposition.

Chukwu et al.[27] extensively experimented and documented almost all aspects of the previously mentioned TAPS blend transportation scenarios using different diluents. However, Igbokwe[28] in particular measured the properties of ternary blends of AKGTL, typical (lighter) ANS oil, and ANS heavy oil. Given the lack of field ANS heavy oil sample, light ends from the conventional ANS oil were distilled off to produce a synthetic heavy oil (matching West Sak API), and three different ternary blends were tested, which basically tracked the gradually dwindling TAPS conventional oil throughput scenario, that is, 70%, 50%, and 30%, respectively. Figure 8.14 shows a composite plot of density and viscosity at 25°C or 77°F for the tested blends and the three different liquid streams, that is, the ANS heavy oil, typical ANS TAPS mix, and AKGTL. As seen in Figure 8.14, the density of ANS TAPS mix oil lies between the heavy oil and the AKGTL, and the case is the same with viscosity, albeit the ANS oil and AKGTL viscosities are a bit closer compared to the heavy oil, and reduction is in orders of magnitude. A complete data set of both density and viscosity as a function of various temperatures is included in Igbokwe.[28] However, as an example, if we focus on the point value of viscosity at 25°C; clearly, such high viscosities

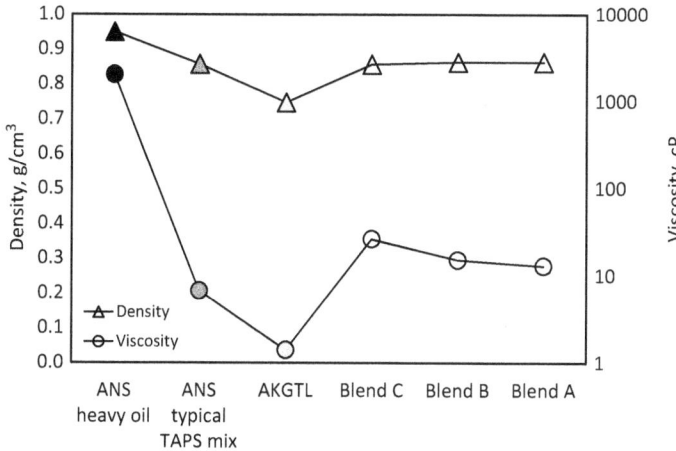

FIGURE 8.14 Density and viscosity of three individual streams and their ternary blends at 25°C or 77°F. Blend % in terms of ANS heavy oil, AKGTL, and ANS typical TAPS mix are as follows. **A**: 10–20–70, **B**: 15–35–50, and **C**: 25–45–30. Plot constructed based on data in Igbokwe.[28]

(~2,000 cP) would pose significant operational problems for TAPS, if only ANS heavy oil were to be the only throughput, since TAPS was designed to transport a relatively lighter and low viscosity oil. Now if we consider the viscosity of diverse blends then despite the fact that the ANS heavy oil controls the overall blend viscosity, the resultant blend viscosity is in the vicinity of the typical TAPS mix. Therefore, these results indicate the necessity to dilute the ANS heavy oil stream (with AKGTL) before it can be successfully transported through TAPS.

Although the use of something relatively as light as AKGTL is feasible from the standpoint of achieving TAPS fluid mechanics specifications, the chemical compatibility still needs to be evaluated. There are two analytical tests that can be typically conducted in the laboratory to evaluate the chemical compatibility of a diluent such as AKGTL with a heavy oil, especially when the blend(s) have a long residence time as is the case in TAPS. These tests are, namely, (1) total asphaltenes using a method such as IP143 and (2) flocculation onset titration stability. The first test was conducted using AKGTL, which resulted in 8.16 g of asphaltene/100 cc of ANS heavy oil (more than twice the amount precipitated by a precipitant such as n-heptane). The flocculation onset titration was, however, conducted with n-heptane (commonly used) that resulted in a small but positive value of the intercept which means that asphaltenes will be present in the ANS heavy oil as stable solutions. The specifics of these tests can be found in Igbokwe.[28] In summary, the AKGTL option provides an enthusiastic argument in favor of ANS heavy oil resource development, and AKGTL, existing ANS mix oil, and ANS heavy oil have the potential for commingled flow through TAPS. However, fluid mechanics as well as chemical compatibility, as shown here, for any diluent should be considered or evaluated prior to deployment in long-distance pipeline transportation.

8.6.3 DRAs

Crude oil or any liquid hydrocarbon products flowing through a pipeline will experience pressure drop in the flow direction owing to frictional energy losses. As expressed by the Weisbach or Darcy–Weisbach frictional pressure drop equation,[29] the drop in pressure is directly proportional to the flow rate (velocity). DRAs are polymeric additives that are injected into long-distance pipelines to reduce the frictional resistance, vis-à-vis pressure drop, when the flow is in the turbulent regime. Therefore, DRAs are also sometimes known as friction reducers.[30] Other definitions of DRAs include the increase in "pumpability" of a fluid caused by a small amount of additive in the fluid.[31] Although the application of DRAs in crude oil pipelines is not routine, and the largest and more common use of oil-soluble DRAs is in the pipeline transportation of refined oils;[30] the most successful use of DRA is in the case of TAPS, which carries a blend of various oils produced on the North Slope of Alaska. In fact, the first commercial injection of DRA into TAPS at pump station 1 (PS1) was on July 1, 1979.[32]

Since drag reduction occurs only in turbulent flow[33] and given the fact that for the most part liquid flows through the petroleum pipeline in a turbulent regime, DRAs can perform very well in most pipelines.[30] For example, in the case of TAPS, with a flow rate of approximately 500,000 bbl/day, average oil gravity of 25°API, average viscosity of 10 cP, and diameter of 48 in. (used as inside diameter as an approximation), the following Reynolds number will be in the fully turbulent region:

$$\text{TAPS Re} = \frac{DU\rho}{\mu} = \frac{48\times0.0254\,(\text{m})\times0.788\,(\text{m}/\text{sec})\times904.1\left(\text{kg}/\text{m}^3\right)}{10\times10^{-3}\left(\text{Nsecm}^2\right)} \approx 86{,}877$$

where $\gamma = 141.5/(25 + 131.5) = 0.9041$ or $\rho = 0.9041$ g/cm³ = 904.1 kg/m³ and $U = [(500{,}000 \times 5.615)/(24 \times 60 \times 60)]/[(\pi/4) \times 4^2] = 2.586$ ft/sec or 0.788 m/sec.

As mentioned earlier, the first commercial application of a DRA to increase the throughput in TAPS began during 1979; however, the DRA experiments conducted in TAPS that were reported by Burger et al.[34] in 1982 ultimately led to the decision for full-scale additive use. The polymeric DRA used in the TAPS experiments was a Conoco Drag Reducer (CDR), which is a solution of a high molecular weight polymer (10.3 wt%) in a hydrocarbon solvent. The efficiency of a DRA is mathematically expressed by the following equations defined by Lescarboura et al.[33] in 1971.

Theoretically, at a given flowrate,

$$\%\text{DR} = \frac{\Delta P - \Delta P_{\text{DRA}}}{\Delta P} \times 100 \tag{8.10}$$

$$\%\text{IT} = \left[\left(\frac{1}{1-0.01\times\%\text{DR}}\right)^{0.55} - 1\right] \times 100 \tag{8.11}$$

where, %DR is percentage drag reduction, ΔP is the base or without-DRA pressure drop in the line, ΔP_{DRA} is the pressure drop in the line using a DRA, and %TI is

Pump Stations	$T_{avg.}$, deg F	5ppm DRA %DR	10ppm DRA %DR	15ppm DRA %DR	20ppm DRA %DR
PS1-3	125.0	12.5	20.0	26.0	27.5
PS6-7	93.5	10.0	15.0	17.5	20.0
PS7-8	89.5	2.5	10.0	13.0	16.3

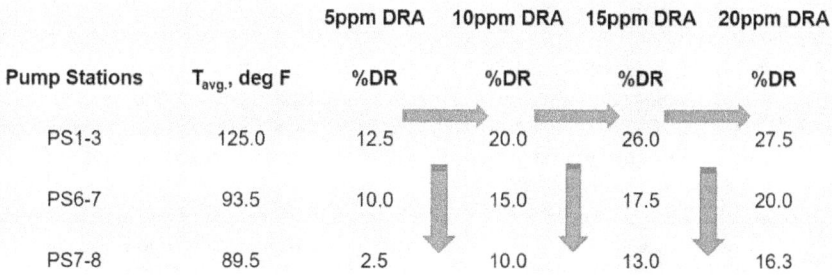

FIGURE 8.15 Effect of temperature and dosage of DRA in the TAPS field experiments. The table and the interpretation compiled/prepared based on the data in Burger et al.[34]

percentage throughput increase. Equation 8.10 is consistent in that if ΔP_{DRA} is the same as ΔP, that is, a DRA not used or not effective, then the %DR is 0, and purely theoretically if ΔP_{DRA} is 0 then %DR is 100. Consequently, if %DR is 0, then %TI also is 0, and again, hypothetically, if %DR is 100, then %TI is infinity. From the standpoint of preceding equations, it should be noted that DRAs are used to increase the pipeline throughput as much as possible without exceeding the safe pressure limit for the pipeline.[33] In other words, any increase in the throughput due to DRAs is a function of the pipeline–pump system. Figure 8.15, which actually is a table, but presented as a figure given the directional arrows, summarizes the results of the field experiments of Burger et al.[34] As seen in this interpretation, the %DR is directly proportional to the DRA ppm dosage, but for the same concentration, it reduces as temperature goes down (oil viscosity increases). Since TAPS is not heated (only insulated), the oil does cool from inlet to outlet. These data and the interpretation are based on actual field experiments conducted by Burger et al.[34] following lab flow loops tests. Note that the ΔP values before DRA treatment were calculated[34] while the post treatment values were measured. With an average %DR of 20 from the data in Figure 8.15, the %TI is a little more than 13%, which is basically what was observed.

REFERENCES

1. Stalkup, F. I. (1983). *Miscible Displacement*. Society of Petroleum Engineers, New York.
2. Dandekar, A., Andersen, S. I. A., & Stenby, E. (1999). Solid Organic Deposition During Gas Injection Studies – Part I. *AIChE's 1999 Spring National Meeting*.
3. Jon Steinar Gudmundsson. (2018). *Flow Assurance Solids in Oil and Gas Production*. CRC Press/Balkema.
4. Dandekar, A. Y. (2013). *Petroleum reservoir rock and fluid properties*. CRC Press, Boca Raton.
5. Ahmed, T. (2010). *Reservoir Engineering Handbook*. Gulf Professional Publishing, Cambridge, MA.
6. Dake, L. P. (1983). *Fundamentals of Reservoir Engineering*, Vol. 8 (1st Edition). Elsevier Science Ltd., Amsterdam

7. Leontaritis, K. J. (1998, February 18). Asphaltene Near-wellbore Formation Damage Modeling. *SPE Formation Damage Control Conference.* https://doi.org/10.2118/39446-MS

8. Abul K. M. & Jamaluddin, T. W. N. (1995). *Process for removing and preventing near wellbore damage due to asphaltene precipitation.* United States Patent (19), 5425422.

9. Kabir, C. S., & Jamaluddin, A. K. M. (2002). Asphaltene Characterization and Mitigation in South Kuwait's Marrat Reservoir. *SPE Production & Facilities,* 17(4), 251–258. https://doi.org/10.2118/80285-PA

10. Ortiz, R., Perez, C., Sánchez, O., Aybar, U., Tellez, F., Mujica, L., Aguilar, J., Andrade, A., Resendiz, T., Camarillo, L., & Thompson, M. (2017). Asphaltene-Prevention Work Flow Enhances Oil Production in High-Temperature Fractured Carbonate Reservoirs. *SPE Production & Operations,* 32(4), 476–490. https://doi.org/10.2118/178956-PA

11. Sutton, G. D., & Roberts, L. D. (1974). Paraffin Precipitation During Fracture Stimulation. *Journal of Petroleum Technology,* 26, 997–1004. https://doi.org/10.2118/4411-PA

12. Adialalis, S. (1982). *Investigation of Physical and Chemical Criteria as Related to the Prevention of Asphalt Deposition in Oil Well Tubings.* MSc Thesis, Imperial College of the University of London.

13. Leontaritis, K. J., & Ali Mansoori, G. (1988). Asphaltene deposition: a survey of field experiences and research approaches. *Journal of Petroleum Science and Engineering,* 1(3), 229–239. https://doi.org/10.1016/0920-4105(88)90013-7

14. Haskett, C. E., & Tartera, M. (1965). A Practical Solution to the Problem of Asphaltene Deposits-Hassi Messaoud Field, Algeria. *Journal of Petroleum Technology,* 17(4), 387–391. https://doi.org/10.2118/994-PA

15. Tuttle, R. N. (1983). High-Pour-Point and Asphaltic Crude Oils and Condensates. *Journal of Petroleum Technology,* 35(6), 1192–1196. https://doi.org/10.2118/10004-PA

16. Thawer, R., Nicoll, D. C. A., & Dick, G. (1990). Asphaltene Deposition in Production Facilities. *SPE Production Engineering,* 5(4), 475–480. https://doi.org/10.2118/18473-PA

17. Al-Kafeef, S. F., Al-Medhadi, F., & Al-Shammari, A. D. (2005). A simplified method to predict and prevent asphaltene deposition in oilwell tubings: Field case. *SPE Production & Facilities,* 20(2), 126–132. https://doi.org/10.2118/84609-PA

18. Kargarpour, M. A., & Dandekar, A. (2016). Analysis of asphaltene deposition in Marrat oil well string: a new approach. *Journal of Petroleum Exploration and Production Technology,* 6(4), 845–856. https://doi.org/10.1007/s13202-015-0221-7

19. Zettlitzer, M. (2000, October 24). Successful field application of chemical flow improvers in pipeline transportation of highly paraffinic crude oil in Kazakhstan. *SPE European Petroleum Conference.* https://doi.org/10.2118/65168-MS

20. Unknown author, Alyeska Pipeline Service Co. Trans Alaska Pipeline System Flow Assurance Overview, April 2021, downloaded June 30, 2021. https://www.alyeska-pipe.com/wp-content/uploads/2021/04/PP-Flow-Assurance-4-21-B.pdf

21. Burger, E. D., Perkins, T. K., & Striegler, J. H. (1981). Studies of Wax Deposition in the Trans Alaska Pipeline. *Journal of Petroleum Technology,* 33(6), 1075–1086. https://doi.org/10.2118/8788-PA

22. Lavenson, D. M., Venkatesan, R., Young, M. K., & Kebert, N. (2017, May 1). The Importance of Periodic Reassessment of Operational Strategies for Wax Management. *Offshore Technology Conference.* https://doi.org/10.4043/27532-MS

23. Wayne Frenier, Murtaza Ziauddin, R. V. (2010). *Organic Deposits in Oil and Gas Production.* Society of Petroleum Engineers.

24. Tarek Ahmed. (2016). *Equations of State and PVT Analysis.* Elsevier, Amsterdam. https://doi.org/10.1016/C2014-0-00119-0

25. Davidson, M. R., Dzuy Nguyen, Q., Chang, C., & Rønningsen, H. P. (2004). A model for restart of a pipeline with compressible gelled waxy crude oil. *Journal of Non-Newtonian Fluid Mechanics*, 123(2), 269–280. https://doi.org/10.1016/j.jnnfm.2004.09.007

26. Timmcke, M. D. (2002). *Rapid evaluation of the gel strength of GTL products during a prolonged trans-Alaska pipeline shutdown* [University of Alaska Fairbanks]. http://hdl.handle.net/11122/6355

27. Godwin A. Chukwu, Santanu Khataniar, Shirish Patil, A. D. (2007). *Operational Challenges in Gas-To-Liquid (GTL) Transportation Through Trans Alaska Pipeline System (TAPS)*. US DOE Report.

28. Igbokwe, C. G. C. (2006). *Experimental investigation on the transportation of commingled blends of gas-to-liquid (GTL) products and Alaskan heavy crude oil through the Trans-Alaska Pipeline System (TAPS)*. http://hdl.handle.net/11122/5843

29. White, F. (2009). *Fluid Mechanics* (7th Edition). McGraw Hill, New York, NY.

30. Kelland, Malcolm A. (2014). *Production Chemicals for the Oil and Gas Industry*. CRC Press, Boca Raton.

31. Savins, J. G. (1964). Drag reduction characteristics of solutions of macromolecules in turbulent pipe flow. *Society of Petroleum Engineers Journal*, 4(3), 203–214. https://doi.org/10.2118/867-PA

32. http://alyeska.wpengine.com/wp-content/uploads/2021/03/FactBookMarch2021.pdf. Downloaded, July 13, 2021.

33. Lescarboura, J. A., Culter, J. D., & Wahl, H. A. (1971). Drag Reduction with a Polymeric Additive in Crude Oil Pipelines. *Society of Petroleum Engineers Journal*, 11(3), 229–235. https://doi.org/10.2118/3087-PA

34. Burger, E. D., Munk, W. R., & Wahl, H. A. (1982). Flow increase in the trans alaska pipeline through use of a polymeric drag-reducing additive. *Journal of Petroleum Technology*, 34(2), 377–386. https://doi.org/10.2118/9419-PA

9 Fluid Mechanics in Flow Assurance

9.1 INTRODUCTION

A fluid is something that can simply be defined as matter that flows from point A to B. Since the solid phase does not flow or move by itself, *fluid* obviously refers to either vapor or liquid, perhaps appropriately called as less and more dense phases. Consequently, these terms are used somewhat interchangeably. *Fluid mechanics* is the study of fluids that are either at rest (statics) or in motion (dynamics).[1] This basic definition of fluid mechanics clearly allows us to recognize the relevance in various flow assurance scenarios. The foremost objective of fluid mechanics in flow assurance is to determine flowline pressure gradient, which basically allows us to find out the pressure at a certain point (distance) in the flow infrastructure. Of course, for non-isothermal systems (common in flow assurance) what needs to be done in parallel is thermal modeling that basically results in the flowline temperature gradient. The pressure and temperature gradient calculations together are sometimes referred to as "thermo-hydraulic" modeling. The calculated pressure and temperature, in turn, can then be compared with the phase behavior of a given fluid stream, i.e., hydrate forming conditions, wax appearance, and asphaltene onset. An appropriate operational action may then be taken such as either increasing or decreasing pressure and temperature to stay away from problematic fluid conditions.

Fluid flow in production infrastructure manifests in different forms and complexities ranging from simple single phase to three phases – gas, oil, and water – and, in some cases, four phases (3 fluids + solid). However, notwithstanding, the three main elements that dictate the pressure gradient are (1) friction, (2) elevation change (+ or −), and (3) fluid acceleration (+ or −). Of course, in static cases both friction and fluid acceleration are zero since the fluid is at rest and not flowing. For example, in a stagnant fluid column in a vertical well or a riser, the only element contributing to the pressure gradient will be elevation change (see the example in Section 9.3). The application of pressure gradient calculation for a flowing single phase and multiphase are covered in Sections 9.4 and 9.7, respectively.

9.2 MECHANICAL ENERGY BALANCE EQUATION

The mechanical energy balance equation relates the pressure gradient to the three elements alternatively stated (1) frictional (friction), (2) hydrostatic (elevation), and (3) kinetic (fluid acceleration). In this discussion of the mechanical energy balance equation, we assume single-phase flow. The pressure form of the mechanical energy balance equation can be mathematically stated as[2]

$$\left(\frac{dP}{dL}\right)_T = \left(\frac{dP}{dL}\right)_F + \left(\frac{dP}{dL}\right)_H + \left(\frac{dP}{dL}\right)_A \qquad (9.1)$$

where $\left(\frac{dP}{dL}\right)_T$ is the total pressure drop over differential length dL and $\left(\frac{dP}{dL}\right)$ with subscripts F, H, and A are frictional, hydrostatic, and acceleration gradients, respectively. The hydrostatic gradient is the simplest to determine because it is dependent only on the fluid density and the deviation angle. However, the density is a variable, along the wellbore or the tubing length given its dependence on pressure, temperature, and the gas deviation factor in the case of gas wells (see Section 9.3). Of course, in the case of horizontal flow, $\left(\frac{dP}{dL}\right)_H$ is automatically zero. The $\left(\frac{dP}{dL}\right)_A$ acceleration term results in a pressure drop in the direction of velocity increase when velocity changes occur (e.g., changing cross-sectional areas). Therefore, it is zero for a constant cross-sectional area and negligible for incompressible fluids (high-pressure gases and liquids).[2, 3] Guo et al.[3] state that $\left(\frac{dP}{dL}\right)_A$ is normally negligible for both gas and oil pipelines.

The frictional pressure gradient, $\left(\frac{dP}{dL}\right)_F$ is, however, much more computationally involved, given the dependence on the flow regime, that is, laminar, transient, and turbulent, which is based on a distinguishing, dimensionless Reynolds number (Re),

$$Re = \frac{DU\rho}{\mu} \qquad (9.2)$$

where D is the pipeline inside diameter (ID), U is the fluid velocity, ρ and μ are the fluid density and viscosity, respectively. Any consistent set of units can be used to render Re dimensionless; for example, if D is in m, U is in m/sec, ρ is in kg/m^3, and μ is in Nsec/m^2, then in $Re = \dfrac{m \times m / \sec \times kg / m^3}{N \sec / m^2}$, all units cancel out (note that kgm/sec^2 = N).

The frictional pressure gradient equation as originally proposed by Weisbach in 1850 is[1]

$$\left(\frac{dP}{dL}\right)_F = \frac{f_D \rho U^2}{2D} \qquad (9.3)$$

where f_D is the dimensionless parameter called the Darcy friction factor (Darcy also is credited with the well-known Darcy's law and the unit of Darcy for permeability adopted by the petroleum industry in his honor). Sometimes f_D also is known collectively as the Darcy–Weisbach friction factor. Later, Fanning proposed the so-called Fanning friction factor, which is one fourth of the Darcy–Weisbach friction factor ($f_D = 4f_F$). Equation 9.3 can be written accordingly as shown and is the form used in this book:

$$\left(\frac{dP}{dL}\right)_F = \frac{2f_F\rho U^2}{D} \tag{9.4}$$

with fluid density ρ in kg/m³, velocity U in m/sec, and pipe ID in m; $\left(\frac{dP}{dL}\right)_F$ will have units of (N/m²)/m (kg/m³ × m²/sec² × 1/m; kgm/sec² = N/m²/m), which can be easily converted to oilfield units of psi/ft (N/m²/m = (14.7/101.325 × 10³)/3.2808 = 4.4219 × 10⁻⁵ psi/ft).

The friction factor is related to or dependent on the Reynolds number. Following equations are commonly used in estimating the friction factor. For laminar region (Re ≤ 2000), $f_F = 16/\text{Re}$. When Re is bracketed between 2,000–4,000, the flow regime is transient or unstable for which the friction factor is estimated by the Colebrook equation:[4]

$$\frac{1}{\sqrt{f_F}} = -4\log_{10}\left[\frac{1.255}{\text{Re}\sqrt{f_F}} + \frac{\varepsilon/D}{3.7}\right] \tag{9.5}$$

Equation 9.5 cannot be solved explicitly for f_F but it can be iterated by shifting the right-hand side to the left and equating it to zero. The parameter ε in Equation 9.5 is called the absolute pipe roughness, and the ratio ε/D is known as the relative roughness. Moody[5] presented a chart of ε/D versus the pipe diameter for various materials that can be used to obtain the relative roughness. Note that the absolute roughness of a given pipe material is constant, but the relative roughness is inversely proportional to the pipe diameter. However, a manufacturer-provided ε value is always preferred. In the absence of both, ε of 0.0005–0.0006 in. can be used as a reasonable substitute in flow assurance calculations.

In a fully turbulent region (Re >> 4000), the Chen[6] equation that allows explicit calculation of f_F is recommended:

$$f_F = \frac{1}{4}\left\{\frac{1}{-2\text{LOG}_{10}\left[\dfrac{\varepsilon/D}{3.7065} - \dfrac{5.0452}{\text{Re}}\text{LOG}_{10}\left(\dfrac{(\varepsilon/D)^{1.1098}}{2.8257} + \left(\dfrac{7.149}{\text{Re}}\right)^{0.8981}\right)\right]}\right\}^2 \tag{9.6}$$

9.3 SHUT-IN WELLHEAD PRESSURE CALCULATIONS AND RELATIONSHIP WITH FLOW ASSURANCE

In principle, the mechanical energy balance equation derived earlier can be easily applied to determine the shut-in wellhead pressure in the case of a stagnant column of gas or oil in the production tubing. The stagnant column of fluid may be a result of planned or unplanned production shutdown. The determined shut-in wellhead pressure at the corresponding temperature provides an important data point from a flow assurance perspective. For example, are these conditions conducive to the formation of hydrates (gas well) or wax (oil well)? As an example, this particular scenario is depicted in Figure 9.1, which shows a stagnant gas column in a deep vertical offshore well. The same can be used to depict an oil well. What is described in Figure 9.1 is somewhat self-explanatory in that the pressure in the tubing will decrease (cooling down), with well shut-in time and decreasing depth with the lowest value at the wellhead, owing to the low mudline temperature. This reduction in temperature also will likely result in the formation of free water due to condensation. Taken as a whole the system poses a situation in which hydrates may form and plug the tubing.

In the case of a static or stagnant gas column, as shown in Figure 9.1, the kinetic energy and friction terms are zero since there is no flow and can be eliminated from the mechanical energy balance equation. Therefore, Equation 9.1 simplifies to a hydrostatic head of gas, $P = \rho_G g X$, where P is the pressure at a certain point corresponding or due to vertical distance X, ρ_G is gas density, and g is acceleration due to gravity. With units of density in kg/m^3, gravitational acceleration in m/sec^2, and X in m, P will be in N/m^2 ($kg/m^3 \times m/sec^2 \times m$; $kg \times m/sec^2 = N$). Note that in the case of a slanted well, the deviation angle (θ) also is included such that $P = \rho_G g \cos(\theta) X$; for a vertical well $\theta = 0°$; that is, $\cos(0) = 1$. A very rudimentary calculation of the shut-in wellhead pressure is as follows. If the bottomhole pressure at a depth of

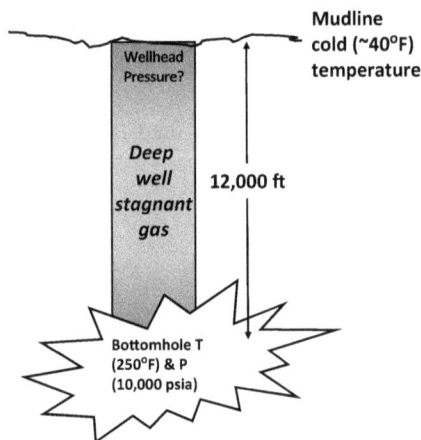

FIGURE 9.1 Stagnant 0.7 gravity gas in a deep vertical offshore well.

12,000 ft (3657.6 m) is recorded as 10,000 psia (68,928.57 kN/m^2), ρ_G is 333 kg/m^3 and g is 9.81 m/sec^2 then the shut-in wellhead pressure, $P_{SIWH} = 68,928.57 \times 10^3 - 333 \times 9.81 \times 3657.6 = 56,980.17$ kN/m^2 or approximately 8268 psia. For field units, the same result can be obtained by using the customary gradient of 0.43353 psi/ft and ρ_G in g/cm^3; $P_{SIWH} = 10,000 - 0.43353 \times 12,000 \times 0.333 = {\sim}8268$ psia.

It should, however, be noted that a more representative calculation of hydrostatic head gets somewhat complicated due to the large ΔT between the bottomhole and wellhead, pressure changes along the wellbore, which alters the gas deviation or compressibility factor, Z, which, in turn, affects the gas density ($\rho_G = PMW_G/ZRT$, where P and T are pressure and temperature, MW_G is the gas molecular weight, and R is the universal gas constant). These uncertainties may result in errors in the calculated P_{SIWH}. One practical approach suggested by Notz[7] is segmentation, that is, to divide the wellbore into small segments of 500 or 1,000 ft. Establish T versus depth, assuming linear relationship and starting with the bottom of the well, and get the density of the gas (Z factor–based) for that segment to calculate the incremental change in pressure going up the wellbore (such as the calculation shown earlier); and eventually ending at the mudline. This methodology helps minimize the aforementioned errors/uncertainties. The method suggested by Notz[7] is illustrated for the scenario shown in Figure 9.1 for a 0.7 gravity gas. As a first step, temperatures at various depths are estimated from the linear relationship plotted in Figure 9.2. The calculation sequence then starts with the Z factor and the gas density at the known bottomhole pressure (10,000 psia) and temperature (250°F). This density is then used in calculating the pressure at the next segment, that is, 11,000 ft since the spacing used is 1,000 ft. For the 0.7 gravity gas ($MW_G = 0.7 \times 28.97 = 20.28$), the Z factor is 1.41, which results in $\rho_G = 10,000 \times 20.28/1.41 \times 10.732 \times (250 + 460) = 18.87$ lb/ft^3 or

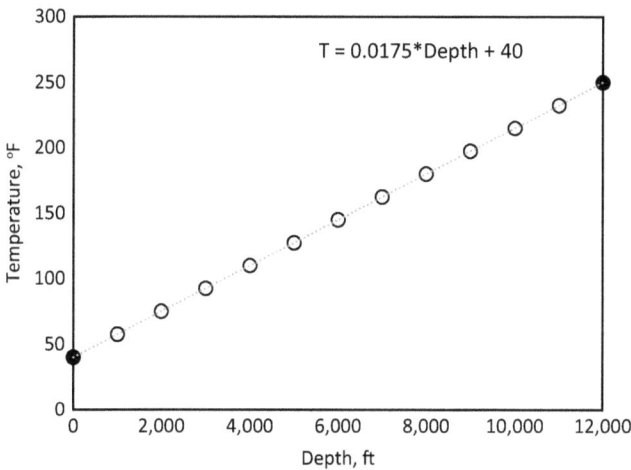

FIGURE 9.2 Depth vs. temperature linear relationship for use in P_{SIWH} calculation example. The two solid circles are mudline and bottomhole temperatures, respectively.

γ_g	T_{pc} (deg R)	P_{pc} (psia)	T, deg F	P, psia	T_{pr}	P_{pr}	Calculated Z	Z FUNCTION = 0
0.7000	377.5900	663.3360	250.0000	10000	1.879472	15.07532	1.41	1.3427E-06

FIGURE 9.3 Screenshot of the Z-factor calculation using the Dranchuk–Abou–Kassem method.[8]

0.3023 g/cm³. The Z factor and other values are calculated by the Dranchuk–Abou–Kassem method[8] (note that the required pseudo-critical pressure and temperature are estimated from the gas gravity–based based empirical correlations[9]). Figure 9.3 shows the screenshot of this Z factor calculation. Using the gas density of 0.3023 g/cm³, pressure at 11,000 ft = 10,000 – 0.43353 × 1,000 × 0.3023 = 9,869 psia. The calculated 9,869 psia and 232.5°F is used to obtain the subsequent Z factor and the gas density, and the pressure at 10,000 ft. This is the calculation sequence followed until the wellhead. The overall results are presented below in Table 9.1, which shows a shut-in wellhead pressure of 8283 psia. Although the difference in the calculated shut-in wellhead pressure using the segmentation method and an average density appears to be small, it may have important consequences from the standpoint of potential hydrate formation if valid or appropriate operating conditions so warrant.

Besides the segmentation method there are other techniques that also can be employed to determine the shut-in wellhead pressure, namely, the average temperature and Z factor[10] and a more rigorous Poettmann[11] method, respectively. Note, however, that these techniques are usually employed to calculate the shut-in bottomhole pressure; however, the equations can be easily rearranged or reversed and expressed in terms of the shut-in wellhead pressure. Both the methods are based on the same principle of the mechanical energy balance equation, and similar to the segmentation method focus on the determination of the pressure- and temperature-dependent gas Z

TABLE 9.1

Shut-In Wellhead Pressure Calculation Using the Segmentation Method of Notz[7]

Depth, ft	Temperature, °F	Z factor	ρ_G, g/cm³	Pressure, psia
0	40.0	1.363	0.3679	8,283
1,000	57.5	1.366	0.3613	8,440
2,000	75.0	1.370	0.3549	8,593
3,000	92.5	1.374	0.3487	8,745
4,000	110.0	1.377	0.3428	8,893
5,000	127.5	1.381	0.3371	9,039
6,000	145.0	1.385	0.3316	9,183
7,000	162.5	1.390	0.3263	9,325
8,000	180.0	1.394	0.3211	9,464
9,000	197.5	1.398	0.3162	9,601
10,000	215.0	1.402	0.3114	9,736
11,000	232.5	1.406	0.3068	9,869
12,000	250.0	1.410	0.3023	10,000

factor and, consequently, the density. The rearranged equations for the average temperature and Z factor method are

$$P_{\text{SIWH}} = \frac{P_{\text{SIBH}}}{\text{EXP}(S/2)}; S = \frac{0.0375\gamma_G L \cos(\theta)}{\bar{Z}\bar{T}} \tag{9.7}$$

where P_{SIBH} is the shut-in bottomhole pressure, γ_G is gas gravity, L is the wellbore length in ft, and $\bar{Z}$ and $\bar{T}$ are the average Z factor and temperature (°R), respectively. If the well is vertical, then θ is 0°. Since P_{SIWH} is obviously unknown, calculations involve an iterative process, starting with an initial guess of

$$P_{\text{SIWH}} = \frac{P_{\text{SIBH}}}{\left[1 + 0.000025L \cos(\theta)\right]} \tag{9.8}$$

For the aforementioned solved example, the initial guess of $P_{\text{SIWH}} = 10,000/(1 + 0.000025 \times 12,000) = 7692$ psia or an arithmetic average pressure of $(10,000 + 7692)/2 = 8846$ psia. The arithmetic average temperature, $\bar{T} = (250 + 40)/2 = 145°F$ or 605°R. Using the Dranchuk–Abou–Kassem method[8] the average Z factor, $\bar{Z} = 1.351$. So all the calculated values can now be used in obtaining the first iteration value, $S = (0.0375 \times 0.7 \times 12,000)/(1.351 \times 605) = 0.3855$, and $P_{\text{SIWH}} = 10,000/\exp(0.3855/2) = 8,247$ psia. The first iteration P_{SIWH} is then used to obtain the new arithmetic average pressure and remaining calculations repeated to check the difference between the iterated P_{SIWH} and the starting value, which, in this case, is 8280 psia; that is, there is a difference of 33 psi. Therefore, the calculations proceed to the next iteration until convergence is achieved (fourth iteration, final $P_{\text{SIWH}} = 8,282$ psia).

For deeper gas wells, Lee and Wattenbarger[12] recommend alternative methods that are more rigorously correct, such as the method developed by Poettmann.[11] The computation procedure is outlined by the following equations modified from Lee and Wattenbarger:[12]

$$\int_{P_{P_r},\text{SIBH}}^{P_{P_r},\text{SIWH}} \frac{Z dP_{P_r}}{P_{P_r}} = -\frac{0.01875\gamma_G L \cos(\theta)}{\bar{T}} \tag{9.9}$$

The preceding integral is broken down into parts choosing an arbitrary limit of 0.2 to represent the lowest values of pressure (because any given pressure = 0.2 × critical pressure):

$$\int_{P_{P_r},\text{SIBH}}^{0.2} \frac{Z dP_{P_r}}{P_{P_r}} + \int_{0.2}^{P_{P_r},\text{SIWH}} \frac{Z dP_{P_r}}{P_{P_r}} = -\frac{0.01875\gamma_G L \cos(\theta)}{\bar{T}} \tag{9.10}$$

$$\int_{0.2}^{P_{P_r},\text{SIWH}} \frac{Z dP_{P_r}}{P_{P_r}} = \int_{0.2}^{P_{P_r},\text{SIBH}} \frac{Z dP_{P_r}}{P_{P_r}} - \frac{0.01875\gamma_G L \cos(\theta)}{\bar{T}} \qquad (9.11)$$

Poettmann[11] presented tabulated values for a wide range of integrals with a lower limit of 0.2. As seen from the earlier equations, the method still requires the use of an arithmetic average temperature but does allow the Z factor to vary with pressure (all expressed by reduced values, e.g., $P_{P_r, \text{SIWH}}$, which is the pseudo-reduced shut-in wellhead pressure). Knowing the P_{SIBH}, the pseudo-reduced shut-in bottomhole pressure is calculated from the critical pressure of the gas mixture (tables in Poettman[11] provide the value of this integral). Then the second term on the right-hand side is calculated, and finally, a value of the left-hand side integral is explicitly obtained and the upper limit determined from the table, which basically gives the $P_{P_r, \text{SIWH}}$. The pseudo-reduced shut-in wellhead pressure is multiplied by the critical pressure of the gas mixture, resulting in the desired P_{SIWH}. This method does not require any iterations. Other methods such as Cullender and Smith[13] are even more rigorous (and computationally somewhat tedious) because they make no simplifying assumptions for the variation of either temperature or Z factor in the wellbore.[12]

9.4 SINGLE-PHASE FLOWLINE/PIPELINE HYDRAULICS

The mechanical energy balance equation and its pressure form described in Section 9.2 can be relatively easily applied to compute the flowing pressure drop for given conditions and pipe geometries. The same equations can be reverse engineered to size larger diameter flowlines and smaller diameter tubing that transport production fluids and production chemicals or inhibitors, respectively. The pressure drop calculation approach can also be applied to estimate the pressure across a pipeline that contains a layer of wax deposit (along the pipe walls) that is scrapped by a pig ("pig" is an acronym for pipeline inspection gauge). All the aforementioned applications are demonstrated by numerical examples in the following four subsections.

9.4.1 Pressure Drop Calculations

Figure 9.4 shows the schematic representation of an offshore production system, along with the relevant data that will be utilized in the numerical examples. It is important to recognize the fact that a single-phase steady state is assumed in all calculations. If the oil is undersaturated, then this assumption is fairly reasonable and much more so in the case of many deepwater Gulf of Mexico fields that exhibit a very large degree of undersaturation.[14] However, in the case of production chemicals/inhibitors, it is safe to assume single-phase conditions. It is also important to understand the fact that despite the single phase, the fluid density, and the viscosity that influence the hydraulic calculations do vary with flow or pressure and temperature, making the calculations challenging and complex. Therefore, single-phase property-based calculations should be considered as estimates, at best, in the absence of other credible values. Something else that can also be done to alleviate the uncertainty is by segmenting the length in smaller sections, such as the example shown in

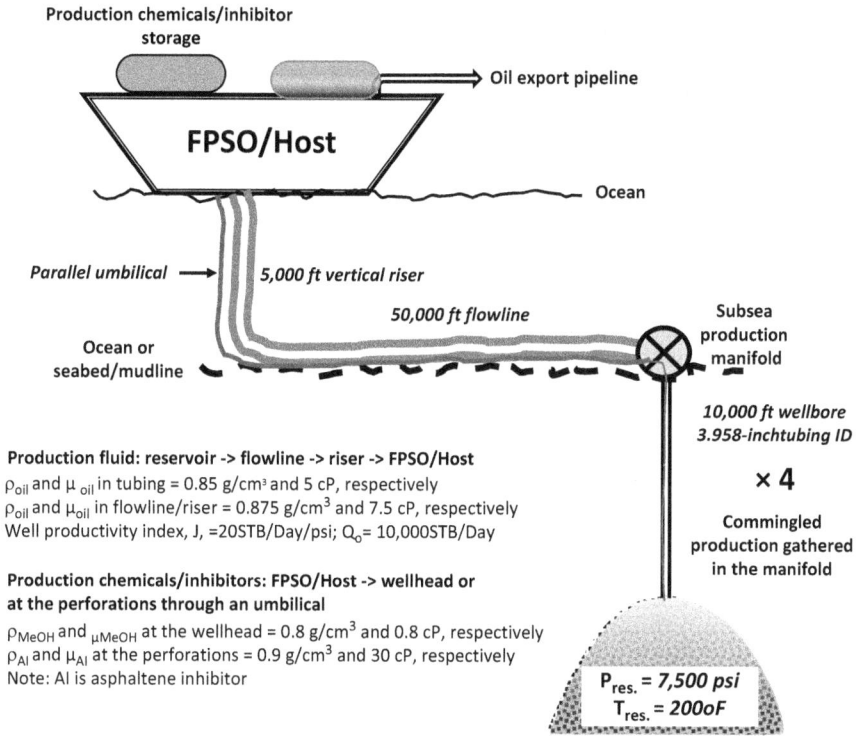

FIGURE 9.4 Schematic representation of an offshore production system for numerical calculations elucidated in Sections 9.4.1 through 9.4.3. Note that the figure is a sketch of textual exercises by Notz.[7]

Section 9.2. Note that in Figure 9.4 the oil properties in the tubing and flowline–riser system are assigned to reflect the oil becoming somewhat dense and more viscous due to the temperature drop.

The pressure drop calculation shown here is that of the wellbore or tubing. First, using the well productivity index, J, the flowing bottomhole pressure can be calculated as

$$J = \frac{Q_O}{\left(P_{\text{reservoir}} - P_{\text{FBHP}}\right)} \tag{9.12}$$

$$P_{\text{FBHP}} = P_{\text{reservoir}} - Q_O / J = 7{,}500 - 10{,}000 / 20 = 7{,}000\,\text{psia}$$

Note that for simplifying the calculations, stock tank barrel (STB) has been used (Q_O or oil rate per producer = 10,000 STB/day). The calculated P_{FBHP}, which is a result of the porous media permeability (Darcy equation), will experience a pressure drop due to the hydrostatic and frictional loss, leading to the flowing wellhead pressure, P_{FWHP} for a well or producer. Using the 10,000-ft wellbore depth and an oil density of 0.85 g/cm³ (see Figure 9.4),

$$\text{Hydrostatic loss} = 10,000 \times 0.43353 \times 0.85 = 3,685 \, \text{psi}.$$

Equation 9.4 can next be used to calculate the frictional loss, which requires the Reynolds number to determine the flow regime and thus the friction factor:

$$Re = \frac{0.1005\text{m} \times 2.3184\text{m} / \sec \times 850\text{kg} / \text{m}^3}{0.005\text{Nsec} / \text{m}^2} = 39,610$$

(10,000 STB/day is converted to velocity using the cross-sectional area for 3.958-in. tubing and other appropriate unit conversions to bring all variables in a consistent set of units).

Assuming an absolute roughness, ε, of 0.0006 in., the roughness ratio, $\varepsilon/D = 0.00015$, and the Re can now be used to set up the Chen equation for f_F since the flow is fully turbulent:

$$f_F = \frac{1}{4} \left\{ \frac{1}{-2\text{LOG}_{10}\left[\frac{0.00015}{3.7065} - \frac{5.0452}{39,610}\text{LOG}_{10}\left(\frac{0.00015^{1.1098}}{2.8257} - \left(\frac{7.149}{39,610} \right)^{0.8981} \right) \right]} \right\}^2$$

$$\cong 0.0056$$

$$\text{Frictional loss} = \frac{2 \times 0.0056 \times 850\text{kg} / \text{m}^3 \times 2.3184^2 \text{m}^2 / \sec^2}{0.1005\text{m}}$$

$$\cong 512\text{N} / \text{m}^2 / \text{m or}$$
$$\approx 0.0226 \, \text{psi} / \text{ft}$$

$\therefore$ for a 10,000-ft wellbore, frictional loss = $0.0226 \times 10,000 = 226$ psi.

The hydrostatic and frictional losses can now be added and subtracted from the P_{FBHP} value to obtain the $P_{FWHP} = 7,000 - 3,685 - 226 = 3,089$ psia.

9.4.2   SIZING OF FLOWLINES

The sizing of the flowline–riser system is generally a function of total throughput and the required arrival pressure at the topsides. Note that there are obviously many other design parameters, such as the fluid being sour, pressure, temperature, depths, insulation, and others, but essentially, the fluid mechanics plays an important role in the determination of the ID of the flowline–riser system. For example, if the host arrival pressure for topsides processing facilities is 1,000 psi for the system depicted in Figure 9.4, then the calculations in Section 9.4.1 are basically reverse engineered. The previously calculated flowing wellhead pressure now becomes the inlet pressure for the flowline, and the required 1,000 psi the outlet pressure; that is, the "pressure drop" over the total length of 55,000 ft is known. However, the ID that will fulfill the pressure drop requirements (and other mechanical constraints, etc.) needs to be iterated since the diameter appears in all friction loss calculations.

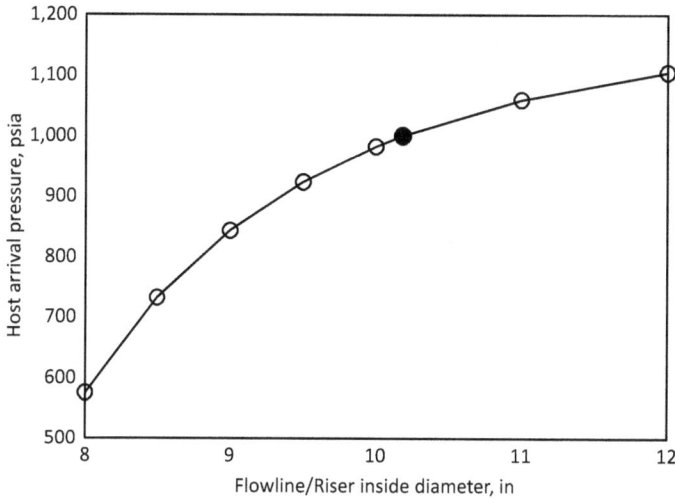

FIGURE 9.5 Host arrival pressure versus flowline/riser inside diameters. Inlet pressure is the calculated P_{FWHP} of 3,089 psia (see Section 9.4.1). The solid circle marker indicates the required arrival pressure achieved by a 10.183-in. ID.

Using the known values and assuming different IDs, a plot of IDs assumed versus the arrival pressure can be constructed as shown in Figure 9.5. Note that the calculations follow the same basic methodology outlined in Section 9.4.1. As seen in Figure 9.5 a precise ID of 10.183 in. satisfies the requirements. However, an odd size such as this is likely not available, and thus, the closest standard size is selected, which, in this case, would be 10 in.

9.4.3 SIZING OF INHIBITOR TUBING

Almost all production chemicals or inhibitors are transported through relatively much smaller ID tubes or lines that travel through a subsea umbilical. The umbilical is basically a fairly large diameter pipe that carries the chemical lines, fiber optic cables, electronics cables, sensors, and so on. The delivery of a given inhibitor is location specific, for example, methanol or wax/paraffin inhibitor at the wellhead and an asphaltene inhibitor or scale inhibitor at the perforations. These chemicals are usually stored in tanks on the host from where they are injected or pumped downhole. A significant advantage is the gain in the hydrostatic head. The dosage rate is dependent on the amount of a given inhibitor that would be needed to continue solids-free operations. The same conceptual pressure drop calculations shown previously can be utilized to determine the inhibitor line sizes and their multiples if one large diameter tube is impractical, and a certain number of smaller lines or tubes are preferred for delivering the same inhibitor dose. Typically, the dosage rates or volumes required for methanol can be up to two orders of magnitude greater than that for scale, paraffin, and asphaltene inhibitors.

The inhibitor tube sizing calculations are similar to the flowline–riser ID calculations shown in Section 9.4.2. That is, they are iterative. Just the way the arrival pressure is known or required for the flowline–riser, a certain arrival or delivery pressure also is needed either at the wellhead or at the perforations, for the respective inhibitors. A somewhat variable parameter also is the inlet pressure, which is typically the discharge pressure of the pump at the host that will deliver the inhibitor at a given spot. Let us now consider an asphaltene inhibitor (AI) with the stated properties and as shown in Figure 9.4 for delivery at the perforations. For the numerical calculations, we will assume a dosage rate of 0.3 gallons/min, 5,000-psi pump discharge pressure at the host, and 8,000-psi arrival pressure at the perforations. For a tube of 0.484-in. ID and the aforementioned values, the Reynolds number is

$$Re = \frac{0.0123m \times 0.1579m / sec \times 900kg / m^3}{0.03Nsec / m^2}$$

$$= 58.5 (1 \text{ gallon} = 0.003785m^3;$$

$$cross - sectional\ area = (\pi / 4) \times 0.0123^2)$$

The calculated Re value in the laminar region means $f_F = 16/58.5 = 0.2734$, which can be used in calculating the frictional loss in the entire 65,000 ft (5,000-ft vertical riser + 50,000-ft flowline + 10,000-ft wellbore), since the tube carrying the AI is considered parallel to the flow system shown in Figure 9.4.

$$Frictional\ loss = \frac{2 \times 0.2734 \times 900kg / m^3 \times 0.1579^2 m^2 / sec^2}{0.01235m}$$

$$\cong 993\,N / m^2 / m\ or$$

$$\approx 0.0439\,psi / ft$$

$\therefore$ for 65,000 ft, frictional loss = 0.0439 × 65,000 = 2,854 psi

The hydrostatic "gain" over 15,000 ft (5,000-ft vertical riser + 10,000-ft vertical wellbore) = 15,000 × 0.43353 × 0.9 = 5853 psi. The outlet pressure or the arrival pressure at the perforations = pump discharge pressure – frictional loss + hydrostatic gain = 5,000 – 2854 + 5,853 = 7,999 psi, which is close enough to the required 8,000 psi. If the preceding calculation is now repeated for an assumed tube ID of 0.5418 in. and a pump discharge pressure of 4,000 psi, then the outlet pressure will also be approximately 8,000 psi. Basically, any reasonable combination of larger tube diameter and lower discharge pressure or vice versa will likely satisfy the precise requirements (e.g., 0.45-in. ID and 6,000 psi discharge pressure as shown in Figure 9.6).

Similar calculations can be carried out for methanol, however, without the 10,000-ft vertical wellbore since the delivery point is wellhead or tree, or in other words frictional loss over 55,000 ft and hydrostatic gain over 5,000 ft. For example, a single tube of 1.6-in. ID can deliver 50 gallons/min methanol at the wellhead at about 3,100 psi (slightly higher than the P_{FWHP}) by using a pump that discharges at 4,400 psi. Alternatively, six 0.75-in. ID tubes with a flow rate of approximately 8.3 gallon/min and a pump discharge pressure of 5,600 psi can deliver methanol at the wellhead at approximately 3,100 psi to collectively meet the total dosage rate of 50 gallons/min.

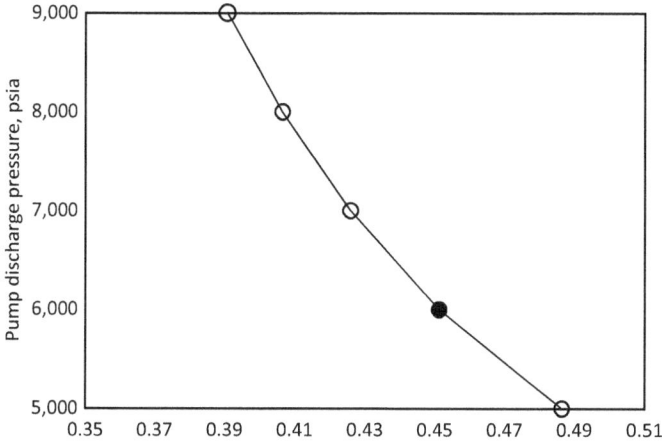

FIGURE 9.6   Pump discharge pressure versus AI inhibitor line ID for 0.3-gallon/min dosage and 8,000 psi perforation arrival pressure. The solid circle marker indicates the combination that will likely satisfy the process requirement (see Section 9.4.3).

9.4.4   ESTIMATING THE PRESSURE DROP IN A LONG-DISTANCE PIPELINE CONTAINING A WAX LAYER

The concept of Notz[7] in one of the exercise problems is extended for estimating the pressure drop in a 48-in.-diameter long-distance (800-mile) Trans-Alaska Pipeline System (TAPS) that transports Alaska North Slope crude oil that is being pigged for deposited wax. Figure 9.7 shows the conceptual schematic of the thin 0.6-mm wax layer scraped by a pig. Part of the data used in this numerical problem are adopted from Notz;[7] that is, the apparent viscosity of 100,000 cP for the wax deposit, while the pig velocity of 3.5 ft/sec and wax density of 0.85 g/cm^3 are similar to the values used by Notz.[7]

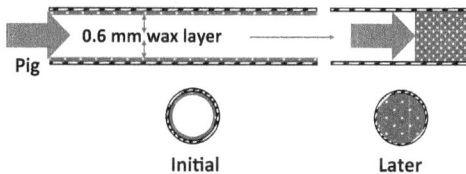

FIGURE 9.7   Conceptual schematic of a wax layer in the 48-in.-diameter TAPS being scraped by a pig.

Using the TAPS ID and the wax layer thickness of 0.6 mm, the effective cross-sectional area is calculated as follows:

$$\text{TAPS cross} - \text{sectional area} = (\pi / 4) \times (47/12)^2 = 12.048\,\text{ft}^2,$$

and the cross – sectional area with reduced TAPS
ID (due to wax layer) $= (\pi / 4)$
$$\times \left[(47/12) - 2 \times 0.001969 \right]^2 = 12.024\,\text{ft}^2$$

∴ Effective cross-sectional area = 12.048 – 12.024 = 0.024 ft², where the 0.6-mm wax thickness is equal to 0.001969 ft. Note that the TAPS wall thickness varies, but an average value of 0.5 in. is used in the calculations where 48 in. is the TAPS outside diameter.

The effective cross-sectional area attributed to wax equates to (0.024 × 800 × 5,280)/5.615 bbl of total wax ≅18,055 bbl (1 mile = 5,280 ft and 1 bbl = 5.615 ft³). The length of wax, assuming the wax-layer thickness, forms a cylinder having a diameter as TAPS ID = 18,055 × 5.615/(π/4)(47/12)² = 8,414 ft (see schematic in Figure 9.7).

Next, using the pig velocity of 3.5 ft/sec and other aforementioned data, the pressure drop can be estimated as follows:

$$\text{Re} = \frac{1.1938\,\text{m} \times 1.0668\,\text{m}/\text{sec} \times 850\,\text{kg}/\text{m}^3}{100\,\text{Nsec}/\text{m}^2} \cong 11,$$

which means fully laminar region, that is, $f_F = 16/11 = 1.45$, which is used to compute the frictional loss:

$$\text{Frictional loss} = \frac{2 \times 1.45 \times 850\,\text{kg}/\text{m}^3 \times 1.0668^2\,\text{m}^2/\text{sec}^2}{1.1938\,\text{m}}$$
$$\cong 2,345\,\text{N}/\text{m}^2/\text{m}\ \text{or}$$
$$\approx 0.104\,\text{psi}/\text{ft}$$

The calculated frictional loss is multiplied by the effective length of 8,414 ft to estimate the pressure drop, which is 0.104 × 8,414 = 875 psi. If TAPS is assumed to be horizontal, then the hydrostatic loss can be considered as zero. However, in reality, that is not the case because the pipeline goes through several elevation gains and losses over the 800-mile distance, which may offset the hydrostatic gradient, to some extent, and is the underlying assumption in these conceptual calculations. Thus, the calculated frictional loss means a pressure drop of about 875 psi, or in other words, if the inlet pressure is 1,000 psi, then the outlet will be 1,000 – 875 = ~125 psi.

9.4.4.1 Estimate of Pressure to Restart a Pipeline Filled with Gelled Oil

When a fluid in a pipeline gels, much pressure is required to restart the flow. Neglecting pressure losses and the elevation effects, the pressure drop required to restart can be estimated using the equation proposed by Perkins and Turner:[15]

$$\Delta P = \frac{4\tau_s L}{D} \tag{9.13}$$

where τ_s is the gel strength or the yield strength (usually reported in dyne/cm²), L is pipe length, and D is pipe ID. If both L and D are in same consistent set of units then ΔP is in dyne/cm². Alternatively, if τ_s is in dyne/cm², L is in miles (typical units), and D is in in. (typical units), then the calculated ΔP from Equation 9.14 is in psi:

$$\Delta P = 3.6758 \frac{\tau_s L}{D} \tag{9.14}$$

or

$$\frac{\Delta P}{L} = 3.6758 \frac{\tau_s}{D} \tag{9.15}$$

Gel strength is typically measured in a Brookfield-type viscometer. Chukwu et al.[16] reported several gel-strength values for Alaska North Slope oil blend flowing through the TAPS. The measured values are a function of a particular temperature achieved after a cold ramp so as to mimic a cooldown after the pipeline shutdown for some time. For example, τ_s values for TAPS blend (at the time of their study) are 5.4 and 823.6 dyne/cm² at 20°F and –20°F, respectively. This means that if TAPS were to shut down and the oil temperature reduce to 20°F and –20°F, then the "cold" restart pressure drop per mile from Equation 9.15 would be 0.422 and 64.4, respectively. Clearly, as seen from the data and ΔP calculations, the lower the temperature, greater the gel strength and, consequently, the restart pressure.

9.5 SUPERFICIAL AND ACTUAL VELOCITIES

In a typical multiphase scenario, the flow patterns, regimes, and, ultimately, the pressure drops depend on gas and liquid velocities. The superficial gas and liquid velocities in a typical two-phase flow are simply defined by the ratio of volumetric flow rate and the line cross-sectional area, A:

$$U_{sL} = Q_L / A; U_{sG} = Q_G / A \tag{9.16}$$

where U_{sL} and U_{sG} are superficial liquid and gas velocities, respectively, and Q_L and Q_G are the liquid and gas volumetric flow rates, respectively.

However, in reality, the pipe cross section will be occupied by both phases, thereby reducing the actual cross-sectional area available for the respective phases, thus resulting in considerably high actual velocities:

$$U_L = Q_L / A_L; U_G = Q_G / A_G \tag{9.17}$$

where U_L and U_G are the actual or in situ liquid and gas velocities, respectively, and A_L and A_G are the cross-sectional areas available for liquid and, gas respectively. In a two-phase flow that involves gas and liquid, the lighter gas phase moves faster than the liquid phase or slips by,[2] resulting in a difference in the actual velocities known as slip or slip velocity: $U_S = U_G - U_L$.

9.5.1 LIQUID HOLDUP

The individual areas A_L and A_G can be used to define the volume fractions or holdup:

$$H_L = \frac{A_L}{A_L + A_G} = \frac{A_L}{A} \tag{9.18}$$

$$H_G = \frac{A_G}{A_L + A_G} = \frac{A_G}{A} \tag{9.19}$$

where H_L and H_G are the liquid and gas holdup, respectively; however, the latter is a misnomer[2] because it is the liquid that has a lower velocity and thus is held up, whereas the gas is not. The A_L and A_G can be substituted in Equation 9.17 as $H_L \times A$ and $H_G \times A$ in order to relate the superficial and actual velocities:

$$U_{sL} = H_L U_L; U_{sG} = H_G U_G \tag{9.20}$$

9.6 FLOW PATTERNS AND REGIMES

Superficial velocities of the gas and oil (liquid) phases are typically used in constructing the flow regime maps.[17] A dimensionless number known as the Froude number has been suggested by Shell Global Solutions[18] that depicts consistent flow regime maps for two-phase horizontal and vertical flows.[17] The Froude number for gas and liquid phases is defined by the following equations:[18]

$$F_{rG} = U_{sG} \sqrt{\frac{\rho_G}{(\rho_L - \rho_G) g D}} \tag{9.21}$$

$$F_{rL} = U_{sL} \sqrt{\frac{\rho_L}{(\rho_L - \rho_G) g D}} \tag{9.22}$$

TABLE 9.2
Froude Numbers for the Gas and Liquid (Oil) for the Data of Burke and Kashou[19]

In Situ Gas Rates		In Situ Oil Rates								
Vol., mmcf/d	Mass, kg/sec	Vol., bbl/d	Mass, kg/sec	ρ_G, kg/m³	ρ_L, kg/m³	Pipe ID, m	U_{sG}, m/sec	U_{sL}, m/sec	F_{rG}	F_{rL}
1.133	1.761	5,709	9.025	4.742	859	0.1828	14.15	0.40	0.79	0.30
1.133	1.761	5,709	9.025	4.742	859	0.1668	16.99	0.48	0.99	0.38

where F_r is Froude number, U_s is superficial velocity, ρ is density, g is acceleration due to gravity, D is pipe ID, and subscripts G and L denote gas and liquid phases, respectively. Any consistent set of units renders F_r dimensionless; for example, if U_s is in m/sec, g is in m/sec² and D is in m then $F_r = $ (m/sec) $\times$ (sec²/m²)$^{0.5}$ = unitless, because the density ratio in the same units is automatically dimensionless. Note that the liquid phase density may represent combined oil and water. Unlike using superficial velocities alone, the inclusion of phase densities in Froude number supposedly captures the pressure effect, in part,[17] in the determination of flow regimes.

The application of Froude number flow regime maps is illustrated by the following example. Using the data of Burke and Kashou,[19] the F_{rG} and F_{rL} values are calculated for the two pipe IDs as shown in Table 9.2. The mass and volumetric flow rates reported by the authors have been used in the pertinent density and superficial velocity calculations, respectively. The Froude numbers from Table 9.2 are then superimposed on the flow regime maps for horizontal and vertical flows to identify the flow regimes. As seen in Figure 9.8, the F_{rG} and F_{rL} values place the cases in Table 9.2 on the boundary of annular dispersed flow and slug flow regimes for the horizontal orientation and annual dispersed flow for the vertical orientation, respectively. Also depicted in Figure 9.8 is the hypothetical case included in Gudmundsson,[17] which places the Froude numbers just inside the slug flow for horizontal orientation and on the interface of annual dispersed flow and slug flow for the vertical orientation.

Gudmundsson[17] has stated that the pipe ID (and the orientation) is the most influential or controlling parameter that dictates the flow regime prevailing in a flowline. However, the slug flow regime is the most undesired in the flow systems since it gives rise to unsteady hydrodynamic behavior that may be a detriment to the topsides equipment. Therefore, it is rather important to know the flow regime that may exist in a given system. A bracketed range of F_{rL} ($0.1 < F_{rL} < 10$) and $F_{rG} < 1$ have been suggested for a quick evaluation for the occurrence of slug flow regime.[17]

9.7 MULTIPHASE PRESSURE DROP PREDICTION

In principle, the same mechanical energy balance (Equation 9.1) can be used to determine the pressure drop for a multiphase flow. For example, the respective terms of pressure gradient in Equation 9.1 can be termed or denoted as "mixture" or simply "m", which appears to be deceptively simple because two or more fluids compete for

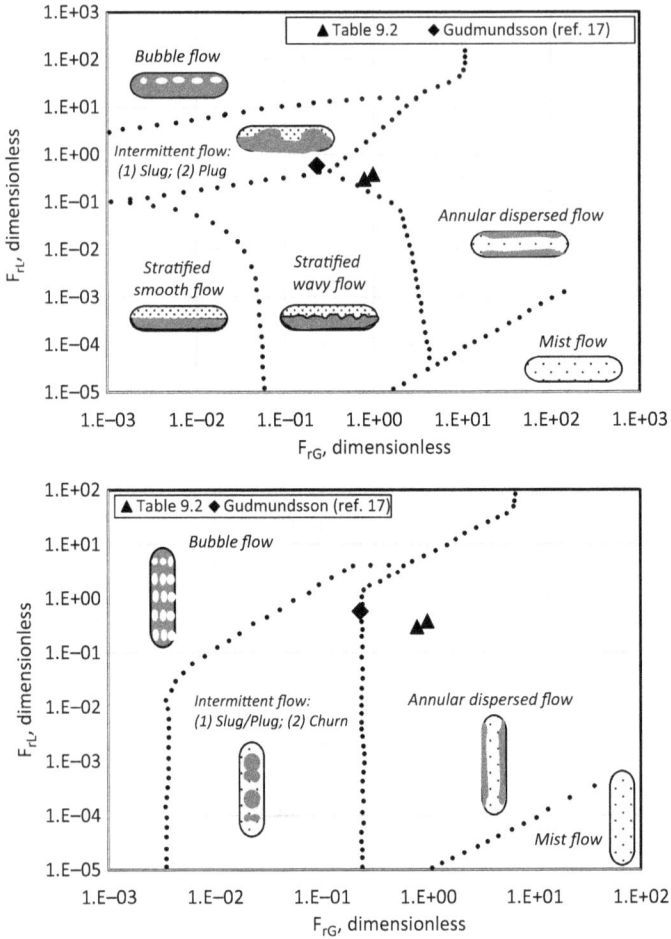

FIGURE 9.8 Digitized flow regime maps[18] for horizontal and vertical flow based on gas and liquid Froude numbers that includes the example in Table 9.2 and the hypothetical case in Gudmundsson.[17]

the same area, holdup of liquid, in situ velocities, flow patterns, volume fractions, and slip, which influence the computations. Basically, three different approaches are available for multiphase pressure drop estimation, namely, (1) the homogeneous model, (2) empirical models, and (3) mechanistic models. All three are covered in the following subsections with the help of a solved numerical example.

9.7.1 HOMOGENEOUS MODEL

This approach is the simplest because of the primary assumption that the multiphase mixture behaves like a well-mixed or homogeneous single phase. Some type of averaging (mass, volume) is used to obtain mixture density, viscosity following which the computational procedure is identical to the single-phase calculations shown previ-ously. The following equations are needed:[2]

$$G_m = U_{sG}\rho_G + U_{sL}\rho_L \tag{9.23}$$

$$x = \frac{U_{sG}\rho_G}{U_{sG}\rho_G + U_{sL}\rho_L} \tag{9.24}$$

$$\mu_m = x\mu_G + (1-x)\mu_L \tag{9.25}$$

$$\rho_m = \frac{1}{\left(\dfrac{x}{\rho_G} + \dfrac{1-x}{\rho_L}\right)} \tag{9.26}$$

$$R_{em} = \frac{DG_m}{\mu_m} \tag{9.27}$$

where G_m is the mixture mass flux in kg/m²-sec; x is the gas mass (flux) fraction (dimensionless); μ_m and ρ_m are mixture viscosity and density, respectively, in cP (converted to Nsec/m² in calculations for unit consistency) and kg/m³, respectively; and R_{em} is the mixture Reynolds number directly expressed in terms of the mixture mass flux, which is basically the product of velocity and density. Note that the liquid phase in these equations can be construed as one liquid phase (e.g., oil + water). The calculated R_{em} value can then be used to obtain the correct f_F (dependent on lami-nar, transition, and turbulent) and thus the frictional pressure gradient. The mixture density is used to obtain the hydrostatic loss. These calculations are similar to those shown previously for single phase. However, in the case of multiphase pressure drop, fluid properties, and the fraction of a given phase continuously change with position, vis-à-vis temperature and pressure in the flow system. This, to some extent, com-plicates the pressure drop calculations even in the homogeneous model. The meth-odology thus entails calculating the total pressure gradient at a particular position, such as the wellhead and then proceeding to the next segment in terms of depth and eventually reaching the bottomhole. Note that the prior segment calculated pressure gradient is used in each subsequent depth. For example, if the wellhead pressure is X psia and at that position the pressure gradient is Y psi/ft, then the pressure Z ft below the wellhead is equal to $(X + YZ)$ psia. Accordingly, the entire distance of interest can be segmented for arriving at the final position, as illustrated by the solved numerical example.

9.7.1.1 Numerical Example

We use the example included in Hasan and Kabir[2] for a 5,151-ft vertical wellbore with a 2.99-in. ID tubing, and production data as follows: oil rate = 1,140 STB/day; gas–oil ratio (GOR) = 450 scf/STB; $\gamma_G = 0.8$; $\gamma_L = 0.9158$ (23°API), and $P_{FWHP} = 505$ psig. The solution basically illustrates the pressure traverse computations over the entire 5,151 ft in a top-down mode starting with the 505-psig wellhead pressure and the given data. The properties of the produced gas and oil (gravities and GOR) indicate that the reservoir fluid is a typical black oil, meaning standard correlations such as Standing's[20] can be used reliably to obtain the needed fluid properties at various positions, vis-à-vis temperature and pressure.

Note that in this particular numerical exercise, the temperature is not provided. Therefore, the following approach is used to back calculate the flowing wellhead temperature first from some of the given data. The gas gravity and P_{FWHP} of 505 psig are used to obtain the Z factor at various temperatures from 105–120°F (considered typical values of presumed wellhead temperatures) using the Dranchuk–Abou–Kassem[8] method. Next, the gas densities and presumed wellhead temperatures are calculated and compared with the 2.19 lb/ft³ value.[2] At 107°F, these densities are

same, $\rho_G = \dfrac{(505+14.7) \times 0.8 \times 28.97}{0.9036 \times 10.732 \times (107+460)} \cong 2.19 \text{lbm} / \text{ft}^3$, thereby indicating a flow-

ing wellhead temperature of 107°F. This is the most definitive manner in which this back calculation can be done because the fundamental real gas equation is used. On the other hand, the stated oil density[2] of 55.042 lb/ft³ and the Standing's[20] R_s and B_o correlations indicate a flowing wellhead temperature of 119°F. This discrepancy between the two back-calculated temperatures is ascribed to the empiricism of R_s and B_o correlations. Therefore, 107°F is used in all the computations for this particular numerical example.

However, since fluid properties are dependent on both temperature and pressure, it is also necessary to obtain a temperature profile for the entire 5,151-ft vertical wellbore for segmentation and subsequent pressure traverse calculations. First, under the assumption of sandface temperature being about the same as that of the formation temperature in a flowing well,[2] the flowing bottomhole temperature can be estimated by utilizing a geothermal gradient; consequently, a flowing bottomhole temperature of 178°F is a reasonable value (75°F surface temperature + 0.02°F/ft × 5,151 ft), where 0.02°F/ft is the geothermal gradient.[17] Typically, when flowing fluid and formation temperatures (based on geothermal gradient) are profiled versus depth, both converge at the well bottomhole but bifurcate as one approaches the wellhead; in other words, the ΔT between the produced fluid and formation steadily increases from a value of 0 to something high (e.g., 107 − 75 = 32°F in this particular example). Basically the geothermal gradient or formation temperature versus depth is a straight line, whereas the flowing fluid temperature versus depth is somewhat curved. It should, however, be noted that in a static case, the formation and fluid temperatures would nearly equal after some time has elapsed, given the no-flow condition. Although much more elaborate and mechanistic approaches[2] can be adopted for the temperature distribution calculation, we use the earlier simplified methodology to estimate (best educated guess) the temperatures over the 5,151-ft vertical wellbore in the absence of temperature data. Figure 9.9 depicts the comparison between the geothermal gradient and the flowing fluid temperature over the vertical wellbore. Note that a 25°F ΔT between the two is assumed at 2,500 ft to represent the curvature in the flowing fluid temperature.

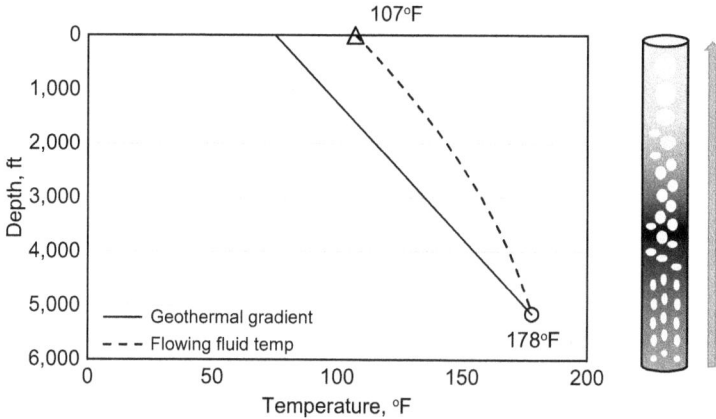

FIGURE 9.9 Comparison of geothermal gradient and flowing fluid temperature versus depth for the numerical problem solved in Section 9.7.1.1.

Using 178°F as the formation temperature, the GOR of 450 scf/STB and the gas and oil gravities, respectively, results in a bubble point of 2,584 psia from the Standing correlation. This value is construed to be reasonable/consistent when compared with the 2,120-psia measured bottomhole temperature,[2] indicating that the oil might be undersaturated in the reservoir but obviously splitting into two phases as it experiences a pressure drop while moving up the tubing, that is, resulting in a two-phase flow. For the first calculation of pressure gradient at the wellhead position, the values of R_s and B_o are necessary (as well as at other subsequent positions), which are

$$R_s \text{ at } 505\,\text{psig and } 107°F = 0.8 \left[\frac{(0.055 \times 519.7 + 1.4)10^{0.0125 \times 23}}{10^{0.00091 \times 107}} \right]^{1.205}$$

$$= 81.6\,\text{scf} / \text{STB and}$$

$$B_o = 0.9759 + 0.00012A^{1.2} = 1.0493\,\text{res.bbl} / \text{STB},$$

$$\text{where } A = 81.6 \left(\frac{0.8}{0.9158} \right)^{0.5} + 1.25 \times 107 = 210.04$$

The preceding values are calculated from Standing's[20] correlation functional forms found in Whitson and Brulé.[21] Similarly, the oil density at these conditions is calculated as

$$\rho_O = \frac{62.432 \times 0.9158 + 0.0136 \times 0.8 \times 81.6}{1.0493} = 55.33 \frac{\text{lb}}{\text{ft}^3} \text{ or } 886.29\,\text{kg} / \text{m}^3$$

Note that the preceding value differs somewhat from the 55.042 lb/ft^3 computed by Hasan and Kabir[2] but is reasonable given the temperature uncertainty.

The volumetric gas and oil flow rates at the prevailing conditions are computed next using $B_g = 0.02827 \dfrac{0.9036 \times (107 + 460)}{519.7} = 0.0278 \, ft^3 / scf, B_o, R_s$, and the given production data,

$$Q_O = \frac{1,140 \times 1.0493 \times 5.615 \times 0.3048^3}{24 \times 60 \times 60} = 0.0022 \, m^3 / sec$$

$$Q_G = \frac{1,140 \times (450 - 81.6) \times 0.0278 \times 0.3048^3}{24 \times 60 \times 60} = 0.0038 \, m^3 / sec$$

and

$$U_{sL} = \frac{0.0022}{\dfrac{\pi}{4} \left(\dfrac{2.99 \times 2.54}{100} \right)^2} = 0.486 \, m / sec, \; U_{sG} = \frac{0.0038}{\dfrac{\pi}{4} \left(\dfrac{2.99 \times 2.54}{100} \right)^2} = 0.847 \, m / sec$$

Subsequently, using Equations 9.23 through 9.27, all the needed parameters for frictional and hydrostatic loss can be obtained:

$$G_m = 0.847 \times 35.09 + 0.486 \times 886.29$$
$$= 460.4 \, kg / m^2 - sec,$$
$$x = (0.847 \times 35.09) / 460.4 = 0.0645.$$

Gas viscosity is obtained using the Lee–Gonzalez[22] correlation commonly employed by PVT laboratories,[21]

$\mu_G = A_1 \times 10^{-4} \exp(A_2 \times 0.03509^{A3}) = 0.0117 \, cP$, which is lower than the 0.019 cP reported by Hasan and Kabir.[2]

$$A_1 = \frac{(9.379 + 0.016 \times 0.8 \times 28.97)(107 + 460)^{1.5}}{209.2 + 19.26 \times 0.8 \times 28.97 + (107 + 460)} = 107.67;$$

$$A_2 = 3.448 + \left(\frac{986.4}{107 + 460} \right) + 0.01009 \times 0.8 \times 28.97 = 5.42, \; and$$

$A_3 = 2.447 - 0.2224 \times 5.42 = 1.24.$

For oil viscosity, we adopted a much simpler yet quite representative correlation of Abu-Khamsin and Al-Marhoun:[23]

$$\mu_O = \exp(-2.652294 + 8.484462 \times 0.88629^4) = 13.24 \, cP$$

The mixture density, viscosity and R_{em} is calculated next:

$$\rho_m = \frac{1}{\dfrac{0.0645}{35.09} + \dfrac{1 - 0.0645}{886.29}} = 345.5 \, kg / m^3 \; and$$

$$\mu_m = 0.0645 \times 0.0117 + (1 - 0.0645) \times 13.23 = 12.37\,cP$$

$$R_{em} = \frac{\left(\dfrac{2.99 \times 2.54}{100}\right) \times 460.4}{12.37 \times 10^{-3}} \cong 2,824 \text{, resulting in } f_F \text{ of } 0.0111 \text{ from the Colebrook}$$

equation,[4] with $\varepsilon/D = 0.0006/2.99 \cong 0.0003$

Subsequently, the frictional and hydrostatic losses are calculated:

$$\text{Frictional loss} = \frac{4.4219 \times 10^{5} \times 2 \times 0.0111 \times 460.4^2}{345.5 \times \left(\dfrac{2.99 \times 2.54}{100}\right)} \cong 0.008\,psi/ft$$

$$\text{Hydrostatic loss} = 345.5 \times 9.81 \times 4.4219 \times 10^{-5} = 0.1499\,psi/ft$$

$$\text{Total pressure loss} = 0.008 + 0.1499 = 0.1578\,psi/ft$$

Note that although in the preceding we refer to the friction and hydrostatic terms as "losses", however, in reality, these could be construed as "gains" in top-down calculations but losses in bottom-up calculations. Sometimes, a negative and a positive sign is associated with loss and gain or is generically referred to as pressure drop, but we use the term *loss* throughout with the aforementioned caveat. For the subsequent segment (200 ft below the wellhead), the pressure is calculated as equaling $519.7 + 0.1578 \times 200 \cong 551$ psia. Then, using this pressure and the estimated temperature at this depth (see Figure 9.9), the gas Z factor and all other parameters are calculated to arrive at the total pressure gradient at 200 ft to proceed to the next segment and ultimately reaching the depth of 5,151 ft. Detailed computations are shown in Table 9.3 that ends with the calculated flowing bottomhole pressure, P_{FBHP} of 1,707 psia, which compares favorably with the computed value of $1,786^2$ psia, despite the temperature uncertainty noted earlier.

9.7.2 Empirical Models – Hagedorn and Brown

The modified Hagedorn–Brown (mHB)[2, 3, 24] correlation is one of the empirical models developed on the basis of the original model – as pointed out by Hasan and Kabir,[2] proprietary commercial software typically applies modifications to the original version, which is seldom transparent to the user. The HB correlation is primarily designed for vertical systems (typically vertical wellbores but conceptually applicable to risers) and requires all the fluid properties used in the homogeneous model. However, the in situ average density and viscosity at a given position is calculated on the basis of liquid holdup and gas volume fraction, respectively. Additionally, gas–liquid surface/interfacial tension is needed for calculating the various dimensionless numbers. Similar to the homogeneous model, the solution for the same numerical example using mHB method is shown in the following.

TABLE 9.3

Detailed Pressure Traverse Calculations Using the Homogeneous Model for the Vertical System Numerical Example (additional columns continued)

Depth, ft	Temperature, °F	Pressure, psia	Z factor	B_g, ft³/scf	R_s, scf/STB	B_o, res. bbl/STB
0	107	520	0.9036	0.0279	81.6	1.0493
200	111	551	0.9004	0.0264	86.4	1.0534
400	115	584	0.8972	0.0250	91.4	1.0576
600	119	618	0.8939	0.0237	96.6	1.0618
800	123	653	0.8907	0.0225	102.0	1.0661
1,000	127	689	0.8875	0.0214	107.6	1.0704
1,200	130	727	0.8842	0.0203	113.5	1.0747
1,400	134	766	0.8810	0.0193	119.5	1.0792
1,600	137	806	0.8777	0.0184	125.7	1.0836
1,800	140	847	0.8745	0.0175	132.2	1.0882
2,000	144	889	0.8713	0.0167	138.9	1.0928
2,200	147	933	0.8682	0.0160	145.8	1.0974
2,400	150	978	0.8650	0.0152	152.9	1.1021
2,600	152	1,024	0.8620	0.0146	160.3	1.1069
2,800	155	1,071	0.8589	0.0139	167.8	1.1118
3,000	158	1,120	0.8560	0.0134	175.7	1.1167
3,200	160	1,169	0.8531	0.0128	183.7	1.1216
3,400	162	1,220	0.8503	0.0123	192.0	1.1267
3,600	165	1,271	0.8477	0.0118	200.5	1.1318
3,800	167	1,324	0.8451	0.0113	209.3	1.1369
4,000	169	1,378	0.8426	0.0109	218.3	1.1422
4,200	171	1,433	0.8403	0.0105	227.5	1.1475
4,400	172	1,489	0.8381	0.0101	237.0	1.1528
4,600	174	1,546	0.8360	0.0097	246.7	1.1583
4,800	176	1,603	0.8341	0.0093	256.7	1.1638
5,000	177	1,662	0.8323	0.0090	266.9	1.1693
5,151	178	1,707	0.8384	0.0089	274.8	1.1736

Q_G, m³/sec	Q_O, m³/sec	U_{sG}, m/sec	U_{sL}, m/sec	ρ_G, kg/m³	ρ_O, kg/m³	G_m, kg/m²-sec	x (dimensionless)
0.0038	0.0022	0.847	0.486	35.09	886.29	460.417	0.0645
0.0036	0.0022	0.791	0.488	37.08	883.64	460.418	0.0637
0.0033	0.0022	0.739	0.490	39.14	881.00	460.418	0.0628
0.0031	0.0022	0.690	0.492	41.28	878.36	460.418	0.0619
0.0029	0.0022	0.645	0.494	43.50	875.74	460.418	0.0610
0.0027	0.0022	0.603	0.496	45.79	873.13	460.419	0.0600
0.0026	0.0023	0.563	0.498	48.17	870.52	460.419	0.0590
0.0024	0.0023	0.527	0.500	50.63	867.93	460.419	0.0579
0.0022	0.0023	0.492	0.502	53.17	865.34	460.420	0.0568
0.0021	0.0023	0.459	0.504	55.79	862.77	460.420	0.0557
0.0019	0.0023	0.429	0.506	58.49	860.20	460.420	0.0545
0.0018	0.0023	0.400	0.508	61.28	857.65	460.421	0.0533
0.0017	0.0023	0.374	0.510	64.15	855.10	460.421	0.0520
0.0016	0.0023	0.348	0.513	67.10	852.57	460.422	0.0508
0.0015	0.0023	0.324	0.515	70.13	850.04	460.422	0.0494

(Continued)

TABLE 9.3 (Continued)
Detailed Pressure Traverse Calculations Using the Homogeneous Model for the Vertical System Numerical Example (additional columns continued)

Q_G, m³/sec	Q_O, m³/sec	U_{sG}, m/sec	U_{sL}, m/sec	ρ_G, kg/m³	ρ_O, kg/m³	G_m, kg/m²-sec	x (dimensionless)
0.0014	0.0023	0.302	0.517	73.25	847.53	460.422	0.0481
0.0013	0.0024	0.281	0.519	76.44	845.02	460.423	0.0466
0.0012	0.0024	0.261	0.522	79.71	842.53	460.423	0.0452
0.0011	0.0024	0.242	0.524	83.06	840.05	460.424	0.0437
0.0010	0.0024	0.224	0.527	86.48	837.57	460.424	0.0422
0.0009	0.0024	0.208	0.529	89.98	835.11	460.425	0.0406
0.0009	0.0024	0.192	0.531	93.54	832.66	460.425	0.0390
0.0008	0.0024	0.177	0.534	97.17	830.22	460.426	0.0373
0.0007	0.0024	0.163	0.536	100.86	827.79	460.426	0.0356
0.0007	0.0024	0.149	0.539	104.62	825.37	460.427	0.0339
0.0006	0.0025	0.136	0.542	108.43	822.96	460.427	0.0321
0.0006	0.0025	0.128	0.544	110.38	821.14	460.428	0.0307

ρ_m, kg/m³	μ_G, cP	μ_O, cP	μ_m, cP	R_{em}	f_F	Hydr. loss, psi/ft	Fric. loss, psi/ft	Total press. loss, psi/ft
345.48	0.0117	13.24	12.38	2,824	0.0111	0.1499	0.0080	0.1578
360.07	0.0116	12.44	11.64	3,003	0.0109	0.1562	0.0075	0.1637
374.75	0.0118	11.69	10.96	3,191	0.0107	0.1626	0.0071	0.1696
389.49	0.0119	11.00	10.32	3,388	0.0105	0.1690	0.0067	0.1756
404.28	0.0121	10.36	9.73	3,594	0.0103	0.1754	0.0063	0.1817
419.08	0.0122	9.76	9.18	3,810	0.0102	0.1818	0.0060	0.1878
433.87	0.0123	9.21	8.67	4,035	0.0100	0.1882	0.0057	0.1939
448.62	0.0125	8.69	8.19	4,270	0.0098	0.1946	0.0054	0.2000
463.31	0.0126	8.21	7.74	4,516	0.0097	0.2010	0.0051	0.2061
477.90	0.0128	7.76	7.33	4,772	0.0095	0.2073	0.0049	0.2122
492.38	0.0129	7.34	6.94	5,039	0.0094	0.2136	0.0047	0.2183
506.71	0.0131	6.95	6.58	5,317	0.0092	0.2198	0.0045	0.2243
520.86	0.0133	6.58	6.24	5,606	0.0091	0.2259	0.0043	0.2303
534.82	0.0134	6.24	5.92	5,906	0.0090	0.2320	0.0041	0.2361
548.54	0.0136	5.92	5.62	6,218	0.0088	0.2380	0.0040	0.2419
562.03	0.0138	5.61	5.35	6,542	0.0087	0.2438	0.0038	0.2476
575.24	0.0139	5.33	5.08	6,877	0.0086	0.2495	0.0037	0.2532
588.16	0.0141	5.07	4.84	7,225	0.0085	0.2551	0.0036	0.2587
600.78	0.0143	4.82	4.61	7,585	0.0084	0.2606	0.0034	0.2641
613.07	0.0145	4.59	4.39	7,957	0.0083	0.2659	0.0033	0.2693
625.02	0.0147	4.37	4.19	8,342	0.0082	0.2711	0.0032	0.2744
636.63	0.0149	4.16	4.00	8,740	0.0081	0.2762	0.0031	0.2793
647.87	0.0151	3.97	3.82	9,150	0.0080	0.2810	0.0030	0.2841
658.75	0.0153	3.79	3.65	9,573	0.0079	0.2858	0.0030	0.2887
669.26	0.0155	3.62	3.49	10,010	0.0078	0.2903	0.0029	0.2932
679.39	0.0157	3.45	3.34	10,459	0.0077	0.2947	0.0028	0.2975
685.66	0.0158	3.34	3.24	10,808	0.0076	0.2974	0.0028	0.3002

The mHB method does include consideration of the prevailing flow regime but only for bubble flow in that separate equations/correlations are needed for computing the liquid holdup and the Reynold's number is based on in situ average liquid velocity.[3] This is, in fact, one of the modifications proposed to address the deficiency of the original HB correlations. However, a simpler modification is to set the liquid holdup (H_L) to the no-slip liquid holdup (C_L) when the correlations incorrectly estimate $H_L < C_L$. This happens to be the case at a depth of 1,400 ft for the solved numerical example (see Table 9.4).

The application of the mHB method begins with the calculation of four different dimensionless numbers, namely the liquid velocity number, gas velocity number, pipe diameter number, and the liquid viscosity number. Each of these requires the value of gas–liquid (oil) interfacial tension at a particular position (wellhead in this case to begin the calculations). Ideally, the most reliable calculation (like other fluid properties) would be based on compositional estimation methods such as the parachor; however, being consistent with the fact that empirical correlations such as Standings are used, interfacial tension values also are determined from a similar class of empirical correlations. For example, in this particular case, we use the Abdul–Majeed[25] equation to first calculate the dead oil surface tension values as a function of temperature (prevailing) and American Petroleum Institute (API) gravity, followed by a correction factor that incorporates prevailing R_s, eventually leading to the dissolved gas corrected final value for use in the dimensionless numbers. Clearly, the correction is mathematically consistent in that the dead oil surface tension value is the same with corrections when $R_s = 0$. For the wellhead conditions, the dead oil surface tension is

$$\sigma_{OD} = (1.17013 - 1.694 \times 10^{-3} \times 107) \times (38.085 - 0.259 \times 23) = 31.77 \text{ dynes/cm}$$

or mN/m

using $R_s = 81.6$ scf/STB (see Table 9.4),

$$(\sigma_{GO}/\sigma_{OD}) = 0.056379 + 0.94362\exp(-3.8491 \times 10^{-3} \times 81.6) = 0.7456,$$

and

σ_{GO} (or σ_{GL}, i. e., O = L and vice versa) at 107 ° F and 519.7 psia (wellhead conditions) = $31.77 \times 0.7456 = 23.69$ dynes/cm or mN/m.

Liquid velocity number, N_{LU}:

$$N_{LU} = U_{sL}\left(\frac{\rho_L}{g\sigma_{GL}}\right)^{0.25} = 0.486\left(\frac{886.29}{9.81 \times 23.69 \times 10^{-3}}\right)^{0.25} = 3.82$$

Gas velocity number, N_{GU}:

$$N_{GU} = U_{sG}\left(\frac{\rho_L}{g\sigma_{GL}}\right)^{0.25} = 0.847\left(\frac{886.29}{9.81 \times 23.69 \times 10^{-3}}\right)^{0.25} = 6.65$$

Pipe diameter number, N_D:

$$N_D = D\left(\frac{g\rho_L}{\sigma_{GL}}\right)^{0.5} = 0.0759\left(\frac{9.81\times886.29}{23.69\times10^{-3}}\right)^{0.5} = 46$$

Liquid viscosity number, N_L:

$$N_L = \mu_L\left(\frac{g}{\rho_L\sigma_{GL}^3}\right)^{0.25} = 13.24\times10^{-3}\left(\frac{9.81}{886.29\times0.02369^3}\right)^{0.25} = 0.071$$

The preceding four dimensionless numbers can also be calculated with common oilfield units, which are given elsewhere.[3]

Next, the value of the group $\left(\dfrac{N_{GU}N_L^{0.38}}{N_D^{2.14}}\right)$ is calculated: $\left(\dfrac{6.65\times0.071^{0.38}}{46^{2.14}}\right) = 0.000673$.

This is used to obtain, Ψ, known as the HB flow parameter, from the chart that begins at 0.01 and $\Psi = 1$, meaning for values ≤ 0.01 for the group, Ψ is set to 1.0, which is the case here. Guo et al.[3] have replaced the chart with the following correlation for obtaining Ψ, with $X_1 = \left(\dfrac{N_{GU}N_L^{0.38}}{N_D^{2.14}}\right)$:

$$\Psi = 0.91163 - 4.82176X_1 + 1{,}232.25X_1^2 - 22{,}253.6X_1^3 + 116{,}174.3X_1^4 \quad (9.28)$$

The second chart is used to determine the HB parameter C_{NL} from the liquid viscosity number, N_L. Guo et al.[3] have replaced this chart also with the following correlation, with $X_2 = \log(N_L) + 3$,

$$C_{NL} = 10^{\left(-2.69851+0.15841X_2-0.551X_2^2+0.54785X_2^3-0.12195X_2^4\right)} \quad (9.29)$$

For this particular example, $X_2 = \log(0.071) + 3 = 1.851$, and

$$C_{NL} = 10^{\left(-2.69851+0.15841\times1.851-0.551\times1.851^2+0.54785\times1.851^3-0.12195\times1.851^4\right)} = 0.0056$$

The third chart is used for calculating (H_L/Ψ) or the liquid holdup factor based on inverse viscosity number,[2] N_f:

$$N_f = \left(\frac{N_{LU}}{N_{GU}^{0.575}}\right)\left(\frac{P}{P_{atm}}\right)^{0.1}\left(\frac{C_{NL}}{N_D}\right)$$

$$(9.30)$$

$$N_f = \left(\frac{3.82}{6.65^{0.575}}\right)\left(\frac{505+14.7}{14.7}\right)^{0.1}\left(\frac{0.0056}{46}\right) = 0.00022$$

Guo et al.[3] have replaced this chart also with the following correlation, with $X_3 =$ log(N_f) + 6:

$$\left(\frac{H_L}{\psi}\right) = -0.10307 + 0.617771X_3 - 0.63295X_3^2 + 0.29598X_3^3 - 0.0401X_3^4 \quad (9.31)$$

For this particular example, $X_3 = $ log(0.00022) + 6 = 2.35, and

$$\left(\frac{H_L}{\psi}\right) = -0.10307 + 0.617771 \times 2.35 - 0.63295 \times 2.35^2 + 0.29598$$
$$\times 2.35^3 - 0.0401 \times 2.35^4 = 0.472.$$

This means the liquid holdup is $H_L = \Psi(H_L/\Psi) = 1 \times 0.472 = 0.472$.

An arbitrarily defined[2] mixture viscosity is calculated using the following equation based on the liquid holdup:

$$\mu_m = \mu_L^{H_L} \times \mu_G^{(1-H_L)} \quad (9.32)$$

Since this numerical example includes gas and oil as the liquid phase (L = O),

$$\mu_m = 13.24^{0.472} \times 0.0117^{(1-0.472)} = 0.323\,cP$$

The liquid volume fraction, C_L, is computed next for obtaining the no-slip mixture density, ρ_n:

$$C_L = \frac{Q_L}{Q_G + Q_L} = \frac{0.0022}{0.0038 + 0.0022} = 0.365;$$

$$\rho_n = 886.29 \times 0.365 + 35.09 \times (1 - 0.365) = 345\,kg\,/\,m^3$$

The mixture Reynold's number using ρ_n, μ_m, U_m ($U_{sG} + U_{sL}$) and tubing ID is calculated to obtain the friction factor:

$$R_{em} = \frac{0.075946 \times (0.847 + 0.486) \times 345}{0.323 \times 10^{-3}} \cong 108,180$$

Given the high Reynold's number, the Chen[6] equation is used to obtain f_F, which is 0.0047 (since calculation using the Chen equation was shown earlier, it is not repeated here). Finally, the density based on holdup and frictional and hydrostatic losses, respectively, are calculated to determine the total pressure loss:

$$\rho_s = \rho_L(H_L) + \rho_G(1 - H_L) = 886.29 \times 0.472 + 35.09 \times (1 - 0.472) = 437\ kg/m^3$$

$$\text{Frictional loss} = \frac{2f_F\rho_n^2U_m^2}{\rho_s D}$$

(9.33)

$$= \frac{4.4219\times10^{-5}\times2\times0.0047\times345^2\times1.333^2}{436\times2.99\times0.0254} = 0.0027\,\text{psi}/\text{ft}$$

$$\text{Hydrostatic loss} = \rho_s g = 436\times9.81\times4.4219\times10^{-5} = 0.189\,\text{psi}/\text{ft}$$

$$\text{Therefore, total pressure loss} = 0.0027+0.189 = 0.192\,\text{psi}/\text{ft.}$$

Table 9.4 shows the detailed calculation results for the mHB method. The final computed flowing bottomhole pressure value differs by approximately 17% from the measured value as opposed to 15% for Hasan and Kabir's calculations.[2] However, this minor difference given the temperature uncertainties noted earlier is reasonable.

TABLE 9.4

Detailed Pressure Traverse Calculations Using the Modified Hagedorn–Brown Correlation for the Vertical System Numerical Example (additional columns continued)

Depth, ft	Temperature, °F	Pressure, psia	Z factor	B_g, ft³/scf	R_s, scf/STB	B_o, res. bbl/STB
0	107	520	0.9036	0.0279	81.6	1.0493
200	111	558	0.8991	0.02601	87.7	1.0539
400	115	597	0.8949	0.02438	93.8	1.0585
600	119	636	0.8909	0.02294	99.9	1.0631
800	123	675	0.8871	0.02166	106.0	1.0677
1,000	127	715	0.8835	0.02051	112.2	1.0722
1,200	130	754	0.8801	0.01947	118.4	1.0768
1,400	134	794	0.8768	0.01854	124.7	1.0813
1,600	137	835	0.8736	0.01767	131.0	1.0859
1,800	140	876	0.8705	0.01686	137.6	1.0905
2,000	144	919	0.8674	0.01610	144.4	1.0951
2,200	147	964	0.8643	0.01538	151.4	1.0998
2,400	150	1,009	0.8613	0.01471	158.7	1.1046
2,600	152	1,056	0.8583	0.01407	166.1	1.1095
2,800	155	1,104	0.8554	0.01348	173.9	1.1144
3,000	158	1,153	0.8525	0.01291	181.8	1.1194
3,200	160	1,203	0.8498	0.01238	190.0	1.1244
3,400	162	1,254	0.8471	0.01189	198.4	1.1295
3,600	165	1,306	0.8445	0.01142	207.1	1.1347
3,800	167	1,360	0.8421	0.01097	215.9	1.1399
4,000	169	1,414	0.8398	0.01055	225.1	1.1452
4,200	171	1,470	0.8376	0.01016	234.4	1.1506
4,400	172	1,526	0.8355	0.00979	244.1	1.1560
4,600	174	1,584	0.8336	0.00944	253.9	1.1615
4,800	176	1,642	0.8319	0.00910	264.0	1.1671
5,000	177	1,701	0.8303	0.00879	274.4	1.1728
5,151	178	**1,746**	0.8292	0.00856	282.4	1.1771

(Continued)

TABLE 9.4   (Continued)
Detailed Pressure Traverse Calculations Using the Modified Hagedorn–Brown Correlation for the Vertical System Numerical Example (additional columns continued)

Q_G, m³/sec	Q_O, m³/sec	U_{sG}, m/sec	U_{sL}, m/sec	ρ_G, kg/m³	ρ_O, kg/m³	σ_{GO}/σ_{OD}	σ_{GO} (or σ_{GL}), dynes/cm or mN/m
0.0038	0.0022	0.847	0.486	35.09	886.29	0.7456	23.69
0.0035	0.0022	0.777	0.488	37.59	883.43	0.7297	23.18
0.0032	0.0022	0.716	0.490	40.10	880.61	0.7141	22.69
0.0030	0.0022	0.663	0.492	42.62	877.81	0.6988	22.20
0.0028	0.0022	0.614	0.494	45.15	875.06	0.6838	21.72
0.0026	0.0022	0.571	0.497	47.68	872.34	0.6690	21.25
0.0024	0.0023	0.533	0.499	50.21	869.66	0.6545	20.79
0.0023	0.0023	0.497	0.501	52.76	867.03	0.6403	20.34
0.0021	0.0023	0.465	0.503	55.34	864.42	0.6262	19.90
0.0020	0.0023	0.434	0.505	58.00	861.83	0.6120	19.44
0.0018	0.0023	0.406	0.507	60.74	859.24	0.5977	18.99
0.0017	0.0023	0.379	0.509	63.57	856.66	0.5832	18.53
0.0016	0.0023	0.353	0.512	66.49	854.09	0.5687	18.07
0.0015	0.0023	0.329	0.514	69.48	851.53	0.5542	17.61
0.0014	0.0023	0.307	0.516	72.56	848.99	0.5396	17.14
0.0013	0.0023	0.286	0.518	75.72	846.45	0.5251	16.68
0.0012	0.0024	0.266	0.521	78.96	843.92	0.5105	16.22
0.0011	0.0024	0.247	0.523	82.27	841.40	0.4961	15.76
0.0010	0.0024	0.229	0.525	85.66	838.90	0.4817	15.30
0.0010	0.0024	0.212	0.528	89.12	836.40	0.4674	14.85
0.0009	0.0024	0.196	0.530	92.65	833.92	0.4532	14.40
0.0008	0.0024	0.181	0.533	96.24	831.44	0.4391	13.95
0.0008	0.0024	0.166	0.535	99.90	828.98	0.4252	13.51
0.0007	0.0024	0.153	0.538	103.63	826.53	0.4115	13.07
0.0006	0.0024	0.140	0.541	107.41	824.09	0.3979	12.64
0.0006	0.0025	0.127	0.543	111.24	821.66	0.3846	12.22
0.0005	0.0025	0.118	0.545	114.17	819.83	0.3747	11.90

N_{LU}	N_{GU}	N_D	N_L	$N_{GU}N_L^{0.38}/N_D^{2.14}$	Ψ	X_2	C_{NL}	N_f	X_3	(H_L/Ψ)	H_L
3.82	6.65	46	0.071	0.000673	1.00	1.85	0.0056	0.00022	2.351	0.472	0.472
3.85	6.14	46	0.068	0.000598	1.00	1.83	0.0054	0.00023	2.360	0.476	0.476
3.89	5.68	47	0.064	0.000532	1.00	1.81	0.0053	0.00023	2.368	0.480	0.480
3.92	5.28	47	0.061	0.000476	1.00	1.79	0.0051	0.00024	2.375	0.483	0.483
3.96	4.92	48	0.059	0.000428	1.00	1.77	0.0050	0.00024	2.383	0.487	0.487
3.99	4.60	48	0.056	0.000385	1.00	1.75	0.0048	0.00025	2.389	0.490	0.490
4.03	4.30	49	0.054	0.000348	1.00	1.73	0.0047	0.00025	2.396	0.493	0.493
4.07	4.04	49	0.052	0.000315	1.00	1.71	0.0046	0.00025	2.402	0.496	0.496
4.10	3.79	50	0.050	0.000285	1.00	1.70	0.0044	0.00026	2.409	0.499	0.499
4.14	3.56	50	0.048	0.000258	1.00	1.68	0.0043	0.00026	2.415	0.502	0.502
4.18	3.34	51	0.046	0.000234	1.00	1.66	0.0042	0.00026	2.423	0.506	0.506
4.22	3.14	51	0.044	0.000212	1.00	1.65	0.0041	0.00027	2.430	0.509	0.509
4.26	2.94	52	0.043	0.000191	1.00	1.63	0.0041	0.00027	2.438	0.513	0.513
4.31	2.76	52	0.041	0.000173	1.00	1.62	0.0040	0.00028	2.447	0.518	0.518

(Continued)

TABLE 9.4 (Continued)
Detailed Pressure Traverse Calculations Using the Modified Hagedorn–Brown Correlation for the Vertical System Numerical Example (additional columns continued)

N_{LU}	N_{GU}	N_D	N_L	$N_{GU}N_L^{0.38}/N_D^{2.14}$	Ψ	X_2	C_{NL}	N_f	X_3	(H_L/Ψ)	H_L
4.35	2.59	53	0.040	0.000156	1.00	1.60	0.0039	0.00029	2.456	0.522	0.522
4.40	2.42	54	0.039	0.000141	1.00	1.59	0.0038	0.00029	2.465	0.526	0.526
4.44	2.27	54	0.038	0.000127	1.00	1.58	0.0038	0.00030	2.475	0.531	0.531
4.49	2.12	55	0.037	0.000114	1.00	1.56	0.0037	0.00031	2.485	0.536	0.536
4.54	1.98	56	0.036	0.000102	1.00	1.55	0.0036	0.00031	2.496	0.542	0.542
4.60	1.84	56	0.035	0.000092	1.00	1.54	0.0036	0.00032	2.508	0.548	0.548
4.65	1.72	57	0.034	0.000082	1.00	1.53	0.0035	0.00033	2.520	0.554	0.554
4.70	1.59	58	0.033	0.000073	1.00	1.52	0.0035	0.00034	2.533	0.560	0.560
4.76	1.48	59	0.032	0.000065	1.00	1.51	0.0034	0.00035	2.547	0.567	0.567
4.82	1.37	60	0.032	0.000058	1.00	1.50	0.0034	0.00036	2.562	0.575	0.575
4.88	1.26	61	0.031	0.000051	1.00	1.49	0.0034	0.00038	2.577	0.583	0.583
4.94	1.16	62	0.030	0.000045	1.00	1.48	0.0033	0.00039	2.594	0.591	0.591
4.99	1.08	62	0.030	0.000041	1.00	1.48	0.0033	0.00041	2.607	0.598	0.598

μ_G, cP	μ_O, cP	μ_m, cP	C_L	H_L to use	ρ_n, kg/m³	R_{em}	f_F	ρ_s, kg/m³	Fric. loss, psi/ft	Hydr. loss, psi/ft	Total press. loss, psi/ft
0.0117	13.24	0.323	0.365	0.472	345	108,180	0.0047	437	0.0027	0.189	0.192
0.0119	12.38	0.325	0.386	0.476	364	107,698	0.0047	440	0.0026	0.191	0.194
0.0121	11.59	0.325	0.406	0.480	382	107,475	0.0047	443	0.0026	0.192	0.195
0.0122	10.86	0.325	0.426	0.483	399	107,447	0.0047	446	0.0026	0.194	0.196
0.0124	10.20	0.325	0.446	0.487	415	107,557	0.0047	449	0.0026	0.195	0.197
0.0126	9.59	0.324	0.465	0.490	431	107,761	0.0047	452	0.0026	0.196	0.199
0.0127	9.03	0.324	0.484	0.493	446	108,023	0.0047	454	0.0026	0.197	0.200
0.0129	8.52	0.335	0.502	0.502	461	104,373	0.0047	461	0.0025	0.200	0.203
0.0130	8.04	0.367	0.520	0.520	476	95,148	0.0048	476	0.0025	0.206	0.209
0.0132	7.60	0.402	0.538	0.538	490	86,889	0.0049	490	0.0025	0.213	0.215
0.0134	7.19	0.440	0.556	0.556	504	79,497	0.0050	504	0.0024	0.219	0.221
0.0136	6.80	0.480	0.573	0.573	518	72,884	0.0050	518	0.0024	0.225	0.227
0.0138	6.44	0.522	0.591	0.591	532	66,969	0.0051	532	0.0024	0.231	0.233
0.0139	6.10	0.567	0.609	0.609	546	61,679	0.0052	546	0.0023	0.237	0.239
0.0141	5.79	0.614	0.627	0.627	559	56,947	0.0053	559	0.0023	0.243	0.245
0.0143	5.49	0.663	0.645	0.645	573	52,715	0.0054	573	0.0023	0.248	0.251
0.0145	5.21	0.715	0.662	0.662	586	48,929	0.0054	586	0.0023	0.254	0.256
0.0147	4.95	0.768	0.680	0.680	598	45,541	0.0055	598	0.0023	0.259	0.262
0.0149	4.71	0.823	0.697	0.697	610	42,508	0.0056	610	0.0023	0.265	0.267
0.0152	4.48	0.879	0.714	0.714	622	39,791	0.0057	622	0.0022	0.270	0.272
0.0154	4.27	0.936	0.730	0.730	634	37,358	0.0057	634	0.0022	0.275	0.277
0.0156	4.06	0.994	0.747	0.747	645	35,177	0.0058	645	0.0022	0.280	0.282
0.0158	3.87	1.053	0.763	0.763	656	33,221	0.0059	656	0.0022	0.285	0.287
0.0161	3.70	1.111	0.779	0.779	667	31,466	0.0060	667	0.0022	0.289	0.291
0.0163	3.53	1.170	0.795	0.795	677	29,891	0.0060	677	0.0022	0.294	0.296
0.0166	3.37	1.228	0.810	0.810	687	28,478	0.0061	687	0.0022	0.298	0.300
0.0168	3.26	1.271	0.822	0.822	694	27,505	0.0061	694	0.0022	0.301	0.303

9.7.3 MECHANISTIC MODELS – VERTICAL FLOW

Empirical pressure drop models such as the mHB and Beggs and Brill[26] (covered in Section 9.7.4.1) can potentially result in inaccuracies when applied to flow conditions outside the bounds of the experimental database on which they are developed. Therefore, a theoretically sound approach is the use of the so-called mechanistic models that use the governing equations based on flow patterns for determining the liquid holdup and pressure drop.

If we consider the case of an undersaturated oil and a vertical wellbore, different flow patterns would exist at various depths owing to the large variation in temperature and pressure, vis-à-vis fluid properties. As illustrated by Hasan and Kabir,[2] at the bottomhole, a single-phase liquid may exist, which may transition to a bubble flow with gas evolution due to pressure reduction (basically past the oil bubble point at the prevailing temperature). However, a further reduction in pressure with decreasing depth will result in additional gas liberation (this is a continual process depending on the overall fluid composition and solubilities of relatively lighter components), in other words, decreasing R_s, which will be conducive to the concatenation of evolved gas (bubbles) becoming sufficiently large to occupy a significant portion of the pipe cross section. This is basically what is termed *slug flow*, and the flow pattern shift is referred to as *bubbly–slug transition*. The decreasing depth may lead to what is known as churn flow following the breakdown of larger bubbles in the preceding slug flow. The movement of gas in this flow regime is often chaotic, changing character with time.[2] This shift in the flow pattern is called a slug–churn transition. Quite often, the slug and churn flow are grouped together on flow pattern maps (e.g., Figure 9.8 for vertical flow). The fourth transition from churn (or slug) flow leads to annular flow, which can be described by the gas flowing through the pipe center and the liquid flowing upward along the pipe wall, a bit like the liquid film "wetting" the pipe wall. Although in principle the flow pattern shifts in a vertical well may be (single phase) → bubbly → slug → churn → annular, the dominant flow regime(s) will be primarily dictated or influenced by the phase envelope of a given oil in conjunction with the PT conditions prevalent in the wellbore, or in other words, it is a matter of how "deep" these conditions enter in the two-phase region of the phase envelope. For instance, if the oil is highly undersaturated, then only single phase, bubbly, and perhaps partly slug flow patterns may exist.

In mechanistic models, basically two approaches can be employed to delineate the flow pattern to determine holdup and eventually the pressure drop. One approach is based on flow pattern maps such as the one depicted in Figure 9.8 based on gas and liquid Froude numbers. The other alternative and flexible approach is based on individual equations that are based on criteria that are convenient for spreadsheet-type computations. Notwithstanding the approach used, special emphasis is always placed on the bubbly → slug flow transition given its importance in flow assurance. Hasan and Kabir[2] have stated that the individual transition criteria (equation-based) is probably the most reliable approach available today. We will use the same numerical example as before to illustrate the application of one of the mechanistic models to calculate the pressure traverse. However, before that is done, the use of a flow pattern map is shown in Figure 9.10 to get an idea of the flow regime. For this purpose, we use the U_{sG}, U_{sL}, ρ_G and ρ_L values from Table 9.4 to obtain the F_{rG} and F_{rL} values for

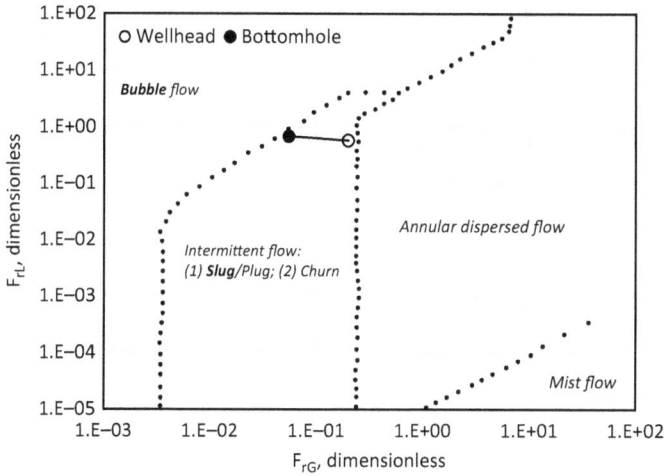

FIGURE 9.10 Froude numbers based on the data shown in Table 9.4 for the numerical example, plotted on the digitized flow regime map,[18] indicating a *possible* transition from bubble flow (bottomhole)-> slug/churn flow (wellhead) -> (*annular flow*).

bottomhole to wellhead conditions. Froude number calculations were shown previously; therefore, they are not repeated here. As seen in Figure 9.10, the bottomhole conditions appear to be on the boundary (or transition) between the bubbly flow and slug flow, moving to slug/churn flow area and eventually ending near the boundary between slug/churn flow and annular flow for the wellhead conditions.

The application of individual transition criteria requires the use of equations that basically result in (gas) velocity limits for establishing the flow pattern transition. Once that is determined, a suitable (mechanistic) model can be employed to obtain the liquid holdup, mixture density, viscosity, and the like for frictional and hydrostatic losses, respectively. For the wellhead conditions, we first calculate $U\infty$, known as the terminal rise velocity of a single bubble, followed by Taylor bubble:[2]

$$
\begin{aligned}
U_\infty &= 1.53 \left[\frac{g \sigma_{GL} \left(\rho_L - \rho_G \right)}{\rho_L^2} \right]^{0.25} \\
&= 1.53 \left[\frac{9.81 \times 23.69 \times 10^{-3} \left(886.29 - 35.09 \right)}{886.29^2} \right]^{0.25} \\
&= 0.193 \, \text{m/sec}
\end{aligned}
\tag{9.34}
$$

$$
\begin{aligned}
U_{\infty T} &= 0.345 \left[\frac{g D \left(\rho_L - \rho_G \right)}{\rho_L} \right]^{0.5} \\
&= 0.345 \left[\frac{9.81 \times 0.0759 \times \left(886.29 - 35.09 \right)}{886.29} \right]^{0.5} \\
&= 0.292 \, \text{m/sec}
\end{aligned}
\tag{9.35}
$$

Subsequently, we calculate three different transitional superficial gas velocities, namely, U_t^{BS}; U_t^{SC} and U_t^{CA}, where the BS, SC and CA superscripts denote bubbly–slug, slug–churn, and churn–annular, respectively.

$$\begin{aligned} U_t^{BS} &= 0.429 U_{sL} + 0.357 U_\infty \\ &= 0.429 \times 0.486 + 0.357 \times 0.193 = 0.277 \, \text{m/sec} \end{aligned} \tag{9.36}$$

$$\begin{aligned} U_t^{SC} &= 12.19 \left(1.2 U_{sL} + U_{\infty T} \right) \\ &= 12.19 \times \left(1.2 \times 0.486 + 0.292 \right) = 10.666 \, \text{m/sec} \end{aligned} \tag{9.37}$$

$$\begin{aligned} U_t^{CA} &= 3.1 \left[\frac{g \sigma_{GL} \left(\rho_L - \rho_G \right)}{\rho_G^2} \right]^{0.25} \\ &= 3.1 \left[\frac{9.81 \times 23.69 \times 10^{-3} \times \left(886.29 - 35.09 \right)}{35.09^2} \right]^{0.25} \\ &= 1.963 \, \text{m/sec} \end{aligned} \tag{9.38}$$

Equation 9.37 is from Kaya et al.[27] and others are provided in Hasan and Kabir.[2] For this particular numerical example, we have adopted a somewhat simplified velocity-based approach, which identifies the prevailing flow pattern via elimination. The transition criteria are as follows. For bubbly flow $U_{sG} < U_t^{BS}$; slug flow $U_{sG} > U_t^{BS}$; churn flow $U_{sG} > U_t^{SC}$; annular flow $U_{sG} > U_t^{CA}$. The preceding calculated values (0.847 > 0.277) denotes slug flow at the wellhead conditions. This process is applied to every segment in the subsequent calculations, eventually reaching the bottomhole conditions. Once the flow pattern is determined, the ensuing calculation steps to determine the frictional and hydrostatic pressure losses are similar to those shown previously.

For the slug flow identified at the wellhead conditions, the liquid holdup is calculated as follows, where 1.2 is the fixed value of a flow coefficient:[2]

$$\begin{aligned} H_L &= 1 - \frac{U_{sG}}{1.2 U_m + \bar{U}_\infty} \\ &= 1 - \frac{0.847}{1.2 \times \left(0.847 + 0.486 \right) + 0.264} \\ &= 0.546 \end{aligned} \tag{9.39}$$

where $\bar{U}_\infty$ is the average rise velocity of small bubbles, mathematically expressed as[2]

$$\begin{aligned} \bar{U}_\infty &= U_\infty \left(1 - e^{-U_t^{BS}/U_{sG}} \right) + U_{\infty T} e^{-U_t^{BS}/U_{sG}} \\ &= 0.193 \times \left(1 - e^{-0.277/0.847} \right) + 0.292 e^{-0.277/0.847} \\ &= 0.264 \, \text{m/sec} \end{aligned} \tag{9.40}$$

The H_L value is used next to calculate the mixture density and viscosity, respectively.

$$\rho_m = 0.546 \times 886.29 + (1 - 0.546) \times 35.09 = 499.5 \, kg / m^3$$

$$\mu_m = 0.546 \times 13.24 + (1 - 0.546) \times 0.0117 = 7.227 \, cP$$

Note that Hasan and Kabir[2] use gas volume fraction in calculating μ_m; we chose to use H_L instead to be somewhat consistent and given the significantly low contribution of friction to the total pressure drop. Next, the mixture Reynold's number is

$$R_{em} = \frac{0.0759 \times (0.847 + 0.486) \times 499.5}{7.227 \times 10^{-3}} = 6{,}996$$

Using Chen's[6] friction factor correlation, $f_F = 0.0085$, and frictional pressure loss is

$$\text{Frictional loss} = \frac{2 f_F \rho_m U_m^2}{D}$$

$$= \frac{2 \times 0.0085 \times 499.5 \times 1.333^2 \times 4.4219 \times 10^{-5}}{0.0759}$$

$$= 0.0088 \, psi / ft$$

$$\text{Hydrostatic loss} = \rho_m g = 499.5 \times 9.81 \times 4.4219 \times 10^{-5} = 0.2167 \, psi / ft$$

Therefore, total pressure loss equals 0.0088 + 0.2167 = 0.2255 psi/ft at the well-head and pressure at 200 ft below = 519.7 + 0.2255 × 200 = 565 psia. Table 9.5 shows the detailed calculations for this numerical example using the mechanistic[2] model. As noted in the table, a bubbly flow pattern occurs at a depth of 2,400 ft and 1,141

TABLE 9.5

Detailed Pressure Traverse Calculations Using the Hasan and Kabir[2] Mechanistic Model for the Vertical System Numerical Example (additional columns continued)

Depth, ft	Temperature, °F	Pressure, psia	Z factor	B_g, ft³/scf	R_s, scf/STB	B_o, res. bbl/STB	Q_G, m³/sec	Q_O, m³/sec
0	107	520	0.9036	0.02790	81.6	1.0493	0.0038	0.0022
200	111	565	0.8979	0.02567	88.9	1.0544	0.0035	0.0022
400	115	611	0.8924	0.02374	96.4	1.0596	0.0031	0.0022
600	119	659	0.8871	0.02204	104.1	1.0648	0.0028	0.0022
800	123	708	0.8818	0.02053	112.1	1.0701	0.0026	0.0022
1,000	127	758	0.8767	0.01918	120.3	1.0756	0.0024	0.0023
1,200	130	810	0.8717	0.01796	128.7	1.0810	0.0022	0.0023

(Continued)

TABLE 9.5   (Continued)

Detailed Pressure Traverse Calculations Using the Hasan and Kabir2 Mechanistic Model for the Vertical System Numerical Example (additional columns continued)

Depth, ft	Temperature, °F	Pressure, psia	Z factor	B_g, ft³/scf	R_s, scf/STB	B_o, res. bbl/STB	Q_G, m³/sec	Q_O, m³/sec
1,400	134	863	0.8670	0.01687	137.3	1.0866	0.0020	0.0023
1,600	137	916	0.8624	0.01589	146.1	1.0922	0.0018	0.0023
1,800	140	971	0.8580	0.01500	155.2	1.0979	0.0017	0.0023
2,000	144	1,027	0.8538	0.01419	164.4	1.1037	0.0015	0.0023
2,200	147	1,083	0.8498	0.01345	173.8	1.1095	0.0014	0.0023
2,400	*150*	*1,141*	*0.8461*	*0.01278*	*183.3*	*1.1153*	*0.0013*	*0.0023*
2,600	152	1,199	0.8426	0.01217	193.0	1.1212	0.0012	0.0024
2,800	155	1,258	0.8394	0.01161	202.8	1.1271	0.0011	0.0024
3,000	158	1,317	0.8365	0.01109	212.8	1.1331	0.0010	0.0024
3,200	160	1,377	0.8337	0.01061	223.0	1.1391	0.0009	0.0024
3,400	162	1,438	0.8313	0.01017	233.3	1.1451	0.0008	0.0024
3,600	165	1,500	0.8290	0.00976	243.9	1.1512	0.0008	0.0024
3,800	167	1,562	0.8271	0.00938	254.6	1.1574	0.0007	0.0024
4,000	169	1,625	0.8254	0.00903	265.4	1.1636	0.0006	0.0024
4,200	171	1,689	0.8239	0.00870	276.5	1.1698	0.0006	0.0025
4,400	172	1,753	0.8227	0.00839	287.7	1.1761	0.0005	0.0025
4,600	174	1,818	0.8217	0.00810	299.1	1.1824	0.0005	0.0025
4,800	176	1,883	0.8210	0.00784	310.6	1.1887	0.0004	0.0025
5,000	177	1,948	0.8206	0.00759	322.3	1.1951	0.0004	0.0025
5,151	178	**1,998**	0.8204	0.00741	331.3	1.2000	0.0003	0.0025

U_{sG}, m/sec	U_{sL}, m/sec	ρ_G, kg/m³	ρ_O, kg/m³	σ_{GO}/σ_{OD}	σ_{GO}, mN/m	U_∞, m/sec	$U_{\infty T}$, m/sec	U_t^{BS}, m/sec	U_t^{SC}, m/sec	U_t^{CA}, m/sec	Flow Pattern
0.847	0.486	35.09	886.29	0.7456	23.69	0.193	0.292	0.277	10.666	1.963	Slug
0.765	0.488	38.09	883.23	0.7266	23.08	0.191	0.291	0.278	10.694	1.868	Slug
0.692	0.491	41.18	880.17	0.7075	22.48	0.190	0.291	0.278	10.722	1.782	Slug
0.629	0.493	44.37	877.10	0.6884	21.87	0.189	0.290	0.279	10.751	1.702	Slug
0.572	0.496	47.64	874.03	0.6693	21.26	0.188	0.290	0.280	10.779	1.627	Slug
0.521	0.498	51.00	870.96	0.6503	20.66	0.186	0.289	0.280	10.808	1.559	Slug
0.476	0.501	54.43	867.89	0.6314	20.06	0.185	0.288	0.281	10.838	1.495	Slug
0.435	0.503	57.95	864.84	0.6126	19.46	0.183	0.288	0.281	10.867	1.435	Slug
0.398	0.506	61.54	861.79	0.5940	18.87	0.182	0.287	0.282	10.897	1.379	Slug
0.365	0.508	65.19	858.76	0.5757	18.29	0.180	0.286	0.283	10.927	1.326	Slug
0.334	0.511	68.91	855.74	0.5576	17.72	0.179	0.286	0.283	10.957	1.277	Slug
0.307	0.514	72.69	852.74	0.5398	17.15	0.177	0.285	0.284	10.988	1.231	Slug
0.281	*0.517*	*76.52*	*849.76*	*0.5223*	*16.59*	*0.176*	*0.284*	*0.284*	*11.018*	*1.187*	*Bubbly*
0.258	0.519	80.36	846.81	0.5054	16.06	0.174	0.283	0.285	11.049	1.146	Bubbly
0.237	0.522	84.25	843.89	0.4887	15.53	0.173	0.283	0.286	11.079	1.108	Bubbly
0.217	0.525	88.18	840.98	0.4724	15.01	0.171	0.282	0.286	11.110	1.071	Bubbly
0.199	0.528	92.16	838.10	0.4564	14.50	0.170	0.281	0.287	11.141	1.036	Bubbly
0.182	0.530	96.16	835.24	0.4408	14.00	0.168	0.280	0.288	11.172	1.003	Bubbly
0.166	0.533	100.20	832.41	0.4255	13.52	0.166	0.279	0.288	11.204	0.972	Bubbly

(Continued)

TABLE 9.5 (Continued)

Detailed Pressure Traverse Calculations Using the Hasan and Kabir2 Mechanistic Model for the Vertical System Numerical Example (additional columns continued)

U_{sG}, m/sec	U_{sL}, m/sec	ρ_G, kg/m³	ρ_O, kg/m³	σ_{GO}/σ_{OD}	σ_{GO}, mN/m	U_∞, m/sec	$U_{\infty T}$, m/sec	U_t^{BS}, m/sec	U_t^{SC}, m/sec	U_t^{CA}, m/sec	Flow Pattern
0.151	0.536	104.26	829.60	0.4106	13.04	0.165	0.278	0.289	11.235	0.942	Bubbly
0.137	0.539	108.34	826.81	0.3961	12.58	0.163	0.278	0.289	11.266	0.914	Bubbly
0.124	0.542	112.43	824.05	0.3820	12.13	0.162	0.277	0.290	11.298	0.887	Bubbly
0.112	0.545	116.54	821.32	0.3682	11.70	0.160	0.276	0.291	11.330	0.861	Bubbly
0.101	0.548	120.66	818.61	0.3548	11.27	0.159	0.275	0.291	11.362	0.837	Bubbly
0.090	0.551	124.78	815.92	0.3419	10.86	0.157	0.274	0.292	11.394	0.813	Bubbly
0.080	0.553	128.91	813.27	0.3293	10.46	0.155	0.273	0.293	11.426	0.790	Bubbly
0.073	0.556	132.02	811.28	0.3200	10.17	0.154	0.272	0.293	11.451	0.774	Bubbly

$\bar{U}_\infty$, m/sec	H_L	ρ_m, kg/m³	μ_G, cP	μ_O, cP	μ_m, cP	R_{em}	f_F	Fric. loss, psi/ft	Hydr. loss, si/ft	Total press. loss, psi/ft
0.264	0.546	499.5	0.0117	13.24	7.227	6996	0.0085	0.0088	0.2167	0.2255
0.261	0.567	517.0	0.0119	12.32	6.984	7043	0.0085	0.0080	0.2243	0.2323
0.257	0.587	533.8	0.0121	11.47	6.739	7118	0.0085	0.0074	0.2315	0.2389
0.254	0.607	549.9	0.0123	10.69	6.492	7217	0.0084	0.0068	0.2385	0.2453
0.250	0.626	565.3	0.0125	9.96	6.247	7338	0.0084	0.0063	0.2452	0.2515
0.246	0.645	580.0	0.0127	9.30	6.004	7480	0.0083	0.0059	0.2516	0.2575
0.242	0.663	594.1	0.0129	8.68	5.765	7643	0.0083	0.0055	0.2577	0.2632
0.238	0.681	607.4	0.0131	8.12	5.532	7825	0.0082	0.0051	0.2635	0.2686
0.234	0.698	620.1	0.0133	7.60	5.306	8025	0.0082	0.0048	0.2690	0.2738
0.229	0.714	632.1	0.0135	7.11	5.086	8242	0.0081	0.0046	0.2742	0.2788
0.225	0.730	643.5	0.0138	6.67	4.874	8477	0.0081	0.0043	0.2791	0.2834
0.220	0.745	654.2	0.0140	6.26	4.670	8728	0.0080	0.0041	0.2838	0.2879
0.215	*0.752*	*657.9*	*0.0142*	*5.88*	*4.425*	*9006*	*0.0079*	*0.0039*	*0.2854*	*0.2893*
0.210	0.767	668.2	0.0145	5.53	4.246	9288	0.0079	0.0037	0.2899	0.2936
0.206	0.782	677.9	0.0147	5.21	4.075	9585	0.0078	0.0035	0.2941	0.2976
0.201	0.796	687.1	0.0149	4.91	3.911	9896	0.0077	0.0034	0.2981	0.3015
0.196	0.809	695.7	0.0152	4.64	3.754	10,221	0.0077	0.0033	0.3018	0.3051
0.191	0.822	703.9	0.0155	4.38	3.604	10,561	0.0076	0.0032	0.3053	0.3085
0.186	0.835	711.6	0.0157	4.14	3.461	10,915	0.0075	0.0031	0.3087	0.3117
0.182	0.847	718.8	0.0160	3.92	3.325	11,283	0.0075	0.0030	0.3118	0.3147
0.177	0.859	725.5	0.0163	3.72	3.195	11,664	0.0074	0.0029	0.3147	0.3176
0.173	0.870	731.9	0.0165	3.53	3.071	12,058	0.0073	0.0028	0.3175	0.3203
0.169	0.882	737.8	0.0168	3.35	2.954	12,465	0.0073	0.0027	0.3201	0.3228
0.165	0.892	743.4	0.0171	3.18	2.842	12,884	0.0072	0.0026	0.3225	0.3251
0.161	0.903	748.7	0.0174	3.03	2.735	13,316	0.0072	0.0026	0.3248	0.3273
0.158	0.913	753.5	0.0177	2.88	2.634	13,760	0.0071	0.0025	0.3269	0.3294
0.156	0.920	757.0	0.0179	2.78	2.561	14,104	0.0071	0.0025	0.3284	0.3309

psia up to the bottomhole depth of 5,151 ft. The only difference in the bubbly flow is the calculation of liquid holdup:

$$H_L = 1 - \frac{U_{sG}}{1.2U_m + U_\infty}$$

(9.41)

At the depth of 2,400 ft,

$$H_L = 1 - \frac{0.281}{1.2 \times (0.281 + 0.517) + 0.176}$$
$$= 0.752$$

With H_L calculated from Equation 9.41, all the other computations are similar to the ones for slug flow. The final bottomhole pressure of 1,998 psia differs from Hasan and Kabir's calculation by less than 2.5%, which again is fairly reasonable and acceptable given the temperature uncertainty and the differences originating from fluid property calculation approaches.

9.7.4　Empirical and Mechanistic Models – Deviated Systems

The term *deviated* primarily refers to flow situations that are offset from the vertical in that not at 90° angle. This could mean horizontal wells, those that are deviated from 90°, and more complicated cases such as flowlines in offshore installations, following the bathymetry resulting in incline and decline in the orientation. The primary difference between vertical and deviated systems is the hydrostatic head, which is the major contributor in the former. In general, the methodology for determining the pressure at a given position in the deviated flow system is similar to vertical flow in that the idea is to calculate the contribution of frictional and hydrostatic components by recognizing the prevailing flow regimes. Similar to vertical flow, empirical and mechanistic approaches are commonplace, which are covered in this section by demonstrating their application to a numerical example (shown only for empirical).

9.7.4.1　Empirical Models – Beggs and Brill

Perhaps the most well-known empirical model that applies to any flow orientation is that of Beggs and Brill,[26] which can be used not just for wellbores but flowlines as well. The numerical example that we use for demonstrating the method is from Hasan and Kabir.[2] As done previously, a detailed breakdown of all calculations is shown. Again, the wellhead temperature information is not provided, which has been estimated from the given data using a similar methodology, as was done in the earlier example. This particular 6,241-ft deep (2.43-in. tubing ID) well is deviated by an angle of 17.5° from the vertical (72.5° from the horizontal) and produces a 33° API oil and 0.6 gravity gas with a GOR of 447 scf/STB. The rate reported[2] is supposedly the total liquid rate of 2,922 STB/D with a 20.02% water cut. Much of the calculations are similar; however, the presence of water in the flow stream requires calculating the composite liquid properties, such as ρ_L, σ_{GL} and μ_L, which are based on B_w,

ρ_W, and σ_{GW}. In this particular example, the water properties were determined from the empirical correlations in McCain.[28] At the given wellhead pressure of 361 psig[2] and the estimated 151°F temperature, $B_W = 1.0224$ res. bbl/STB, $\rho_W = 978.09$ kg/m³, $\sigma_{GW} = 59.8$ mN/m, and $\mu_W = 0.40$ cP. Since the salinity information is unknown, these values represent pure water properties at the stated conditions. The composite liquid properties are

$$\rho_L = 818.65 \times (1 - 0.2002) + 978.09 \times 0.2002 = 850.57 \, \text{kg} / \text{m}^3;$$

$$\sigma_{GL} = 22.26 \times (1 - 0.2002) + 59.80 \times 0.2002 = 29.78 \, \text{mN} / \text{m};$$

$$\mu_W = 3.19 \times (1 - 0.2002) + 0.40 \times 0.2002 = 2.63 \, \text{cP}.$$

Q_L and Q_G and U_{sL} and U_{sG} at the wellhead conditions are

$$Q_L = \frac{\left[2922 \times (1 - 0.2002) \times 1.0593 + 2922 \times 0.2002 \times 1.0224\right] \times 5.615 \times 0.3048^3}{24 \times 60 \times 60}$$
$$= 0.0057 \text{m}^3 / \sec$$

$$U_{sL} = \frac{0.0057}{\dfrac{\pi}{4} \times \left(\dfrac{2.43 \times 2.54}{100}\right)^2} = 1.890 \, \text{m} / \sec$$

$$Q_G = \frac{\left[2922 \times (1 - 0.2002) \times (447 - 53.6) \times 0.04446\right] \times 0.3048^3}{24 \times 60 \times 60} = 0.0134 \text{m}^3 / \sec$$

$$U_{sG} = \frac{0.0134}{\dfrac{\pi}{4} \times \left(\dfrac{2.43 \times 2.54}{100}\right)^2} = 4.478 \, \text{m} / \sec$$

The no-slip liquid holdup $C_L = \dfrac{1.890}{1.890 + 4.478} = 0.297$ and Froude number,

$$N_{Fr} = \frac{U_m^2}{gD} = \frac{(1.890 + 4.478)^2}{9.81 \times \dfrac{2.43 \times 2.54}{100}} = 67.$$

Next, four different boundary parameters are calculated from C_L to determine the prevailing flow pattern, which actually is based on their map for horizontal systems (corrected for deviated systems, as shown later). These are, $L_1 = 316 \times 0.297^{0.302} = 219$; $L_2 = 0.0009252 \times 0.297^{-2.4684} = 0.0185$; $L_3 = 0.1 \times 0.297^{-1.4516} = 0.583$; $L_4 = 0.5 \times 0.297^{-6.738} = 1792$. The calculated C_L, L_1–L_4 and N_{Fr} values indicates an intermittent flow pattern at the wellhead conditions, based on Table 9.6.

TABLE 9.6

Beggs and Brill[26] Flow Pattern Determination Criteria Based on C_L, L_1–L_4 and N_{Fr}

Flow regime	Criterion 1	OR	Criterion 2
Segregated	$C_L < 0.01$ and $N_{Fr} < L_1$		$C_L \geq 0.01$ and $N_{Fr} < L_2$
Transition	$C_L \geq 0.01$ and $L_2 < N_{Fr} \leq L_3$		–
Intermittent	$0.01 \leq C_L < 0.4$ and $L_3 < N_{Fr} \leq L_1$		$C_L \geq 0.4$ and $L_3 < N_{Fr} \leq L_4$
Distributed	$C_L < 0.4$ and $N_{Fr} \geq L_1$		$C_L \geq 0.4$ and $N_{Fr} > L_4$

The liquid holdup for horizontal orientation ($0°$ inclination) – since Beggs and Brill method basis is horizontal orientation – is calculated based on C_L, N_{Fr} and the flow pattern–specific parameters,[26] a, b, and c (value shown directly in calculations):

$$
\begin{aligned}
H_L\left(0°\right) &= a\, C_L^b \Big/ N_{Fr}{}^c \\
&= 0.845 \times \left(0.297^{0.5351} \Big/ 67^{0.0173}\right) \\
&= 0.410
\end{aligned}
\tag{9.42}
$$

The value of $H_L(0°)$ is corrected for the inclination angle using the following equation:

$$
H_L\left(\alpha°\right) = 1 + C\left[\sin\left(1.8\alpha\right) - \frac{1}{3}\sin^3\left(1.8\alpha\right)\right]
\tag{9.43}
$$

where $\alpha = 90° - 17.5° = 72.5°$ and C is a coefficient determined from C_L, N_{LU}, N_{Fr}, and flow pattern–specific "$d - g$" parameters in Beggs and Brill,

$$
\begin{aligned}
C &= \left(1 - C_L\right)\ln\left(dC_L{}^e N_{LU}{}^f N_{Fr}{}^g\right) \\
&= \left(1 - 0.297\right) \times \mathrm{Ln}\left(2.96 \times 0.297^{0.305} \times 13.9^{-0.4473} \times 67^{0.0978}\right) \\
&= -0.036
\end{aligned}
\tag{9.44}
$$

note that the calculation of N_{LU} is identical to the one shown previously.

Therefore, $H_L\left(\alpha°\right) = 1 + \left(-0.036\right)\left[\sin\left(1.8 \times 72.5\right) - \frac{1}{3}\sin^3\left(1.8 \times 72.5\right)\right] = 0.978$

and $H_L = H_L(0°) \times H_L(\alpha°) = 0.410 \times 0.978 = 0.401$.

The no-slip density, ρ_n, and viscosity, μ_n, are calculated using C_L to obtain the no-slip Reynold's number, R_{en}, and the no-slip friction factor, f_{Fn}:

$$
\rho_n = 850.57 \times 0.297 + 16.49 \times \left(1 - 0.297\right) = 264 \ \mathrm{kg/m^3}
$$

$$\mu_n = 2.63 \times 0.297 + 0.012 \times (1 - 0.297) = 0.788 \, \text{cP}$$

$$R_{en} = \frac{0.06172 \times (4.478 + 1.890) \times 264}{0.788 \times 10^{-3}} = 131,670$$

The two-phase or mixture friction factor, f_{Fm}, is obtained from

$$f_{Fm} = f_{Fn} \times \exp(s) \tag{9.45}$$

where

$$s = \frac{LN(y)}{-0.0523 + 3.182LN(y) - 0.8725\left[LN(y)\right]^2 + 0.01853\left[LN(y)\right]^4} \tag{9.46}$$

with $y = C_L/H_L^2 = 0.297/0.401^2 = 1.844$,

$$s = \frac{LN(1.844)}{-0.0523 + 3.182LN(1.844) - 0.8725\left[LN(1.844)\right]^2 + 0.01853\left[LN(1.844)\right]^4} = 0.390$$

and $f_{Fm} = 0.0046 \times \exp(0.390) = 0.0068$ (f_{Fn} of 0.0046 is from the Chen equation), resulting in a frictional loss of 0.105 psi/ft using the no-slip density of 264 kg/m³. This calculation is similar to what was shown earlier; hence, it is not repeated here.

The mixture density with slip, ρ_{ms}, is calculated to obtain the hydrostatic loss:

$$\rho_{ms} = 850.57 \times 0.401 + 16.49 \times (1 - 0.401) = 351 \, \text{kg/m}^3;$$

$$\text{Hydrostatic loss} = 351 \times 9.81 \times \sin(72.5°) \times 4.4219 \times 10^{-5} = 0.145 \, \text{psi/ft.}$$

Therefore, the total pressure loss equals $0.105 + 0.145 = 0.250$ psi/ft and the pressure at 200 ft below the wellhead is 426 psia. Note that at a depth of 2,200 ft or 941 psia, the flow pattern becomes distributed, which is logical given the higher pressure or in other words relatively less free gas fraction. This flow regime has a subcategory of bubble flow[26] and is obviously akin to the scenario discussed earlier in the context of mechanistic models for vertical flow. Again, detailed calculations that end at 6,241 ft are shown below in Table 9.7. The calculated values of pressure traverse are nearly identical to those reported by Hasan and Kabir;[2] as a matter of fact, they are a bit closer to the measured data despite the uncertainty in the wellhead temperature.

9.7.4.2 Mechanistic Models

Similar to the mechanistic models for vertical flow, the ones for deviated flow too are based on fundamental laws. Following the prediction of flow regime, empirical closure relationships are still required to complete the calculations. The formulations of the mechanistic models generally differ based on the applications, that is, wellbores and pipelines (for the most part, anything other than the tubular). For example, Hasan

TABLE 9.7

Detailed Pressure Traverse Calculations Using the Beggs and Brill[26] Model for the Deviated System Numerical Example (additional columns continued)

Depth, ft	Temp., °F	Pres., psia	Z factor	B_g, ft³/scf	R_s, scf/STB	B_o, res. bbl/STB	B_w, res. bbl/STB	Q_G, m³/sec	Q_L, m³/sec
0	151	376	0.9671	0.04446	53.6	1.0593	1.0224	0.0134	0.0057
400	157	476	0.9602	0.03519	69.0	1.0679	1.0240	0.0102	0.0057
800	162	576	0.9540	0.02911	84.9	1.0766	1.0255	0.0081	0.0057
1,200	167	678	0.9482	0.02478	101.3	1.0855	1.0270	0.0066	0.0058
1,600	172	782	0.9429	0.02153	118.1	1.0945	1.0284	0.0054	0.0058
2,000	176	888	0.9381	0.01900	135.5	1.1035	1.0296	0.0045	0.0059
2,400	180	998	0.9336	0.01692	154.0	1.1130	1.0308	0.0038	0.0059
2,800	183	1,114	0.9293	0.01517	173.7	1.1228	1.0318	0.0032	0.0059
3,200	187	1,233	0.9255	0.01372	194.2	1.1328	1.0327	0.0027	0.0060
3,600	189	1,355	0.9222	0.01250	215.4	1.1430	1.0335	0.0022	0.0060
4,000	192	1,479	0.9193	0.01146	237.3	1.1534	1.0342	0.0018	0.0061
4,400	194	1,604	0.9170	0.01057	260.0	1.1640	1.0348	0.0015	0.0061
4,800	196	1,732	0.9151	0.00980	283.4	1.1748	1.0352	0.0012	0.0062
5,200	198	1,862	0.9137	0.00913	307.5	1.1858	1.0355	0.0010	0.0062
5,600	199	1,993	0.9127	0.00853	332.4	1.1970	1.0357	0.0007	0.0063
6,000	200	2,125	0.9123	0.00801	358.1	1.2084	1.0357	0.0005	0.0063
6,241	200	2,206	0.9122	0.00772	374.0	1.2154	1.0357	0.0004	0.0063

U_{sG}, m/sec	U_{sL}, m/sec	ρ_G, kg/m³	ρ_o, kg/m³	ρ_w, kg/m³	ρ_L, kg/m³	σ_{GO}/σ_{OD}	σ_{GO}, mN/m	σ_{GW}, mN/m	σ_{GL}, mN/m
4.478	1.890	16.49	818.65	978.09	850.57	0.824	22.26	59.80	29.78
3.405	1.903	20.84	813.93	976.55	846.48	0.780	21.06	58.79	28.62
2.698	1.917	25.19	809.25	975.09	842.45	0.737	19.90	57.84	27.50
2.193	1.930	29.60	804.63	973.71	838.48	0.695	18.78	56.93	26.42
1.812	1.943	34.07	800.05	972.43	834.56	0.655	17.70	56.07	25.38
1.515	1.957	38.60	795.52	971.24	830.70	0.616	16.65	55.26	24.38
1.269	1.971	43.36	790.96	970.16	826.84	0.578	15.61	54.47	23.39
1.061	1.985	48.36	786.34	969.18	822.94	0.540	14.58	53.73	22.42
0.888	2.000	53.47	781.74	968.32	819.09	0.503	13.59	53.02	21.49
0.741	2.015	58.68	777.18	967.56	815.29	0.468	12.65	52.36	20.60
0.615	2.030	63.99	772.65	966.92	811.54	0.435	11.74	51.73	19.75
0.506	2.045	69.38	768.15	966.40	807.84	0.403	10.89	51.15	18.95
0.410	2.061	74.84	763.70	965.99	804.19	0.373	10.08	50.61	18.20
0.326	2.077	80.37	759.28	965.70	800.60	0.345	9.32	50.11	17.49
0.250	2.093	85.96	754.89	965.54	797.06	0.319	8.61	49.65	16.83
0.182	2.110	91.60	750.55	965.49	793.58	0.294	7.94	49.24	16.21
0.144	2.120	95.02	747.94	965.52	791.50	0.280	7.56	49.00	15.86

C_L	N_{Fr}	L_1	L_2	L_3	L_4	Flow Pattern	$H_L(0°)$	N_{LU}	C	$H_L(\alpha)$	H_L
0.297	67.0	219	0.0185	0.583	1791.6	Interm.	0.410	13.9	-0.036	0.978	0.401
0.359	46.5	232	0.0116	0.443	501.6	Interm.	0.457	14.1	-0.023	0.986	0.450
0.415	35.2	242	0.0081	0.358	186.4	Interm.	0.496	14.3	-0.015	0.991	0.492
0.468	28.1	251	0.0060	0.301	83.3	Interm.	0.531	14.6	-0.010	0.994	0.528

(Continued)

TABLE 9.7 (Continued)
Detailed Pressure Traverse Calculations Using the Beggs and Brill[26] Model for the Deviated System Numerical Example (additional columns continued)

C_L	N_{Fr}	L_1	L_2	L_3	L_4	Flow Pattern	$H_L(0°)$	N_{LU}	C	$H_L(\alpha)$	H_L
0.517	23.3	259	0.0047	0.260	42.4	Interm.	0.562	14.8	−0.006	0.996	0.560
0.564	19.9	266	0.0038	0.230	23.8	Interm.	0.590	15.0	−0.004	0.998	0.589
0.608	*17.3*	*272*	*0.0032*	*0.206*	*14.2*	*Distr.*	*0.670*	*15.3*	*0.000*	*1.000*	*0.670*
0.652	15.3	278	0.0027	0.186	9.0	Distr.	0.703	15.5	0.000	1.000	0.703
0.693	13.8	283	0.0023	0.170	5.9	Distr.	0.733	15.8	0.000	1.000	0.733
0.731	12.5	287	0.0020	0.158	4.1	Distr.	0.761	16.1	0.000	1.000	0.761
0.767	11.6	292	0.0018	0.147	3.0	Distr.	0.786	16.3	0.000	1.000	0.786
0.802	10.8	296	0.0016	0.138	2.2	Distr.	0.810	16.6	0.000	1.000	0.810
0.834	10.1	299	0.0014	0.130	1.7	Distr.	0.832	16.9	0.000	1.000	0.832
0.864	9.5	302	0.0013	0.124	1.3	Distr.	0.853	17.2	0.000	1.000	0.853
0.893	9.1	305	0.0012	0.118	1.1	Distr.	0.872	17.4	0.000	1.000	0.872
0.921	8.7	308	0.0011	0.113	0.9	Distr.	0.890	17.7	0.000	1.000	0.890
0.936	8.5	310	0.0011	0.110	0.8	Distr.	0.900	17.9	0.000	1.000	0.900

μ_G, cP	μ_O, cP	μ_W, cP	μ_L, cP	ρ_n, kg/m³	μ_n, cP	R_{en}	f_{Fn}
0.0120	3.19	0.40	2.63	264	0.788	131,670	0.0046
0.0122	2.92	0.39	2.41	317	0.873	118,979	0.0047
0.0124	2.68	0.37	2.22	365	0.929	111,775	0.0047
0.0127	2.47	0.36	2.05	408	0.965	107,623	0.0048
0.0129	2.28	0.35	1.89	448	0.986	105,388	0.0048
0.0131	2.11	0.34	1.76	485	0.995	104,460	0.0048
0.0134	1.95	0.34	1.63	520	0.996	104,404	0.0048
0.0136	1.81	0.33	1.51	553	0.990	105,064	0.0048
0.0139	1.68	0.33	1.41	584	0.978	106,368	0.0048
0.0141	1.56	0.32	1.31	612	0.962	108,195	0.0048
0.0144	1.45	0.32	1.22	638	0.943	110,459	0.0047
0.0147	1.35	0.32	1.15	661	0.921	113,090	0.0047
0.0150	1.26	0.32	1.07	683	0.898	116,036	0.0047
0.0153	1.18	0.32	1.01	703	0.874	119,252	0.0047
0.0156	1.11	0.32	0.95	721	0.850	122,704	0.0047
0.0159	1.04	0.32	0.90	738	0.826	126,361	0.0046
0.0161	1.00	0.32	0.87	747	0.811	128,652	0.0046

y	s	f_{Fm}	ρ_{ms}, kg/m³	Fric. loss, psi/ft	Hydr. loss, psi/ft	Total press. loss, psi/ft
1.844	0.390	0.0068	351	0.105	0.145	0.250
1.769	0.385	0.0069	393	0.088	0.162	0.251
1.716	0.382	0.0069	427	0.077	0.177	0.254
1.678	0.380	0.0070	457	0.069	0.189	0.258
1.648	0.378	0.0070	483	0.063	0.200	0.263
1.625	0.377	0.0070	505	0.058	0.209	0.267
1.354	0.364	0.0069	568	0.054	0.235	0.289
1.319	0.363	0.0069	593	0.051	0.245	0.296

(Continued)

TABLE 9.7   (Continued)
Detailed Pressure Traverse Calculations Using the Beggs and Brill[26] Model for the Deviated System Numerical Example (additional columns continued)

y	s	f_{Fm}	ρ_{ms}, kg/m³	Fric. loss, psi/ft	Hydr. loss, psi/ft	Total press. loss, psi/ft
1.289	0.363	0.0069	615	0.048	0.254	0.302
1.263	0.363	0.0068	634	0.046	0.262	0.308
1.241	0.363	0.0068	652	0.044	0.270	0.313
1.221	0.364	0.0068	668	0.042	0.276	0.318
1.204	0.365	0.0068	682	0.041	0.282	0.323
1.189	0.366	0.0068	695	0.039	0.287	0.327
1.175	0.368	0.0067	706	0.038	0.292	0.330
1.163	0.370	0.0067	716	0.037	0.296	0.334
1.156	0.371	0.0067	722	0.037	0.299	0.335

Note: actual calculations are on the basis of a 200-ft segment; however, a truncated table is presented.

and Kabir's[29] model for deviated wellbore systems and Xiao et al.'s[30] model for (two-phase) flow in pipelines. However, Petalas and Aziz[31] model is construed as applicable to both wellbores and pipelines, and others include Gomez et al.[32] Somewhat similarly, the "empirical" Beggs and Brill[26] model also is versatile (and unified) in that it applies to any systems and inclinations. Additionally, the Beggs and Brill[26] model appears to have stood the test of time, which is evident from the comparison of Xiao et al.'s[30] mechanistic model and the former for a database of more than 400 points. Four out of the five statistical parameters (includes average percentage error and absolute deviation) defined by Xiao et al.[30] are comparable with Beggs and Brill,[26] which basically points to the fact that a great deal of sophistication may not always be necessary within this class of steady-state pressure drop prediction methods.

9.8   TRANSIENT MODELS

Transient models, sometimes referred to as dynamic models, belong to a totally different class compared to those covered earlier. Pipeline operations are inherently transient[33] such as start-up, shutdown, changing inlet and outlet flows, pigging, pumps and compressors switching on and off, control set points change, fluid compositions, and properties change (GOR increasing, etc.). These facts would seem to indicate that a useful flow model should be a transient model; that is, it should solve the **time-** (and **position-**)dependent flow (and heat) equations. Steady-state models, on the other hand, do not include time dependency but still are widely used to design pipelines and estimate the flow and pressure drop.

For a transient model, a single pressure equation that includes all three phases can be constructed as follows, which is modified from Ellul et al.[34] and Bendiksen et al.,[35] respectively:

$$
\left[\frac{f_G}{\rho_G} \left(\frac{\partial \rho_G}{\partial P} \right)_{T,R_s} + \frac{1-f_G-f_W}{\rho_O} \left(\frac{\partial \rho_O}{\partial P} \right)_{T,R_s} + \frac{1-f_G-f_O}{\rho_W} \left(\frac{\partial \rho_W}{\partial P} \right)_{T,R_s} \right] \frac{\partial P}{\partial t}
$$

$$
= -\frac{1}{A\rho_G} \left(\frac{\partial A f_G \rho_G V_G}{\partial z} \right) - \frac{1}{A\rho_O} \left(\frac{\partial A f_O \rho_O V_O}{\partial z} \right) - \frac{1}{A\rho_W} \left(\frac{\partial A f_W \rho_W V_W}{\partial z} \right) \quad (9.47)
$$

$$
+ G_G \frac{1}{\rho_G} + G_O \frac{1}{\rho_O} + G_W \frac{1}{\rho_W}
$$

where gas mass fraction, R_s is defined as

$$
R_s = \frac{m_G}{m_G + m_O + m_W} \quad (9.48)
$$

In the preceding equation, G, O, and W are the gas, oil, and water phases, respectively; ρ_G, ρ_O, and ρ_W are densities of the respective phases, which are functions of pressure (P), temperature (T), and composition; f_G, f_O, and f_W are volume fractions; A is pipe cross-sectional area; t is time; V_G, V_O, and V_W are velocities of the respective phases; G_G, G_O, and G_W are mass sources; and z is the length coordinate. Equation 9.47 together with the momentum equations for the three phases are discretized and solved simultaneously for pressure and phase velocity. Of course, this does not lend itself to hand or spreadsheet-type calculations and is available in most commercial simulators. Modisette and Modisette[33] state that steady-state models or approximations are often much easier to implement than full transient models given the fact that in the former, the solutions are not subject to numerical instabilities that transient models must avoid. A remarkable classical analog of the transient model is in reservoir simulation where the objective is to solve for pressure and saturation as a function of position and time.

Ge and Rasheed[36] presented a modeling case study for wet gas transport in a flowline having condensate and water along with gas. Owing to the changing pressure and temperature the phase behavior alters and thus liquid loading continuously varies (bbl of condensate and water/mmscf of gas). This is a scenario that requires reliable ΔP and liquid holdup predictions so that a proper slug catcher can be sized. In this case study, condensate and water loading vary from 10–100 and 1–50 bbl/mmscf gas at rates of 25–300 mmscfd for transport through a 32-in., 35-mile flowline (changing elevation profile) traversing onshore as well as offshore. The flowline exit conditions are fixed at 200 psig and 50°F, respectively. The authors compared pressure drop and liquid holdup predictions for various models that included Beggs and Brill and a dynamic or transient model. Both the pressure drop and liquid holdup predictions of transient model and the Beggs and Brill[26] correlation are quite comparable for superficial gas velocities of 15 ft/sec and above; however, below this particular cutoff the latter results in significant underpredictions.

9.9 SLUGGING AND WATER HAMMER AND ITS MITIGATION

Both slugging (severe and more commonly encountered) and water hammer are flow assurance challenges that are unrelated to any solid deposition. Slugging is basically a result of variations in the flow stream, such as sudden surges or disproportionate amount of gas and liquid in a given space (e.g., flowline or a riser), whereas water hammer is a result of sudden stoppage of liquid flow. The practical aspects of both are covered in the following sections.

9.9.1 SLUGGING

Slugging is basically a transient state fluid mechanics problem, which is more common or obvious and serious (from health, safety, and environmental standpoint) in deepwater flowline–riser combinations given the terrain, bathymetry, and large gradients. The denser phase liquid collects or builds up at the "low spots" in the flow infrastructure and thus basically separates from the gas, which is referred to as slugging. Obviously, the separated gas has a tendency to carry these liquid slugs that may lead to flooding at the receiving end, which is typically designed to handle a certain gas–liquid ratio or rate, thus upsetting the system. The obvious consequences of slugging are unstable pressures in the system, significant impact on frictional ΔP, and cyclic flow conditions. Slugging is commonly attributed to shutdown/restart; low flow rates, low GORs, high water cuts; and decreasing reservoir or wellhead pressures. Sarica et al.[37] schematically represented severe (terrain) slugging in a flowline–riser system, where the flowline in the direction of flow is offset from the horizontal by about −20°, thereby forming a low point at the end where it meets the vertical riser. The four sequential steps or stages are as follows: (a) slugging onset, that is, low gas and liquid velocities giving rise to stratified flow; (b) higher hydrostatic pressure of the liquid phase compared to the upstream gas, that is, liquid column acts like a "seal" on the incoming gas phase; (c) gas phase pressure is greater than the column hydrostatic or in other words, liquid slug is getting pushed out of riser (surging the liquid volume in the separator), gas expands, gas velocity increases; and (d) liquid film unswept from the riser starts falling back or liquid accumulates again, that is, back to step (a).

Typical considerations for slugging mitigation/containment include a slug catcher on the top side (may be uneconomical for offshore, logistically inconvenient); a slug catcher on the seabed, downhole, or subsea separation of gas and liquid and then their separate flow through different risers; pigging; gas lift (analogous to artificial lift in vertical wells) at the riser base to lighten the liquid column; horizontal or a 2–5°-upward-slope flowline section immediately ahead of the riser (may not always be feasible given the bathymetry) to eliminate the "sump" effect;[37] high flow rates; and backpressure or choking upstream of the separator or riser top (at the expense of reduced flow rates) and mixing devices at the riser base. Table 9.8 shows the different proposed techniques/ideas and their pros and cons, which indicates that methods such as subsea separation are gaining traction with successful reported cases.

Sarica et al.[37] extensively studied severe slugging in a laboratory-based test facility that mimicked the flowline–riser system in which they tested the performance of

TABLE 9.8

Suggested Slugging Elimination Methods and their Pros and Cons

Technique	Pros and Cons/Other Comments
Backpressure	Reduces production capacity
Gas lift	Joule Thomson (JT) cooling possible, may promote hydrate, wax
Choking (common) – akin to backpressure	Careful choking required to achieve the least back pressure
Gas lift and choking combo	Viable method
Riser base pressure control with a surface control valve	Somewhat similar to choking
Pipe in Pipe type (concentric smaller pipe insertion in riser)	Concerns of gas lift also apply here in principle; intrusive (cannot pig the line)
Multiphase riser base lift (divert part of production from nearby source)	System specific, feasible for limited cases
Adding foams to "bind" the gas and oil	
Subsea separation (getting more traction – state of the art)	Viable solution, no back pressure imposed, requires separate flowlines (and a pump for liquid), economics? Marlim (Brazil); Pazflor (Angola); Troll (Norway); Tordis (Norway)

gas lift and surfactant in eliminating slugging. They evaluated the efficacy of these mitigation measures via "Elimination (slugging) Performance Index" and Modified Elimination (slugging) Performance Index. These indices basically reflect the pressure drops in the system. Their test facility results indicate that some combination of gas injection and surfactant appears to be the best in slugging elimination. Xing and Yeung[38] reported results from a computational fluid dynamics (CFD) simulation of backpressure and choking of the riser mimicked in a laboratory test facility somewhat similar to Sarica et al.[37] Their results suggest that the riser ΔP stabilizes due to back pressurizing, but this obviously increases the riser base pressure.

9.9.1.1 Slug Catchers

A slug catcher is basically a process vessel, a bit like a separator, having a primary function of handling excess liquid due to slugging (terrain, hydrodynamic, and pigging) that a separator alone cannot handle. Slug catchers essentially stabilize the process and their basic design mimics the separator and is typically placed upstream of a production separator. Some of the important factors to consider in a slug catcher are size (e.g., design basis being transient modeling), slug frequency, location in the flow infrastructure, and type (conventional or pipe). The volumes of liquids produced under steady-state conditions, transients (start-ups, etc.), and pigging are reviewed and compared with a "rule of thumb" method for sizing, which is 3–5 times the riser volume.[19]

Other slug catchers include a small-scale pipe-fitting slug catcher that is suitable for small 150–200-bbl slugs and is economical given the size and footprint. These are essentially designed as pipe and fittings rather than a typical pressure vessel. Somewhat similar in principle are the harp- or finger-type slug catchers. Márquez et al.[39] provided a step-by-step methodology they used in designing a slug catcher for the heavy oil in Orinoco belt, Venezuela. For a 16° API oil, and rates of about 15,000 bbl/day liquid and 8,500 mmscfd gas, their design was for four fingers, each 20-in. in diameter and 26 ft in length, which was about 25% cheaper than the conventional slug catcher.

As the name suggests, slug frequency models are used to determine the frequency of slugs. These are defined as the number of liquid slugs (no discrimination per se in terms of oil/water) passing by a certain point along the pipeline over time. Obviously, slug flow, and thus its frequency, is important because it is one of the most observed characteristics in multiphase transport. For the most part it is particularly problematic because of the intermittent nature due to the unique phase distribution. Clearly, slug frequency prediction is important in flowline designs and separation facilities and is also needed in mechanistic models to accurately predict pressure drop and liquid holdup. Most empirical models consider the effect of operations, geometrical, and mechanistic parameters on slug frequency; note that Gokcal et al.'s[40] model specifically includes a (high) viscosity effect. The functional forms of the Shea et al.[41] and Gokcal et al.[40] slug frequency empirical correlations, respectively, follow.

$$f_s = \frac{0.47 U_{sL}^{0.75}}{D^{1.2} L_p^{0.55}} \tag{9.49}$$

$$f_s = 2.623 \frac{1}{N_f^{0.612}} \frac{U_{sL}}{D} \tag{9.50}$$

$$N_f = \frac{D^{3/2} \sqrt{\rho_L (\rho_L - \rho_G) g}}{\mu_L} \tag{9.51}$$

where f_s is slug frequency/sec; U_{sL} is superficial liquid velocity in m/sec; D is pipe or flowline inside diameter in m; L_p is pipe or flowline length measured from pipe inlet in m; N_f is dimensionless inverse viscosity number; ρ_L and ρ_G are liquid and gas densities, respectively, in kg/m³; g is acceleration due to gravity in m/sec²; and μ_L is liquid viscosity in Pa sec.

Candelier and Papot[42] compared engineering software predictions and a West African deepwater field data that also included slugging analysis. Field data on slugging frequency can also be deduced from pressure variations recorded at the topsides, which are representative of slug size due to change in the hydrostatic gradient in the riser; the minimum indicating slug at the base and the maximum indicating its removal or arrival at topsides. Figures 9.11 and 9.12 show the comparisons.

9.9.2 WATER HAMMER

This is a phenomenon typically caused by abrupt or sudden valve closure that brings the flowing liquid upstream of the valve to a sudden halt, causing the pressure to rapidly surge. Water hammer also is sometimes referred to as "hydraulic shock," having consequences such as a loud bang or noise-induced vibrations, an enlargement effect on the pipe, a rupture of the pipeline, or an implosion caused by vacuum

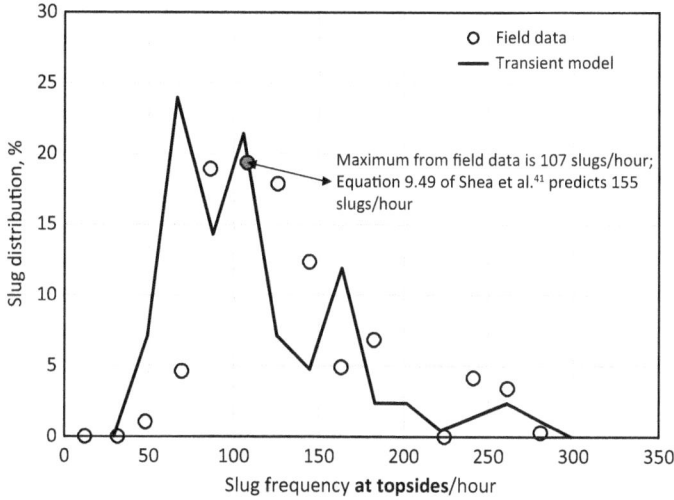

FIGURE 9.11   Slugging analysis comparison of field data, transient model, and Shea et al.'s[41] slug frequency correlation. Plot constructed based on data presented in Candelier and Papot.[42]

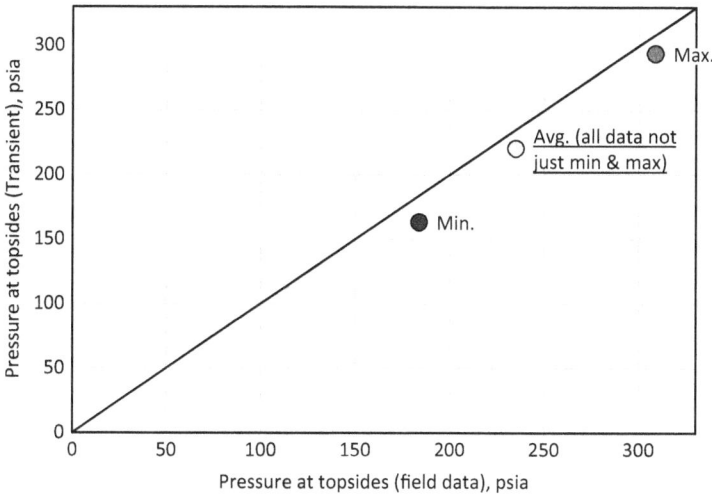

FIGURE 9.12   Transient and field topsides pressure comparison to deduce slug frequency. Plot constructed based on data presented in Candelier and Papot.[42]

creation on the downstream side of the valve as this liquid will continue to flow. Han et al.[43] simulated multiphase fluid hammer effects in start-up and shut-in for a generic offshore production system that included a single wellbore, flowline–riser, and a test separator to basically study the sensitivity of various parameters. Their main conclusions that may have some practical implications are as follows. Hammer effect is a strong function of fluid compressibility (or implicitly GOR), vis-à-vis high compressibilities (gas or any less dense phase) will tend to "dampen" the pressure oscillation. Hammer effects generally decrease with volume, for example, long flowlines. A wellbore storage–type effect and the reservoir (again large volume) will "absorb" the hammers and will not reach the separator. A shut-in/start-up valve placed at the bottomhole rather than the platform (near the separator) may minimize water hammer effects because, again, the hydraulic shock will likely be "absorbed" by the reservoir.

REFERENCES

1. White, F. (2009). *Fluid Mechanics* (7th Edition). McGraw Hill.
2. Hasan, A., & Kabir C. (2002). *Fluid Flow and Heat Transfer in Wellbores*. Society of Petroleum Engineers.
3. Guo, B., Lyons, W. C., & Ghalambor, A. (2007). *Petroleum Production Engineering, A Computer-Assisted Approach*. Elsevier. https://doi.org/10.1016/B978-0-7506-8270-1. X5000-2
4. Colebrook, C. (1939). Turbulent Flow in Pipes, with Particular Reference to the Transition between the Smooth and Rough Pipe Laws. *Journal of the Institution of Civil Engineers*, 11, 133–156.
5. Moody, L. F. (1944). Friction Factors for Pipe Flow. *ASME Transactions*, 66, 671–684.
6. Chen, N. H. (1979). An Explicit Equation for Friction Factor in Pipe. *Industrial & Engineering Chemistry Fundamentals*, 18(3), 296–297. https://doi.org/10.1021/i160071a019
7. Notz, P. (2009). *Flow Assurance for Offshore Production*. PetroSkills. Houston, TX
8. Dranchuk, P. M., & Abou-Kassem, H. (1975). Calculation of Z Factors for Natural Gases Using Equations of State. *Journal of Canadian Petroleum Technology*, 14(3). https://doi.org/10.2118/75-03-03
9. Sutton, R. P. (1985, September 22). Compressibility Factors for High-Molecular-Weight Reservoir Gases. In *SPE Annual Technical Conference and Exhibition*. https://doi.org/10.2118/14265-MS
10. Smith, R. V. (1950). Determining Friction Factors for Measuring Productivity of Gas Wells. *Journal of Petroleum Technology*, 2(03), 73–82. https://doi.org/10.2118/950073-G
11. Poettmann, F. H. (1951). The Calculation of Pressure Drop in the Flow of Natural Gas Through Pipe. *Journal of Petroleum Technology*, 3(11), 317–326. https://doi.org/10.2118/115-G
12. Lee, J., & Wattenbarger, R. A. (1996). *Gas Reservoir Engineering*. Society of Petroleum Engineers.
13. Cullender, M. H., & Smith, R. V. (1956). Practical Solution of Gas-Flow Equations for Wells and Pipelines with Large Temperature Gradients. *Transactions of the AIME*, 207(1), 281–287. https://doi.org/10.2118/696-G
14. Dindoruk, B., & Christman, P. G. (2004). PVT Properties and Viscosity Correlations for Gulf of Mexico Oils. *SPE Reservoir Evaluation & Engineering*, 7(6), 427–437. https://doi.org/10.2118/89030-PA
15. Perkins, T. K., & Turner, J. B. (1971). Starting Behavior of Gathering Lines and Pipelines Filled with Gelled Prudhoe Bay Oil. *Journal of Petroleum Technology*, 23, 301–308. https://doi.org/10.2118/2997-PA
16. Chukwu, G.A.; Khataniar, S.; Patil, S.L. and Dandekar, A. (2007). *Operational challenges in gas-to-liquid (GTL) transportation through Trans Alaska Pipeline System (TAPS)*. https://www.osti.gov/servlets/purl/919007

17. Jon Steinar Gudmundsson. (2018). *Flow Assurance Solids in Oil and Gas Production.* CRC Press/Balkema.

18. Shell Global Solutions International. (2007). *Gas/Liquid Separators – Type Selection and Design Rules, Design and Engineering Practice, DEP 31.22.05.11, 100 pp.*

19. Burke, N. E., & Kashou, S. F. (1996). Slug-Sizing/Slug-Volume Prediction: State of the Art Review and Simulation. *SPE Production & Facilities*, 11(03), 166–172. https://doi.org/10.2118/30902-PA

20. Standing, M. B. (1947). A Pressure-Volume-Temperature Correlation for Mixtures of California Oils and Gases. *Drilling & Production Practice*, 275–287.

21. Whitson, C.H. & Brulé, M. R. (2000). *Phase Behavior.* Society of Petroleum Engineers Inc.

22. Lee, A. L., Gonzalez, M. H., & Eakin, B. E. (1966). The Viscosity of Natural Gases. *Journal of Petroleum Technology*, 18(8), 997–1000. https://doi.org/10.2118/1340-PA

23. Abu-Khamsin, S. A., & Al-Marhoun, M. A. (1991). Development of a New Correlation for Bubblepoint Oil Viscosity. *Arabian J. Sci. & Eng.*, 16(2A).

24. Hagedorn, A. R., & Brown, K. E. (1965). Experimental Study of Pressure Gradients Occurring During Continuous Two-Phase Flow in Small-Diameter Vertical Conduits. *Journal of Petroleum Technology*, 17(4), 475–484. https://doi.org/10.2118/940-PA

25. Abdul-Majeed, G. H., & Abu Al-Soof, N. B. (2000). Estimation of gas–oil surface tension. *Journal of Petroleum Science and Engineering*, 27(3), 197–200. https://doi.org/10.1016/S0920-4105(00)00058-9

26. Beggs, D. H., & Brill, J. P. (1973). A Study of Two-Phase Flow in Inclined Pipes. *Journal of Petroleum Technology*, 25(5), 607–617. https://doi.org/10.2118/4007-PA

27. Kaya, A. S., Sarica, C., & Brill, J. P. (2001). Mechanistic Modeling of Two-Phase Flow in Deviated Wells. *SPE Production & Facilities*, 16(3), 156–165. https://doi.org/10.2118/72998-PA

28. McCain Jr., W. D. (1989). *Properties of Petroleum Fluids* (2nd Edition). PennWell.

29. Hasan, A. R., & Kabir, C. S. (1988). Predicting Multiphase Flow Behavior in a Deviated Well. *SPE Production Engineering*, 3(4), 474–482. https://doi.org/10.2118/15449-PA

30. Xiao, J. J., Shoham, O., & Brill, J. P. (1990, September 23). A Comprehensive Mechanistic Model for Two-Phase Flow in Pipelines. In *SPE Annual Technical Conference and Exhibition*. https://doi.org/10.2118/20631-MS

31. Petalas, N., & Aziz, K. (2000). A Mechanistic Model for Multiphase Flow in Pipes. *Journal of Canadian Petroleum Technology*, 39(6). https://doi.org/10.2118/00-06-04

32. Gomez, L. E., Shoham, O., Schmidt, Z., Chokshi, R. N., & Northug, T. (2000). Unified Mechanistic Model for Steady-State Two-Phase Flow: Horizontal to Vertical Upward Flow. *SPE Journal*, 5(3), 339–350. https://doi.org/10.2118/65705-PA

33. Modisette, J. L., & Modisette, J. P. (2001, October 17). Transient and Succession-of-Steady-States Pipeline Flow Models. In *PSIG Annual Meeting*.

34. Ellul, I. R., Saether, G., & Shippen, M. E. (2004, October 20). The Modeling of Multiphase Systems Under Steady-State and Transient Conditions - A Tutorial. In *PSIG Annual Meeting*.

35. Bendiksen, K. H., Maines, D., Moe, R., & Nuland, S. (1991). The Dynamic Two-Fluid Model OLGA: Theory and Application. *SPE Production Engineering*, 6(2), 171–180. https://doi.org/10.2118/19451-PA

36. Ge, J., & Rasheed, M. A. (2009, May 12). Wet Gas Pipeline Liquid Holdup and Pressure Calculation by Different Calculation Methods. *PSIG Annual Meeting*.

37. Sarica, C., Yuan, G., Shang, W., Pereyra, E., & Kouba, G. (2015). Feasibility and Evaluation of Surfactants and Gas Lift in Combination as a Severe-Slugging-Suppression Method. *Oil and Gas Facilities*, 4(4), 78–87. https://doi.org/10.2118/170595-PA

38. Xing, L., & Yeung, H. (2010, June 2). CFD Modelling of Severe Slugging in Pipeline-Riser System. In *7th North American Conference on Multiphase Technology*.

39. Márquez, J., Manzanilla, C., & Trujillo, J. (2010). A Conceptual Study of Finger-Type Slug Catcher for Heavy-Oil Fields. *SPE Projects, Facilities & Construction*, 5(4), 192–199. https://doi.org/10.2118/122829-PA

40. Gokcal, B., Al-Sarkhi, A. S. S., Sarica, C., & Al-Safran, E. M. M. (2010). Prediction of Slug Frequency for High-Viscosity Oils in Horizontal Pipes. *SPE Projects, Facilities & Construction*, 5(3), 136–144. https://doi.org/10.2118/124057-PA

41. Shea, R. H., Eidsmoen, H., Nordsveen, M., Rasmussen, J., Xu, Z. G., & Nossen, J. (2004). Slug frequency prediction method comparison. In *4th North American Conference on Multiphase Technology*, 227–237. https://www.scopus.com/inward/record.uri?eid=2-s2.0-3042703210&partnerID=40&md5=ba9d3fc7b77e0a99aa13b1d5d46c5d76

42. Candelier, C., & Papot, F. (2015, June 10). Comparisons Between Engineering Software Predictions and West of Africa Deepwater Field Data Including Thermal Performances. In *17th International Conference on Multiphase Production Technology*.

43. Han, G., Ling, K., Khor, S. H., Zhang, H., & Thakur, R. K. (2013). Simulation of Multiphase Fluid-Hammer Effects During Well Startup and Shut-In. *Oil and Gas Facilities*, 2(6), 68–77. https://doi.org/10.2118/160049-PA

10 Flow Assurance Case Studies and Engineer's Toolbox

10.1 INTRODUCTION

In this final chapter, we first cover notable field case studies that involve the four organic solids, that is, hydrates, waxes, asphaltenes, and naphthenates and inorganic solids such as scales and fines. This is followed by a somewhat unique or emergent issue that ties flow assurance and hydraulic fracturing. Since hydrates are perhaps the most notorious from a health, safety, and environmental (HSE) standpoint, these aspects are also included, which is followed by perspectives on flow assurance economics. The chapter concludes with a summary of flow assurance technologies and their maturity levels, and finally, the flow assurance engineer's toolbox that attempts to address questions, such as (1) Where does the engineer start – or what is the workflow? and (2) What tools are available for the engineer to evaluate and solve operational problems?

10.2 COMPILATION OF FIELD CASE STUDIES ON SOLIDS

These case studies are those reported in the literature, and their discussion is basically organized around (1) the solid(s) problem in question, (2) how the problem was identified, (3) the solid(s) risk mitigation strategy, (4) the effectiveness of the treatment or changes in operational parameters, and (5) the follow-up if any.

10.2.1 Hydrates

Clearly, in deepwater fields, a favorable combination of light-end alkanes, produced water, cold ambient temperatures, and high pressures almost always exist, resulting in the formation of hydrates, which can agglomerate and form blockages in the flow infrastructure, ultimately leading to production deferrals.[1] Typical remediation strategies include pressure reduction (depressurization), temperature increase (addition of heat), and the use of thermodynamic inhibitors, such as methanol and or other chemicals such as anti-agglomerants. However, deploying any of these strategies in deep water is particularly difficult due to logistical challenges and access limitations. The first case study discussed on hydrates is for a Gulf of Mexico (GOM) asset in water depths of about 4,500 ft (dry tree). This is a good example of a paradigm shift in hydrate risk management in that it is a pivot from complete avoidance of hydrate formation toward efficient management of hydrate to avoid blockages[1] in the flow infrastructure.

DOI: 10.1201/9781003091301-10

Similar to the standard practice of hydrate avoidance, this particular GOM asset was originally designed as such. However, with maturity, water cut increased, and wells were placed on artificial (gas) lifts, rendering the original philosophy difficult to implement. This was one of the primary motivations to experiment by pushing the conventional boundaries of hydrate management under the understanding[1] that "hydrate formation does not necessarily lead to a hydrate blockage". In Kurup et al.'s[1] quest for the new hydrate (risk) management strategy, the first step was fundamental hydrate phase equilibrium curves. They predicted hydrate PT conditions for varying gas–oil ratio (GORs) using pure water rather than the produced 3% saline water in order to obtain a conservative estimate (note that thermodynamically salinity shifts the hydrate equilibrium curve, i.e., depending on the salt concentration there is an inhibitor effect). As described in previous chapters, the hydrate PT curve separates the hydrate forming and nonforming conditions – PT conditions on the left and right represent favorable and unfavorable hydrate-forming conditions, respectively, while the curve itself signifies equilibrium between the gas, water, and hydrate phases. If the operating temperature is on the left side of the equilibrium curve, then at any given pressure, the difference between the equilibrium temperature and the operating temperature is called subcooling, which is basically the ΔT in the Hammerschmidt equation for estimating the methanol wt% required for inhibition.[2] Although any subcooling is conducive to hydrate formation, it is the magnitude of the ΔT that is of significance as far as the severity is concerned, that is, higher ΔTs leading to a relatively rapid hydrate formation and consequently high probability of blockage. Figure 10.1 shows a plot of GOR versus subcooling at a typical shutdown pressure of 4,500 psia

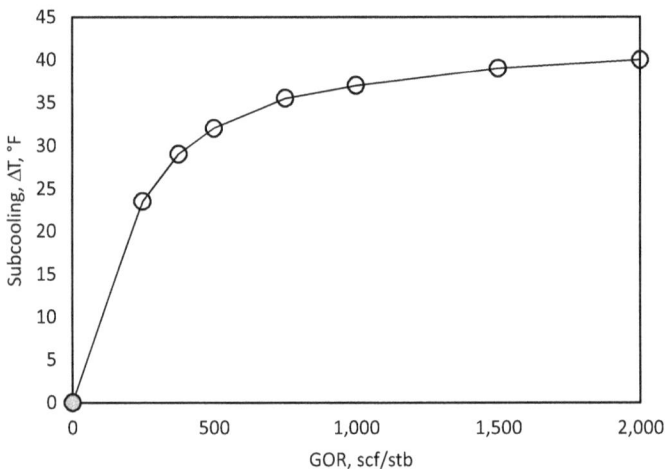

FIGURE 10.1 Plot of GOR versus subcooling ΔT at a shutdown pressure of 4,500 psia. The shutdown temperature is 40°F (basically mudline/seawater temperature at the stated depth); consequently, the first data point on the graph means GOR and ΔT are 0 scf/stb and 0°F, respectively. Note that the plot is constructed based on the hydrate equilibrium curves and other data in Kurup et al.[1]

(shutdown temperature being 40°F used as *operating temperature* to estimate the subcooling for a worst-case scenario). As seen in this figure, the ΔT is directly proportional to GOR (note that 250 scf/STB is the original asset GOR); in other words, this means operating with gas lift alters the fluid composition, thus shifting the hydrate PT curves to the right. For the original GOR, the hydrate equilibrium point of 4,500 psia and 63.5°F means a subcooling of 23.5°F (63.5–40); however, for the highest GOR of 2,000 scf/STB this value is as high as 40°F, which means considerable hydrate risk. This situation is somewhat of a dichotomy in that gas lift is necessary to efficiently lift the production, but it comes with a potentially negative consequence of considerable hydrate blockage. The subsequent methodology proposed by the authors to address this issue for this particular asset is summarized in the following.

Kurup et al.[1] describe the hydrate management philosophy for this particular asset in the form of a bowtie with two barriers: (1) preventative and (2) mitigating. The preventative barriers include tubing/riser insulation, dead oil displacement (DOD), low-dosage hydrate inhibitors (LDHI), and low GOR. Since fluids are warm enough during steady-state flow, the conditions are below the hydrate equilibrium curve, that is, an insulating effect; however, in the case of a shutdown, the fluids will experience a gradual drop in the temperature right below the mudline, which may be conducive to hydrate formation (basically a somewhat curved PT path approaching or crossing the given hydrate equilibrium curve). It should be noted, however, that this will not happen instantaneously but is defined by what is known as the cooldown time. In the case of a shutdown, bullheading or DOD is a reliable method to prevent hydrate formation; for shorter durations, the production fluids are displaced below the mudline, and for longer durations, the displacement is even further below the surface-controlled subsurface safety valve and into the formation. In case of planned short shutdowns, LDHIs can be used to pretreat production fluids; however, high water cuts (WCs) and hydraulics are some of the limitations. The effect of low GORs is somewhat self-explanatory from Figure 10.1. In case the preventative barriers fail and result in hydrate blockages, then mitigating barriers or remediation options are considered such as depressurization, hot oil circulation, methanol, and coiled tubing interventions.

Since fluids start to cool following a shutdown, hydrate safe out measures need to be implemented within the cooldown times. Another term that is defined by the authors[1] is the "no touch time", which is the difference between the cooldown time and the execution time for hydrate safe-out measures. The cooldown times can be predicted based on multiphase transient simulations using the field parameters such as flow rates, steady-state temperatures, fluid properties, and others. Figure 10.2 shows the predicted cooldown times for the four wells, of which two are placed on gas lift. As seen in Figure 10.2 (and Figure 10.1 that depicts the subcooling), wells placed on gas lift result in cooldown times of less than 1 hour, which is operationally impractical, if not difficult, and points to the level of higher risk in gas-lifted wells.

The hydrate management strategy for this asset basically evolved from a combination of hydrate phase equilibria calculations and historical episodes of unplanned shutdowns that exceeded the cooldown times. Since the hydrate volume (%) plays an important role in the extent of blockage, vis-à-vis obstruction for flow, the authors first computed this for various GORs and water cuts at the shut-in conditions of 4,500 psia and 40°F. For all the tested GORs, the hydrate volume fraction first

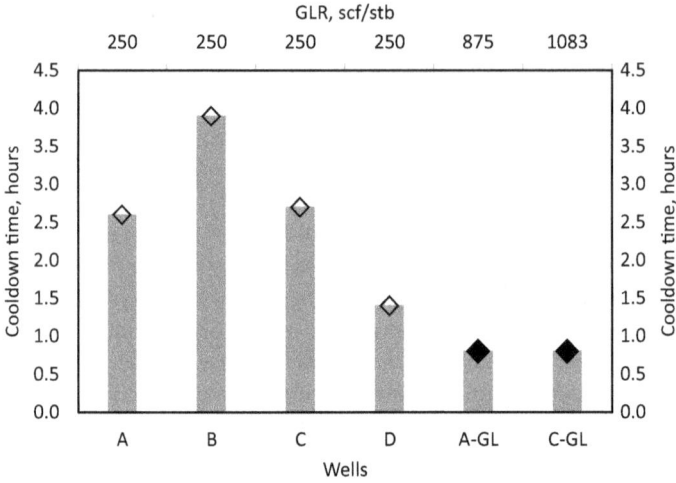

FIGURE 10.2 Effect of gas lift (GLR) on cooldown times. Note that the non-gas-lifted (GL) wells assume a GL(O)R of 250 scf/stb. Plot is constructed based on the tabulated data in Kurup et al.[1]

increased with water cut, reached a peak, and then continually decreased. Clearly, the hydrate volume fraction versus WC relationship is basically governed by the amounts of gas and water present, specifically the WC (limiting reaction component) in reaching the peak and the hydrocarbon components (limiting reaction component) in the reducing hydrate volume fraction. Figure 10.3 plots the ratio of GOR and WC versus the maximum or peak hydrate volume fraction for the cases studied by the authors.[1]

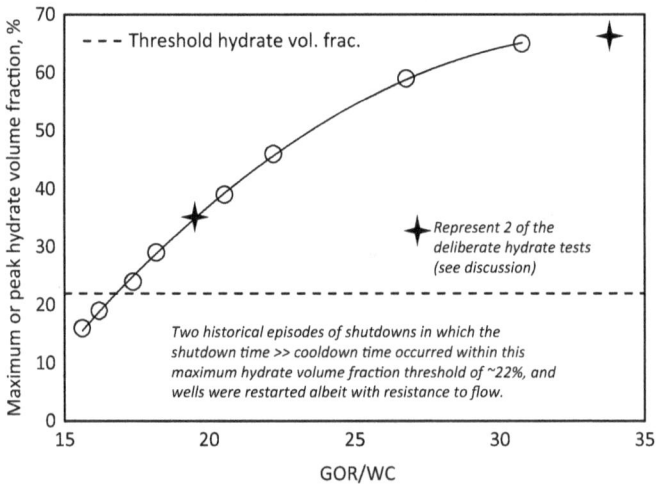

FIGURE 10.3 Maximum hydrate volume fraction as a function of GOR and WC ratio. Plot is constructed based on digitized data from Kurup et al.[1]

Also, shown in the inset is the "threshold" or limiting value of the maximum hydrate volume fraction, within which the two historical episodes of well restarts (after resistance to flow) fall. The authors state that in these two cases, the hydrate safe out measures could not be implemented, which obviously resulted in the shutdown times well exceeding the cooldown times, yet the wells were restarted. This analysis thus indicates that there exists a "safe operating window" for the GOR and WC combination that gives a hydrate volume fraction less than about 22%; that is, hydrates obviously do form but not enough to form a full blockage in the tubing. Based on this inference, it was decided to conduct deliberate hydrate tests to essentially push the boundaries of hydrate management.

The deliberate tests were conducted on well C, which had the highest WC (50%) and was on gas lift. The tested gas-lift rates and GORs ranged from 2–3.5 mmscfd and 1,021–1,706 scf/stb. The first test at 2 mmscfd and 1.5-h shutdown before DOD showed a 1:1 correspondence between the monitored downhole and wellhead pressures, respectively, for the DOD, meaning good communication and no indication of blockage. The test carried out at 3.5 mmscfd and 5 hours' shutdown before DOD showed a 500–1000 psi spike, within an hour of DOD, in the wellhead pressure, while the downhole pressure remained fairly flat. However, a 1:1 correspondence was observed much later in the test. The initial period thus indicates resistance to displacement, with displacement still feasible at 5 h. The GOR and WC ratio of these two tests are mapped in Figure 10.3, which basically indicates that even at higher hydrate volume fractions, it is possible to restart the wells with DOD within a 5-h time frame. In summary, the hydrate management strategy for this asset has been relying on a correlation that combines hydrate phase equilibria and operating parameters, with one caveat that operating closer to limits does come with a risk of hydrate blockage.

Although, for the most part, the hydrate phase equilibria are fundamentally the same, operating companies have to, however, pivot the hydrate management strategies depending on the conditions and circumstances. For example, the case study discussed earlier on pushing the boundaries may not be universally applicable, such as in greenfield development. Uribe et al.[3] describe a novel hydrate management strategy in a high-risk hydrate environment in a fast-track tie-in GOM development. The fast-track project means about 1.5 years from concept to first oil-leveraging existing subsea infrastructure. The challenges included flowline lengths and water depths exceeding 10,000 ft and 5,000 ft, respectively; tie-in being a single flexible flowline, limiting insulation; low reservoir temperature (150°F); continuous hydrate inhibition not possible from an operating expenses standpoint; low oil price environment; and some degree of uncertainty since analog field fluid information was used in the design due to the fast track. This necessitated a novel hydrate and other flow assurance strategy, which basically consisted of intermittent (instead of continuous) methanol and LDHI use upstream of choke and gooseneck during shutdown and start-up and hydrate remediation in general. First oil was achieved within a year of drilling exploration well.

10.2.2 WAXES

From a phase behavior standpoint, wax is generally a problem manifested by low temperatures and the loss of light ends. The low-temperature threat always exists in deep waters, and the loss of light ends is basically governed by the degree of under-saturation at a given temperature. In some cases, however, the wax precipitation and deposition problem may get quite convoluted and somewhat mixed up due to other unknown interference and interactions. This was basically the case with the two well (C-2 and C-3) wet gas and gas condensate subsea Coulomb field, in 7,500 ft of water depths, tied back to the NaKika platform in the GOM via a single 27-mile, 8-in. bare flowline.[4] The condensate-to-gas ratios (CGRs) of fluids from C-2 and C-3 are 65 bbl/mmscf (lean) and 200 bbl/mmscf (rich), respectively, or about 15,000 scf/stb and 5,000 scf/stb, which are typical of a wet gas and a gas condensate. The design basis included continuous injection of monoethylene glycol (MEG) for hydrate inhibition and "wax in place" strategy under the assumption that the paraffin inhibitor (PI) would keep the wax deposition rate low and not constrain production.[4] However, it was known after the fact that the fluid in well C-3 was significantly rich in wax compared to the well C-2 used in the design basis. The chronology of events reported by Manfield et al.[4] are very telling for understanding the problem in this particular field. Initially, with just the C-2 well flowing, the flowline pressure drop was more or less steady, which surged to approximately 4,000 psi over a 1-month period, with well C-3 coming online. Given this high ΔP, it was decided to shut in C-3, with C-2 continuing, which again resulted in the stabilization of ΔP.

The authors[4] carried out subsequent laboratory work and utilized analogous experience from the Mensa subsea development that led to the hypothesis that the formation and accumulation of a highly viscous phase (emulsion of condensate, MEG, wax) or a wax slurry that is stable at the low ambient temperatures encountered in the subsea flowline caused elevated pressure drops. Consequently, various mitigation options were considered by the team, and it was determined that heat retention or keeping the fluids warm in the flowline would have a high probability of reducing the flowline ΔP. However, among other challenges, such as ineffective PI and the fact that the riser could not be insulated, the bare flowline was perhaps the most crucial. This predicament, however, led to a somewhat novel solution based on steady-state thermo-hydraulic modeling, which recommended burial of the flowline to alleviate the heat loss. The flowline burial was executed after about 2 years of the initial start-up. The burial operation was completed in two passes, each beginning close to the touchdown point, that is, at the end of the line nearest the host. An underwater trenching tool deployed from a support vessel placed the flowline in the first pass in a channel some 6–8 ft deep, covered with 1–3 ft of backfill, whereas the second pass provided 6–8 ft of backfill over the flowline. The effect of burial was realized soon enough, which the authors[4] termed "immediate gratification". Prior to burial, C-2 was producing at approximately 68 mmscfd, with a flowline ΔP of 3,900 psi, and within 1 week of the burial, the C-2 choke was fully opened and production reached 102 mmscfd, with a corresponding flowline ΔP of only 2,700 psi. However, the most notable was well C-3, which was returned to production and gradually ramped up, and after 1 month, gas production was about 110 mmscfd, whereas liquid production increased from 4,800–15,600 bbl/d, and the flowline ΔP stabilized at about 3,500 psi. In addition to the well production data, the differences in the colors of the collected liquid samples provided further evidence of the efficacy of the flowline burial in that the initial samples were opaque, containing a large amount of wax, whereas the later ones were almost transparent, exhibiting typical stabilized condensate color. Finally, the effect of flowline cleanup was also corroborated by a rather dramatic decrease in the liquid viscosity from more than 500 cP to 2 cP (both at 4°C and 10 s^{-1}). An important (universal) message this paper[4] conveys, however, is the importance of good-quality representative fluid samples and property data early in the project development stages and retention of heat to prevent rapid cooling of the produced stream; in other words, an insulated flowline may prove to be a worthwhile expense and valuable insurance against potential production deferrals.

Conventional techniques for combating paraffin wax deposition problems include chemicals (inhibitors, dispersants), hot oiling or water circulation, and mechanical methods, such as pigging and scraping. However, other nontraditional methods such as the use of thermochemicals that generate heat due to the exothermic reaction are also starting to emerge. Qamruzzaman et al.[5] documented successful application of thermochemicals in addressing what appears to be a near-wellbore paraffin deposition in five wells in Western India. In this particular case, thermal stimulation is achieved via an exothermic reaction between aqueous solution of equimolar NH_4Cl and $NaNO_2$. The reaction rate and time (delay time) taken to reach the peak temperature is controlled by a suitable catalyst and the pH of the solution such that the heat release is targeted where it is needed in the (near) wellbore or in the pipeline.[5] For application, the two tanks store a mixture of water, $NaNO_2$ and an additive and water, NH_4Cl and acetic acid (catalyst), respectively, which is pumped by two separate pumps at the same flow rate; the solutions mix at the wellhead and react per the following,[5] which is controlled by the catalyst and the pH:

$$NaNO_2 + NH_4Cl \rightarrow NaCl + N_2 + 2H_2O; \Delta H = -309\,kJ/mole$$

The supporting laboratory tests and the mathematical modeling carried out by Qamruzzaman et al.[5] show that the delay time (25–30 min) is controlled by the catalyst and the pH. The volume of the formulation required can be basically estimated from a combination of the near-wellbore pore volume and the production casing between the tubing shoe and the perforation bottom. For the five wells on which this treatment was carried out, the reaction mixture was pumped with a margin of 5–10 min before achieving the peak temperature. This was followed by a soaking time of 60–90 min, and the well(s) flowed back to monitor production. Although not stated as such explicitly by the authors, the thermochemical stimulation is essentially akin to a huff-and-puff process that uses steam for heavy oil EOR. The ratio of post- and pretreatment oil rates reported by the authors ranged from a low of 1.2 to a high of 6.6, with the sustainability of the gains being 2–7 months.

Singh et al.[6] recommended a unique solution to combat the wax deposition problem in an arctic environment on the Alaska North Slope, which obviously comes with an added challenge of a nearly continuous permafrost up to a depth of 1,800–2,000 ft. They proposed the use of vacuum insulated tubing (VIT) for wax control, which would automatically also address the subsidence issue by maintaining the integrity of the permafrost – that is truly serving a dual purpose. From a heat transfer standpoint, the concept of VIT is quite straightforward in that the key benefit being superior thermal insulation, mainly owing to the vacuum in the annular space between the two

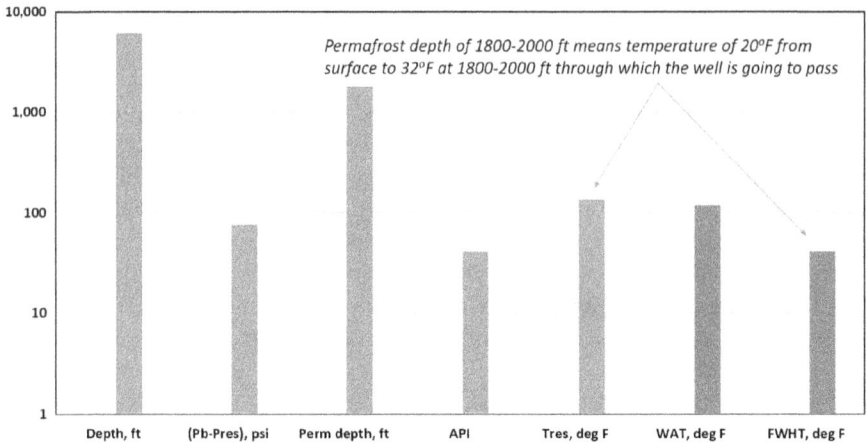

FIGURE 10.4 Waxy oil and other data for the VIT candidate well. Note the 40+° API oil gravity and WAT that far exceeds the FWHT. Plot constructed based on data from Singh et al.[6]

concentric pipes. Essentially, the vacuum acts as a thermal barrier that keeps the produced (waxy) oil warm in the tubing while not allowing any heat from the produced oil to transfer to the continuous permafrost layer. The bar chart in Figure 10.4 shows the key data for the problem well, which was the candidate for VIT application.[6] So, how did the solution proposed by the authors actually evolve? This is shown by the following points:

- Heating to a higher temperature is not an option because of the delicateness of the permafrost, which if melted will lead to serious subsidence and well failure issues
- Use of inhibitors not considered – logistical challenges, degradation of performance, etc.
- VIT was selected for thermal management (heat to be contained in the fluid)
- Selected conditions for assessing the efficacy of VIT thermo-hydraulic modeling were 650 BOPD; GL(O)R of 1500 scf/STB and FWHP of 300 psi. Water production is minimal.
- VITs through the entire depth showed doubling of original flowing wellhead temperature (FWHT; useful in reducing the wax deposition rate) but still 30°F below WAT
- Basically, this means wax management or mitigation – wax-free production cannot be achieved.

The authors[6] also conducted flow loop experiments to determine wax deposition rates, combined with heat transfer calculations based on which monthly deposition thickness versus depth were established. This showed an increase from 0 at a depth of 6,000 ft depth to 10 mm/month to the top. If this deposit is not removed, then it

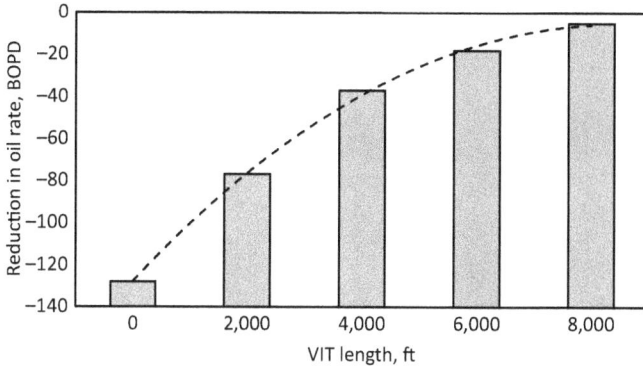

FIGURE 10.5   Reduction in oil rate for various VIT lengths, based on "wax-free" oil rate of 650 BOPD. Note the somewhat less pronounced gain beyond 4,000-ft VIT. In other words, despite the increasing length, oil rate does not reach 650 BOPD. Plot constructed based on predicted production decline in Singh et al.[6]

will actually reduce the deposition rate a little bit because this will supposedly act as a barrier between the permafrost and the flowing fluid in the well. Figure 10.5 shows the loss of oil rate (based on predictions) after 1 month due to wax precipitation. The operator (ConocoPhillips) adopted a cautious and balanced approach considering the capital expense of VIT installation, cost of wax remediation jobs, loss of revenue due to production deferral, and long lead time in getting VITs. This basically means that there is no point in directly going for the highest VIT length, because as seen in Figure 10.5, although the reduction in oil rate continues to improve; beyond 4,000 ft, the gain is not very significant. Thus, the final recommendation was 4,000 ft of VIT length. This is an ideal example of field data, reservoir engineering, lab work, and modeling in arriving at an "optimum" solution.

10.2.3   ASPHALTENES

Altemeemi et al.[7] presented an interesting case study of asphaltene deposition formation damage in a carbonate reservoir rock in northern Kuwait. Although the primary focus was formation damage, it was reported that the asphaltenes precipitated out of the liquid phase and deposited in the production liner and tubing,[7] obviously making it a flow assurance challenge as well. The fluid in the subject reservoir is a typical volatile oil with an American Petroleum Institute (API) gravity in the high 30s, a viscosity of less than 0.5 cP, and a relatively high GOR and bubble point. Fairly comprehensive asphaltene onset pressure (AOP) measurements were also conducted for this particular oil, especially within a temperature range of 234.5 to 190°F (reservoir temperature is 241°F) given the fact that the oil will cool as it flows from the reservoir to the wellhead, and it is critical to assess the precipitation tendency along the tubing. Following quality control of various AOP measurements at the reservoir temperature, the authors concluded that the AOP for the Magwa-Marrat reservoir fluid is 5,600 ± 500 psia. Figure 10.6 shows the key

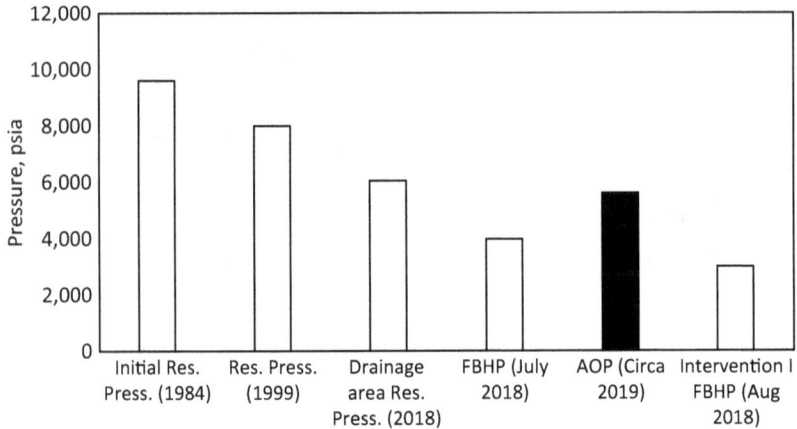

FIGURE 10.6   Pressure chronology of the Magwa-Marrat well with respect to the asphaltene onset pressure. Bar chart constructed based on numerical data in Altemeemi et al.[7]

pressure data for this particular reservoir along with the AOP, which clearly demonstrates the well flowing bottomhole pressure well below the AOP, making it easy to comprehend that asphaltene deposition was the formation damage mechanism, also leading to tubing deposition.

The first intervention in August 2018 that was carried out consisted of tubing cleanout and coiled tubing jetting job using diesel and toluene. The authors compared the multirate tests in July 2018 and August 2018, respectively, which indicated that the intervention marginally improved the well performance but still showed significant damage; the slope of the skin factor–versus–oil rate post intervention was slightly smaller than the one before. Given the only slight improvement from the first intervention, a second much more comprehensive (see details in reference 7) intervention was embarked upon in October 2018. One of the main components and the process involved the use and squeezing of an aromatic solvent into the formation. Multirate tests were re-performed, following this intervention, which indicated significant improvement in the well performance; the skin–versus–oil rate plot showed negative skin and a nearly horizontal line. The authors reported continuous monitoring (up until April 2019 – about 6 months had elapsed) via multirate tests that confirmed the sustenance of the second intervention. This case is a good example of solving the asphaltene deposition problem at its "root" or "in situ" in order to produce asphaltenic fluids at Flowing Bottom Hole Pressure (FBHP)s below the AOPs. Additionally, the operating experience of these authors is somewhat similar to the Ula field in Norway;[8] that is, if the liquid velocity is high enough, then that may be adequate to carry the precipitated asphaltenes to flow out and not deposit.

Misra et al.[9] documented asphaltene deposition problems in an onshore carbonate field in the United Arab Emirates. At first the production strings were getting plugged with solid scales, which were cleaned with acid washing, while others showed the presence of an organic composition rich in asphaltenes. The analyzed solids were characterized for composition that indicated 63+% asphaltenes. The two main origins of this asphaltene deposition were suspected to be a naturally occurring tar mat and a rich-gas injection pilot, both located in the vicinity of the severely affected parts of the reservoir. The operating company approached the problem by considering (1) the field assessment in defining the extent of the problem; (2) a characterization of the collected deposit; (3) a crude oil assessment from the standpoint of colloidal instability index (CII), de Boer plot, and n-C7 titration; (4) the role of the tar-mat layer and rich-gas injection; (5) AOP measurements; and (6) finally screening potential asphaltene dispersants and inhibitors to address the problem. The oil in question is basically a fairly high API gravity (little less than 40°) and having average asphaltene percentage ranging between 0.3% to about 2%. The process of asphaltene assessment is perhaps best described in Table 10.1.

Clearly, the (somewhat) empirical assessment methods in Table 10.1 are not very definitive or conclusive, but one of the stronger (initial) suspicions of tar mat (in) stability of the asphaltenes in the oil and due to rich-gas injection is eliminated because several wells close to the gas injectors did not exhibit asphaltene problems but other distant wells did, which is the reason the authors[9] concluded that gas injection may not be the primary contributor to the (in)stability of asphaltenes. Basically, after eliminating the preceding possibilities, Misra et al.[9] came up with a "preferred"

TABLE 10.1
Summary of Asphaltene Stability Assessment for a Circa 40° API Oil from a Carbonate Oilfield in the United Arab Emirates

Assessment Method	Parameter Value for the Oil or Analog	Stability Outcome	Criterion or Value
Resin/Asphaltene ratio (R/A)	3–26	Stable	R/A > 2.8
Colloidal Instability Index (CII)	1.16–1.40	*Unstable*	CII > 0.9
de Boer plot	0.62–0.63 g/cm³ (in situ density) and 2,200 psia bubble point	*Uncertain*	Data point bracketed between severe and no problem
n-C7 titration	2–5 cm³ n-C7/cm³ oil	*Mildly unstable*	>2.5 cm³ n-C7
Tar mat zone core and extracts	High asphaltene content in tar mat	Marked as "stable"	Extremely low φ and k in tar mat zone; asphaltene migration unlikely
Live oil AOP measurements at 249°F and 130°F	Not detected visually or by near infrared	Construed as "stable"	Authors state bubble point and AOP proximity as the reason for "non-detection"
Live oil AOP with CO₂ or rich gas	AOPs detected, rapid increase with gas mole%	*Potential asphaltene instability*	Rapid increase in AOPs

Note: ϕ = porosity and k = permeability; asphaltene onset pressure (AOP) measurements are on another sample (analog) that has shown asphaltene deposition in tubing. Table compiled based on the data and information in Misra et al.[9]

explanation and attributed the observed asphaltene precipitation to variations in oil compositions via successive oil influxes from two source rocks with wide distribution of mixing ratios, which, in principle, is akin to the blending of incompatible fluids in a pipeline or flowline. A noteworthy operational observation made by the authors is similar to the Magwa-Marrat[7] and Ula[8] fields, respectively, in that asphaltene blockage is retarded at high production rates. Finally, in order to address the asphaltene deposition problem, the operating company resorted to traditional solutions in that they evaluated various asphaltene dispersants/inhibitors and recommended one for tubing cleanout to prevent deposition. The screening was based on a dispersion parameter that used light transmittance through the sample containing the inhibitor and the control.

10.2.4 NAPHTHENATES

Naphthenates originate from naphthenic acid in crudes and form heavier calcium naphthenate (CaN) and lighter sodium naphthenate (NaN).[10] The Ca and Na obviously originate from formation water. Gudmundsson[10] describes naphthenates as "amphiphilic" given the fact that one end has an affinity to water and the other end to hydrocarbons. The physical appearance of naphthenates (calcium) is paste or gel-like deposits that tend to harden with exposure to air.[10, 11] The major naphthenate issue is in the production of acidic crudes; hence, acidity of crude oil is used to gauge the severity of naphthenate problem. However, mechanistic understanding of (calcium) naphthenate deposition is still very limited. For that reason, there appear to be no "screening criteria", "rule of thumb", or "prediction models" in terms of an index, phase envelope, and so on to determine if naphthenates would form. A simple keyword search of "naphthenate" in www.onepetro.org results in only 1600 publications as opposed to nearly 19,000 on hydrates, indicating the elusiveness of naphthenates and perhaps the lack of understanding of their behavior, which is primarily the reason for limited reports of field experiences. Although Junior et al.[11] state that to date CaN and NaN problems have been encountered in almost all oil-producing regions in the world.

As stated by Gudmundsson,[10] naphthenate deposits may come as a surprise in both new and mature field developments due to lack of reliable screening methods to assess the risk a priori. The common spot for naphthenates to occur is typically in topsides, for example, separators and heat exchangers, and in the moving elements, such as pumps. This was precisely what was reported early in the life of a deepwater oilfield in Angola.[11] The Gimboa field crude has an average $24°$ API gravity and relatively high $24°C$ pour point. The formation water contains 10,920 and 900 mg/liter Na and Ca, respectively, and an unusually high 31.2 mole% CO_2. According to Gudmundsson,[10] based on field data reported for some oils and the elemental analysis, higher Na should lead to NaN and higher Ca to CaN, respectively. However, the naphthenate deposit in Gimboa were identified[11] as CaN despite the lower Ca content compared to Na. Junior et al.[11] state that the given type of API gravity tends to be susceptible to CaN, but a moderate total acid number (TAN) of 0.65 mg/g potassium hydroxide (KOH) tend to form NaN instead of CaN. The reasoning provided by the authors[11] is CO_2. Under downhole conditions, the high CO_2 results in very acidic solutions, which inhibits naphthenate salt formation; however, the pressure drop owing to production results in the liberation of CO_2 (solubility proportional to pressure), making the water more basic, thus facilitating the formation of CaN. The formation of CaN in Gimboa began with the water cut, which implicitly means no reported issues for dry oil. The CaN initially manifested as solids and emulsion formation at the interface in process separators, bulk oil treaters and hydrocyclones,[11] subsequently plugging and fouling (heat exchangers) much of the surface processing facility, including the instrumentation. Following repeated production deferrals, the severity of the problems caused by CaN increased, which included oil becoming more viscous and the (sticky) paste-like CaNs bonding with any and all particulate matter, such as scale, sand, and formation fines.[11] Although the precise timing of the severity is not reported, CaN onset was approximately 5 months after first oil. The various mitigation measures and supporting analysis, which became the top priority, that were reported by the authors are best elucidated by the following bullet points:

- The CaNs were trapped by filtration at key points in the processing facility, such as outlets of inlet separator, production separator, dehydrator, desalter
- The trapped CaN solids dissolved in a solution of xylene and hydrochloric acid, basically separating in a top oily layer and a somewhat opaque water layer. Note that given the amphiphilic nature of naphthenates, an organic, as well as inorganic, solvent is necessary.
- Initially some glacial acetic acid (GAA) was used in conjunction with de-emulsifiers to reduce the formation of naphthenates. GAA reduces the pH of the aqueous phase and dissolves the already formed CaN solids.
- Physical parameters are, of course, pressure and temperature – high pressure means much CO_2 remaining in solution, thus reducing the pH, whereas temperature exceeding 65°C exacerbates CaN formation.
- Other combinations are removing as much water as quickly as possible from the front end of the process, varying the pressure and temperature to maintain as much CO_2 as possible in the solution, and injecting GAA into the produced water to keep low pH levels.
- Water cut can also be construed as a measure given the observations in Gimboa in that the CaN problem worsened with increasing water cut; however, beyond a 50% value, there was a decline in the CaN.
- Some of the preceding measures resulted in a somewhat temporary solution in that the CaN solids started to build up again, thus necessitating a more robust treatment.
- Although theoretically maintaining a high enough pressure is beneficial, it may not always be practically feasible.
- Ultimately, the authors report the use of an aggressive combination of lower temperature and a commercial CaN inhibitor injection to bring the solids loading to a minimum.

Among the multitude of chemical components found in crude oils, naphthenic acids (NA) are one of them, and therefore, the severity of naphthenate problems can be assessed by measuring the acidity of the crude oil.[10] Obviously, since hydrocarbons

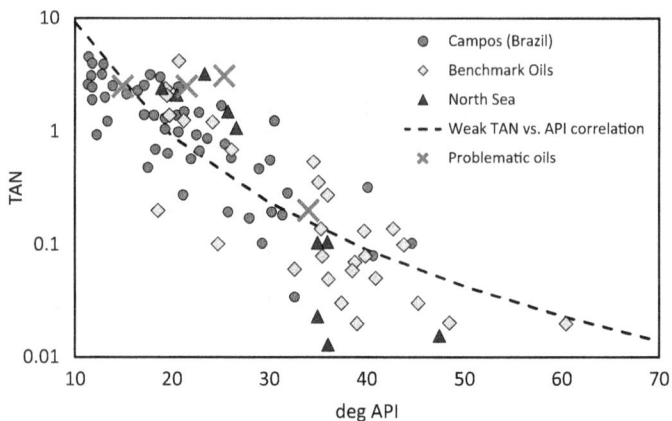

FIGURE 10.7 Empirical evidence of high TAN–low API and vice versa for different crude oils and a suggested correlation. Plot constructed based on various data sources in Gudmundsson.[10]

are not aqueous, they are not amenable to direct pH measurements, and hence, a special type of titration[10] is carried out to determine the TAN. For example, the TAN value of 0.65 mg/g of KOH for the Gimboa crude. Based on a comparison of API gravities and TAN values, it has been suggested[10] that heavy crude oil (e.g., due to maturity, biodegradation) is more acidic compared to light crude oil, which has been alternatively stated as high concentration of naphthenic acids are often found in heavy crudes with high TAN.[12] The empirical evidence of this is demonstrated in the TAN-versus-API graph shown in Figure 10.7. Gudmundsson[10] presented an empirical equation based on some literature data that allows the estimation of naphthenic acid wt% from TAN; NA = 0.45(TAN − 0.6), which means NA wt% is zero for a TAN value of 0.6. Naphthenate deposits can, however, occur in high, as well as low, TAN numbers.[10]

Somewhat similar to the experience in Gimboa,[11] Nichols et al.[12] also state the significance of the nature of the NAs found in the crude, the produced brine composition, and operating conditions, such as temperature, pressure, and water cut, all have an effect on the naphthenate risk. However, given the more than 1000 species of the NAs that are reportedly found in crude oils,[10, 13] the treatment of naphthenates requires customized or field-specific[12] approaches to combat the problem. Nichols et al.[12] presented a CaN case study in a complex production system, in the Campos Basin, handling mixed oils from different reservoirs on a single floating production storage and offloading (FPSO) separator. The oils in question are five different crudes; three are high TAN (2–2.2 mg/g KOH), and two are low TAN (0.8–1.1 mg/g KOH) that produce formation waters with calcium concentrations of 1,840 mg/liter and 15,000 mg/liter, respectively. The collected deposits were analyzed and showed the presence of high-molecular-weight acids, which is characteristic of NAs (the reader is to referred to Gudmundsson[10] for the chemistry aspects). The use of a flow loop, vis-à-vis ΔP measurement versus time, is a fairly well-accepted standard laboratory methodology to assess the flow assurance risk posed by solids (especially paraffin waxes, asphaltenes, and scale) and subsequently evaluate the efficacy of inhibitors. In principle, this is basically what the authors[12] used to investigate the effects of mixing different combinations of the oils and brines to identify those with the greatest naphthenate risk. The flow loop tests were conducted[12] in two modes using two different rigs – first one was for the determination of main scale of interest, calcium carbonate (brine alone, no oil), and the second one, which is a bit like flow through than loop, was primarily for the determination of naphthenate and inhibitors. The operating temperature and pressure in both rigs were, however, common at 85°C and 5 bars (185°F and 72.5 psi) to mimic the FPSO separator conditions. The scaling rig is equipped with a 1/16th-in. outer diameter and 1,000-mm-long stainless steel coil in which the test brine flow rate is 10 ml/min. The naphthenate rig is operated with a 20% WC, which represents the FPSO separator. Test oil and brine are basically mixed in a cell at a speed of 13,500 rpm, following which the homogenized mixture passes through two filters placed in series – a coarse filter and a fine 7-μm sintered metal filter – across which the ΔP is measured. The outlet of the fine filter is hooked to a transparent separator for visual observation of phases. The results obtained in these flow experiments can be best summarized or presented in a tabular format, as in Table 10.2, which will benefit the reader.

Based on the earlier summary and details provided by the authors,[12] some specific conclusions can be drawn or lessons can be learned. These include (1) both oil A and

---

TABLE 10.2

Summary and Results of Timed Flow Experiments Conducted at 85°C and 5 Bars (185°F and 72.5 psi) for Campos Basin Fluids

Test Fluids	Composition/ Concentration/ TAN	Flow Assurance Risk	ΔP	Assessment
High Ca brine	Containing 100 mg/liter HCO$_3^-$	Scaling tendency	Stable 1.5 psi over ~120 min	No blocking
High Ca brine	Containing 585 mg/liter HCO$_3^-$ (field level)	Scaling tendency	Spiking to 8.5 psi in ~20 min	Rapid scaling
High TAN oil A	With high Ca 100 mg/liter HCO$_3^-$, 2.0 mg/g KOH TAN	Scaling/CaN	Spiking to 100 psi in ~ 10 min	Scale (likely due to high shear) + CaN
High TAN oil A	No HCO$_3^-$, 2.0 mg/g KOH TAN	CaN	Spiking to 100 psi in ~ 15 min	CaN solids being formed
Low TAN oil D	With high Ca 100 mg/liter HCO$_3^-$, 0.8 mg/g KOH TAN	Scaling/CaN	Steady rise to ~40 psi in 60 min (stable thereafter)	Scale (likely due to high shear) but no CaN
Low TAN oil D	No HCO$_3^-$, 0.8 mg/g KOH TAN	CaN	Steady rise to ~10 psi in 45 min (stable thereafter)	No CaN solids

(Continued)

TABLE 10.2   (Continued)
Summary and Results of Timed Flow Experiments Conducted at 85°C and 5 Bars (185°F and 72.5 psi) for Campos Basin Fluids

Test Fluids	Composition/ Concentration/ TAN	Flow Assurance Risk	ΔP	Assessment
High TAN oil A	With high Ca brine but no HCO_3^-, 2.0 mg/g KOH TAN	CaN	Spiking to ~55 psi in 55 min	CaN solids being formed
High TAN oil A + 200 ppm selected CNI	With high Ca brine but no HCO_3^-, 2.0 mg/g KOH TAN	CaN	Stable < 10 psi over ~60 min	No blocking; CNI effective
High TAN oil B + 25 ppm SI	Various high and low Ca brines with HCO_3^-, 2.2 mg/g KOH TAN	CaN (in conjunction with HCO_3^-)	Rapid rise to ~100 psi within 30 min	CaN solids blocking the fine filter
High TAN oil B + 25 ppm SI	Various high and low Ca brines without HCO_3^-, 2.2 mg/g KOH TAN	CaN (without HCO_3^-)	Compared to oil A, relatively much gradual increase	Milder CaN potential compared to oil A
High TAN oil B + 25 ppm SI + 200 ppm selected CNI	With high Ca 585 mg/liter HCO_3^-, 2.2 mg/g KOH TAN	CaN	Rising to same ΔP as without CNI; albeit with a time lag	CNI slows the formation of CaN solids but ineffective in preventing them

Note: CNI is CaN inhibitor, and SI is scale inhibitor. Table prepared based on the data and information or discussion in Nichols et al.[12]

B are crudes with nearly the same high-TAN values, yet their propensity for CaN risk differs; (2) although a certain CaN inhibitor (CNI) qualifies or is effective for oil A, the efficacy is not universal, in that it does not perform at the same level in case of oil B; (3) in blended production systems such as the one in the Campos Basin, it is possible that a particular oil represents the most extreme or severe conditions of CaN risk, which may actually be offset by mixing with another crude (of lower risk); and (4) the specific type of NAs is perhaps more important than the reported bulk or overall concentration. Nevertheless, the preceding type of comprehensive workflow can be very useful in the assessment of the flow assurance risks (and ultimately in shortlisting a potential CNI) posed by naphthenates in a diverse set of operating parameters, especially the commingling of different (TAN) crudes and brines.

10.2.5   SCALE

The formation of scale or precipitation is basically a result of exceeding the solubility limits of some of the components or salts in the aqueous phase. At least relative to asphaltenes, scale chemistry is somewhat better defined or known. For instance, there are mainly two groups[14] of salts present; one is higher solubility (NaCl, KCl, $CaCl_2$), and the other is lower solubility such as $BaSO_4$, $CaCO_3$ (also referred to as calcite), $SrSO_4$, and $CaSO_4$. Clearly, given the lower solubility of the second group, salts that are more likely to precipitate and deposit, commonly referred to as scale, are $CaSO_4$, $BaSO_4$, $SrSO_4$, and $CaCO_3$. As far as produced water is concerned, it may have a self-scaling tendency, for example, the calcium carbonate, such as in the case of the Campos Basin, or a common household example being the lime scale or $CaCO_3$ in the teakettles. The scale may also form when incompatible waters mix (seawater and formation water), producing a scaling brine that may result in carbonate or sulfate scale. Obviously, the formation of scale has operational and economic impacts, such as reduced production or deferrals, scaling of artificial lift equipment (electrical submersible pumps or ESPs), well shut-ins, fouling of processing equipment, and even well abandonment in extreme cases.

In principle, the overall scale management strategy is somewhat similar to other solids such as asphaltenes. The two main elements are monitoring or surveillance and treatment (as needed). Monitoring may involve scale prediction, tubing checks, production performance, and water and scale analysis, whereas treatment may be mechanical (coiled tubing, reaming, etc.) or chemical, such as (acetic or hydrochloric acid) scale dissolvers. However, the most cost-effective method for dealing with scale is to prevent the scale before it forms.[15] The five steps[15] in scale inhibition strategy are (1) evaluate scaling conditions (location and type); (2) conduct laboratory testing; (3) screen a suitable scale inhibitor (SI; typically acid-based) based on effectiveness, water chemistry, and temperature; (4) design a dosage plan; and, finally, (5) field deployment. The application of SIs can be continuous or via squeeze treatment; however, some of the important prerequisites are (1) that the SI is added upstream of where the deposition occurs;[15] (2) SI is released at enough concentration to be effective; and (3) treatment lifespan is long enough to be economically viable.[16]

The exercise/case study[17] covered here, for a new field development that will produce from a single reservoir using primary depletion initially and via artificial lift later (ESP, gas lift), is intended to provide a workflow for selecting or screening a particular type of SI. A neighboring field, which contains similar formation water, is known to suffer from $CaCO_3$ scale and has led to negative experiences due to poor selection of SIs, thus requiring chemical dissolver treatments. Therefore, the operator is erring on the side of extreme caution. The total dissolved solids (TDS) in the new field formation water is fairly high (~130,000 mg/liter), and the calcium concentration is 6,900 mg/liter. The initial reservoir pressure and temperature are 4,000 psi and 190°F. In the SI screening, anticipated production scenarios for five different stages of field life are considered, which range from early life to year 15. As a first step, standard industry software is utilized to predict the potential calcite deposit (in mg/liter of produced water) at various spots and field life stages. Figure 10.8 shows the growth of calcite deposit at all spots with field maturity. If the deposits in year 10 and 15 are compared, the rates stabilize a bit, however, the slight increase is attributed to the increase in the water cut with time. A somewhat standard ensuing operational practice is to invite chemical vendors to submit products for independent third-party testing in order to screen SIs. These tests typically consist of (visual)

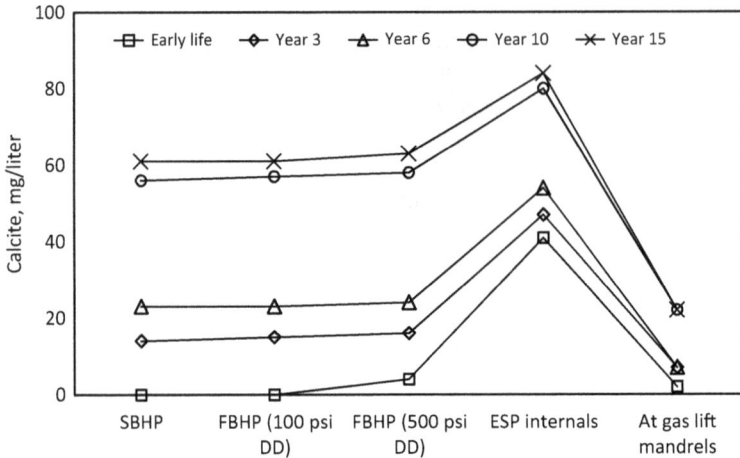

FIGURE 10.8 Evolution of calcite with time at various locations. Note: SBHP = static bottomhole pressure; FBHP = flowing bottom hole pressure; ESP = electrical submersible pump; and DD = drawdown. Plot constructed based on tabulated predicted values in reference 17.

compatibility on mixing, as well as with the metallurgy, plastics, elastomers, and others that are typically encountered in the oilfield; (dynamic) flow loop tests; core floods to assess the (formation damage) impact on end point oil and water permeabilities; and pore volumes (PV) to minimum inhibitor concentration (MIC). In this particular case, the four products according to their generic chemistry are (1) sulfonated copolymer, (2) acid phosphonate, (3) neutralized phosphonate, and (4) phosphate ester. As far as dosages up to a concentration of 100 ppm in the formation water at 190°F are concerned, all four products show a clear appearance after 16 h of testing; however, at higher concentrations of 10,000 and 100,000 ppm, product 2 has a hazy appearance, whereas product 3 results in a precipitate. All four SIs resulted in compatibility with plastics, elastomers, and metals (except mild steel and aluminum).[17] Metal and plastic/elastomer compatibilities were tested using corrosion rates and mass changes, respectively. The results of the jar test, SI dynamic scale loop test, and core floods (for changes in end point relative permeabilities and PV to MIC) are summarized in Table 10.3. Corefloods (9% porosity, no other data available) are designed to assess the suitability for squeeze application.

In Table 10.3, the relative permeability ratios of close to 1 basically indicate no damage, whereas value far less than 1 (shown by the italics) point to potential damage caused by the squeezed SI. The PV to MIC is primarily indicative of the longevity of the SI. Although the acid and the neutralized phosphonates result in higher PVs to MIC, the end point relative to permeability ratios for both are far less than 1. Relatively, the return permeabilities in the case of the neutralized phosphonate are much better compared to the acid phosphonate; it is the only SI that results in a sudden spike in ΔP and produces a precipitate, albeit at much more elevated concentrations. Clearly, the two phosphonate-based products 2 and 3 can be eliminated. Now if the performance of products 1 and 4 are compared, the former is superior from the

TABLE 10.3
Summary of SI Screening Tests at 190°F

Scale Inhibitor	Jar test – SI Required for 95% Inhibition	Dynamic Scale Loop Pressure Drops for 3 ppm Dosage	$\dfrac{k_{ro} @ S_{wi}(\text{after})}{k_{ro} @ S_{wi}(\text{before})}$	$\dfrac{k_{rw} @ S_{or}(\text{after})}{k_{rw} @ S_{or}(\text{before})}$	PV to MIC
Sulfonated copolymer	4 ppm	Fairly flat ΔP	1.01	1.00	*500 PV*
Acid phosphonate	2.5 ppm	Fairly flat ΔP	*0.45*	*0.55*	890 PV
Neutralized phosphonate	3 ppm	*Sudden spike in ΔP at 50 min*	*0.85*	*0.73*	810 PV
Phosphate ester	**2 ppm**	**Fairly flat ΔP**	**0.95**	**0.79**	**715 PV**

Note: k_{ro} and k_{rw} are end point oil and water relative permeabilities, respectively, and S_{wi} and S_{or} are irreducible water and residual oil saturation, respectively. Italics indicate values far less than 1. The table prepared based on the data and information from reference 17.

standpoint of no damage (both relative permeability ratios ~1); however, it does result in the lowest (500) PV to MIC and requires 4 ppm (highest among all 4) for 95% inhibition as shown by the jar tests. This means product 1 can also be eliminated and ultimately product 4, which is phosphate ester, is recommended (values in bold in Table 10.3) for squeeze treatment based on the overall performance to provide control of the scaling risk over the field life. Note that the relatively lower (715) PV to MIC and the somewhat marginal loss of return permeability is a bit of compromise considering other performance indicators for the other three SI products. This example of SI screening methodology, which may also be applicable to other solids inhibitors, demonstrates the importance of recommendation that should be based on multiple assessments or comprehensive tests rather than relying on one or two, which may be misleading, for example, choosing acid phosphonate, solely based on the highest (890) PV to MIC.

Guraieb et al.[18] recently reported case studies on novel extended reach (ER) SIs to combat $CaCO_3$ and $CaSO_4$ scaling risk in the Permian Basin. The SI deployment is squeeze treatment–based, which the authors[18] state as being effective in preventing scale formation in wells for extended periods, hence the name ER. However, in some cases, these treatments may last for a limited duration, thus requiring repeated squeezes and negatively impacting the production and economics. The authors reported on the success of a phosphonate-based ER SI for scale treatment, compared to an incumbent phosphonate-based SI. Coreflood PVs to MIC (authors refer to this as the minimum effective concentration, MEC) and the SI adsorbed per gram of core material (somewhat congruent to PV to MIC, because the higher the adsorption, the higher the likelihood of larger PVs to MIC). The ER SI showed over 10,000 PV of flow with SI concentration remaining above the MIC, which was approximately 2–3 times the incumbent SI dropping below the MIC. The adsorption of the ER SI was reported to be about 6–12 times the incumbent. In subsequent field trials, the ER SI treatment volumes were targeted to achieve a 5-ft radial penetration. The average squeeze treatment lifetime of the ER SI was demonstrated to be about 2–3 times longer than the incumbent in the three fields (~70 wells) in which the comparisons were carried out. This example or case study simply points out the developments taking place in not only the efficacy of the SI but also the longevity.

The first instance of production loss attributed to scale in the carbonate wells of South Marlim field, Campos Basin, Brazil, occurred in July 2014, which was treated by remote acidification (acetic and hydrochloric).[19] The scale deposition mechanism hypothesized by the authors is water evaporation due to the high salinity (110,000 mg/liter) formation water, with elevated levels of calcium, sodium, and chloride, coming in contact with the dry gas from gas lift. Consequently, the most probable[19] scale suspects were calcium carbonate ($CaCO_3$) and halite (NaCl). Although acids are considered as an effective method to treat scale, logistics, economics, sustained efficacy, and HSE concerns are some of the main challenges. Following the acid treatment in early 2018, the scale problem resurfaced within about 1.5 months, resulting in nearly 42% reduction in the oil rate. For this reason, a somewhat unique and conceptually simple approach in combating scale was proposed and implemented, as recently reported by Martins and Freiman[19] in which fresh water is used to remove the scale. The application of fresh water involves soaking of the production tubing under the assumption that it would dissolve (at least some of it) the scale. However, there are two main challenges: one is logistics, that is, ready availability of needed volumes of fresh water, especially in offshore environment, and the second is a solids-specific risk in that fresh water may be thermodynamically conducive to the formation of hydrates. Note that the salts present in water act as thermodynamic inhibitors, shifting the hydrate equilibrium curve up. Both challenges were overcome[19] in this particular case because fresh water is produced by the FPSO itself, and by adjusting the operating conditions and taking advantage of the layout (avoidance of liquid accumulation), the hydrate risk could be mitigated. The freshwater deployment procedure as described by the authors[19] is somewhat like a "huff-n-puff" in that the fresh water is bullheaded in the formation and tubing, and there is a 4-hour soak period, following which the well is restarted. In one of the wells, the oil rate doubled post the freshwater treatment, and two other wells showed a similar result. A lab analysis conducted on the injected fresh water and the return water after soaking, confirmed halite dissolution.[19] However, following the initial success, the frequency of these treatments increased considerably, which was the reason SI was added to the fresh water in soaking operations,[19] thereby widening the span between treatments. With the freshwater soaking treatment, a saving of US$700,000 compared with the remote acidification, on the basis of four operations is realized.[19] The authors report positive results in mitigating scale deposition by using fresh water, without hydrate occurrences, since 2018 and given the publication year of 2021, the method appears to be sustainable.

10.2.6 MULTIPLE ORGANIC AND INORGANIC SOLIDS

For the most part, flow assurance risks are typically related to a given type of predominant solid, i.e., hydrate, or paraffin wax or asphaltene or naphthenates, such as the specific cases discussed above. That means flow assurance risk (simultaneously) posed by multiple solids may not be very commonplace. The deepwater Stones field in the GOM is somewhat unique in that not only were multiple solid threats identified, but there were also other unique challenges, such as very rough terrain, a furrowed sea bottom, and its proximity to an escarpment.[20] The field was discovered in 2005,[21] which contains an estimated reserve of 2.5 MMBOE (million barrels of oil equivalent). With reservoir depth around 26,500 ft below sea level, 17,000 ft below the mud line, and water depth of 9,500 ft, the field is a record holder in deepwater. The reservoir is characterized as high-pressure, high temperature (HPHT); low GOR; having a lower grade API gravity; and having a somewhat higher viscosity than typical GOM fluids.[20] The oil is shipped via FPSO, whereas gas is exported by a lateral pipeline (faces ~3,000 ft of elevation change) that ties into a gathering system. Clearly, flow assurance challenges are obvious given the extremes and compounded by a lack of good fluid samples.[20] Start-up or production in the first phase of Stones development commenced in the fall of 2016 and shortly after that, Hoffman et al.'s[20] paper was published in May 2017; however, since then there have been no documented (in the open literature) instances to date of production deferrals due to flow assurance issues. This is a good indication or assumption that the adopted[20] flow assurance combat strategy, as summarized in Table 10.4, continues to be effective, which perhaps serves as an example of a flow assurance engineers "toolbox". Given the lack of good fluid samples, consequently, the slightly conservative approach adopted, especially with respect to hydrate and scale, may also be a contributing

TABLE 10.4
Flow Assurance Threats and Combat Strategies in the Stones Field

Type of Solids	Contributing/ Uncertainty Factors for Solids	Location in the Flow Infrastructure			
		Wellbore	Tree	Flowline/ Riser	Topsides
Hydrate	Uncertainty in produced water salinity	MeOH bullheading for unplanned and planned shutdowns	MeOH flush for shutdowns and MeOH injection for cold start-up	DOD for shutdowns and hot oiling for cold startup	Not identified
Wax	Elevated cloud point and critical wax deposition temperature	Not identified	Hot oiling (also for cold start-up) and/or paraffin inhibitor at tree		Not identified
Scale	Co-deposition of $CaCO_3$ and $BaSO_4$	SI downhole	Not identified		SI
Asphaltene	N/A	Not identified	Kept as contingency	Not identified	
Naphthenate	N/A		Not identified		

Table modified after Hoffman et al.[20]

factor for sustained production. Note that the team also considered other ancillary flow assurance risks such as corrosion (due to the presence of CO_2), erosion, and emulsion, which were deemed to be somewhat typical in that corrosion inhibitor and demulsifier were recommended, mainly at the topsides.

10.2.7 SAND MIGRATION AND FLOW

Subsurface formations can be consolidated or unconsolidated; typically, it is the latter types that are prone to the migration of sands, thus causing flow assurance challenges. Therefore, sand can be considered as an unwanted by-product of oil and gas production. Salama[22] has alternatively stated that sand production may be a certainty in reservoirs that have relatively low formation strengths (less than 1,000 psi). Thus, sand production typically falls in the realm of geomechanics and/or well completions, which is beyond the scope of this book. Additionally, sand is something that is not part of phase behavior, which is basically the case with all the other solids; its presence in the flow infrastructure along with the other solid phases manifests a very peculiar flow assurance problem. The migrated or produced sand can settle (below a certain minimum velocity) in flowlines and various processing equipment (especially in the low points), and if it does flow together with the produced fluids, then it commonly leads to metal erosion,[10] thus becoming a matter of HSE for oilfield installations. Gudmundsson[10] also states that the sands present in the flow system may act as induction or seeding sites for nucleation of common flow assurance solids. The erosion rate is a function of factors, such as flow velocity, sand production rate, sand size and sand shape; material characteristics; and flow system geometry (bends, tees etc.).[22] Primarily, three different mitigation measures for sand are prescribed,[22] namely, (1) for first oil or from day 1, complete the production wells with downhole sand exclusion systems, such as sand screens; (2) work over to recomplete the well later by installing sand exclusion systems at the onset of sand production under the assumption that for the most part, sand production will not begin as soon as the well is put on production; and (3) "manage" the sand, without any mechanical exclusion systems, by designing facilities to handle the produced sand. The last option is preferred in many cases because the exclusion systems may pose the risk of well plugging and mechanical damage.[22] However, this requires sand erosion prediction models, which are necessary for establishing tolerable sand production rates, and, in turn, allowable well flow rates.[22] A somewhat commonly used equation for determining the solids free erosion velocity limit is the API Recommended Practice (RP)14E,[23, 24] expressed by the following:

$$U_e = \frac{C}{\sqrt{\rho_m}} \tag{10.1}$$

where U_e is the erosional velocity in ft/sec, C is an empirical coefficient, and ρ_m is the fluid mixture density at prevailing flow conditions in lbm/ft^3.

Since the validity of Equation 10.1, especially for sand-laden fluids, has been challenged,[24] Salama[24] suggested an alternative equation that has the following functional form:

$$U_e = S \frac{D\sqrt{\rho_m}}{\sqrt{W}} \tag{10.2}$$

Note that using the units of m/sec for U_e; kg/m³ for ρ_m; mm for pipe diameter D, and kg/day for sand production rate; the value of S factor is 0.05 for pipe bends. Consequently, using $S = 0.05$ in Equation 10.2, $U_e = \frac{D\sqrt{\rho_m}}{20\sqrt{W}}$, which is the recommended[24] correlation for establishing the erosional velocity limits for sand-laden oil and gas production. Another practical equation developed by Salama[22, 24] for predicting the erosion rate in terms of sand production is as follows:

$$ER = \frac{1}{S_m} \frac{WU_m^2 d}{D^2 \rho_m} \tag{10.3}$$

where ER is the erosion rate in mm/year; S_m is geometry factor (numerical value of 5.5 for pipe bends[24]); U_m and ρ_m is fluid mixture velocity and density in m/sec and kg/m³, respectively; and d is sand size in μm. Typical sand grain size is 250 μm in conventional oil and gas reservoirs; however, above 400 μm the effect of d becomes negligible, and thus, a limit of 400 is used in Equation 10.3 for $d > 400$.

The application of Equations 10.2 and 10.3 can be illustrated by the following numerical example. For a certain two-phase flow that also is accompanied by sand production, $U_{sL} = 20$ m/sec, $U_{sG} = 5$ m/sec, $\rho_L = 850$ kg/m³, $\rho_G = 50$ kg/m³, $W = 2$ kg/day, $d = 200$ μm, and $D = 50$ mm. The given data are used to first calculate the mixture density from the phase densities and the respective superficial velocities: $\rho_m = (\rho_L U_{sL} + \rho_G U_{sG})/U_m = (850 \times 20 + 50 \times 5)/(20 + 5) = 690$ kg/m³. Substituting the previously calculated ρ_m and other given data, $U_e = \frac{50\sqrt{690}}{20\sqrt{2}} = 46.4$ m / sec ; clearly, if everything else remains the same but the sand (co)production rate goes up, then the U_e value will decrease, which is to be expected, vis-à-vis the higher the rate of sand, the more stringent is the erosional velocity threshold and vice-versa. The erosion rate can also be calculated by direct substitution of the given data in Equation 10.3,

$$ER = \frac{1}{5.5} \frac{2 \times 25^2 \times 200}{50^2 \times 690} = 0.026 \text{ mm / year}, \text{ which is well below the acceptable or}$$

tolerable erosion rate of the order of 0.1 mm/year.[24]

10.3   FLOW ASSURANCE AND HYDRAULIC FRACTURING

The rapid growth in the US domestic oil and gas production in recent years is primarily attributed to what is known as unconventional reservoirs in regions such as Bakken, the Permian Basin, and Eagle Ford. These unconventional reservoirs are generally characterized as such due to the low rock permeability that can be in the range of nano Darcies (nD) as opposed to milli Darcies (mD) in the case of typical conventional systems, that is, orders of magnitude difference in permeability.

Given such low permeabilities, a common oilfield practice is to stimulate the rock by hydraulic fracturing to enable the fluids to flow into the production well(s). Typically, it is the "cold" fracturing fluid used that is the reason for somewhat unique flow assurance issues that mainly pertain to wax. In fact, way back in 1974, Sutton and Roberts[25] published the first paper on paraffin precipitation during fracture stimulation, which is ubiquitous in the current times due to the boom in the unconventionals. In particular, this is a major concern for the shale oils in the aforementioned regions given the fact that they tend to be highly paraffinic[26] and thus have high WATs (up to a carbon number of 100), although sweet and light with API gravities in the range of approximately 40–45°. Therefore, there is always a persistent risk of deposition in the pore spaces, resulting in positive skin, as well as in the wellbore, when temperatures drop below the WAT. Mahmoudkhani et al.[26] classified or graded the wax deposit hierarchy according to the molecular weights (MWs) as follows: higher MW or carbon number wax deep in the fractures, mixed-MW wax in the horizontal section of the wellbore that is proximal to the fractures, medium-MW wax near the base of the vertical section of the wellbore, and, finally, lower MW wax near the wellhead. Clearly, from a phase behavior standpoint, this gradation is consistent with the temperature profile at these respective (problem) locations. This is also evident from the paraffin-distribution high-temperature gas chromatography (HTGC)[26] of samples from the wellhead ($\sim C_{10}$–C_{30}) and downhole spots ($\sim C_{30}$–C_{70}), respectively. Given such diversity of wax, the authors recommend the use of multi-chemistry or a "cocktail" instead of single chemistry to target a certain carbon number range. For example, the proportions of carbon number groups can be calculated per the HTGC, and the same can be applied in selecting a blend. For wax mitigation, the authors mention the dual-purpose use of porous ceramic proppants as a chemical (inhibitor) delivery system as well as a proppant to keep the fractures open.

10.4 HSE CONSIDERATIONS OF FLOW ASSURANCE

Given the fact that the flow interruptions or blockage caused by gas hydrates can be sudden or instantaneous; for the most part, HSE considerations of flow assurance pertain to gas hydrates. The primary reason behind this is the characteristics of gas hydrates because dissociation of one unit volume of hydrate can result in as much as 180 volumes of gas, meaning a potentially very rapid buildup of pressure in a confined space. According to Sloan,[2] hydrate plug lengths ranging from 25 to as much as 200 ft can move at velocities between 60 and 270 ft/sec, which will almost be like a torpedo or a projectile. Sloan[2] reported more than 30 cases of accidents related to hydrate blockages resulting in loss of revenue and/or life. Of these cases, two report loss of life, remediation time ranging between 1–2 days to over a month and loss of revenue from as little as US$4,000 to millions of dollars. Both from HSE as well as economics standpoints, Sloan's[2] quote is very noteworthy and apt: *"a gallon of prevention is worth a mile of cure"*. In summary, the best practice is to prevent the formation of hydrate plugs through better design and operating practices. For example, staying in the safe hydrate-free region or using inhibitors, which may be expensive but are overshadowed by the significance of HSE. Table 10.5 provides a summary of HSE incidences related to hydrate blockage.

TABLE 10.5

Summary of Hydrate Blockage Operational Events and their Role in Health, Safety, and the Environment

Field Operator	Condition	Cause of Blockage	Extent of Blockage	Downtime	Loss of Life and Revenue	Remedy
Chevron	Sour gas flowline, distant end of the wellhead	Not reported	Not reported	Not reported	Foreman lost life due to line failure because of hydrate plug	Not reported
Gulf oil	Rimbey gas plant start-up	Plug formed in overhead line having preexisting cracks	Not reported	Not reported	Chief operator lost life, plant suffered million $6 in damages	Not reported
Mobil	Sour gas flowline	Not reported	Two locations between wellhead and satellite	3 days	None	Depressurization
Statoil North Sea	Oilfield	Not reported	Topsides	Not reported	None	Heating – which resulted in an exploded line
Elf Aquitaine North Sea	Plug in a riser	Not reported	Plug in a riser	Not reported	None	Discharge end left open to extrude plug – expelled into the ocean (platform rose by 8 in.)
GOM	1400 psi, 65°F, free water, gas, and condensate	Cold temperature in export line	8–10 ft	3 days	$40,000	Depressurization, MeOH injection in future
Wyoming	120°F and 360 psi, but ambient −20°F	Line wrapped with heating tape – heat input lowered for cost	Not reported	1 day	Not reported	Depressurization + heat + MeOH

(Continued)

TABLE 10.5 (Continued)

Summary of Hydrate Blockage Operational Events and their Role in Health, Safety, and the Environment

Field Operator	Condition	Cause of Blockage	Extent of Blockage	Downtime	Loss of Life and Revenue	Remedy
Chevron – Oklahoma	Sales gas line – T&P drops after choke	Cold ambient temperature	Not reported	½ day	Not reported	Depressurization + heat + MeOH
Chevron – Canada	Gas gathering line; gas saturated with water vapor	Extended shut-in period, 37–41°F ambient temperature	300 ft in a 15-mile line	2 days	Not reported	Welding rig used to apply electric current directly to the affected section
Lasmo – North Sea	Single flowline between 2 satellite wells (no R/T pigging)	Combined wax + hydrate problem caused due to high WAT, uninsulated line and cold seawater temperature	1.2-mile section blocked	2 years' worth of struggle	Not reported	Heat, MeOH and PIs used, affected section replaced – field abandoned due to remaining reserves
Texaco – GOM	Gas export line, 8000 bbl/d liquids, and 70 mmscf/d gas	Gas not dehydrated – water vapor condensed	U bend (bottom of the riser)	2–3 days	Not reported	MeOH injection
Texaco – Wyoming	Gas wells – flowing wellhead conditions of 2000 psi, 52–56°F, 0.8–1.4 mmscf/d, water 2–4 bbl/d	MeOH already injected at a rate of 30 gal/d	Intermittent hydrate problems	None	None	Solution of kinetic hydrate inhibitor (KHI), MeOH injected instead of pure MeOH; represented 50% cost savings. KHI effective in several cases by Texaco

Occidental – North Sea	5,000 psi and 86°F inlet cooling down to 35°F	25wt% MeOH injection conc. maintained	Intermittent hydrate problems – with MeOH rate drops	4–15 days	Not reported	Depressurization
Amoco – North Sea	70-mile-long export line	Line not pigged for 3 months (free water conc. low but accum.)	30 mile from inlet – line completely blocked	2 months to remove hydrate blockage and 3 months of lost production	Million $0.5 for hydrate remedy and $5.5 for lost production	Two sided depressurization, i.e., up- and downstream of the plug
Offshore – South America	Production well – large gas, condensate rates and 6% WC	1-day shut-in to collect buildup data	Shut-in conditions cooled the well fluid forming hydrates	13 days lost production	Not reported	Coiled tubing string and hot glycol

Table compiled and prepared based on the documented cases in Sloan.[2]

10.5   PERSPECTIVES ON FLOW ASSURANCE ECONOMICS

Detailed flow assurance economics is a rather vast topic, which is beyond the scope of this book. Therefore, only perspectives (in terms of numbers) on flow assurance that are, for the most part, related to operational costs are included from the standpoint of completeness (see the following bulleted points). Basically, the operational cost is mostly controlled by the chemical dosage rate; for example, if methanol (MeOH) injection is required at a rate of X gallons/day for a certain pipeline, flowline, and so on, then the (operational) price of MeOH per gallon can be directly utilized to calculate the daily cost or expenditures, since MeOH is generally not recovered and thus considered as spent. This can, in turn, be used to express the cost of flow assurance chemicals per stock tank barrel of oil or (mm)scf of gas, as appropriate. Although average cost of straight chemicals such as MeOH, MEG, and xylene tend to be somewhat low; specialized chemistries used in the various inhibitors (kinetic hydrate, asphaltene, scale, corrosion, paraffin) are the reason they tend to be the most expensive. For example, MeOH price is about US$1/gallon, whereas a kinetic hydrate inhibitor (KHI) may be as much as 7–10 times costly. It is, however, important to note that solely the consideration of price of the chemical alone may be quite misleading. A case in point is the comparison between MeOH and a KHI; although MeOH is very inexpensive, the dosage rate tends to be very high as opposed to a ppm level in the case of KHI, or in other words the gallons and gallons of MeOH multiplied by its (low) price and vice versa in the case of a KHI. Essentially, the dosage rate and price tends to offset each other.

- "Annually, a significant operating expense equal to hundreds of millions of US dollars is devoted to hydrate prevention with half spent on inhibition while offshore operations additionally spend approximately US$1 million per mile on insulation of subsea pipelines to prevent hydrates". Jassim et. al.[27]
- A typical example shared by an industry expert on cleaning pig operations of 24-in., 15-mile-long pipeline would cost about $250,000 plus a disposal fee of approximately $30,000, done before running a smart pig (another $100,000). Total pigging cost of about $380,000. If this is extrapolated to all US pipelines at one time then this would amount to $50+ billion. www.rbenergy.com.[28]
- According to US Senate testimony on the Trans Alaska Pipeline System (TAPS), it would cost US$2 million to run a smart pig operation for the entire length of 800 mile (including the staff time, data analysis, and transit time). www.senate.gov.[29]
- Per BP Pipelines report from 2011, a complete replacement of TAPS insulation (better R value, 2–4 times the existing) would cost US$800,000 to $1,000,000/mile – this is mainly for wax control and retaining heat.
- A Statoil study in 2006 for a long-distance pipeline for 55,000 STB/day oil flow rate over a 7-year period reported that cost of pigging and wax inhibitor versus pigging alone would vary by a factor of 3–4 (pure pigging is a cheaper alternative).

- According to Statoil, 2% of production losses in 2007 were due to scale amounting to US$469 million (oil price of $60/bbl).[30]
- Sloan[2] also provides two rules of thumb for heat-managed hydrate control: US$1.5 million/mile for overall heat transfer coefficient of 0.3 BTU/hr-ft-°F (8–10-in.-diameter pipe).

10.6 SUMMARY OF FLOW ASSURANCE TECHNOLOGIES AND PRACTICAL CONSIDERATIONS

Kondapi and Moe[31] provide conclusions from an evaluation of 30 different state-of-the-art flow assurance technologies according to their maturity levels, such as (1) embryonic, (2) emerging, (3) matured, and (4) aging. They categorized the technologies into five different categories: (1) thermal, (2) chemical, (3) hardware or facilities, (4) operational, and (5) software. Thermal consisted of heating, insulation, direct heating, electrically heated pipe in pipe, and cold flow; chemical consisted of inhibitors for hydrate, waxes, asphaltenes, scale, flow improvers, and the like; hardware or facilities consisted of mostly subsea-type separation, boosting, and others; operational meant pigging, staying in safe zones, depressurization, hot oiling, and so on; and software meant real-time flow assurance advisor to alert operating personnel. The distribution of technologies according to their maturity levels is shown in a bar graph in Figure 10.9. As expected, the use of chemicals is the most mature technologies.

10.6.1 OPERATING COMPANY PERSPECTIVES

Since it is the operating companies that typically have to tackle all the flow assurance challenges that are posed by the various solids, each of them has its own

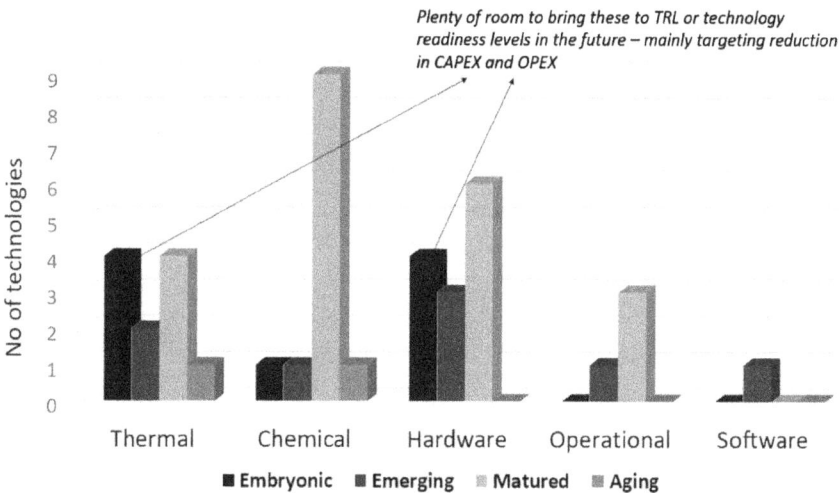

FIGURE 10.9 Summary of flow assurance technologies and their maturity levels. Bar graph constructed based on data from Kondapi and Moe.[31]

"operating perspectives" which basically reflect the importance given to different solids. Summaries of some of these from major companies are included in the following such that the reader, vis-à-vis another operator, may be able to form an informed opinion or decision when handling flow assurance challenges.

The British Petroleum (BP) perspective on measurement and modeling of asphaltenes, as presented by Gonzalez,[32] focuses on an important practical question as far as the representativity from the actual field performance and addressing the problem is concerned. The two main uncertainties outlined by Gonzalez[32] are the asphaltene definition and precipitation versus deposition. The industry and, for the most part, much of the (academic) literature tends to focus on precipitation or AOP or phase change rather than deposition, the understanding of which is under development.[32] Consequently, the asphaltene inhibitors (AI) are designed to target precipitation instead of deposition. The key takeaway messages[32] are the improvement of laboratory techniques using representative fluid samples (surface samples not recommended), field observation and surveillance, and model validation against field and lab data.

According to Ramanathan,[33] (Shell's) philosophy on deepwater design and its flow assurance strategy (mainly from solids-prevention perspective) are significant components of project design – obviously, increased complexity means higher capital expense or CAPEX. Perhaps for that reason, there appears to be somewhat of a paradigm shift and that "risk-based" flow assurance, supposedly based on operational experience, could be an enabler of system design simplification, thus making (marginal) projects competitive. The self-explanatory concept is illustrated using the hydrate equilibrium curve shown in Figure 10.10. An important element of risk-based flow assurance is risk mapping,[33] which is constructed on the basis of solids, operational events, time, and location (see Figure 10.11). For a specific asphaltene management example, it is suggested[33] that it is a balance of three combination options, namely, (A) continuous chemical inhibitor injection and solvent soak, (B) no chemical injection and solvent soak, and (C) coated (not painted but molecularly bonded) tubing and solvent soak. Note that regardless of the options, varying degrees of deposition will likely take place; thus, a solvent soak will be needed.[33] The lifetime cost analysis comparison of these three options for eight wells shown in Figure 10.12 indicates that the coated flowlines with solvent soak is the least expensive. Finally, other guidelines/requirements[33] as far as flowline insulation thickness or U value is concerned are as follows: (1) for hydrates, host arrival temperature > hydrate equilibrium

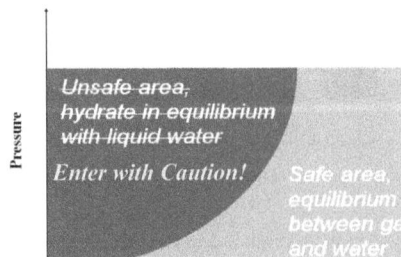

FIGURE 10.10 Illustration of "risk-based" flow assurance strategy.

	Months		
	1 2 3 4 5 6 7 8 9 10 11 12 13 14 15 16 17 18 19 21		
Operational event	*Gas lift starts*	*Water BT*	*WC > 10%*
Solids			
Asphaltene	*Impact on wells, subsea and host*		
Hydrate	*Low risk initially, impact on wells, subsea*		
Wax	*Low risk initially, impact on wells, FL; solvent effect and lighter components decrease*		
Scale	*Low risk initially, impact on wells, FL*		

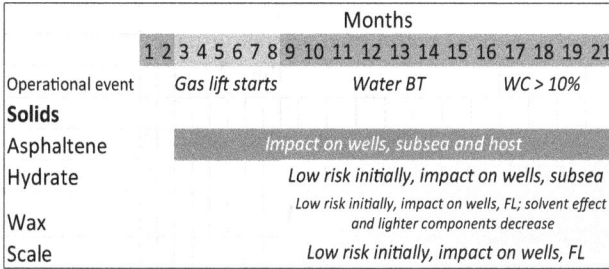

FIGURE 10.11 Flow assurance "risk mapping". The darkening of the color shade with field maturity for hydrate, wax, and scale indicates increasing risk of impacting production/ integrity, which may need mitigation. The onset of asphaltene risk with the start of gas lift is attributed to the oil being potentially unstable as far as asphaltenes are concerned. Modified after Ramanathan.[33]

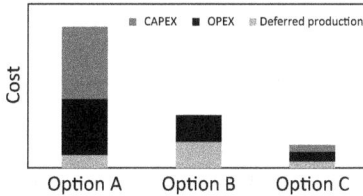

FIGURE 10.12 Lifetime cost analysis comparison of eight wells for three different asphaltene management options (see Section 10.6.1). Note that the actual costs are not provided; thus, they have been interpreted strictly for the purpose of (qualitative/quantitative) comparison. Modified after Ramanathan.[33]

and cooldown time of 12–15 hours to secure the system; (2) for paraffin or wax, the host arrival temperature > WAT; (3) limiting design factor in insulation thickness is enabling of reel lay; (4) cost-wise, a certain insulation thickness that would meet the >WAT requirement is in the range of about US$3–4 million; and (5) for the most part, WAT > hydrate equilibrium will tend to satisfy both requirements.

Vey recently, Chinnappan et al.[34] reported on the successful use of glass-reinforced epoxy (GRE)–lined tubing for prevention of asphaltene deposition in problematic Kuwaiti oil wells. This is conceptually similar to what has been suggested by Shell in that the idea is to alter the internal tube surface, which will prevent the deposition of asphaltenes. However, there is a fundamental difference between the two approaches; the former uses a lining, whereas molecular bonding is used in the latter. Based on the documented results in Chinnappan et al.,[34] it can be construed that, going forward, this would be the operating philosophy of Kuwait Oil Company (KOC) in most of their high API gravity Jurassic oilwells. The salient features of the GRE application, based on Chinnappan et al.'s[34] paper, are summarized here. The subject wells produce light (nearly 40°API) crudes, which contain only 1–3 wt% asphaltene content and the AOP ranges between 4,000–5,000 psi. Given the depleted reservoir pressure down to about 6,000 psi, owing to a lack of any other pressure support, AOP occurring in the tubing (and asphaltenes eventually depositing) can be readily realized. The authors report (and present visual evidence) a reduction in the tubing inside diameter from 2.75 in. to less than 2 in. due to asphaltene deposition. Many mechanical and chemical trials were attempted; however, none of them proved promising for field-wide implementation, which was the motivation to implement GRE-lined tubing given their ability to retard, and even eliminate, scale nucleation and deposition by virtue of its smooth surface (surface roughness to 0.00011 in.). The performance criteria used in evaluating the efficacy of GRE-lined tubing included pre- and post-FWHP as well as FWHT. For example, the 0.75 psi/day FWHP decline rate of preinstallation dropping down to only 0.19 psi/day in one of the first test wells. The authors also report the asphaltene cleaning frequency of approximately 3 jobs/year reducing to nil when GRE was implemented. Although the cost of GRE-lined tubing is nearly 3.5 times that of the normal tubing, the time to recover the incremental cost or the payback period is only about a year (based on the basic economic analysis). Finally, the authors report an annual operational savings of US$519,750 per well, which comes with the added intangible benefits of the method being environmentally friendly since it eliminates the use of highly flammable and toxic aromatics.

The following are Total's perspective on deepwater as presented by Candelier and Papot:[35]

- Hydrates are among the worst fluid chemistry issues; however, they are not a problem during normal operations but mainly during transients (start-up and shutdown).
- Design principle slogan of some sorts seems to be stay free/away from hydrate formation
- Design recommendations heavily rely on the accuracy of the multiphase flow modeling used during design phases; validation is important and achieved by internal experimental database and by comparison with field data when available.
- Key parameters of interest are liquid holdup, flow pattern prediction, slug frequency and characterization, pressure drop, and temperature profile

10.6.2 SUBSEA PROCESSING

An emergent technology as far as deepwater is concerned is subsea processing, which is basically either subsea separation and/or subsea boosting. As the name suggests, the idea is to separate the production stream subsea or somewhat in situ and then bring it up (separately) to the FPSO, whereas *boosting* simply refers to raising the pressure. The potential gains subsea processing may offer are (1) reduced topsides constraints and footprint; (2) simplified flow assurance management (free water removed); (3) prediction of pressure and temperature distribution in flowlines and risers simplified (~single phase); (4) minimize/eliminate slugging issues; (5) in situ

"diversion" of separated fluids, that is, gas and water reinjected for pressure maintenance, and oil straight into the flowlines and risers; and, finally, (6) cost reductions
and thus enhancement of economics in deepwater. Some examples of deepwater subsea processing are Pazflor[36] (offshore Angola) and Marlim[37] (offshore Brazil). The
heavy and viscous oil Pazflor[36] development is reported as the world's first largest
subsea separation system, which is in 2,000–4,000 ft water depth. The actual subsea separation unit (SSU) is 9 m in height and 3.5 m in diameter, operates at about
300 psi and 140°F, is rated to process 110,000 STB/D of oil and 35 mmscf/d gas,
and the "separated" gas and liquid free flow through respective risers. The Marlim[37]
development is at a water depth of 870 m with a subsea footprint of 29 m × 11 m,
and the subsea-separated water is reinjected. The system functionalities are as follows: the first separation stage is sand removal, followed by separation in vertically
arranged pipes in a "harp" shape (somewhat like a forked slug catcher) for free gas
separation, that is, a gas–liquid separator; next is pre-separation of oil and water in
a 60-m-long U-shaped tube pipe separator; oil–water separation is completed in an
outlet vessel; separated gas recombines with oil flows through the flowline to FPSO
via the riser (2.4 km total length); separated water is further processed through a
water desander and two hydro-cyclones to bring it to injection quality; finally, a
centrifugal pump boosts the water pressure for reinjection, passing through an online
oil-in-water monitor.

10.6.3 WAXPHALTENES

Typically, paraffin waxes and asphaltenes are dealt with individually from a flow
assurance standpoint. However, depending on the fluid composition and in particular
the strong affinity between paraffin wax and asphaltenes may lead to coprecipitation
of what is termed as "waxphaltenes". The occurrence of waxphaltenes is somewhat
rarely reported in the open literature; for example, a search of this term in www.
onepetro.org results in only five hits. These coprecipitated complexes actually pose
challenges as far as inhibition (more broadly operating strategy) is concerned, thus
necessitating a right inhibitor chemistry that can basically tackle both solids. Punase
et al.[38] reported a waxphaltene case study for oil from deepwater reservoir from West
Africa. In this particular case, it is relatively easy to comprehend the formation of
waxphaltenes based on the oil characteristics. The oil is somewhat light, having an
API gravity of 29° and room temperature viscosity of about 23 cP (rapidly increases
with cooling), a high saturate content, and a consequently high WAT of about 39°C.[38]
Besides these wax conducive oil characteristics, the saturates–aromatics–resins–
asphaltenes analysis results in a fairly high CII of 1.53 (reflects lower stability of
asphaltenes).[38] Clearly, both are favorable in terms of waxphaltenes; the authors[38]
state that further characterization analyses highlighted heavier paraffinic components
to have very high affinity toward the asphaltene clusters and creating waxphaltene
precipitation issues. This presents a unique opportunity[38] to revisit the way product
development is performed allowing chemical treatment strategy to be adopted and
aligned based on actual deposit (waxphaltene) characteristics.

10.7 FLOW ASSURANCE ENGINEER'S TOOLBOX

As far as the practical aspects of flow assurance are concerned, an important question or two arise: (1) Where does the (flow assurance) engineer start? or (2) What could be in the (flow assurance) engineer's toolbox? Although some of the case studies included in this chapter and the perspectives do provide some guidance to specific flow assurance problems, the two questions are certainly not trivial and hence do not have a straightforward answer. One primary reason is the multidisciplinary (both science and engineering) nature of flow assurance, with many different (specialized) activities intersecting each other, which is akin to a jigsaw puzzle or a Venn diagram, as illustrated in Figure 10.13. For example, production chemistry can guide the conditions at which hydrates are likely to form based on the gas composition – this information needs to be conveyed to the thermohydraulic modeler and production operations. Therefore, we have made an attempt to provide a framework or a logical workflow, that is self-explanatory (see Figure 10.14), in the context of flow assurance as a discipline, that can potentially serve as a general guideline. Notwithstanding the proffered approach if a direct or explicit solution is not found; then, as rightly stated by Di Lullo,[39] *"we can always invent it!"* Finally, in closing, having stated the popular adage "prevention is better than cure", it is extremely important to be mindful (and, of course, compliant) of all HSE issues while ensuring production sustainability or system operability.

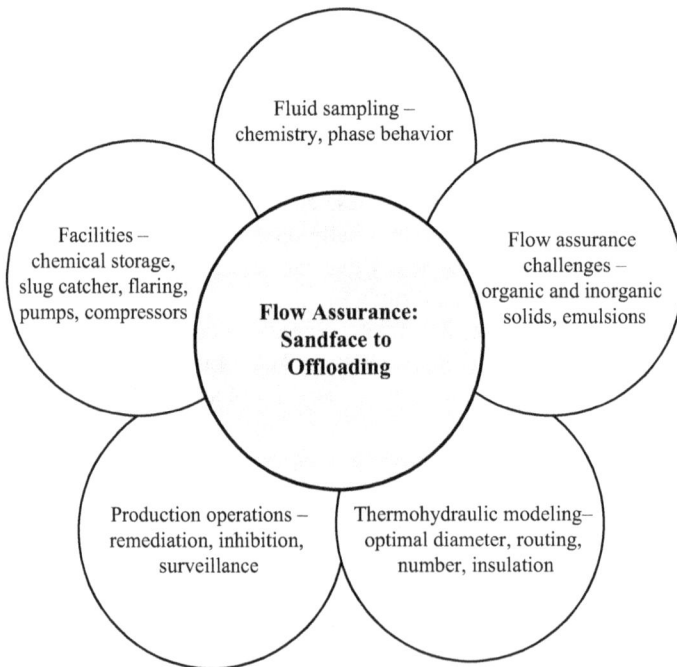

FIGURE 10.13 Intersecting multidisciplinary flow assurance activities.

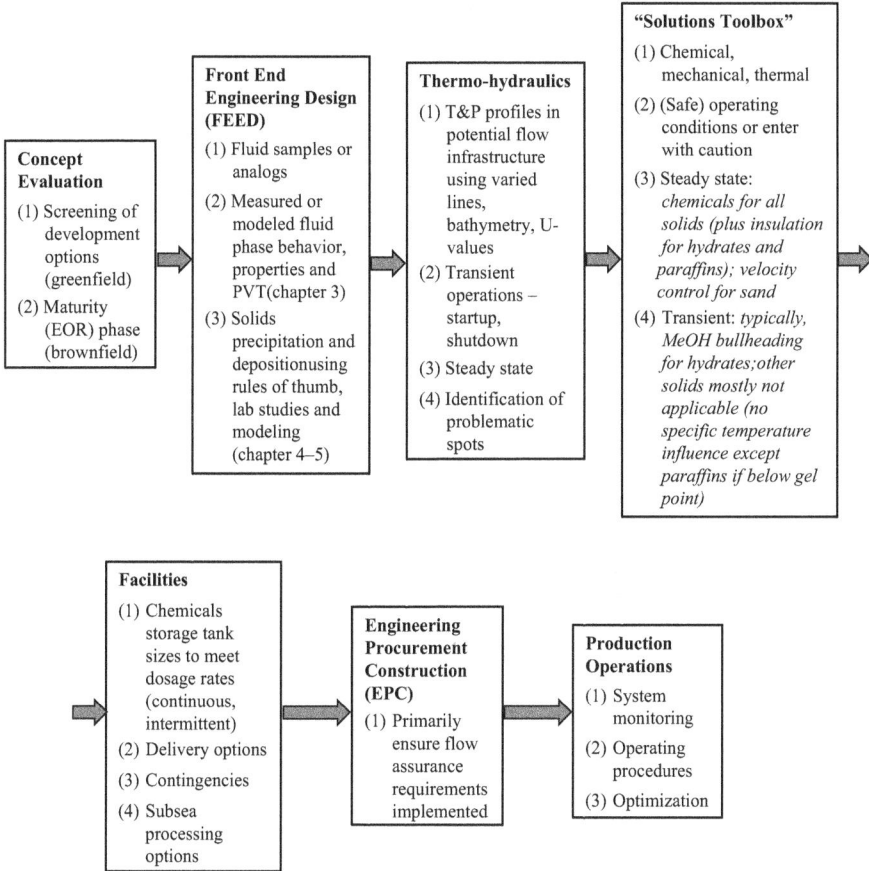

FIGURE 10.14 Generic framework or workflow that a flow assurance engineer can use.

REFERENCES

1. Kurup, A. S., Hernandez, O., Idstein, T., Zamora, C. A., Greenly, L., & Anderson, J. (2017, May 1). Pushing Conventional Boundaries of Hydrate Management in a Dry Tree Facility. *Day 2 Tue, May 02, 2017.* https://doi.org/10.4043/27780-MS
2. Sloan, E. D. J. (2000). *Hydrate Engineering.* Society of Petroleum Engineers.
3. Uribe, O., McDermott, C., & Hudson, A. (2017, May 1). A Novel Hydrate Strategy in a High Risk Hydrate Environment for a Fast Track Tie-in GoM Development. *Day 2 Tue, May 02, 2017.* https://doi.org/10.4043/27916-MS
4. Manfield, P., Stone, B., Kwan, T., Zimmerman, E., & Devlin, P. (2007, April 30). Wax-On, Wax-Off: Understanding and Mitigating Wax Deposition in a Deepwater Subsea Gas/Condensate Flowline. *All Days.* https://doi.org/10.4043/18834-MS
5. Qamruzzaman, M., Khan, M., Roy, D. C., Raman, R., & Arora, S. (2020, November 12). Combating Production Loss Due to Paraffin Deposition: A Success Story of Thermochemical Stimulation in Western India. *Day 1 Tue, November 17, 2020.* https://doi.org/10.2118/202457-MS

6. Singh, P., Walker, J. A., Lee, H. S., Gharfeh, S. G., Thomason, W., & Blumer, D. (2007). An Application of Vacuum Insulation Tubing for Wax Control in an Arctic Environment. *SPE Drilling & Completion*, 22(02), 127–136. https://doi.org/10.2118/111006-PA

7. Altemeemi, B., Gonzalez, F. A., Gonzalez, D. L., Jassim, S., Snasiri, F., Al-Nasheet, A., Al-Mansour, Y., Ali, A., & Sheikh, B. (2020, February 12). Overcoming Formation Damage in a Carbonate Reservoir Rock due to Asphaltene Deposition. *Day 1 Wed, February 19, 2020*. https://doi.org/10.2118/199264-MS

8. Thawer, R., Nicoll, D. C. A., & Dick, G. (1990). Asphaltene Deposition in Production Facilities. *SPE Production Engineering*, 5(4), 475–480. https://doi.org/10.2118/18473-PA

9. Misra, S., Abdulla, D., Bazuhair, M., Aboukshem, A., Stenger, B., & Katheeri, A. Al. (2011, September 25). Management of Asphaltenes Deposition in a Giant Carbonate Onshore Oil Field, Abu Dhabi, U.A.E. *All Days*. https://doi.org/10.2118/140278-MS

10. Jon Steinar Gudmundsson. (2018). *Flow Assurance Solids in Oil and Gas Production*. CRC Press/Balkema.

11. Junior, J., Borges, L., Carmelino, C., Hango, P., Milliken, J. D., & Asomaning, S. (2013, April 8). Calcium Naphthenate Mitigation at Sonangol's Gimboa Field. *All Days*. https://doi.org/10.2118/164069-MS

12. Nichols, D. A. A., Rosário, F. F. F., Bezerra, M. C. M. C. M., Gorringe, S. E. E., Williams, H. L. L., & Graham, G. M. M. (2014, May 14). Calcium Naphthenates in Complex Production Systems - Evaluation and Chemical Inhibition Challenges. *Day 2 Thu, May 15, 2014*. https://doi.org/10.2118/SPE-169756-MS

13. Shafizadeh, A., McAteer, G. & Sigmon, J. (2003). *High Acid Crudes*. Presentation., Crude Oil Quality Group Meeting, New Orleans.

14. Pedersen, K. S., Christensen, P. L., & Shaikh, J. A. (2015). Phase behavior of petroleum reservoir fluids (2nd ed.). CRC Press, Taylor & Francis Group.

15. Henrichsen, C. (2010). Keynote on How to Prevent the Loss of Oil and Gas Production Caused by Scale Deposits. SPE Applied Technology Workshop: Managing Scale and Asphaltene. Unpublished Materials.

16. Stalker, R. (2010). Effect of Near Wellbore Fractures on Scale Formation and Treatment. SPE Applied Technology Workshop: Managing Scale and Asphaltene. Unpublished Materials.

17. Jordan, M. (2010). Workshop Exercise on Scale Management. SPE Applied Technology Workshop: Managing Scale and Asphaltene. Unpublished Materials.

18. Guraieb, P., Tomson, R., Courville, A., Nikam, V., Kennedy, J., Nelson, R., & Sears, B. (2021). Novel Scale Inhibitor Extends Treatment Lifetimes in Permian EOR. *Journal of Petroleum Technology*, 73(08), 38–40. https://doi.org/10.2118/0821-0038-JPT

19. Martins, P. H. de A., & Freiman, B. M. (2021, August 9). Scale Removal In A Deepwater Oil Production Well Using Fresh Water. *Day 3 Wed, August 18, 2021*. https://doi.org/10.4043/31300-MS

20. Hoffman, J., Clausing, K., Robinson, S., Subramanian, P., & Zummo, A. (2017, May 1). The Stones Project: Subsea, Umbilical, Riser and Flowline Systems. *Day 2 Tue, May 02, 2017*. https://doi.org/10.4043/27569-MS

21. https://www.shell.com/about-us/major-projects/stones/_jcr_content/par/relatedtopics.stream/1473156161876/16b40cdb9b716348a54ce6f0ecf66a21fcc4dd1d/stones-project-overview.pdf. Retrieved August 19, 2021.

22. Salama, M. M. (2000). Sand Production Management. *Journal of Energy Resources Technology*, 122(1), 29–33. https://doi.org/10.1115/1.483158

23. API (2000), API RP 14E Recommended Practice for Design and Installation of Offshore Production Platform Piping Systems, 5th Edition, American Petroleum Institute, Washington, DC.

24. Salama, M. M. (1998, May 4). An Alternative to API 14E Erosional Velocity Limits for Sand Laden Fluids. *All Days*. https://doi.org/10.4043/8898-MS

25. Sutton, G. D., & Roberts, L. D. (1974). Paraffin Precipitation During Fracture Stimulation. *Journal of Petroleum Technology*, 26(9), 997–1004. https://doi.org/10.2118/4411-PA

26. Mahmoudkhani, A., Feustel, M., Reimann, W., & Krull, M. (2017, April 3). Wax and Paraffin Control by Fracturing Fluids: Understanding Mode of Actions and Benefits of Water-Dispersible Wax Inhibitors. *Day 3 Wed, April 05, 2017*. https://doi.org/10.2118/184594-MS

27. Jassim, E. I., Abdi, M. A., & Muzychka, Y. (2008, May 5). A CFD-Based Model to Locate Flow-Restriction Induced Hydrate Deposition in Pipelines. *All Days*. https://doi.org/10.4043/19190-MS

28. www.rbenergy.com; retrieved July 2017.

29. www.senate.gov; retrieved July 2017.

30. Fleming, N. (2010). Nanotechnology – Engineered Solutions to Downhole Scale Management. SPE Applied Technology Workshop: Managing Scale and Asphaltene. Unpublished Materials.

31. Kondapi, P., & Moe, R. (2013, May 6). Today's Top 30 Flow Assurance Technologies: Where Do They Stand? *All Days*. https://doi.org/10.4043/24250-MS

32. Gonzalez, D. (2019). Understanding of Asphaltene Precipitation and Deposition. Asphaltene measurements and modeling: Do they represent actual field performance and help fixing the problem? SPE Workshop. Unpublished Materials.

33. Ramanathan, K. (2019). SPE Luncheon on Risk Based Flow Assurance Strategy for Deepwater Developments. Unpublished Materials.

34. Edappillikulangara Chinnappan, R., Telang, M., Quttainah, R., Radhakrishnan, G., Fernandes, A., & Rajendran, K. (2021, March 16). First Time Worldwide Application of Glass Reinforced Epoxy Lined Tubing for Prevention of Asphaltene Deposition on Tubing in Oil Wells – A Case Study from Kuwait. *Day 3 Thu, March 25, 2021*. https://doi.org/10.2523/IPTC-21441-MS

35. Candelier, C., & Papot, F. (2015, June 10). Comparisons Between Engineering Software Predictions and West of Africa Deepwater Field Data Including Thermal Performances. *17th International Conference on Multiphase Production Technology*.

36. Bon, L. (2009, September 8). Pazflor: A World Technology First in Subsea Separation. *All Days*. https://doi.org/10.2118/123787-MS

37. Orlowski, R., Euphemio, M. L. L., Euphemio, M. L., Andrade, C. A., Guedes, F., Tosta da Silva, L. C., Pestana, R. G., de Cerqueira, G., Lourenço, I., Pivari, A., Witka, A., Folhadella, H., Pacheco, L., Kronemberger, S., & Vilela, J. (2012, April 30). Marlim 3 Phase Subsea Separation System - Challenges and Solutions for the Subsea Separation Station to Cope with Process Requirements. *All Days*. https://doi.org/10.4043/23552-MS

38. Punase, A., Aguiar, J., Smith, R., Mazzeo, C., & Wylde, J. (2019, October 25). Importance of Comprehending Waxphaltenes and the Need to Develop New Inhibitors to Prevent its Deposition. *SPE/IATMI Asia Pacific Oil & Gas Conference and Exhibition*. https://doi.org/10.2118/196368-MS

39. Di Lullo, A. (2012). Discover a Career: Flow Assurance. *The Way Ahead*, 08(1), 24–25. https://doi.org/10.2118/0112-024-TWA

Index

Page numbers in *italics* refer to figures and **bold** refer to tables.

For Product Safety Concerns and Information please contact our EU
representative GPSR@taylorandfrancis.com
Taylor & Francis Verlag GmbH, Kaufingerstraße 24, 80331 München, Germany

www.ingramcontent.com/pod-product-compliance
Lightning Source LLC
Chambersburg PA
CBHW060752220326
41598CB00022B/2404